Flexible Multibody
Dynamics

Flexible Multibody Dynamics

Efficient Formulations with Applications

Second Edition

Arun K. Banerjee

CRC Press
Taylor & Francis Group
Boca Raton London New York

CRC Press is an imprint of the
Taylor & Francis Group, an **informa** business

Second edition published 2022
by CRC Press
6000 Broken Sound Parkway NW, Suite 300, Boca Raton, FL 33487-2742

and by CRC Press
2 Park Square, Milton Park, Abingdon, Oxon, OX14 4RN

© 2022 Arun K. Banerjee

First edition published by Wiley in 2016

CRC Press is an imprint of Taylor & Francis Group, LLC

Library of Congress Cataloging-in-Publication Data

Names: Banerjee, Arun K., author.
Title: Flexible multibody dynamics : efficient formulations with applications /
Arun Banerjee.
Description: Second edition. | Boca Raton, FL : CRC Press, [2022] |
Includes index. | Identifiers: LCCN 2021046654 (print) | LCCN 2021046655 (ebook) | ISBN
9781032139197 (hbk) | ISBN 9781032139289 (pbk) | ISBN 9781003231523
(ebk)
Subjects: LCSH: Machinery, Dynamics of. | Multibody systems--Mathematical
models.
Classification: LCC TJ173 .B235 2022 (print) | LCC TJ173 (ebook) | DDC
621.8/11--dc23
LC record available at https://lccn.loc.gov/2021046654
LC ebook record available at https://lccn.loc.gov/2021046655

ISBN: 978-1-032-13919-7 (hbk)
ISBN: 978-1-032-13928-9 (pbk)
ISBN: 978-1-003-23152-3 (ebk)

DOI: 10.1201/9781003231523

Typeset in Times
by Deanta Global Publishing Services, Chennai, India

Alpona, Chhoto, and Buddho,...

and

Professor Thomas Kane, from whose magnificent books I learned to do dynamics, and whose expectation "to do many interesting problems together", is what started me on the path and inspired me to move on, to develop a body of research that this book contains.

Contents

Preface

This book is based on my published research on deriving computationally efficient equations of motion of multibody systems with flexible components. It reflects my work of over 35 years done mostly at Lockheed Missiles and Space Company, and partially at Martin-Marietta and Northrop Companies. The cover of the book depicts two examples of flexible multibody systems: the Galileo spacecraft, a satellite that deployed a truss-like probe to study the planet Jupiter, and the Hubble Space Telescope, for which a computer code based on an algorithm given in this book was used for simulation at Lockheed. Other examples of multibody systems with flexible components are robotic manipulators, helicopters, spacecraft with articulated reflectors, the space shuttle deploying a tethered subsatellite, and a ship reeling out a cable connecting it to a vehicle performing sea-floor mine search, and flexible rockets. Deriving equations of motion for such systems, and solving them to predict motion, are basic steps in their design.

In this book I choose to use Kane's method of deriving equations of motion for two reasons: its efficiency over other methods in reducing the labor of deriving the equations, and the simplicity of the final equations obtainable by a choice of variables that the method allows. However, the contribution of the book goes beyond a direct formulation using Kane's equations, to more computationally efficient algorithms that involve small matrices to invert, like the order-n formulation and the block-diagonal formulation for systems with flexible components. A major contribution of the book is in compensating for errors of premature linearization, inherent with the use of vibration modes of structures in large overall motion, by using geometric stiffness due to inertia loads.

A strong feature of the book is the application of the theory to solve complex problems. An introductory background material on dynamics and vibrations required of readers is given in Chapter I. In Chapter 1, I explain Kane's method, first with a simple example and then for a more complex problem of the dynamics of a three-axis controlled spacecraft with fuel slosh modeled as a spherical pendulum. Kane's method of the direct linearization of the equation of motion, without deriving the non-linear equations, is given next. The method of compensation of errors due to premature linearization, by adding geometric stiffness due to inertia loads, is shown by a simple example. Kane's equations with undetermined multipliers are presented next, for a system with motion constraints. In the Appendix to chapter 1, a guideline for choosing variables that simplify equations of motion is provided; another topic, of sliding impact dynamics with friction at the contact surface, is presented for the ejection of a nose cap pulling a packed parachute.

In Chapter 2, Kane's method is used to derive non-linear dynamical equations for deploying a satellite connected to the space shuttle by a 100 km long tether, and doing station-keeping and retrieval. Chapters 3 and 4 deal with large overall motion of beams and plates that illustrate the application of Kane's method of direct linearization. Chapter 5 gives a derivation of equations of large overall motion of an arbitrary flexible body, with a method of redeeming prematurely linearized equations by adding motion-induced geometric stiffness. Chapter 6 incorporates the motion-induced geometric stiffness into a dense-matrix formulation of the dynamics of a system of flexible bodies in large overall motion. Chapter 7 provides review material from structural dynamics, based mainly on the book by Craig, with some additional work done at Lockheed on component modes and modal truncation vectors. Chapter 8 presents an efficient algorithm for dynamical equations with block-diagonal mass matrices that was used for the Hubble and Next Generation Space telescopes; the algorithm is then extended to systems with structural loops, comparing results from the theory with test data for an antenna deployment. Chapter 9 illustrates the power of efficient motion variables in a block-diagonal mass algorithm, for systems with multiple structural loops. Chapter 10 presents an order-n algorithm for modeling n spring-connected rigid rods, to simulate large bending of beams in large overall motion, and shows that the solution is as accurate as a non-linear finite element solution, while being computationally much faster than the finite element method. A code in FORTRAN, perhaps an archaic language, is given for the order-n formulation as an Appendix at the end of the book. Chapter 11 uses a variable-n order-n algorithm for deploying/retracting a boom from a spacecraft, and deployment and retraction of a cable from a ship to an underwater maneuvering vehicle. Chapter 12 is on modeling the dynamics of a flexible rocket. Chapter 13 deals with the problem of large amplitude fuel slosh for various fill-fractions, giving detailed results for a spacecraft undergoing large overall motion.

In writing the book, I was helped by Dr Paul Mitiguy, on the efficient choice of motion variables; by Professor Arun Misra at McGill on the formation flying of tethered satellites; by Dr John Dickens, formerly of Lockheed, on modal truncation vectors and geometric stiffness issues, and for providing his structural dynamics codes; and by Professor William Singhose of the Georgia Institute of Technology on electrodynamic input shaping for tethers. I thank my Lockheed colleagues, Dr Ron Dotson, manager, Structural Dynamics group, who gave me a freehand to do research; in particular, I want to thank my Lockheed colleague, Mark Lemak, who developed a workhorse code based on the algorithms given here and produced the results given in Chapters 6 and 8–10. I am grateful to my Lockheed colleague, David Levinson, whose appreciation of my flexible body dynamics work at Lockheed was a morale booster. I am also grateful to my friend of many years, Dr Tushar Ghosh, for his insightful snippets of analysis in the book.

About the Author

Dr Arun K. Banerjee retired as a principal research scientist from the Lockheed Research Laboratory in Palo Alto, CA, USA. Previously, he worked for Martin-Marietta Aerospace and Northrop Services. He was an associate editor of the *AIAA Journal of Guidance, Control, and Dynamics*, 1995–2001. His honors include an invitation from the European Space Agency to give a state-of-the-art lecture, titled "Dynamics of Multibody Systems with Flexible Components", and giving invited short courses from this book at the Beijing Institute of Technology and the Indian Space Research Organization. He has contributed numerous journal publications on flexible multibody dynamics, and presented papers at international conferences in Venice and Tokyo, and at the International Congress of Theoretical and Applied Mechanics, Warsaw, 2004, and in Adelaide, 2008.

Background Material on Dynamics and Vibrations

I.1 INTRODUCTION

The objective of this book is to enable the reader to *derive equations of motion* of complex systems consisting of many interconnected flexible bodies. The first part of this task is kinematics, where one writes a set of differential equations that describe how the configuration of a body changes with time, without reference to the masses and forces involved in it. The second part is to *do* dynamics, where one derives a set of differential equations, the solution of which predicts how the body will move under the action of forces and torques on the bodies. Regarding the formulation of the equations of motion, we choose to follow Kane's method, because it (1) requires the least amount of labor among all methods to derive the equations, and (2) produces the simplest form of the final equations, with an appropriate set of motion variables. We will expand on this later.

We include here the key kinematical entities involved in Kane's method. We present the background material necessary for describing the kinematics. All of the materials given in this chapter are based on Refs. [1, 2]. We also review the basic theory of vibration of a structure, represented by its mass and stiffness matrices that lead to the mode shapes and frequencies of a flexible body component of a multibody system; we also discuss a method of reduction of coordinates in this context. A good reference for structural dynamics is Ref. [3].

I.2 DIRECTION COSINE MATRIX

If a rigid body B, or the reference frame B for deformations of a flexible body, rotates with respect to a frame or rigid body A, the orientation and angular velocity of B with respect to A can be described in several ways. We first describe this kinematics in terms of Euler angles and the associated direction cosine matrix. In the most general case, this direction matrix can be obtained by three successive rotations. Let A and B be characterized by three body-fixed orthogonal unit vectors $\mathbf{a}_1$, $\mathbf{a}_2$, $\mathbf{a}_3$ and $\mathbf{b}_1$, $\mathbf{b}_2$, $\mathbf{b}_3$, respectively, be initially aligned, and then undergo three successive rotations as follows: for a "body three 1-2-3 rotation" [2], body B undergoes a rotation by θ_1 about $\mathbf{a}_1$ to form a triad $\mathbf{a}_1', \mathbf{a}_2', \mathbf{a}_3'$, from which a second rotation by θ_2 forms a different triad $\mathbf{a}_1'', \mathbf{a}_2'', \mathbf{a}_3''$, and finally a third rotation by θ_3 about $\mathbf{a}_3''$ to get to the final orientation $\mathbf{b}_1, \mathbf{b}_2, \mathbf{b}_3$. One can express the direction cosine matrix of B with respect to A, written as $^A C^B$, such that: $\{A\} = [^A C^B]\{B\}$, and obtained by three *successive matrix multiplications*, as follows:

$$\begin{Bmatrix} \mathbf{a}_1 \\ \mathbf{a}_2 \\ \mathbf{a}_3 \end{Bmatrix} = \begin{bmatrix} c_2 c_3 & -c_2 s_3 & s_2 \\ s_1 s_2 c_3 + s_3 c_1 & -s_1 s_2 s_3 + c_3 c_1 & -s_1 c_2 \\ -c_1 s_2 c_3 + s_3 s_1 & c_1 s_2 s_3 + c_3 s_1 & c_1 c_2 \end{bmatrix} \begin{Bmatrix} \mathbf{b}_1 \\ \mathbf{b}_2 \\ \mathbf{b}_3 \end{Bmatrix} \qquad (\text{I}.1)$$

$$\Rightarrow \{A\} = [^A C^B]\{Bl\}$$

$$\mathbf{a}_2 = (s_1 s_2 c_3 + s_3 c_1)\mathbf{b}_1 + \left(-s_1 s_2 s_3 + c_3 c_1\right)\mathbf{b}_2 + \left(-s_1 c_2\right)\mathbf{b}_3;$$

Note : pick elements from row

$$\mathbf{b}_2 = (-c_2 s_3)\mathbf{a}_1 + (-s_1 s_2 s_3 + c_3 c_1)\mathbf{a}_2 + (c_1 s_2 s_3 + c_3 s_1)\mathbf{a}_3;$$

Note : pick elements from columns

Here, $c_1, s_1, c_2, s_2, c_3, s_3$ stand for $\cos(\theta_1)$, $(\sin\theta_1)$, $\cos(\theta_2)$, $(\sin\theta_2)$, $\cos(\theta_3)$, $(\sin\theta_3)$, respectively. There are 12 different forms of the directions matrices, $^A C^B$ depending on the sequence of body-3 rotations [1-2-3; 2-3-1; 3-1-2; 1-3-2; 2-1-3; 3-2-1]; or body-2 rotations [1-2-1; 1-3-1; 2-1-2; 2-3-2; 3-1-3; 3-2-3]. If the body rotation occurs such that the body B rotates by θ_1, about a_1, next by θ_2 about a_2, and then by θ_3 about a_3, then a "space-three 1-2-3" direction cosine matrix can be constructed [2]. The explicit forms of these 12 direction matrices are given in the Appendix of Ref. [2]. One important property of the direction cosine matrices is that

$$\left[^A C^B \right]^{-1} = \left[^A C^B \right]^T \qquad (\text{I}.2)$$

We will see later that Euler angle rates are related to angular velocity components and for any sequence of rotation, these rates can become singular, in which case one has to choose a different rotation sequence.

I.3   FUNDAMENTAL THEOREM ON DIFFERENTIATION OF A VECTOR IN TWO FRAMES

The word, differentiation of a vector, is meaningless unless one specifies which frame the vector is differentiated in. Let $\mathbf{z}$ be *any* vector, and let N be the Newtonian, i.e., inertial frame, and B a body frame. Then, the differentiation of $\mathbf{z}$ in frame N is related to the differentiation of $\mathbf{z}$ in frame B by the following rule:

$$\frac{^N d\mathbf{z}}{dt} = \frac{^B d\mathbf{z}}{dt} + {^N}\boldsymbol{\omega}^B \times \mathbf{z} \qquad (I.3)$$

Here, ${^N}\boldsymbol{\omega}^B$ is the angular velocity of B in N. As an example, let the central angular momentum $\mathbf{h}^{B/B^*}$ of a rigid body about the mass center B^* of B, and the torque due to a couple on B, $\mathbf{T}^B$, be represented and expressed as follows:

$$^N\boldsymbol{\omega}^B = \omega_1 \mathbf{b}_1 + \omega_2 \mathbf{b}_2 + \omega_3 \mathbf{b}_3$$

$$\mathbf{h}^{B/B^*} = I_1 \omega_1 \mathbf{b}_1 + I_2 \omega_2 \mathbf{b}_2 + I_3 \omega_3 \mathbf{b}_3 \qquad (I.4)$$

$$\mathbf{T}^B = T_1 \mathbf{b}_1 + T_2 \mathbf{b}_2 + T_3 \mathbf{b}_3$$

Here $\mathbf{b}_1$, $\mathbf{b}_2$, $\mathbf{b}_3$ are basis vectors fixed in B, and I_1, I_2, I_3 are the centroidal principal moments of inertia of B about B^*. Then, use of Eq. (I.4) in Eq. (I.3) leads to the equations, known as Euler's dynamical equations:

$$I_1 \dot{\omega}_1 - (I_2 - I_3)\omega_2 \omega_3 = T_1$$

$$I_2 \dot{\omega}_2 - (I_3 - I_1)\omega_3 \omega_1 = T_2 \qquad (I.5)$$

$$I_3 \dot{\omega}_3 - (I_1 - I_2)\omega_1 \omega_2 = T_3$$

A rigorous definition of the angular velocity of B in N is given by

$$^N\boldsymbol{\omega}^B = \mathbf{b}_1 \left(\frac{^N d\mathbf{b}_2}{dt} \cdot \mathbf{b}_3 \right) + \mathbf{b}_2 \left(\frac{^N d\mathbf{b}_3}{dt} \cdot \mathbf{b}_1 \right) + \mathbf{b}_3 \left(\frac{^N d\mathbf{b}_1}{dt} \cdot \mathbf{b}_2 \right) \qquad (I.6)$$

While this definition [1] is quite abstract, it is actually very useful, as we will see. The intuitive definition of angular velocity as a rotation rate, ${^N}\boldsymbol{\omega}^B = \Delta\boldsymbol{\theta}/\Delta t$, where $\Delta\boldsymbol{\theta}$ is the small angle rotation vector in time Δt as $\Delta t \to 0$, does not hold for large angle rotations. For large rotations, ${^N}\boldsymbol{\omega}^B$ is a function of the sines and cosines of the angles and depends on the rotation sequence.

Now selecting any rotation sequence, like body three 1-2-3 as in Eq. (I.1), one can use the direction cosine matrix, replacing the frame A by N, i.e., differentiating in N and multiplying by $\mathbf{b}_i$ ($i = 1,2,3$), as required in Eq. (I.6) to get the following [2]:

$$\begin{Bmatrix} \omega_1 \\ \omega_2 \\ \omega_3 \end{Bmatrix} = \begin{bmatrix} c_2 c_3 & s_3 & 0 \\ -c_2 s_3 & c_3 & 0 \\ s_2 & 0 & 1 \end{bmatrix} \begin{Bmatrix} \dot{\theta}_1 \\ \dot{\theta}_2 \\ \dot{\theta}_3 \end{Bmatrix} \qquad (1.7)$$

Or its inverse relation:

$$\begin{Bmatrix} \dot{\theta}_1 \\ \dot{\theta}_2 \\ \dot{\theta}_3 \end{Bmatrix} = \begin{bmatrix} c_3/c_2 & -s_3/c_2 & 0 \\ s_3 & c_3 & 0 \\ -c_3s_2/c_2 & s_3s_2/c_2 & 1 \end{bmatrix} \begin{Bmatrix} \omega_1 \\ \omega_2 \\ \omega_3 \end{Bmatrix} \tag{I.8}$$

Note that there is a division by $\cos(\theta_2)$, and the Euler angle rate equation becomes singular when $\theta_2 = \pi/2$. This is the case with any rotation sequence: with division by either $\cos(\theta_2) = 0$ or $\sin(\theta_2) = 0$, depending on the rotation sequence. If this happens, one just chooses another sequence for which $\cos(\theta_2) \neq 0$ or $\sin(\theta_2) \neq 0$.

The addition theorem for angular velocities for n successive rotations from $N \to B_1 \to B_2 \to \dots \to B_{n-1} \to B_n$ is

$$^{N}\omega^{B_n} = {}^{N}\omega^{B_1} + {}^{B_1}\omega^{B_2} + \dots + {}^{B_{n-1}}\omega^{B_n} \tag{I.9}$$

I.4   QUATERNIONS

A commonly used set of parameters, called quaternions, which should be called more appropriately [2] "Euler parameters", describe the orientation of a body, without the feature of singularity in their rate equations. In 1776, Euler established a theorem that "every change in the relative orientation of two rigid bodies A and B can be produced by a simple rotation of B in A". Consider the sketch in Figure I.1.

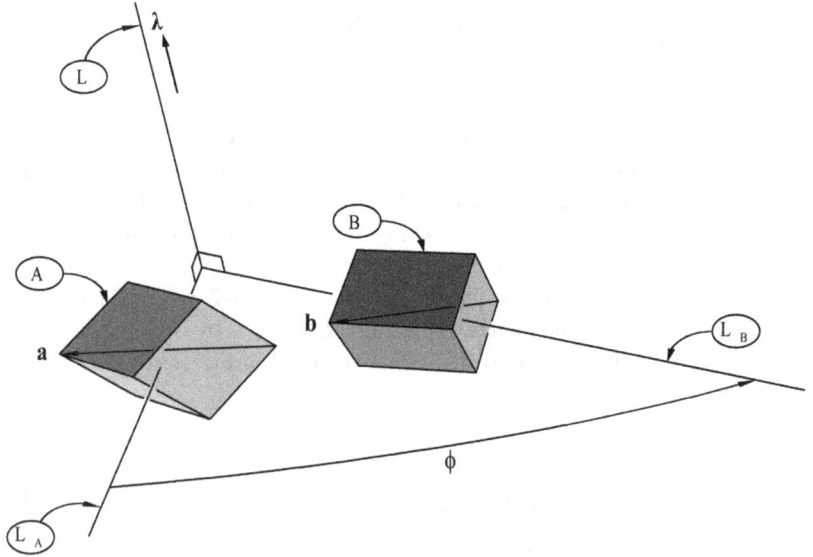

FIGURE I.1   Relative orientation of two rigid bodies A and B shown by a simple rotation of B in A.

As required for *simple rotations*, the unit vector λ, which must remain fixed in both frames A and B, and ϕ is the angle of rotation involved, one can define a Euler vector $\mathbf{q}$ and four scalar quantities, denoted q_1, q_2, q_3, q_4, and called quaternions, as follows:

$$\mathbf{q} = \lambda \sin \frac{\phi}{2} \tag{I.10}$$

$$q_i = \mathbf{q}.\mathbf{a}_i = \mathbf{q}.\mathbf{b}_i \quad (i = 1,2,3); \, q_4 = \cos \frac{\phi}{2} \tag{I.11}$$

Quaternions are not independent of each other because Eqs. (I.10) and (I.11) show

$$q_1^2 + q_2^2 + q_3^2 + q_4^2 = 1 \tag{I.12}$$

The three most-used relations involving quaternions are as follows [2]:

1. Elements of the direction cosine matrix from A to B are

$$^A C^B = \begin{bmatrix} 1-2(q_2^2+q_3^2) & 2(q_1q_2-q_3q_4) & 2(q_1q_3+q_2q_4) \\ 2(q_2q_1+q_3q_4) & 1-2(q_3^2+q_1^2) & 2(q_2q_3-q_1q_4) \\ 2(q_3q_1-q_2q_4) & 2(q_3q_2+q_1q_4) & 1-2(q_1^2+q_2^2) \end{bmatrix} \tag{I.13}$$

Conversely, given a direction cosine matrix $^A C^B$, one can find the quaternions:

$$q_1 = \frac{C_{32}-C_{23}}{4q_4}$$

$$q_2 = \frac{C_{13}-C_{31}}{4q_4} \tag{I.14}$$

$$q_3 = \frac{C_{21}-C_{12}}{4q_4}$$

$$q_4 = 0.5\sqrt{1+C_{11}+C_{22}+C_{33}}$$

2. Quaternion rates and angular velocity components do not give rise to a singularity. This is the main reason for the use of quaternion.

$$\begin{Bmatrix} \dot{q}_1 \\ \dot{q}_2 \\ \dot{q}_3 \\ \dot{q}_4 \end{Bmatrix} = \frac{1}{2} \begin{bmatrix} q_4 & -q_3 & q_2 \\ q_3 & q_4 & -q_1 \\ -q_2 & q_1 & q_4 \\ -q_1 & -q_2 & -q_3 \end{bmatrix} \begin{Bmatrix} \omega_1 \\ \omega_2 \\ \omega_3 \end{Bmatrix} \tag{I.15}$$

3. Successive rotations: When a body A is commanded to attain a desired orientation B, but acquires only the orientation of $\bar{B}$, "error quaternions", used in feedback control, are computed by the formula for successive rotations from frame A to frame $\bar{B}$, to frame B, as follows:

$$\begin{Bmatrix} q_{1e} \\ q_{2e} \\ q_{3e} \\ q_{4e} \end{Bmatrix} = \begin{bmatrix} q_4 & q_3 & -q_2 & -q_1 \\ -q_3 & q_4 & q_1 & -q_2 \\ q_2 & -q_1 & q_4 & -q_3 \\ q_1 & q_2 & q_3 & q_4 \end{bmatrix} \begin{Bmatrix} q_{1c} \\ q_{2c} \\ q_{3c} \\ q_{4c} \end{Bmatrix} \tag{I.16}$$

Details of the derivation of Eqs. (I.13)–(I.16) are given in Ref. [2].

I.5   FOUR BASIC THEOREMS IN KINEMATICS

1. Consider two points P and Q fixed on a rigid body B. Then, the velocity of P in N is related to the velocity of Q in N as follows:

$$^N\mathbf{v}^P = {}^N\mathbf{v}^Q + {}^N\boldsymbol{\omega}^B \times \mathbf{p}^{QP} \tag{I.17}$$

Here, $\mathbf{p}^{QP}$ is the position vector from Q to P, on B.

2. Acceleration of P in N is related to the acceleration of Q in N as

$$^N\mathbf{a}^P = {}^N\mathbf{a}^Q + {}^N\boldsymbol{\omega}^B \times \mathbf{p}^{QP} + {}^N\boldsymbol{\omega}^B \times ({}^N\boldsymbol{\omega}^B \times \mathbf{p}^{QP}) \tag{I.18}$$

3. Velocity in N of one point P moving on a moving rigid body (or frame B) is given by

$$^N\mathbf{v}^P = {}^N\mathbf{v}^{\bar{B}} + {}^B\mathbf{v}^P \tag{I.19}$$

Here, $\bar{B}$ is a point fixed on B, which is coincident with a point P at the instant under consideration, and $^B\mathbf{v}^P$s the velocity of P on B.

4. Acceleration in N of a point P moving on a rigid body B is given by

$$^N\mathbf{a}^P = {}^N\mathbf{a}^{\bar{B}} + {}^B\mathbf{a}^P + 2{}^N\boldsymbol{\omega}^B \times {}^B\mathbf{v}^P \tag{I.20}$$

In Eq. (I.20), the last term is called the Coriolis acceleration. Application of Eqs. (I.19) and (I.20) occurs in flexible body kinematics, when a material point P in the undeformed body B moves (or vibrates) as B rotates in N. If the deformation at point P located by $\mathbf{p}^{QP}$ is given by a sum of a number ν of vibration modes $\boldsymbol{\Phi}_i$ and modal coordinates q_i, i.e.,

$$\delta = \sum_{i=1}^{\nu} \boldsymbol{\Phi}_i q_i \tag{I.21a}$$

$$^N\mathbf{v}^P = {}^N\mathbf{v}^Q + {}^N\boldsymbol{\omega}^B \times \left[\mathbf{p}^{QP} + \sum_{i=1}^{v}\Phi_i q_i\right] + \sum_{i=1}^{v}\Phi_i\dot{q}_i \qquad (I.21b)$$

$$^N\mathbf{a}^P = \frac{d\,^N\mathbf{v}^P}{dt} = {}^N\mathbf{a}^Q + {}^N\boldsymbol{\alpha}^B \times \left(\mathbf{p}^{QP} + \sum_{i=1}^{v}\Phi_i q_i\right)$$

$$+ \sum_{i=1}^{v}\Phi_i\ddot{q}_i + {}^N\boldsymbol{\omega}^B \times \left[{}^N\boldsymbol{\omega}^B \times \left(\mathbf{p}^{QP} + \sum_{i=1}^{v}\Phi_i q_i\right) + 2\sum_{i=1}^{v}\Phi_i\dot{q}_i\right] \qquad (I.21c)$$

I.6 GENERALIZED COORDINATES AND GENERALIZED SPEEDS

Generalized coordinates describe the instantaneous configuration of a system. The displacements of particle i of a system may be described by variables other than the Cartesian coordinates x_i, y_i, z_i, and these may be lengths, angles, or modal coordinates for the modes of vibration. Let $q_1, q_2, q_3, \ldots, q_n$ be such a general system of coordinates such that

$$x_i = f_i(q_1, q_2, q_3, \ldots, q_n); \quad y_i = g_i(q_1, q_2, q_3, \ldots, q_n);$$
$$z_i = h_i(q_1, q_2, q_3, \ldots, q_n) \qquad (I.22)$$

Then, $q_1, q_2, q_3, \ldots, q_n$ are called generalized coordinates, with the following properties: the displacement of the system is completely described by them and the n-coordinates are completely independent in which case the system is called an n-degree of freedom system, or they are subject to m number of constraints, called *holonomic* constraints, with the degree of freedom of the system being (n − m).

Generalized speeds are scalars, introduced by Kane [1], that characterize the motion of a system; they can be linear speeds, angular speeds, or rates of change of vibration mode coordinates. Kane defines generalized speeds of an n-degree of freedom system in N as

$$u_r = \sum_{s=1}^{n} Y_{rs}\dot{q}_s + Z_r \quad (r = 1, \ldots, n) \qquad (I.23)$$

In Eq. (I.23), Y_{rs}, Z_r are functions of the generalized coordinates $q_1, q_2, q_3, \ldots, q_n$ and time t, and are so chosen that Eq. (I.22) can be solved uniquely for $\dot{q}_1, \dot{q}_2, \dot{q}_3, \ldots, \dot{q}_n$. Equation (I.23) are called the *kinematic differential equations* for S in N. If the generalized speeds u_i are all independent, the system is of degree of freedom n. If, on the other hand, there are m number of motion constraints, the system is called a simple non-holonomic system with the constraint equations given by

$$\sum_{r=1}^{n} A_{sr} u_r + B_s = 0 \quad s = 1, \ldots, m \tag{I.24}$$

and the degree of freedom of the system becomes $n - m$. A good choice of generalized speeds makes the dynamical equations simple.

I.7   PARTIAL VELOCITIES AND PARTIAL ANGULAR VELOCITIES: KEY COMPONENTS IN KANE'S METHOD

The partial velocity of a particle and the partial angular velocity of a rigid body are centrally important vectors in Kane's method. They are described as follows. If $q_1, q_2, q_3, \ldots, q_n$ and $u_1, u_2, u_3, \ldots, u_n$ are generalized coordinates and generalized speeds for a system, respectively, for a body B rotating in a frame N, the angular velocity of B in N can be expressed uniquely as

$$^N\omega^B = \sum_{r=1}^{n} {}^N\omega_r^B u_r + {}^N\omega_t^B \tag{I.25}$$

Similarly, the velocity of a particle P translating in N, can be expressed uniquely as

$$^N\mathbf{v}^P = \sum_{r=1}^{n} {}^N\mathbf{v}_r^P u_r + {}^N\mathbf{v}_t^P \tag{I.26}$$

Kane [1] defines $^N\omega_r^B$ as the r-th partial angular velocity of B in N, and $^N\mathbf{v}_r^P$ as the r-th partial velocity of P in N. With this definition, the r-th partial angular velocity of B and the r-th partial velocity of P can always be found by *inspection* of the expressions for the angular velocity and velocity of B and P, respectively, as the vector coefficients of u_r in the angular velocity and velocity expressions.

Example: If a rigid body B has two points Q and P fixed on it, then Eq. (I.17) applies, and the partial velocities of P and the partial angular velocities of B are as follows. To facilitate the process Eq. (I.17) is repeated here.

$$^N\mathbf{v}^P = {}^N\mathbf{v}^Q + {}^N\omega^B \times \mathbf{p}^{QP}$$

the r-th partial velocities of P and partial angular velocities of B follow from Eqs. (I.26) and (I.25):

$$^N\omega_r^B = 0 \quad r = 1, 2, 3$$

$$^N\mathbf{v}_r^P = \mathbf{b}_{r-3} \times \mathbf{p}^{QP} \quad r = 4, 5, 6 \tag{I.27}$$

$$^N\omega_r^B = \mathbf{b}_{r-3} \quad r = 4, 5, 6_r^B = \mathbf{b}_{r-3} \quad r = 4, 5, 6$$

Let the reference frame B move in large rotation and translation with respect to the reference frame N, and let the body undergo small vibration with respect to frame B. Then, we define $(6 + \nu)$ as follows:

$$^{N}\mathbf{v}^{Q} = u_1\mathbf{b}_1 + u_2\mathbf{b}_2 + u_3\mathbf{b}_3$$

$$^{N}\boldsymbol{\omega}^{B} = u_4\mathbf{b}_1 + u_5\mathbf{b}_2 + u_6\mathbf{b}_3$$

$$\sum_{i=1}^{\nu}\varphi_i^P\dot{q}_i = \sum_{i=1}^{\nu}\varphi_i^P u_{6+i}$$

(I.28)

Then, following Eq. (I.21a), the velocity of P can be written as

$$^{N}\mathbf{v}^{P} = u_1\mathbf{b}_1 + u_2\mathbf{b}_2 + u_3\mathbf{b}_3 + (u_4\mathbf{b}_1 + u_5\mathbf{b}_2 + u_6\mathbf{b}_3)$$

$$\times \left(\mathbf{p}^{QP} + \sum_{i=1}^{\nu}\varphi_i q_{6+i}\right) + \sum_{i=1}^{\nu}\varphi_i u_{6+i}$$

(I.29)

Then, the i-th partial velocity of P in N is the coefficients of u_i in Eq. (I.29):

$$^{N}\mathbf{v}_i^{P} = \mathbf{b}_i \qquad\qquad i = 1,2,3$$

$$^{N}\mathbf{v}_i^{P} = \mathbf{b}_{i-3} \times \left(\mathbf{p}^{QP} + \sum_{i=1}^{\nu}\varphi_i^P\eta_i\right) \quad i = 4,5,6$$

$$^{N}\mathbf{v}_i^{P} = \varphi_{i-6}^{P} \qquad\qquad i = 7,\ldots,6+\nu$$

(I.30)

I.8 DEFINITION OF INERTIA FORCE AND INERTIA TORQUE

1. Inertia force on a particle P is:

$$\mathbf{F}^{*P} = -m^{PN}\mathbf{a}^{P}$$

(I.31)

2. Inertia force and inertia torque for a rigid body B with mass center B*
   are given by:

$$\mathbf{F}^{*B} = -m^{BN}\mathbf{a}^{B*}$$

(I.32)

$$\mathbf{T}^{*B} = -\left[\overline{\mathbf{I}}^{B/B*} \cdot {}^{N}\boldsymbol{\alpha}^{B} + {}^{N}\boldsymbol{\omega}^{B} \times \left(\overline{\mathbf{I}}^{B/B*} \cdot {}^{N}\boldsymbol{\omega}^{B}\right)\right]$$

Here $\overline{\mathbf{I}}^{B/B*}$ is the moment of inertia dyadic of B about B*.

I.9   VIBRATION OF AN ELASTIC BODY: MODE SHAPES, FREQUENCIES, MODAL EFFECTIVE MASS, AND MODEL REDUCTION: THE EIGENVALUE PROBLEM

This section is expounded in greater details in Chapter 7. The following is a brief introduction, based on Ref. [3]. First, we write the equations of free vibration of a system of discrete particles or rigid bodies connected by springs, in matrix form given below, where M is the mass matrix and K is the stiffness matrix of the system:

$$[\mathbf{M}]\{\ddot{\mathbf{x}}\} + [\mathbf{K}]\{\mathbf{x}\} = \{\mathbf{0}\} \tag{I.33}$$

Assume a solution:

$$\{x\} = \{\varphi\}\cos(\omega t - \alpha) \tag{I.34}$$

Substitution of Eq. (I.34) in Eq. (I.33) gives the "algebraic eigenvalue problem" based on

$$[K - \omega^2 M]\{\varphi\} = 0 \tag{I.35}$$

For a non-trivial solution, the following "characteristic equation" must be satisfied:

$$\det([K - \omega^2 M]) = 0 \tag{I.36}$$

The roots of this equation are the eigenvalues, or squared natural frequencies ω_i^2 with

$$0 \le \omega_1^2 \le \omega_2^2, \ldots, \omega_i^2 \le \cdots \le \omega_n^2 \tag{I.37}$$

For any eigenvalue, ω_i^2, there is an eigenvector, ϕ_i, or a vibration mode shape.

$$\varphi_i = \begin{bmatrix} \varphi_{1i} & \varphi_{2i} & .. & .. & .. & \varphi_{ni} \end{bmatrix}^t \tag{I.38}$$

An eigenvector multiplied by a constant is also an eigenvector. The most important property of natural modes is their orthogonality. Proof of this property is started by multiplying Eq. (I.35) for r-th mode ϕ_j and multiplying the result by φ_s^t to get

$$(\phi_s^t K \phi_r) - \omega_r^2 (\phi_s^t M \phi_r) = 0 \tag{I.39}$$

Since K and M are symmetric matrices, Eq. (I.39) can be transposed and written as

$$(\phi_r^t K \phi_s) - \omega_s^t (\phi_r^t M \phi_s) = 0 \qquad (I.40)$$

Eq. (I.38) can be subtracted from Eq. (I.39) to give

$$(\omega_s^2 - \omega_r^2)(\phi_s^t M \phi_r) = 0 \qquad (I.41)$$

For modes with distinct frequencies, $\omega_r \neq \omega_s$, it is necessary that

$$\phi_i^t M \phi_j = 0 \quad i \neq j \qquad (I.42)$$

Equation (I.41) shows that the i-th and j-th modes are orthogonal with respect to the mass matrix, M. Equation (I.41) can be substituted in Eq. (I.38), replacing M with K, to show that, ϕ_i, ϕ_j are also orthogonal to the stiffness matrix, K.

Scaling of mode shapes produces a "normalized" mode shape in respect to the mass matrix.

$$\phi_i^t M \phi_i = \mu_i \rightarrow \hat{\phi}_i = \phi_i / \sqrt{\mu_i} \qquad (I.43)$$

For mode ϕ_r of a flexible body that is attached to a rigid body, μ_r is called the modal effective mass for the mode; μ_r represents the excitability of mode ϕ_r due to shaking of the rigid body (see Chapter 7), and modes of higher effective mass are retained in the model.

I.9.1 REDUCTION OF DEGREES OF FREEDOM

Now we describe a process of reducing the degrees of freedom in a model. We start with Eq. (I.33): $[M]\{\ddot{x}\} + [K]\{x\} = \{0\}$. Here, x is an n × 1 column matrix, and we may seek to reduce the number of states. A physical example [4] is a rotating cantilever beam of length 2L, flexural rigidity EI, and mass mL, represented by a two finite element model with vertical and rotational displacements, for Eq. (I.32), with x as a 4 × 1 column matrix. Let δ_1, θ_1 denote the deflection and section rotation at the mid-point, and δ_2, θ_2 are the deflection and rotation at the free-end, as shown in Figure I.2.

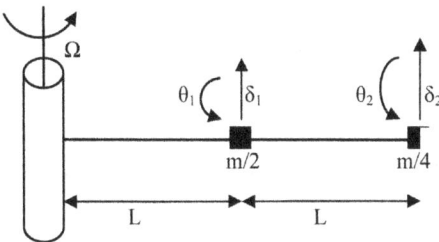

FIGURE I.2 Rotating cantilever beam represented by a two-element lumped mass model.

For convenience we will set $\Omega = 0$. Chapter 3 gives a detailed analysis of a beam attached to a moving base. $\{x\} = \begin{bmatrix} \delta_1 & \theta_1 & \delta_2 & \theta_2 \end{bmatrix}^t$:

$$[K]\{x\} = \frac{EI}{L^3} \begin{bmatrix} 24 & 0 & -12 & 6 \\ 0 & 8 & -6 & 2 \\ -12 & -6 & 12 & -6 \\ 6 & 2 & -6 & 4 \end{bmatrix} \begin{Bmatrix} \delta_1 \\ \theta_1 \\ \delta_2 \\ \theta_2 \end{Bmatrix} \qquad (I.44)$$

Since we are not interested in the states θ_1, θ_2, we seek to eliminate them by a process called *static condensation* [3]. First, we reorder the states to $\{x\} = \begin{bmatrix} \delta_1 & \delta_2 & \theta_1 & \theta_2 \end{bmatrix}^t$. To do this, we reshape the [K] matrix in two steps: (1) do a first transformation by interchanging the third and second columns, and then (2) interchange the second and third rows obtained in the first transformation. This process yields the new [K] matrix for the states in the new order $\{x\} = \begin{bmatrix} \delta_1 & \delta_2 & \theta_1 & \theta_2 \end{bmatrix}^t$:

$$\begin{bmatrix} \hat{K} \end{bmatrix}\{\hat{\delta}\} = \frac{EI}{L^3} \begin{bmatrix} 24 & -12 & 0 & 6 \\ -12 & 12 & -6 & -6 \\ 0 & -6 & 8 & 2 \\ 6 & -6 & 2 & 4 \end{bmatrix} \begin{Bmatrix} \delta_1 \\ \delta_2 \\ \theta_1 \\ \theta_2 \end{Bmatrix} \qquad (I.45)$$

State reduction is completed for the reduced number of states, δ_1, δ_2 by defining the sub-matrices in Eq. (I.45), corresponding to the active (a) and the deleted (d) states:

$$\hat{K}_{aa} = \frac{EI}{L^3} \begin{bmatrix} 24 & -12 \\ -12 & 12 \end{bmatrix} \qquad (I.46)$$

$$\hat{K}_{ad} = \frac{EI}{L^3} \begin{bmatrix} 0 & 6 \\ -6 & -6 \end{bmatrix} \qquad (I.47)$$

$$\hat{K}_{dd} = \frac{EI}{L^3} \begin{bmatrix} 8 & 2 \\ 2 & 4 \end{bmatrix} \qquad (I.48)$$

Reference [3] shows that the reduced stiffness matrix corresponding to the $[\delta_1, \delta_2]^t$ states, which retain the stiffness contribution due to the deleted states, is given by

$$[K_{red}] = [K_{aa} - K_{ad}K_{dd}^{-1}K_{ad}^t] \qquad (I.49)$$

One can use the same process to reduce the mass matrix, which in its original form for a consistent mass two-element beam is [4] as follows, where m is the mass per unit length of the beam. The two transformations interchanging columns and rows provide

$$[M]\begin{Bmatrix} \ddot{\delta}_1 \\ \ddot{\theta}_1 \\ \ddot{\delta}_2 \\ \ddot{\theta}_2 \end{Bmatrix} = \frac{mL}{420} \begin{bmatrix} 312 & 0 & 54 & -13 \\ 0 & 8 & 13 & -3 \\ 54 & 13 & 156 & -22 \\ -13 & -3 & -22 & 4 \end{bmatrix} \begin{Bmatrix} \ddot{\delta}_1 \\ \ddot{\theta}_1 \\ \ddot{\delta}_2 \\ \ddot{\theta}_2 \end{Bmatrix} \qquad (I.50)$$

This process really means

$$\begin{bmatrix} \hat{M} \end{bmatrix} \begin{Bmatrix} \ddot{\hat{\delta}} \end{Bmatrix} = \frac{mL}{420} \begin{bmatrix} 312 & 54 & 0 & -13 \\ 54 & 156 & 13 & -22 \\ 0 & 13 & 8 & -3 \\ -13 & -22 & -22 & 4 \end{bmatrix} \begin{Bmatrix} \ddot{\delta}_1 \\ \ddot{\delta}_2 \\ \ddot{\theta}_1 \\ \ddot{\theta}_2 \end{Bmatrix} \qquad (I.51)$$

Now define and form

$$M_{aa} = \frac{mL}{420} \begin{bmatrix} 312 & 54 \\ 54 & 156 \end{bmatrix};$$

$$M_{ad} = \frac{mL}{420} \begin{bmatrix} 0 & -13 \\ 13 & -22 \end{bmatrix};$$

$$M_{dd} = \frac{mL}{420} \begin{bmatrix} 8 & -3 \\ -22 & 4 \end{bmatrix} \qquad (I.52)$$

$$[M_{red}] = [M_{aa} - M_{ad}M_{dd}^{-1}M_{ad}^{t}] \qquad (I.53)$$

With that, the equations for the rotating cantilever beam in reduced number of states are

$$[M_{red}] \begin{Bmatrix} \ddot{\delta}_1 \\ \ddot{\delta}_2 \end{Bmatrix} + [K_{red}] \begin{Bmatrix} \delta_1 \\ \delta_2 \end{Bmatrix} = \begin{Bmatrix} 0 \\ 0 \end{Bmatrix} \qquad (I.54)$$

Ref. [4] compares the frequencies of this reduced order model to those of an "exact" full order model element model; scaled by a constant multiplier involving on model parameters the first two natural frequencies are follows:

Frequencies	Exact model	Reduced model
ω_1	3.515	3.51
ω_2	22.034	22.22

REFERENCES

1. Kane, T.R. and Levinson, D.A. (eds), (1985), *Dynamics: Theory and Applications*, McGraw-Hill.
2. Kane, T.R., Likins, P.W., and Levinson, D.A. (eds), (1983), *Spacecraft Dynamics*, McGraw-Hill.
3. Craig, R.R. (eds), (1981), *Structural Dynamics: An Introduction to Computer Methods*, Wiley.
4. Thomson, W.T. and Dahleh, M.D. (eds), (1998), *Theory of Vibration*, Prentice Hall.

1 Derivation of Equations of Motion

1.1 AVAILABLE ANALYTICAL METHODS AND THE REASON FOR CHOOSING KANE'S METHOD

In this book, we formulate equations of motion for a system of rigid and flexible bodies undergoing large overall motion, where large overall motion for a flexible body means a small elastic vibration of the body with respect to a frame moving in large rotation and/or translation. Various choices of analytical methods are available for deriving equations of motion, v, such as Newton–Euler methods and methods based on D'Alembert's principle together with the principle of virtual work, Lagrange's equations, Hamilton's equations, Boltzmann–Hamel equations, Gibbs equations, and Kane's equations. The most recent among these is Kane's method, based on a paper published in 1965, by Kane and Wang [1], and the method was given detailed exposition, with extensive applications, by Kane [2], Kane and Levinson [3], and Kane, Likins, and Levinson [4].

In a survey paper, Kane and Levinson [5] took up a fairly complex example, of an 8 degree of freedom (dof) system consisting of a spacecraft containing a four-bar linkage to show the difference between seven analytical methods; an application to a flexible spacecraft was also considered. The conclusion in Ref. [5] was that (i) D'Alembert's method is less laborious than a method using the conservation of momentum, with both methods requiring the introduction and elimination of constraint forces; (ii) Lagrange's equations require no introduction of workless constraint forces, but the labor to derive the equations is prohibitive; (iii) Boltzmann–Hamel equations, also called Lagrange's equations in quasi-coordinates, use variables that simplify the equations of motion but require order n^3 computations for certain terms, for an n dof system, and the process of getting the final equations is formidable; and (iv) Gibbs equations are better than Lagrange's equations using quasi-coordinates, but require one to form terms with n^3 computations for an n dof system. With an exposition of Kane's method, Ref. [5] showed that Kane's method is superior to the other methods, on the basis of two crucial considerations: (i) operational simplicity, meaning *reduced labor* in the derivation of the equations of motion either by hand or in terms of computer operations via symbol manipulation; and (ii) *simplicity* of the final form of the equations, giving rise to a reduction in the computational time, which is achieved by an efficient choice of motion variables that Kane calls generalized speeds. An exposition of Kane's method follows.

DOI: 10.1201/9781003231523-2

1.2   KANE'S METHOD OF DERIVING EQUATIONS OF MOTION

Consider a system of particles and rigid bodies whose *configuration* in N, a Newtonian reference frame (meaning an inertial frame), is characterized by *generalized coordinates*, $q_1, q_2, \ldots, q_n$, for an n dof system. Let $u_1, u_2, \ldots, u_n$ the motion variables, called *generalized speeds* by Kane, be linear combinations of $\dot{q}_1, \dot{q}_2, \ldots, \dot{q}_n$ (where a dot over variable indicates its time-derivative):

$$u_i = \sum_{j=1}^{n} W_{ij}\dot{q}_j + X_i \quad (i = 1, \ldots, n) \tag{1.1}$$

Eq. (1.1) represents the *kinematical differential equations* of the system. Here, W_{ij} and X_i are functions of the n generalized coordinates and time t; W_{ij} and X_i are so chosen that Eq. (1.1) can be uniquely solved for $\dot{q}_1, \dot{q}_2, \ldots, \dot{q}_n$. Typically, pre-scribed motion terms appear in X_i. The angular velocity of any rigid body and the velocity of any material point of the system can *always* be expressed uniquely as a linear function of the generalized speeds, $u_1, u_2, \ldots, u_n$. Thus, for a particle P_k in a system, Kane has shown on p. 46 of Ref. [3] that the velocity of P_k in N, defined as the *inertial time-derivative* of the position vector of P_k from a point O fixed in N, can always be written in a series of terms involving the generalized speeds, and another free of them:

$$^N\mathbf{v}^{P_k} = \frac{^N d\mathbf{p}^{OP_k}}{dt} = \sum_{i=1}^{n} {}^N\mathbf{v}_i^{P_k} u_i + {}^N\mathbf{v}_t^{P_k} \tag{1.2}$$

Here, $^N\mathbf{v}_i^{P_k}$, $^N\mathbf{v}_t^{P_k}$ are vector functions of the generalized coordinates, $q_1, q_2, \ldots, q_n$. In Ref. [2], Kane calls the vector $^N\mathbf{v}_i^{P_k}$, that is, the coefficient of the i-th general-ized speed u_i in Eq. (1.2), the *i-th partial velocity* of the point P_k. Similarly, for a rigid body B_k, the velocity of its mass center B_k^* and the angular velocity of B_k in N for a system can always be expressed [3] as

$$^N\mathbf{v}^{B_k^*} = \sum_{i=1}^{n} {}^N\mathbf{v}_i^{B_k^*} u_i + {}^N\mathbf{v}_t^{B_k^*}$$

$$^N\boldsymbol{\omega}^{B_k} = \sum_{i=1}^{n} {}^N\boldsymbol{\omega}_i^{B_k} u_i + {}^N\boldsymbol{\omega}_t^{B_k} \tag{1.3}$$

Kane calls $^N\mathbf{v}_i^{B_k^*}$ the *i-th partial velocity* of B_k^*, and $^N\boldsymbol{\omega}_i^{B_k}$ the *i-th partial angular velocity* of the body B_k in N. These are vector functions of the generalized coor-dinates. Typically, $^N\mathbf{v}_t^{P_k}$, $^N\mathbf{v}_t^{B_k^*}$, $^N\boldsymbol{\omega}_t^{B_k}$ in Eqs. (1.2) and (1.3) are remainder terms associated with prescribed velocity and angular velocity. Partial velocities and partial angular velocities are crucial items in Kane's method, and throughout this book we will see their central roles in the formulation of equations of motion. Once the velocities of points P_j and of mass centers B_j^*, and the angular velocities

of rigid bodies B_j are expressed in some vector basis fixed in B_j, inertial acceleration of those points and mass centers, as well as angular acceleration of those
bodies can be obtained by differentiating the angular velocity expressions in a
Newtonian reference frame N. This is done by appealing to the *rule for the differentiation of a vector in two reference frames* [3], as

$$
{}^N\mathbf{a}^{P_j} = \frac{{}^N d \, {}^N \mathbf{v}^{P_j}}{dt} = \frac{{}^{B_j} d \, {}^N \mathbf{v}^{P_j}}{dt} + {}^N\boldsymbol{\omega}^{B_j} \times {}^N\mathbf{v}^{P_j}
$$

$$
{}^N\mathbf{a}^{B_j^*} = \frac{{}^N d \, {}^N \mathbf{v}^{B_j^*}}{dt} = \frac{{}^{B_j} d \, {}^N \mathbf{v}^{B_j^*}}{dt} + {}^N\boldsymbol{\omega}^{B_j} \times {}^N\mathbf{v}^{B_j^*} \tag{1.4}
$$

$$
{}^N\boldsymbol{\alpha}^{B_j} = \frac{{}^N d \, {}^N \boldsymbol{\omega}^{B_j}}{dt} = \frac{{}^{B_j} d \, {}^N \boldsymbol{\omega}^{B_j}}{dt}
$$

The first equality in these equations is a definition, and the second equality sign
provides a basic kinematic relationship between the differentiation of a vector in
two reference frames, and it assumes that the frame, B_k, is different from frame N.

Kane's equations of motion are stated in terms of what Kane calls generalized
inertia forces and generalized active forces. For an n degree of freedom system
consisting of NR number of rigid bodies and NP number of particles, the i-th *generalized inertia force* is defined by Kane [3] as the following expression, involving
dot-products of i-th partial velocities with inertia forces, and i-th partial angular
velocities with inertia torques:

$$
F_i^* = -\sum_{j=1}^{NR} [m_j \, {}^N\mathbf{a}^{B_j^*} \cdot {}^N\mathbf{v}_i^{B_j^*} + \{\mathbf{I}^{B_j/B_j^*} \cdot {}^N\boldsymbol{\alpha}^{B_j} + {}^N\boldsymbol{\omega}^{B_j} \times (\mathbf{I}^{B_j/B_j^*} \cdot {}^N\boldsymbol{\omega}^{B_j})\} \cdot {}^N\boldsymbol{\omega}_i^{B_j}]
$$

$$
- \sum_{j=1}^{NP} m_j \, {}^N\mathbf{a}^{P_j} \cdot {}^N\mathbf{v}_i^{P_j} \quad (i = 1,\dots,n) \tag{1.5}
$$

Here, ${}^N\mathbf{a}^{B_j^*}, {}^N\mathbf{a}^{P_j}$ are the Newtonian frame accelerations of the mass centers B_j^* of
the body B_j and particle P_j, respectively; $\mathbf{I}^{B_j/B_j^*}$ is the *inertia dyadic* of B_j about B_j^*;
and ${}^N\boldsymbol{\alpha}^{B_j}$ is the angular acceleration of B_j in N. The i-th *generalized active force*
for this n degree of freedom system is obtained by taking dot-products of partial
velocities of NP number of particles P_j with the forces on P_j, and the forces on NR
number of bodies B_j with centers B_j^*, and of partial angular velocities of B_j with
the torques on B_j, and summing over all bodies and particles:

$$
F_i = \sum_{j=1}^{NR} [\mathbf{F}^{B_j^*} \cdot {}^N\mathbf{v}_i^{B_j^*} + \mathbf{T}^{B_j} \cdot {}^N\boldsymbol{\omega}_i^{B_j}] + \sum_{j=1}^{NP} \mathbf{F}^{P_j} \cdot {}^N\mathbf{v}_i^{P_j} \quad (i = 1,\dots,n) \tag{1.6}
$$

Here, the resultant of all contact and body forces on body B_j is $\mathbf{F}^{B_j^*}$ at B_j^* together with a couple of torque $\mathbf{T}^{B_j}$, and the resultant of external and contact forces on particle P_j is $\mathbf{F}^{P_j}$. Note that all non-working interaction forces are automatically eliminated by taking the sum in Eq. (1.6) over bodies and particles, with actions and reactions canceling, as generalized active forces are formed. Some special cases of generalized active force that are covered by Eq. (1.6) are those due to elastic-dissipative mechanical systems, by forces derivable from a *potential function* $V(q_1,\ldots, q_n, t)$ and dissipative forces derived from a *dissipation function* $D(u_1,\ldots, u_n)$, as follows:

$$F_i^c = -\frac{\partial V}{\partial q_i} - \frac{\partial D}{\partial u_i} \quad (i = 1,\ldots,n) \tag{1.7}$$

These are added to Eq. (1.6) to form the total of generalized active forces F_i in Eq. (1.6).

1.3   KANE'S EQUATIONS OF MOTION

Kane's equations for an n degree of freedom system can now be written by adding up the generalized active and inertia forces, and setting them equal to zero, as

$$F_i + F_i^* = 0, \quad i = 1,\ldots,n \quad \text{or} \quad -F_i^* = F_i, \quad i = 1,\ldots,n \tag{1.8}$$

These equations of motion, Eq. (1.8), together with the kinematical equation of Eq. (1.1), can be written as two sets of n coupled, non-linear, differential equations in matrix form:

$$[M(q)]\{\dot{U}\} = \{C(q, U, t)\} + \{F(q, U, t)\}$$
$$[W(q)]\{\dot{q}\} = \{U - X(q, t)\} \tag{1.9}$$

Here, $M(q)$ is called the n × n "mass matrix"; $C(q, U, t)$ is the n × 1 matrix of "Coriolis and centrifugal inertia forces"; and $F(q, U, t)$ is the n × 1 matrix of "generalized force". Equation (1.9) completely describes the dynamics of the system. The algebra involved in forming Eq. (1.5) and Eq. (1.6) can be quite massive for a complex mechanical system, as may be checked by an analyst deriving equations of motion of such systems by hand. That is why a computerized symbol manipulation code, was developed by Schaechter and Levinson in Ref. [9] to derive the equations of motion. Finally, it should be mentioned that Kane had originally [2] called Eq. (1.8), Lagrange's form of D'Alembert's principle, because just as Lagrange's equations can be derived by dot-multiplying the D'Alembert force equilibrium equations by the components of virtual displacement in a virtual work procedure, Kane obtains his equations by dot-multiplying the D'Alembert equilibrium equations by the partial velocities and partial angular velocities.

1.3.1 SIMPLE EXAMPLE: EQUATIONS FOR A DOUBLE PENDULUM

Figure 1.1 shows a planar double pendulum. Consider the links OP, PQ as mass-less rigid rods, each of length l, with lumped mass m at the end of each rod acted on only by gravity. Configuration of the pendulum is defined by two generalized coordinates q_1, q_2.

In order to use Kane's method, we can choose, as per Eq. (1.1), the following generalized speeds:

$$u_1 = \dot{q}_1; \ u_2 = \dot{q}_1 + \dot{q}_2 \text{ or } \begin{Bmatrix} \dot{q}_1 \\ \dot{q}_2 \end{Bmatrix} = \begin{Bmatrix} u_1 \\ u_2 - u_1 \end{Bmatrix} \tag{1.10}$$

Note that Eq. (1.10) is also the kinematical equation, defining the rates of change of the generalized coordinates in terms of the generalized speeds. The velocity of P in the Newtonian reference frame N can be written in terms of the angular velocity of the link OP in N, $u_1\mathbf{n}_3$, ($\mathbf{n}_3$ being perpendicular to the plane)

$$^N\mathbf{v}^P = u_1\mathbf{n}_3 \times l(\cos q_1 \mathbf{n}_1 + \sin q_1 \mathbf{n}_2) = l\, u_1(-\sin q_1 \mathbf{n}_1 + \cos q_1 \mathbf{n}_2) \tag{1.11}$$

Here, $\mathbf{n}_1$, $\mathbf{n}_2$, $\mathbf{n}_3$ are unit vectors in N, respectively, as shown in Figure 1.1.

The velocity of Q in N is given in terms of the velocity of P in N, by

$$\begin{aligned}
^N\mathbf{v}^Q &= {}^N\mathbf{v}^P + u_2\mathbf{n}_3 \times l[(\cos(q_1 + q_2)\mathbf{n}_1 + \sin(q_1 + q_2)\mathbf{n}_2] \\
&= \mathbf{n}_1 l[-u_1 \sin q_1 - u_2 \sin(q_1 + q_2)] + \mathbf{n}_2 l[u_1 \cos q_1 + u_2 \cos(q_1 + q_2)]
\end{aligned} \tag{1.12}$$

Here, $u_2\mathbf{n}_3$ is the angular velocity in N of the link PQ. Now we form partial velocities of P and Q, coefficients of generalized speeds u_1, u_2, that are obtained by inspection of Eq. (1.11) and Eq. (1.12), as shown in Table 1.1.

Accelerations of P and Q in N are obtained by differentiating the velocity vectors in N:

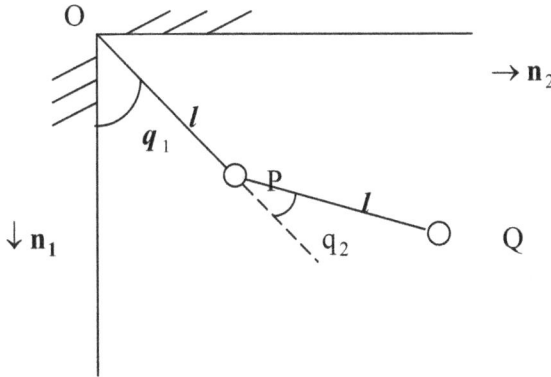

FIGURE 1.1 A planar double pendulum.

TABLE 1.1
**Partial Velocities for the Points P and Q in the Double
Pendulum Example**

r	v_r^P	v_r^Q
1	$l(-\sin q_1 \mathbf{n}_1 + \cos q_1 \mathbf{n}_2)$	$l(-\sin q_1 \mathbf{n}_1 + \cos q_1 \mathbf{n}_2)$
2	0	$l[-\sin(q_1+q_2)\mathbf{n}_1 + \cos(q_1+q_2)\mathbf{n}_2]$

$$^N\mathbf{a}^P = \frac{^N d\, ^N\mathbf{v}^P}{dt} = l[\dot{u}_1(-\sin q_1\mathbf{n}_1 + \cos q_1\mathbf{n}_2) + u_1^2(-\cos q_1\mathbf{n}_1 - \sin q_1\mathbf{n}_2)] \quad (1.13)$$

$$^N\mathbf{a}^Q = \frac{^N d\, ^N\mathbf{v}^Q}{dt} = l[\dot{u}_1(-\sin q_1\mathbf{n}_1 + \cos q_1\mathbf{n}_2) + u_1^2(-\cos q_1\mathbf{n}_1 - \sin q_1\mathbf{n}_2)]$$

$$+l\{\dot{u}_2[-\sin(q_1+q_2)\mathbf{n}_1 + \cos(q_1+q_2)\mathbf{n}_2] + \qquad\qquad (1.14)$$

$$u_2^2[-\cos(q_1+q_2)\mathbf{n}_1 - \sin(q_1+q_2)\mathbf{n}_2]\}$$

Generalized inertia forces in two generalized speeds are formed by using Eq. (1.5):

$$F_1^* = -ml^2[2\dot{u}_1 + \dot{u}_2\cos q_2 - u_2^2\sin q_2] \qquad\qquad (1.15)$$

$$F_2^* = -ml^2[\dot{u}_1\cos q_2 + u_1^2\sin q_2 + \dot{u}_2] \qquad\qquad (1.16)$$

These equations would be more complex had we chosen $u_i = \dot{q}_i, i = 1,2$. The relative simplicity of the form of Eqs. (1.15) and (1.16) is due to an efficient choice of generalized speeds, a guideline for which is described in the Appendix to this chapter. Generalized active forces due to gravity on P and Q are obtained as per Eq. (1.6).

$$F_1 = mg\mathbf{n}_1 \cdot \mathbf{v}_1^P + mg\mathbf{n}_1 \cdot \mathbf{v}_1^Q = -2\,mgl\sin q_1 \qquad\qquad (1.17)$$

$$F_2 = mg\mathbf{n}_1 \cdot\, ^N\mathbf{v}_2^P + mg\mathbf{n}_1 \cdot\, ^N\mathbf{v}_2^Q = -mgl\sin(q_1+q_2) \qquad\qquad (1.18)$$

Substituting Eqs. (1.15) through Eq. (1.18) into Kane's equation, Eq. (1.8), yields the dynamical equations:

$$ml^2[2\dot{u}_1 + \dot{u}_2\cos q_2 - u_2^2\sin q_2] = -2mgl\sin q_1 \qquad\qquad (1.19)$$

$$ml^2[\dot{u}_1\cos q_2 + u_1^2\sin q_2 + \dot{u}_2] = -mgl\sin(q_1+q_2) \qquad\qquad (1.20)$$

These dynamical equations are written in matrix form as

$$\begin{bmatrix} 2ml^2 & ml^2\cos q_2 \\ ml^2\cos q_2 & ml^2 \end{bmatrix}\begin{Bmatrix} \dot{u}_1 \\ \dot{u}_2 \end{Bmatrix} = (ml^2\sin q_2)\begin{Bmatrix} u_2^2 \\ -u_1^2 \end{Bmatrix} - \begin{Bmatrix} 2mgl\sin q_1 \\ mgl\sin(q_1+q_1) \end{Bmatrix} \quad (1.21)$$

Note that the multiplier matrix of the column matrix for the derivatives of the generalized speeds, the so-called mass matrix, is symmetric. Dynamical equations, Eq. (1.21), together with the kinematical equations, Eq. (1.10), complete the equations of motion for the double pendulum.

1.3.2 COMPLEX EXAMPLE: EQUATIONS OF MOTION FOR A SPINNING SPACECRAFT WITH THREE ROTORS, A TANK WITH THRUSTER FUEL SLOSH, AND A NUTATION DAMPER

Consider a more complex spacecraft example, for which equations of motion were derived by the author and reported in Ref. [10]. Figure 1.2 shows a spinning spacecraft with three reaction wheels, a nutation damper, and thruster fuel sloshing in a tank. The spacecraft bus is a rigid body G, with three reaction control rotors W_1, W_2, W_3; sloshing fuel is represented by an offset spherical pendulum with a massless rod with an end point mass m_P; and the nutation damper has a point mass m_Q, at point Q, with $\hat{Q}$ being the location in the spacecraft body G when the nutation damper spring is unstretched. The slosh pendulum attachment point is located from the mass center G* of G by $(-Z_0\mathbf{g}_3)$. Two angles ψ_1, ψ_2 in a body 2-1 rotation orient the pendulum representing fuel slosh. Let σ denote *the nominal length plus the elastic stretch of the spring* at the nutation damper of mass m_Q. This describes a 12 dof system, for which we define the generalized speeds u_i $(i=1,\ldots,12)$ as follows:

$$u_i = {}^N\mathbf{v}^{G^*}\cdot\mathbf{g}_i \quad i=1,2,3 \tag{1.22}$$

$$u_{3+i} = {}^N\boldsymbol{\omega}^G\cdot\mathbf{g}_i \quad i=1,2,3 \tag{1.23}$$

$$u_{6+i} = {}^G\boldsymbol{\omega}^{W_i}\cdot\mathbf{g}_i \quad i=1,2,3 \tag{1.24}$$

$$u_{9+i} = \dot{\psi}_i \quad i=1,2 \tag{1.25}$$

$$u_{12} = \dot{\sigma} \tag{1.26}$$

The velocity of G* follows from Eq. (1.22):

$$^N\mathbf{v}^{G^*} = u_1\mathbf{g}_1 + u_2\mathbf{g}_2 + u_3\mathbf{g}_3 \tag{1.27}$$

The angular velocity of G follows from Eq. (1.23):

$$^N\boldsymbol{\omega}^G = u_4\mathbf{g}_1 + u_5\mathbf{g}_2 + u_6\mathbf{g}_3 \tag{1.28}$$

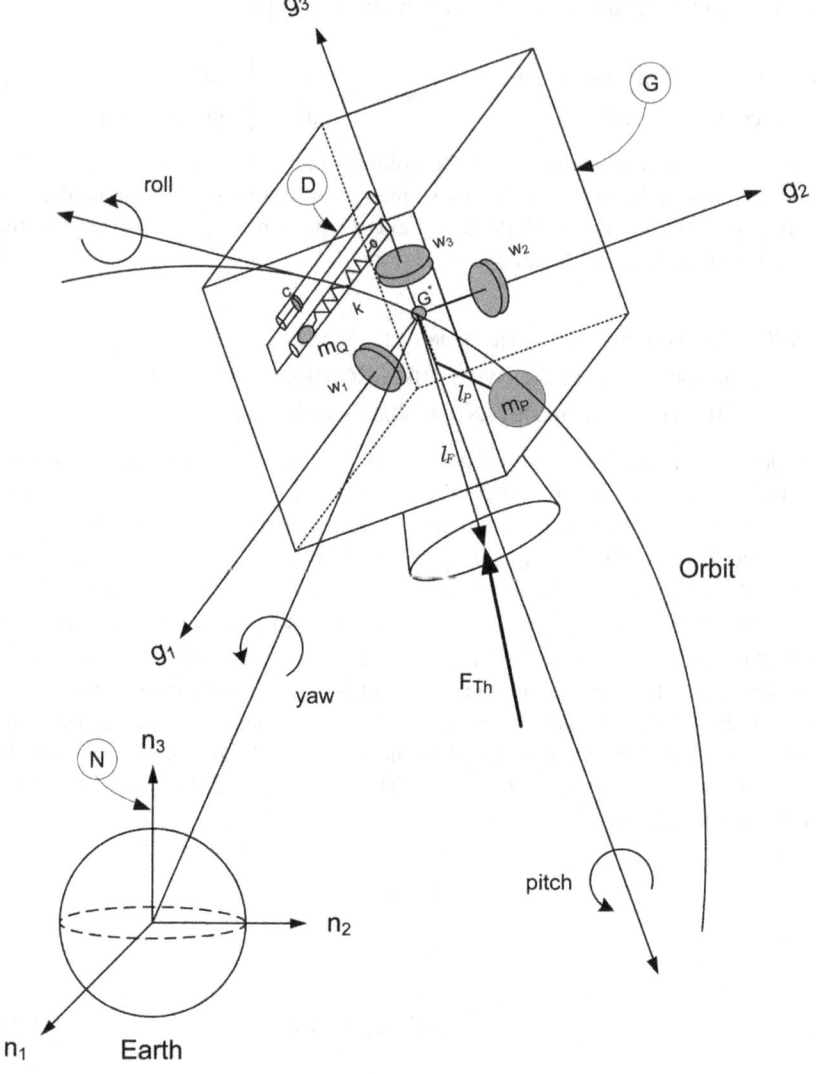

FIGURE 1.2 Spacecraft with three reaction wheels for attitude control, a thruster with thrust fuel sloshing in a tank represented by a spherical pendulum, and a nutation damper; the roll, pitch, and yaw angles shown are *about* the orbit frame, not the body frame.

Angular velocities of the wheels, given by the angular velocity addition theorem [3], follow from Eqs. (1.23) and (1.24) as

$$^{N}\omega^{W_1} = {}^{N}\omega^{G} + {}^{G}\omega^{W_1} = (u_4 + u_7)\mathbf{g}_1 + u_5\mathbf{g}_2 + u_6\mathbf{g}_3 \tag{1.29}$$

$$^{N}\omega^{W_2} = {}^{N}\omega^{G} + {}^{G}\omega^{W_2} = u_4\mathbf{g}_1 + (u_5 + u_8)\mathbf{g}_2 + u_6\mathbf{g}_3 \tag{1.30}$$

$$^N\boldsymbol{\omega}^{W_3} = {}^N\boldsymbol{\omega}^G + {}^G\boldsymbol{\omega}^{W_3} = u_4\mathbf{g}_1 + u_5\mathbf{g}_2 + (u_6 + u_9)\mathbf{g}_3 \tag{1.31}$$

Let $\mathbf{r}_1$, $\mathbf{r}_2$, $\mathbf{r}_3$ be a set of basis vectors fixed on the pendulum rod. For a body 2-1 rotation sequence [4], in angles ψ_1, ψ_2, and noting that the pendulum offset is along $(-z_0\mathbf{g}_3)$ as in Figure 1.2, we have the direction cosine matrix relation:

$$\begin{Bmatrix} \mathbf{g}_1 \\ \mathbf{g}_2 \\ \mathbf{g}_3 \end{Bmatrix} = \begin{bmatrix} c\psi_1 & s\psi_1 s\psi_2 & s\psi_1 c\psi_2 \\ 0 & c\psi_2 & -s\psi_2 \\ -s\psi_1 & c\psi_1 s\psi_2 & c\psi_1 c\psi_2 \end{bmatrix} \begin{Bmatrix} \mathbf{r}_1 \\ \mathbf{r}_2 \\ \mathbf{r}_3 \end{Bmatrix} \tag{1.32}$$

Here, we have denoted $c\psi_1 = \cos\psi_1$, $s\psi_2 = \sin\psi_2$, and so on. The position vector from G* to P is

$$\mathbf{p}^{G*P} = -z_0\mathbf{g}_3 + l_P\mathbf{r}_3 = l_P(s\psi_1 c\psi_2\mathbf{g}_1 - s\psi_2\mathbf{g}_2) + \left(-z_0 + l_P c\psi_1 c\psi_2\right)\mathbf{g}_3 \tag{1.33}$$

Note that the particle P at the end of the pendulum is moving on the moving body G. Hence, the inertial velocity of the particle of mass m_P at P is given by the theorem [3]:

$$^N\mathbf{v}^P = {}^N\mathbf{v}^{G*} + {}^N\boldsymbol{\omega}^G \times \mathbf{p}^{G*P} + \frac{{}^G d\mathbf{p}^{G*P}}{dt} \tag{1.34}$$

Using Eqs. (1.27) and (1.28) in Eq. (1.34) leads to

$$^N\mathbf{v}^P = \mathbf{g}_1[u_1 + u_5(-z_0 + l_P c\psi_1 c\psi_2) + u_6 l_P s\psi_2 + l_P(u_{10}c\psi_1 c\psi_2 - u_{11}s\psi_1 s\psi_2)]$$

$$+ \mathbf{g}_2[u_2 + u_6 l_P s\psi_1 c\psi_2 - u_4(-z_0 + l_P c\psi_1 c\psi_2) - l_P u_{11}c\psi_2] \tag{1.35}$$

$$+ \mathbf{g}_3[u_3 - l_P(u_4 s\psi_2 + u_5 s\psi_1 c\psi_2 + u_{10}s\psi_1 c\psi_2 + u_{11}c\psi_1 s\psi_2)]$$

The nutation damper mass particle position vector in Figure 1.2 is $\mathbf{p}^{G*Q} = \sigma\mathbf{g}_1 + z_Q\mathbf{g}_3$ (here σ denotes *the nominal length plus the elastic stretch of the spring*). Thus, the inertial velocity of Q is

$$^N\mathbf{v}^Q = {}^N\mathbf{v}^{G*} + {}^N\boldsymbol{\omega}^G \times \mathbf{p}^{G*Q} + \frac{{}^G d\mathbf{p}^{G*Q}}{dt} \tag{1.36}$$

$$= \mathbf{g}_1(u_1 + u_5 z_Q + u_{12}) + \mathbf{g}_2(u_2 + u_6\sigma - u_4 z_Q) + \mathbf{g}_3(u_3 - u_5\sigma)$$

With all the velocities and angular velocities expressed, a table of partial velocities and partial angular velocities can be formed from Eqs. (1.27)–(1.31), (1.34), and (1.35), as the coefficients of the generalized speeds in these equations. In the partial velocity table, Table 1.2, the first column refers to the index of the generalized speed, and the other columns describe corresponding partial velocities of G* P, Q, or the partial angular velocities of G, W_1, W_2, W_3.

The inertial angular accelerations of G, W_1, W_2, W_3 are obtained by differentiating the angular velocity expressions in Eqs. (1.28) through (1.31) in the N-frame:

$$^N\boldsymbol{\alpha}^G = \dot{u}_4\mathbf{g}_1 + \dot{u}_5\mathbf{g}_2 + \dot{u}_6\mathbf{g}_3 \tag{1.37}$$

TABLE 1.2

Partial Velocities for the Dynamics of a Spacecraft with Three Reaction Wheels, a Pendulum Representing Fuel Slosh, and a Nutation Damper

r	v_r^{G*}	ω_r^G	$\omega_r^{W_1}$	$\omega_r^{W_2}$	$\omega_r^{W_3}$	v_r^P	v_r^Q
1	$\mathbf{g}_1$	0	0	0	0	$\mathbf{g}_1$	$\mathbf{g}_1$
2	$\mathbf{g}_2$	0	0	0	0	$\mathbf{g}_2$	$\mathbf{g}_2$
3	$\mathbf{g}_3$	0	0	0	0	$\mathbf{g}_3$	$\mathbf{g}_3$
4	0	$\mathbf{g}_1$	$\mathbf{g}_1$	$\mathbf{g}_1$	$\mathbf{g}_1$	$-\mathbf{g}_2(-z_0 + l_P c\psi_1 c\psi_2) - \mathbf{g}_3 l_P s\psi_2$	$-\mathbf{g}_2 z_Q$
5	0	$\mathbf{g}_2$	$\mathbf{g}_2$	$\mathbf{g}_2$	$\mathbf{g}_2$	$\mathbf{g}_1(-z_0 + l_P c\psi_1 c\psi_2) - \mathbf{g}_3 l_P s\psi_1 c\psi_2$	$\mathbf{g}_1 z_Q - \mathbf{g}_3 \sigma$
6	0	$\mathbf{g}_3$	$\mathbf{g}_3$	$\mathbf{g}_3$	$\mathbf{g}_3$	$\mathbf{g}_1 l_P s\psi_2 + \mathbf{g}_2 l_P s\psi_1 c\psi_2$	$\mathbf{g}_2 \sigma$
7	0	0	$\mathbf{g}_1$	0	0	0	0
8	0	0	0	$\mathbf{g}_2$	0	0	0
9	0	0	0	0	$\mathbf{g}_3$	0	0
10	0	0	0	0	0	$\mathbf{g}_1 l_P c\psi_1 c\psi_2 - \mathbf{g}_3 l_P s\psi_1 c\psi_2$	0
11	0	0	0	0	0	$-\mathbf{g}_1 l_P s\psi_1 s\psi_2 - \mathbf{g}_2 l_P c\psi_2 - \mathbf{g}_3 l_P c\psi_1 s\psi_2$	0
12	0	0	0	0	0	0	$\mathbf{g}_1$

$$^N\alpha^{W_1} = \dot{u}_4\mathbf{g}_1 + (\dot{u}_5 + u_6 u_7)\mathbf{g}_2 + (\dot{u}_6 - u_5 u_7)\mathbf{g}_3 + \dot{u}_7\mathbf{g}_1 \tag{1.38}$$

$$^N\alpha^{W_2} = (\dot{u}_4 - u_6 u_8)\mathbf{g}_1 + \dot{u}_5\mathbf{g}_2 + (\dot{u}_6 + u_4 u_8)\mathbf{g}_3 + \dot{u}_8\mathbf{g}_2 \tag{1.39}$$

$$^N\alpha^{W_3} = (\dot{u}_4 + u_5 u_9)\mathbf{g}_1 + (\dot{u}_5 - u_4 u_9)\mathbf{g}_2 + \dot{u}_6\mathbf{g}_3 + \dot{u}_9\mathbf{g}_3 \tag{1.40}$$

Accelerations of G*, P, Q are found by differentiating velocities in the N-frame, yielding

$$^N\mathbf{a}^{G*} = \dot{u}_1\mathbf{g}_1 + \dot{u}_2\mathbf{g}_2 + \dot{u}_3\mathbf{g}_2 + {}^N\omega^G \times {}^N\mathbf{v}^{G*} \tag{1.41}$$

$$^N\mathbf{a}^P = \frac{{}^G d\,{}^N\mathbf{v}^P}{dt} + {}^N\omega^G \times {}^N\mathbf{v}^P \tag{1.42}$$

$$^N\mathbf{a}^Q = \frac{{}^G d\,{}^N\mathbf{v}^Q}{dt} + {}^N\omega^G \times {}^N\mathbf{v}^Q \tag{1.43}$$

The negative of the generalized inertia forces for this example system is now formed following Eq. (1.5) as follows:

$$-F_r^* = \left[\mathbf{I}^{G/G*} \cdot {}^N\boldsymbol{\alpha}^G + {}^N\boldsymbol{\omega}^G \times (\mathbf{I}^{G/G*} \cdot {}^N\boldsymbol{\omega}^G) \right] \cdot \boldsymbol{\omega}_r^G$$

$$+ \sum_{i=1}^{3} \left[\mathbf{I}^{W_i/W_i*} \cdot {}^N\boldsymbol{\alpha}^{W_i} + {}^N\boldsymbol{\omega}^{W_i} \times \left(\mathbf{I}^{W_i/W_i*} \cdot {}^N\boldsymbol{\omega}^{W_i} \right) \right] \cdot {}^N\boldsymbol{\omega}_r^{W_i} \tag{1.44}$$

$$+ m_G \, {}^N\mathbf{a}^{G*} \cdot {}^N\mathbf{v}_r^{G*} + m_P \, {}^N\mathbf{a}^P \cdot {}^N\mathbf{v}_r^P + m_Q \, {}^N\mathbf{a}^Q \cdot {}^N\mathbf{v}_r^Q \quad (r = 1,\dots,12)$$

Here the assumption is that the mass of the three wheels is embedded in the mass of G. The quantities with boldface **I** with a superscript denote the *dyadics* of the moment of inertia about the mass center of the body concerned. Generalized active forces, due to thrust, gravity, thrust torque about G* (due to the thrust line of action *not* passing through G*), wheel torques, spring-damper forces from an unstretched length σ_0 on the nutation damper Q, and viscous torque on the pendulum P are obtained by following Eq. (1.6):

$$F_r = (\mathbf{F}_{th}^{G*} + \mathbf{F}_g^{G*}) \cdot {}^N\mathbf{v}_r^{G*} + \mathbf{T}_{th}^G \cdot \boldsymbol{\omega}_r^G + \sum_{i=1}^{3} \mathbf{T}^{W_i} \cdot {}^G\boldsymbol{\omega}_r^{W_i}$$

$$+ (\mathbf{F}_g^P - \mathbf{F}_v^P) \cdot {}^N\mathbf{v}_r^P + \{\mathbf{F}_g^Q \tag{1.45}$$

$$- [k(\sigma - \sigma_0) + c\dot{\sigma}]\mathbf{g}_1\} \cdot {}^G\mathbf{v}_r^Q \quad (r = 1,\dots,12)$$

Equations (1.44) and (1.45) are put in Kane's dynamical equations, Eq. (1.8).

Finally, for an assumed *body 3-2-1 rotation* in the angles, $\theta_1, \theta_2, \theta_3$, the attitude of G, the kinematical equations relating rotation rates to angular velocity components are [4]

$$\dot{\theta}_1 = (u_5 s_3 + u_6 c_3)/c_2$$

$$\dot{\theta}_2 = u_5 c_3 - u_6 s_3 \tag{1.46}$$

$$\dot{\theta}_3 = s_2 (u_5 s_3 + u_6 c_3)/c_2 + u_4$$

A detailed analysis and simulation results are given in Ref. [8]. First, the well-known result of flat-spin of a spacecraft, initially spinning about the minimum axis $\mathbf{b}_2$, with dissipation provided by the nutation damper, re-orients and spins about the maximum axis of inertia, $\mathbf{b}_1$ (see Figure 1.3A).

Angular velocity component about the $\mathbf{b}_3$ axis after flat-spin behaves similar to that in the top part in Figure 1.3A. Next, we consider the case of a *desired reorientation* behavior of the spacecraft. Let the spacecraft spin initially about the axis of maximum moment of inertia, and then we then spin up a reaction wheel about the intermediate axis of inertia about $\mathbf{b}_3$ as per the following wheel speed function of time:

$${}^N\omega^{W_3} = 0 \qquad\qquad\qquad 0 \le t < 1000$$

$$= 0.9375(t - 1000) \qquad 1000 \le t < 7400 \tag{1.47}$$

$$= 6000 \qquad\qquad\qquad 7400 \le t \le 10000$$

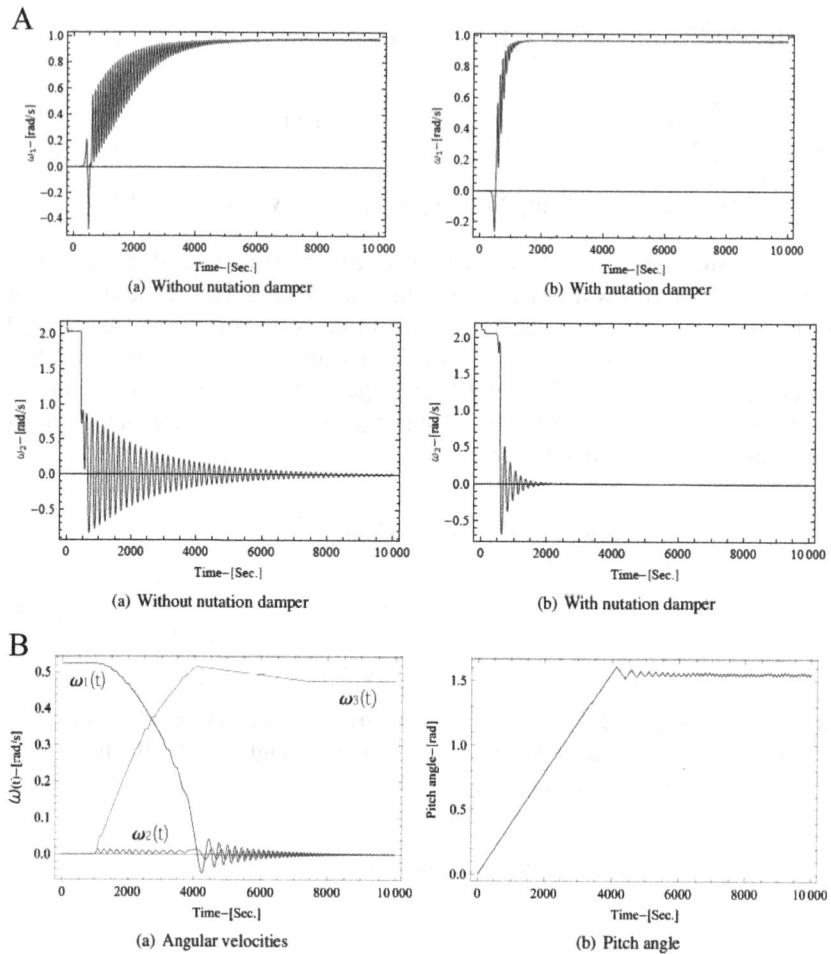

FIGURE 1.3   A: Flat spin of spacecraft. B: Reorientation of spacecraft.

The resulting behavior of the spacecraft angular velocity and pitch angle are shown in Figure 1.3B.

Figure 1.3A shows the flat spin behavior of the spacecraft, as the angular velocity of rotation about minimum axis of moment of inertia (lower two curves for ω_2) changes to rotation about maximum axis of moment of inertia (upper two curves for ω_3). Figure 1.3B shows the reorientation behavior of the spacecraft spinning initially about the maximum inertia axis by spinning up a wheel along the axis of intermediate axis of inertia; angular velocity components and the pitch angle are shown.

Equations (1.5) and (1.6) show that Kane's equations require computing accelerations and angular accelerations, and taking dot-products with partial velocities and partial angular velocities. Also, a good choice of generalized speeds leads to simpler final equations. These two aspects of reduced labor and simplicity of final equations make Kane's equations superior to Lagrange's equations, which are reviewed next.

1.3.3 COMPARISON TO DERIVATION OF EQUATIONS OF MOTION BY LAGRANGE'S METHOD: APPLICATION TO THE SAME COMPLEX EXAMPLE SHOWN IN FIGURE 1.2

Lagrange's equations are widely used as a method for deriving equations of motion. The formulation of equations by this method consists of forming the system kinetic energy function, $K = f(q_1,\ldots,q_n,\dot{q}_1,\ldots,\dot{q}_n)$, where q_i, $i = 1,\ldots,n$ are the n generalized coordinates, and $\dot{q}_i$ are their time-derivatives. Lagrange's equations are given by

$$\frac{d}{dt}\frac{\partial K}{\partial \dot{q}_i} - \frac{\partial K}{\partial q_i} = Q_i - \frac{\partial V}{\partial q_i} - \frac{\partial D}{\partial \dot{q}_i} \quad (i = 1,\ldots,n) \tag{1.48}$$

Here, Q_i are the *generalized active forces* due to forces not derivable from a potential function or a dissipation function; Q_i are evaluated in the same way as in Kane's method if the generalized speeds are chosen as the derivatives of the generalized coordinates. Further, $V(q_1,\ldots,q_n)$ is the potential function, and $D(\dot{q}_1,\ldots,\dot{q}_n)$ is the dissipation function. Equation (1.48) is a road map that one follows to derive equations of motion by Lagrange's method. To illustrate the process, consider the same system shown in Figure 1.2, and start with the kinetic energy function for this system, written as follows:

$$K = (1/2)[m_G \, {}^N\mathbf{v}^{G*} \cdot {}^N\mathbf{v}^{G*} + {}^N\boldsymbol{\omega}^G \cdot \mathbf{I}^{G/G*} \cdot {}^N\boldsymbol{\omega}^G]$$

$$+(1/2)\sum_{i=1}^{3}\left[{}^N\boldsymbol{\omega}^{W_i} \cdot \mathbf{I}^{W_i/W_i*} \cdot {}^N\boldsymbol{\omega}^{W_i} \right] + (1/2)\left[m_P \, {}^N\mathbf{v}^P \cdot {}^N\mathbf{v}^P + m_Q \, {}^N\mathbf{v}^Q \cdot {}^N\mathbf{v}^Q \right]$$

$$\tag{1.49}$$

Here, ${}^N\mathbf{V}^{G*}$, ${}^N\boldsymbol{\omega}^G$, ${}^N\boldsymbol{\omega}^{W_i}$, ${}^N\mathbf{V}^P$, ${}^N\mathbf{V}^Q$ have all been expressed before in terms of generalized speeds, and since Lagrange's equations work with generalized coordinates, rather than generalized speeds, we have to express $u_1,\ldots,u_{12}$ in Eq. (1.49) as $u_i = \dot{q}_i$, $i = 1,\ldots,12$. A choice of roll–pitch–yaw sequence can be made by the analyst; assuming a body 1-2-3 rotation sequence [4], the direction cosine matrix relating the body components of the mass center velocity to their inertial components, and the relation between the body angular velocity components to the Euler angle rates are given below:

$$\begin{bmatrix} u_1 \\ u_2 \\ u_3 \end{bmatrix} = \begin{bmatrix} cq_5cq_6 & -cq_5sq_6 & sq_5 \\ sq_4sq_5cq_6 + sq_6cq_4 & -sq_4sq_5sq_6 + cq_6cq_4 & -sq_4cq_5 \\ -cq_4sq_5cq_6 + sq_6sq_4 & cq_4sq_5sq_6 + cq_6sq_4 & cq_4cq_5 \end{bmatrix}\begin{bmatrix} \dot{q}_1 \\ \dot{q}_2 \\ \dot{q}_3 \end{bmatrix} \tag{1.50}$$

$$\begin{bmatrix} u_4 \\ u_5 \\ u_6 \end{bmatrix} = \begin{bmatrix} cq_5cq_6 & sq_6 & 0 \\ -cq_5sq_6 & cq_6 & 0 \\ sq_5 & 0 & 1 \end{bmatrix}\begin{bmatrix} \dot{q}_4 \\ \dot{q}_5 \\ \dot{q}_6 \end{bmatrix} \tag{1.51}$$

Here, q_1, q_2, q_3 are the inertial position coordinates of the mass center, and q_4, q_5, q_6 are three Euler angles for attitude. Wheel angles, slosh pendulum angles, and damper location are given by

$$u_{6+i} = \dot{q}_{6+i} \quad i = 1, 2, 3$$

$$u_{9+i} = \dot{q}_{9+i} \quad i = 1, 2 \qquad\qquad (1.52)$$

$$u_{12} = \dot{q}_{12} = \dot{\sigma}$$

Equations (1.50)–(1.52) complete the 12 kinematical equations. The generalized coordinates and their rates, $q_i, \dot{q}_i$; $i = 1, \ldots, 12$, can now be substituted in the kinetic energy expression in terms of the generalized coordinates and their time-derivatives. The expression for K in Eq. (1.49) resulting from these substitutions is a formidably complex function, when written explicitly.

$$
\begin{aligned}
K = (1/2)\, m_G &\{ [\dot{q}_1 cq_5 cq_6 + \dot{q}_2 (sq_4 sq_5 cq_6 + sq_6 cq_4) + \dot{q}_3 (-cq_4 sq_5 cq_6 + sq_6 sq_4)]^2 \\
&+ [-\dot{q}_1 cq_5 sq_6 + \dot{q}_2 (-sq_4 sq_5 sq_6 + cq_6 cq_4) + \dot{q}_3 (cq_4 sq_5 sq_6 + cq_6 sq_4)]^2 \\
&+ [\dot{q}_1 sq_5 - \dot{q}_2 sq_4 cq_5 + \dot{q}_3 cq_4 cq_5]^2 \} + (1/2) \{ I_1 (\dot{q}_4 cq_5 cq_6 + \dot{q}_5 sq_6)^2 \\
&+ I_2 (-\dot{q}_4 cq_5 sq_6 + \dot{q}_5 cq_6)^2 + I_3 (\dot{q}_4 sq_5 + \dot{q}_6)^2 \} \\
&+ (1/2) \{ I_1^W [(\dot{q}_4 cq_5 cq_6 + \dot{q}_5 sq_6) + \dot{q}_7]^2 + I_2^W [(-\dot{q}_4 cq_5 sq_6 + \dot{q}_5 cq_6) + \dot{q}_8]^2 \\
&+ I_3^W [(\dot{q}_4 sq_5 + \dot{q}_6) + \dot{q}_9]^2 \} \\
&+ (1/2)\, m_P \{ [\dot{q}_1 cq_5 cq_6 + \dot{q}_2 (sq_4 sq_5 cq_6 + sq_6 cq_4) + \dot{q}_3 (-cq_4 sq_5 cq_6 + sq_6 sq_4) \\
&+ (-\dot{q}_4 cq_5 sq_6 + \dot{q}_5 cq_6)(-z_0 - l_P cq_{11}) - (\dot{q}_4 sq_5 + \dot{q}_6)l_P sq_{10} sq_{11} \\
&+ l_P (u_{11} cq_{10} cq_{11} - u_{10} sq_{10} sq_{11})]^2 \\
&+ [-\dot{q}_1 cq_5 sq_6 + \dot{q}_2 (-sq_4 sq_5 sq_6 + cq_6 cq_4) + \dot{q}_3 (cq_4 sq_5 sq_6 + cq_6 sq_4) \\
&+ (\dot{q}_4 sq_5 + \dot{q}_6)l_P cq_{10} sq_{11} - (\dot{q}_4 cq_5 cq_6 + \dot{q}_5 sq_6)(-z_0 - l_P cq_{11}) \\
&+ l_P (u_{10} cq_{10} sq_{11} + u_{11} sq_{10} cq_{11})]^2 + [\dot{q}_1 sq_5 - \dot{q}_2 sq_4 cq_5 + \dot{q}_3 cq_4 cq_5 \\
&+ l_P (u_4 sq_{10} sq_{11} - u_5 cq_{10} sq_{11} + u_{11} sq_{11})]^2 \} \\
&+ (1/2)\, m_Q \{ [\dot{q}_1 cq_5 cq_6 + \dot{q}_2 (sq_4 sq_5 cq_6 + sq_6 cq_4) + \dot{q}_3 (-cq_4 sq_5 cq_6 + sq_6 sq_4) \\
&+ (-\dot{q}_4 cq_5 sq_6 + \dot{q}_5 cq_6) z_Q + \dot{q}_{12}]^2 \\
&+ [-\dot{q}_1 cq_5 sq_6 + \dot{q}_2 (-sq_4 sq_5 sq_6 + cq_6 cq_4) + \dot{q}_3 (cq_4 sq_5 sq_6 + cq_6 sq_4) \\
&+ (\dot{q}_4 sq_5 + \dot{q}_6)\sigma - (\dot{q}_4 cq_5 cq_6 + \dot{q}_5 sq_6) z_Q]^2 \\
&+ [\dot{q}_1 sq_5 - \dot{q}_2 sq_4 cq_5 + \dot{q}_3 cq_4 cq_5 - (-\dot{q}_4 cq_5 sq_6 + \dot{q}_5 cq_6)\sigma]^2 \}
\end{aligned}
$$

$$(1.53)$$

Shabana [11], in his book on multibody dynamics, has used Lagrange's equations to derive equations of motion for very simple systems. However, for complex systems, Lagrange's equations are too laborious to apply.

For each of the n generalized coordinates q_i, one must (1) form K, and differentiate it with respect to the derivatives of the generalized coordinates, $\dot{q}_i$; (2) differentiate the *resulting expression* with respect to time; and (3) differentiate K with respect to the generalized coordinates q_i. Looking at the expression for K, and using Eq. (1.48), for $q_i = 1,\ldots,12$, it is clear what a formidable task this is! Here, no attempt was made to get a solution using Lagrange's method; not only is the labor prohibitive to an analyst, but even when done by a machine using computer algebra, the resulting set of equations will take an enormous number of terms to evaluate. This is directly in contrast to the simple operations of just taking the *dot-products involved in Kane's formulation* to obtain the equations. Figure 1.3, which shows the results of the system in Figure 1.2, illustrates that while Kane's equations were formed and solved for this complex problem, just to form the equations by Lagrange's method, before the solution is obtained, is itself a massive task. Thus, in terms of required labor alone, Kane's method is superior to Lagrange's method, by far.

Another significant reason for the use of Kane's equations, as has already been seen in the example of the complex spacecraft shown in Figure 1.2, and will be repeatedly seen in later examples in this book, is the relative simplicity of the final form of the equations. This is made possible by the fact that Kane uses generalized speeds rather than derivatives of generalized coordinates as the motion variables. Kane's generalized speeds have been thought of as time-derivatives of "quasi-coordinates". A form of Lagrange's equations for quasi-coordinates, known as Boltzmann–Hamel equations, is far more complex than Eq. (1.48), and is computationally very intensive, as seen below.

1.3.4 BOLTZMANN–HAMEL EQUATIONS

Consider the definition of quasi-coordinates, Eq. (1.1), in a slightly different form and their inverse relation:

$$u_j = \sum_{i=1}^{n} W_{ji}\dot{q}_i + X_j \quad (j = 1,\ldots,n) \tag{1.54}$$

$$\dot{q}_k = \sum_{i=1}^{n} Z_{ki}u_i - Y_k \quad (k = 1,\ldots,n) \tag{1.55}$$

The system kinetic energy can be represented in two ways, using Eq. (1.55), as follows:

$$K = K(q_1,\ldots,q_n,\dot{q}_1,\ldots,\dot{q}_n) = \overline{K}(q_1,\ldots,q_n,u_1,\ldots,u_n) \tag{1.56}$$

This leads to the derivative forms, based on Eq. (1.54), and their time-derivative:

$$\frac{\partial K}{\partial \dot{q}_i} = \sum_{j=1}^{n} \frac{\partial \overline{K}}{\partial u_j} \frac{\partial u_j}{\partial \dot{q}_i} = \sum_{j=1}^{n} \frac{\partial \overline{K}}{\partial u_j} W_{ji} \tag{1.57}$$

$$\frac{d}{dt}\left\{\frac{\partial K}{\partial \dot{q}_i}\right\} = \sum_{j=1}^{n} \left\{\frac{\partial \overline{K}}{\partial u_j}\right\} \dot{W}_{ji} + \sum_{j=1}^{n} \frac{d}{dt}\left\{\frac{\partial \overline{K}}{\partial u_j}\right\} W_{ji} \tag{1.58}$$

Now, $\dot{W}_{ji}$ required in Eq. (1.58) can be written after recognizing the dependence of W_{ji} on the generalized coordinates:

$$\dot{W}_{ji} = \sum_{k=1}^{n} \frac{\partial W_{ji}}{\partial q_k} \dot{q}_k \tag{1.59}$$

It follows from Eqs. (1.58) and (1.59) that

$$\frac{d}{dt}\left\{\frac{\partial K}{\partial \dot{q}_i}\right\} = \sum_{j=1}^{n} \sum_{k=1}^{n} \left\{\frac{\partial \overline{K}}{\partial u_j}\right\} \frac{\partial W_{ji}}{\partial q_k} \dot{q}_k + \sum_{j=1}^{n} \frac{d}{dt}\left\{\frac{\partial \overline{K}}{\partial u_j}\right\} W_{ji} \tag{1.60}$$

Now replacing Eq. (1.57) in Eq. (1.60), one gets

$$\frac{d}{dt}\left\{\frac{\partial K}{\partial \dot{q}_i}\right\} = \sum_{j=1}^{n} \sum_{k=1}^{n} \left\{\frac{\partial \overline{K}}{\partial u_j}\right\} \frac{\partial W_{ji}}{\partial q_k} \left[\sum_{i=1}^{n} Z_{ki} u_i - Y_k\right] + \sum_{j=1}^{n} \frac{d}{dt}\left\{\frac{\partial \overline{K}}{\partial u_j}\right\} W_{ji} \tag{1.61}$$

It follows from Eqs. (1.54) through Eq. (1.56) that

$$\frac{\partial K}{\partial q_i} = \frac{\partial \overline{K}}{\partial q_i} + \sum_{j=1}^{n} \frac{\partial \overline{K}}{\partial u_j} \frac{\partial u_j}{\partial q_i} = \frac{\partial \overline{K}}{\partial q_i} + \sum_{j=1}^{n} \frac{\partial \overline{K}}{\partial u_j} \left\langle \sum_{k=1}^{n} \frac{\partial W_{jk}}{\partial q_i} \dot{q}_k + \frac{\partial X_k}{\partial q_i} \right\rangle \quad i = 1, \ldots, n \tag{1.62}$$

Again, using Eq. (1.55) in Eq. (1.62) gives

$$\frac{\partial K}{\partial q_i} = \frac{\partial \overline{K}}{\partial q_i} + \sum_{j=1}^{n} \frac{\partial \overline{K}}{\partial u_j} \left\langle \sum_{k=1}^{n} \frac{\partial W_{jk}}{\partial q_i} \left[\sum_{i=1}^{n} Z_{ki} u_i - Y_k\right] + \frac{\partial X_k}{\partial q_i} \right\rangle \quad i = 1, \ldots, n \tag{1.63}$$

Using Eqs. (1.61) and (1.63), we can complete the form of Boltzmann–Hamel equations:

$$\sum_{j=1}^{n} \sum_{k=1}^{n} \left\{\frac{\partial \overline{K}}{\partial u_j}\right\} \frac{\partial W_{ji}}{\partial q_k} \left[\sum_{i=1}^{n} Z_{ki} u_i - Y_k\right] + \sum_{j=1}^{n} \frac{d}{dt}\left\{\frac{\partial \overline{K}}{\partial u_j}\right\} W_{ji} - \frac{\partial \overline{K}}{\partial q_i}$$

$$- \sum_{j=1}^{n} \frac{\partial \overline{K}}{\partial u_j} \left\langle \sum_{k=1}^{n} \frac{\partial W_{jk}}{\partial q_i} \left[\sum_{i=1}^{n} Z_{ki} u_i - Y_k\right] + \frac{\partial X_k}{\partial q_i} \right\rangle = F_i \quad i = 1, \ldots, n \tag{1.64}$$

Here, F_i is the generalized active force in the i-th quasi-coordinate, which is formed exactly as in Kane's method using generalized speeds. Note from Eq. (1.64) that one needs to invert the coefficient matrix in the definition of the quasi-coordinates in Eq. (1.54) and compute n^3 number of non-vanishing quantities in the following term T_{BH} in Eq. (1.64):

$$T_{BH} = \sum_{j=1}^{n} \sum_{k=1}^{n} \left[\left\{ \frac{\partial \overline{K}}{\partial u_j} \right\} \frac{\partial W_{ji}}{\partial q_k} \right] \sum_{i=1}^{n} Z_{ki} u_i - \sum_{j=1}^{n} \frac{\partial \overline{K}}{\partial u_j} \sum_{k=1}^{n} \frac{\partial W_{jk}}{\partial q_i} \sum_{i=1}^{n} Z_{ki} u_i \quad (i,j,k = 1,...,n)$$

(1.65)

For the 12 dof system of Figure 1.1, this means forming 1728 such terms after forming the inverse implied in Eq. (1.54). It is evident that the Boltzmann–Hamel equations, Eq. (1.64), are prohibitively labor-intensive to use as a tool in deriving equations of motion. It seems, as has been noted in Ref. [5], that the fact that Boltzmann–Hamel equations are used at all is surprising, except perhaps to check the correctness of equations derived by other methods.

1.3.5 GIBBS EQUATIONS

Gibbs equations, sometimes called Gibbs–Appell equations, see Refs. [6,7,8], use quasi-coordinates and are easier to use than Boltzmann–Hamel equations. Gibbs equations are written in terms of a Gibbs function, G:

$$\frac{\partial G}{\partial \dot{u}_i} = \sum_{j=1}^{n} W_{ji} Q_j \quad i = 1,...,n$$

(1.66)

Here, G is defined for a system of ν particles $P_1,...,P_\nu$, and nb rigid bodies B_i as

$$G = G_P + G_R$$

(1.67)

$$G_P = \frac{1}{2} \sum_{i=1}^{\nu} m_i \left({}^N\mathbf{a}^{P_i} \cdot {}^N\mathbf{a}^{P_i} \right)$$

(1.68a)

$$G_B = \frac{1}{2} \left[\sum_{i=1}^{nb} M_i \, {}^N\mathbf{a}^{B_i^*} \cdot {}^N\mathbf{a}^{B_i^*} + {}^N\alpha^{B_i} \cdot \mathbf{I}^{B_i/B_i^*} \cdot {}^N\alpha^{B_i} + 2 \, {}^N\alpha^{B_i} \cdot {}^N\omega^{B_i} \times \mathbf{I}^{B_i/B_i^*} \cdot {}^N\omega^{B_i} \right]$$

(1.68b)

In Eq. (1.66), Q_j is the generalized force corresponding to the generalized coordinate q_j in Lagrange's equations, i.e., $u_j = \dot{q}_j$, and W_{ji} is obtained from Eq. (1.54) for the general case defining generalized speeds.

First thing to note here is that one has to (1) form the square of the acceleration expression G, which is generally a lengthy expression when rigid bodies are involved, as shown by Eq. (1.68a) and Eq. (1.68b), though in the light of Eq. (1.66) only terms involving derivatives of generalized speeds need to be kept; (2) take the derivatives of G with respect to the derivatives of generalized speeds; (3) get $\dot{q}_i$ in terms of u_i needed to form the right-hand side in Eq. (1.66) by inverting the $n \times n$ matrix in Eq. (1.55), a labor-intensive task: see Ref. [6,7,9]. In the context of the dynamics of multibody systems with flexible components, the overall labor involved in performing these three tasks noted above is formidable.

It has been commented sometimes [6] that Kane's equations are just another form of Gibbs equations. Reviewing the processes involved in Kane's method (requiring dot-products of partial velocities with accelerations, etc.) and Gibbs' method (squaring the accelerations in Gibbs function, and differentiating, etc.), it is seen that such comments are entirely unfounded.

1.3.5.1   Reader's Exercise

Consider the double pendulum problem given before, and derive Lagrange's equation in quasi-coordinates, Eq. (1.64), for it, using Eq. (1.10) as the quasi-coordinates. Observe that one indeed forms n^3 or eight non-vanishing quantities in forming Eq. (1.64).

1.4   KANE'S METHOD OF DIRECT DERIVATION OF LINEARIZED DYNAMICAL EQUATION

It is often necessary to derive linearized dynamical equations, as for example in considering the stability of an equilibrium state, or when some variables are assumed to stay small as in the case of elastic deformation of flexible bodies undergoing large overall motion. Linearized equations can, of course, be derived by first deriving the non-linear equations and then linearizing them by discarding all non-linear terms. A simpler method, of directly deriving the linear equations without deriving the non-linear equations, has been given by Kane in Ref. [3]. This direct method of linearization consists of two steps to be followed in sequence:

1. First, form the partial velocities of all particles and mass centers of rigid bodies, and the partial angular velocities of all rigid bodies from the corresponding non-linear expressions for velocity and angular velocity, and then linearize these non-linear partial velocity and partial angular velocity expressions.
2. Now linearize the velocity and angular velocity expressions, differentiate these to get accelerations and angular accelerations, and form the generalized inertia force and generalized active force expressions by dot-multiplications with the linearized partial velocity and partial angular velocity expressions, discarding all non-linear terms in the process.

Kane and Levinson [3] have illustrated the application of this procedure to a complex problem. Below, we provide an analytical basis for this process. Consider an n degree of freedom system, with a particle P_k whose velocity in a Newtonian frame N is given by Eq. (1.2), and let its i-th partial velocity vector be expanded in a Taylor series in deviations about the equilibrium state of $q_1 = \cdots = q_n = 0;\ u_1 = \cdots = u_n = 0$:

$$\mathbf{v}_i^{P_k} = \sum_{i=1}^{n}\left[\mathbf{f}_i^k(0,\ldots,0,t) + \sum_{j=1}^{n}\left[\frac{\partial \mathbf{f}_i^k}{\partial q_j}\right]_0 q_j + \text{h.o.t.}\right] + {}^N\mathbf{v}_t^{P_k} \qquad (1.69)$$

Here, the derivatives are evaluated about the zero state, $q_1 = \cdots = q_n = 0$, and h.o.t. stands for higher order terms in q's. Because of Eqs. (1.2) and (1.69), the velocity of P_k in N about the equilibrium state becomes

$$\mathbf{v}^{P_k} = \sum_{i=1}^{n}\left[\mathbf{f}_i^k(0,\ldots,0,t) + \sum_{j=1}^{n}\left[\frac{\partial \mathbf{f}_i^k}{\partial q_j}\right]_0 q_j + \text{h.o.t.}\right] u_i + {}^N\mathbf{v}_t^{P_k} \qquad (1.70)$$

Eq. (1.70) is a non-linear expression of the velocity because of the product of q's and u's. Here, the coefficient vector of $u_i, i = 1,\ldots,n$, is the i-th non-linear, partial velocity, and linearizing it in the generalized coordinates, q's, the correctly linearized i-th partial velocity of P_k is

$$^N\tilde{\mathbf{v}}_i^{P_k} = \mathbf{f}_i^k(0,\ldots,0,t) + \sum_{j=1}^{n}\left[\frac{\partial \mathbf{f}_i^k}{\partial q_j}\right]_0 q_j \qquad (1.71)$$

The linearized expression for the velocity of P_k in N is obtained from Eq. (1.71) by ignoring the products of u's and q's, yielding

$$^N\tilde{\mathbf{v}}^{P_k} = \sum_{i=1}^{n}\left[\mathbf{f}_i^k(0,\ldots,0,t)\right] u_i + {}^N\mathbf{v}_t^{P_k} \qquad (1.72)$$

The partial velocity of P_k with respect to the i-th generalized speed, obtained from this linear velocity, Eq. (1.72), is

$$^N\hat{\mathbf{v}}_i^{P_k} = \mathbf{f}_i^k(0,\ldots,0,t) \qquad (1.73)$$

Comparing Eq. (1.73) with Eq. (1.71), we see that the expression for the partial velocity obtained from the linearized velocity, Eq. (1.72), has already lost a term involving q's in Eq. (1.71). This is called an error of premature linearization, and we will see that this makes a crucial difference between correct and incorrect linear equations. A similar analysis can be done to show that the correct partial angular velocity of a rigid body is obtained only by linearizing the non-linear partial angular velocity obtained from the non-linear angular velocity expression.

The foregoing analysis shows that the correctly linearized partial velocity/ partial angular velocity already contains terms involving zeroth and first degree terms in the q's. Thus, in taking dot-products of these quantities with the acceleration/angular acceleration terms, a process involving *products of* q's *and* u's, one has to remove the non-linear terms anyway to get equations linear in q's and u's.

This justifies the second step in Kane's way of directly deriving the linearized equations.

This process of directly deriving linearized equations will be demonstrated here by means of a simple example shown in Figure 1.4, where a massless rod of length L attached to a rotating table by a torsion spring, has a particle P of lumped mass m, at the end of the rod. The table has a prescribed angular velocity $\omega(t)$, and the orientation of the rod is shown by the angle $q_1(t)$. We introduce a generalized speed:

$$u_1 = \dot{q}_1 \tag{1.74}$$

In terms of the basis vectors a_1, a_2 fixed in the table, the non-linear expression for the velocity of the particle P located by the position vector $L[\cos q_1 a_1 + \sin q_1 a_2]$ is

$$^N v^P = \omega(t) R\, a_2 + [\omega(t) + u_1] L [\cos q_1 a_2 - \sin q_1 a_1] \tag{1.75}$$

which may be rewritten in the format of Eq. (1.2) as

$$^N v^P = L\, [\cos q_1 a_2 - \sin q_1 a_1]\, u_1 + \omega(t)\{[R + L\,\cos q_1]a_2 - L \sin q_1 a_1\} \tag{1.76}$$

Using the notations used above, the non-linear partial velocity of P becomes

$$f_1^P = L[\cos q_1 a_2 - \sin q_1 a_1] \tag{1.77}$$

Linearizing this non-linear partial velocity, following Eq. (1.65), we get the correctly linearized partial velocity:

$$^N \tilde{v}_1^P = f_1^P(0) + \left[\frac{\partial f_1^P}{\partial q_1}\right]_0 q_1 = L\, [a_2 - q_1 a_1] \tag{1.78}$$

Now, linearize the velocity, Eq. (1.76), by neglecting all non-linear terms, to get

$$^N \tilde{v}^P = -\omega\, q_1 L\, a_1 + [\omega\,(R+L) + L\, u_1]\, a_2 \tag{1.79}$$

The linearized acceleration of P in N is obtained by differentiating $^N \tilde{v}^P$ in N and removing all non-linear terms in the process:

$$^N \tilde{a}^P = -[\dot{\omega} q_1 + \omega u_1]\, La_1 + [\dot{\omega}(R+L) + L\dot{u}_1]\, a_2$$
$$+\omega\{-\omega q_1 L a_2 - [\omega(R+L) + L u_1]\, a_1\} \tag{1.80}$$

Finally, dot-multiplying the linearized partial velocity, Eq. (1.78), with mass times the acceleration in Eq. (1.80), *and* removing the resulting non-linear terms gives the linearized generalized inertia force. The linearized generalized active force being simply, $(-kq_1)$, Kane's way of directly deriving linearized dynamical equations gives for this system:

$$\dot{u}_1 = -[k/(mL^2)+(R/L)\omega^2]q_1 - (1+R/L)\dot{\omega} \qquad (1.81)$$

Equation (1.81) shows that the natural frequency, $[k/(mL^2)]q_1$, is augmented by $(R/L)\omega^2 q_1$. This is clearly identifiable as centrifugal stiffening, where the effective *stiffness increases with the rotational speed* of the base. Using the method in Section 1.2, it can be checked that the full, non-linear equation of the system in Figure 1.4 is

$$\dot{u}_1 = -k/(mL^2)q_1 - (R/L)\omega^2 \sin q_1 - [1+(R/L)\cos q_1]\dot{\omega} \qquad (1.82)$$

Linearization of this equation in q_1 would indeed give Eq. (1.81), verifying that Eq. (1.81) is the correctly linearized equation for this system.

The above approach of *directly* deriving linearized dynamical equations without deriving the non-linear equations is a special feature of Kane's method that has no analog in Lagrange's method. The method has been extended to special cases of continua like beams and plates in Refs. [12, 13], respectively. The direct linearization method, however, is not extensible to general elastic continua because it is extremely difficult to derive velocity expressions for non-linear deformation for general continua. In search of a general method that does work for arbitrary continua, we will presently discuss, rather paradoxically, a procedure of *premature*

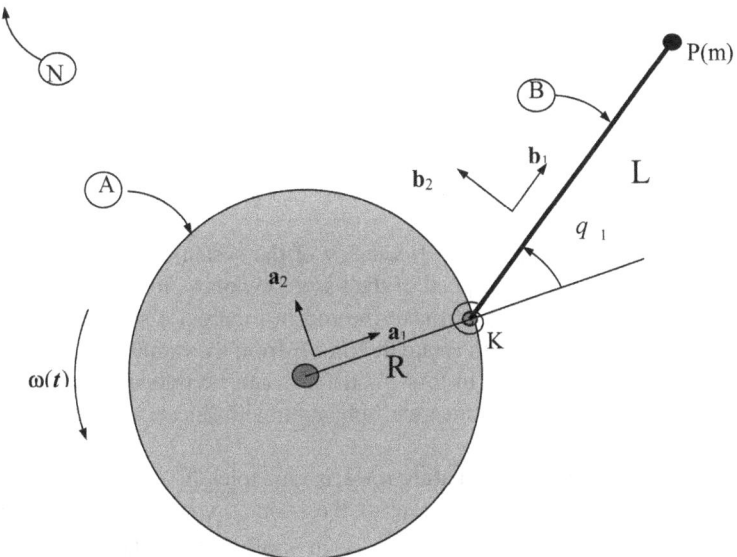

FIGURE 1.4 Elastically restrained bar connected to a table with prescribed rotation.

linearization that actually produces incorrect equations that can, however, be redeemed to get the correct linear equations, by the a posteriori addition of geometric stiffness due to inertia loads [14].

1.5 PREMATURELY LINEARIZED EQUATIONS AND A POSTERIORI CORRECTION BY AD HOC ADDITION OF GEOMETRIC STIFFNESS DUE TO INERTIA LOADS

In directly deriving the linear equations without deriving the non-linear equations, premature linearization occurs when the partial velocities of particles and mass centers of rigid bodies and the partial angular velocities of rigid bodies are obtained from the linearized velocities and angular velocities. Ref. [3] hints at this without going into details. We will first show that premature linearization leads to incorrect equations, with loss of structural stiffness with motion; next, we show how to compensate for the error in stiffness by adding load-dependent geometric stiffness. Again, the ideas are best explained by a simple example. So, we will consider the same example as in Section 1.5. The expression for the *linearized* velocity of P in N here was given in Eq. (1.79),

Prematurely linearized partial velocity with respect to u_1 from this linearized velocity expression is

$$^N \hat{\mathbf{v}}_1^P = L\,\mathbf{a}_2 \tag{1.83}$$

Note that we have already lost a term in Eq. (1.83) from the partial velocity expression, Eq. (1.78). Dot-multiplying this partial velocity expression with mass times the linearized acceleration from Eq. (1.74) would give the prematurely linearized generalized inertia force. The linearized generalized active force being simply, $-kq_1$, dynamical equations for the system with premature linearization can be written after rearranging terms as

$$\dot{u}_1 = -[k/mL^2 - \omega^2]q_1 - (R/L+1)\dot{\omega} \tag{1.84}$$

It shows that the effective natural frequency of the system shown in Figure 1.5 decreases with the rotational speed of the base! Obviously, this is physically not true, as centrifugal force would make the rod in Figure 1.4 stay straight along $\mathbf{a}_1$, that is, it would be stiffer to rotate the link B from the radial line, and so, Eq. (1.84) is incorrect. The error in loss of stiffness can be traced to the incorrect partial velocity in Eq. (1.83), since multiplication with linearized acceleration is common to both approaches.

When is the error due to premature linearization tolerable? It can be seen from Eq. (1.84) that when the rotation rate of the frame, ω, is far smaller than the natural frequency of the structure, that is, $\omega^2 << [k/(mL^2)]$, the error committed is small. Thus, in such situations of relatively very small rotation rates, the indigenous error due to premature linearization may be acceptable. Certain simplifications are then acceptable, and we mention applications of this in Chapter 6.

It may be noted in passing that error due to using linearized velocity also affects other methods such as Lagrange's equation. Thus, computing the kinetic energy for this example problem based on the linearized velocity given by Eq. (1.85), one obtains

$$K = \frac{m}{2}[\omega^2 q_1^2 L^2 + \omega^2(R+L)^2 + L^2\dot{q}_1^2 + 2\omega(R+L)L\dot{q}_1] \qquad (1.85)$$

Upon substitution in Lagrange's equation with the generalized force due to the torsional spring:

$$\frac{d}{dt}\frac{\partial K}{\partial \dot{q}_1} - \frac{\partial K}{\partial q_1} = -kq_1 \qquad (1.86)$$

the reader can check that one gets the same incorrect equation as Eq. (1.84). Ironically, for most general elastic bodies undergoing large overall motion, where vibration modes are used to describe small deformation, one is innately doing premature linearization since vibration modes come from a linear theory. Hence, it is of interest to see if there is a way to redeem the incorrect equation, a posteriori, with some appropriate augmentation of stiffness. In fact, this is possible by appealing to the notion of geometric stiffness due to loads [13]. To illustrate this, we continue with the same example problem.

Cook [15] has shown that for a bar or rod type of structure, geometric stiffness, K_G, due to an axial load, F_a, is as follows, leading to generalized forces as shown:

$$K_G = \frac{1}{2}\frac{F_a}{L}[\delta_1 \quad \delta_2]\begin{bmatrix} 1 & -1 \\ -1 & 1 \end{bmatrix}\begin{Bmatrix} \delta_1 \\ \delta_2 \end{Bmatrix} \rightarrow \begin{Bmatrix} F_1 \\ F_2 \end{Bmatrix} = \begin{Bmatrix} \dfrac{\partial K_G}{\partial \delta_1} \\ \dfrac{\partial K_G}{\partial \delta_2} \end{Bmatrix} \qquad (1.87)$$

Here, δ_1, δ_2 are the transverse end displacements of the bar under axial load, F_a. For the rod in Figure 1.5, the end displacements and load are $\delta_1 = 0$, $\delta_2 = Lq_1$; $F_a = m\omega^2(R+L)$. Now Eq. (1.87) gives

$$K_G = 0.5m\omega^2(R+L)L(q_1)^2 \qquad (1.88)$$

The generalized active force due to the potential energy in geometric stiffness then becomes

$$F_G = -\frac{dK_G}{dq_1} = -m\omega^2(R+L)Lq_1 \qquad (1.89)$$

Adding this term, a posteriori, to the right-hand side of the prematurely linearized equation, Eq. (1.78), produces:

$$mL^2\ddot{u}_1 = -[k + mRL\omega^2]q_1 - mL(R + L)\dot{\omega} \tag{1.90}$$

Equation (1.90) is precisely the same as Eq. (1.81) obtained by direct linearization in the previous section. In conclusion, using a simple example, we have shown here that the error in loss of motion-induced stiffness due to premature linearization can be rectified, a posteriori, by adding geometric stiffness due to the relevant inertia load. This idea is developed more fully for an arbitrary flexible body in Ref. [14].

1.6   KANE'S EQUATIONS WITH UNDETERMINED MULTIPLIERS FOR CONSTRAINED MOTION

For systems with constraints on their motion, the constraints may be of the (holonomic) form:

$$\phi_i(q_1,\ldots,q_n) = 0 \quad i = 1,\ldots,m \tag{1 91}$$

or the non-holonomic, i.e., non-integrable, form stated in terms of generalized speeds [3]:

$$\sum_{j=1}^{n} a_{ij}u_j + b_i = 0 \quad i = 1,\ldots,m \tag{1.92}$$

Kane [1–3] eliminates m of the generalized speeds from Eq. (1.92) to work with (n − m) independent generalized speeds, developing non-holonomic velocities and angular velocities, non-holonomic partial velocities and partial angular velocities, and associated generalized active and inertia forces. See Ref. [3] for details. If the constraints are "workless", that is, their satisfaction involves no work of the constraint forces, the method produces a reduced set of differential equations of motion and is computationally efficient, though the labor involved in solving Eq. (1.92) for a complex system should not be overlooked. However, if the constraint forces are also of interest, then a modified version of Kane's method, called *Kane's method with undetermined multipliers* [16], may well have an advantage from the point of view of computer coding. The method consists of introducing as many extra degrees of freedom as there are constraint forces to be determined, proceeding as though the constraints are violated, and then enforcing the constraints in terms of the augmented set of generalized speeds; the process is described below. Recall that the velocity of a point j or the angular velocity of frame j can always be written as m constraint equations in the n generalized speeds.

$$\sum_{i=1}^{n} \frac{^N\partial \mathbf{v}^j}{\partial u_i} u_i + \mathbf{v}_t^j = 0 \quad j = 1,\ldots,m \tag{1.93}$$

Here, $^N\mathbf{v}^j$ is a vector that stands symbolically for either the j-th particle velocity, or the i-th body frame angular velocity, in the Newtonian frame N. Now Eq. (1.93) can be differentiated in N with respect to time to get the corresponding acceleration form of the constraint equations including the acceleration remainder terms that absorb all terms *not* involving $\dot{u}_j$:

$$\sum_{i=1}^{n} \frac{^N\partial \mathbf{v}^j}{\partial u_i} \dot{u}_i + \mathbf{a}_t^j = 0 \quad j = 1,\ldots,m \tag{1.94}$$

Now, if certain forces and torques are at play to maintain the m number of constraints, given by Eq. (1.93), then if the constraints are relaxed, these forces will do work. Introducing m number of undetermined force measure numbers $\lambda_1,\ldots,\lambda_m$ (representing scalar values of forces acting on points or torques on rigid bodies), generalized active forces due to forces/torques can be evaluated for the augmented set of generalized speeds, in the usual manner. Since constraint forces do no work when the constraints are active, the physical constraint force has to be in the direction of $\mathbf{n}_j$, perpendicular to the partial velocity of j, the point originally constrained, and the generalized force due to the physical constraint force must be of the form given below:

$$F_i^c = \sum_{i=1}^{n} \frac{^N\partial \mathbf{v}^j}{\partial u_i} \cdot \lambda_j \mathbf{n}_j \quad j = 1,\ldots,m \tag{1.95}$$

where $\mathbf{v}^j$ is the velocity of the point (or angular velocity of the frame) j. Equation (1.94) can be written in the following matrix form, where A is an (m × n) matrix, with n the dimension of the augmented state U, and B is a column matrix with m elements:

$$A\dot{U} + B = 0 \tag{1.96}$$

Note that the coefficients λ_j in Eq. (1.95) are exactly the same partial velocity terms corresponding to u_j in Eq. (1.94). Hence, the matrix form of Eq. (1.95) for the generalized forces due to forces obtained by relaxing the constraints, can be written in terms of the A matrix and an (m × 1) matrix of so-called undetermined multipliers, Λ:

$$F^c = A^t\Lambda \tag{1.97}$$

The complete set of equations to solve is given next.

1.6.1 SUMMARY OF EQUATIONS OF MOTION WITH UNDETERMINED MULTIPLIERS FOR CONSTRAINTS

Equations of motion with constraint equations in the augmented set of generalized speeds are of the form:

$$M \dot{U} = C + F + A^t \Lambda \tag{1.98a}$$

$$A \dot{U} + B = 0 \tag{1.98b}$$

From these equations, the constraint forces and the equations of motion are determined:)

$$\Lambda = -(AM^{-1}A^t)^{-1}[AM^{-1}(C+F)+B] \tag{1.99}$$

$$\dot{U} = M^{-1}(C+F+A^t\Lambda) \tag{1.100}$$

Note that Kane's equations with undetermined multipliers have the advantage of computing the forces Λ needed to maintain the constraints.

1.6.2 A SIMPLE APPLICATION

Consider the slider crank shown in Figure 1.5. It is a single degree of freedom system, with the constraint that the vertical displacement of the slider is zero. To maintain this constraint, a vertical force R is applied to the slider. If we want to determine this force R, we relax the constraint, adding a degree of freedom to the two-link system.

The constraint equation follows from the bottom part of Figure 1.4, the actual system.

$$r\sin q_1 + l\sin(q_1 + q_2) = 0 \tag{1.01a}$$

Differentiating the above equation with respect to time *twice*, yields

$$\begin{bmatrix} r\cos q_1 + l\cos(q_1 + q_2) & l\cos(q_1 + q_2) \end{bmatrix} \begin{Bmatrix} \ddot{q}_1 \\ \ddot{q}_2 \end{Bmatrix} + \tag{1.101b}$$

$$\{-r(\dot{q}_1)^2 \sin q_1 - l(\dot{q}_1 + \dot{q}_2)^2 \sin(q_1 + q_2)\} = 0$$

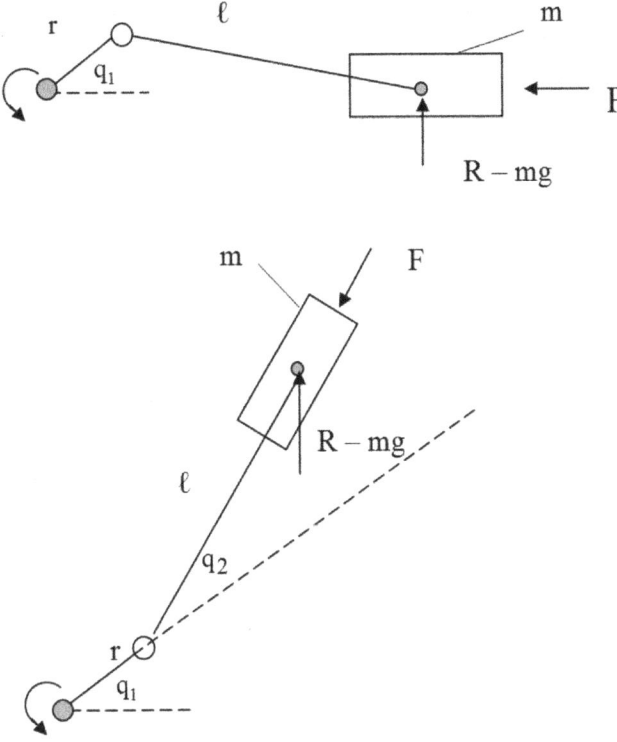

FIGURE 1.5   Slider crank, actual and unconstrained versions.

Defining $U = [\dot{q}_1; \dot{q}_2]$ allows us to write Eq. (1.101) in the particular form:

$$A = \begin{bmatrix} A_{11} & A_{12} \end{bmatrix}; \quad A_{11} = r\cos q_1 + l\cos(q_1 + q_2); \quad A_{12} = l\cos(q_1 + q_2);$$

$$B = \{-r(\dot{q}_1)^2 \sin q_1 - l(\dot{q}_1 + \dot{q}_2)^2 \sin(q_1 + q_2)\}$$

(1.102)

Using Kane's method for the augmented set of generalized speeds for this system, the equations of motion of the slider under external force F_e and gravity and reactive load R, are written in the form of Eq. (1.98a). For simplicity, we assume that the rods are massless, and only the slider has mass m. The equations of motion and the constraint equation in Eq. (1.98b) are of the form:

$$\begin{bmatrix} M_{11} & M_{12} & A_{11} \\ M_{12} & M_{22} & A_{12} \\ A_{11} & A_{12} & 0 \end{bmatrix} \begin{Bmatrix} \ddot{q}_1 \\ \ddot{q}_2 \\ -R \end{Bmatrix} = \begin{Bmatrix} C_1 \\ C_2 \\ 0 \end{Bmatrix} + \begin{Bmatrix} F_{1nr} \\ F_{2nr} \\ -B \end{Bmatrix}$$

(1.103)

Here, F_{1nr}, F_{2nr} are the generalized force, not coming from reactions, but due to gravity and any other external forces, and

$$M_{11} = m\left[r^2 + l^2 + rl\cos q_2 \right]$$

$$M_{12} = m[rl\cos q_2 + l^2]$$

$$M_{22} = ml^2$$

$$C_1 = m(\dot{q}_2^2 + 2\dot{q}_1\dot{q}_2)rl\sin q_2 \qquad\qquad (1.104)$$

$$C_2 = -mrl\dot{q}_1^2 \sin q_2$$

$$F_{1nr} = -Fr\sin q_2 - mg[r\cos q_1 + l\cos(q_1 + q_2)]$$

$$F_{2nr} = -mgl\cos(q_1 + q_2)$$

These equations can be combined in the form of Eq. (1.92) and Eq. (1.93) as

$$
\begin{bmatrix}
m\left[r^2 + l^2 + rl\cos q_2 \right] & m[rl\cos q_2 + l^2] & [r\cos q_1 + l\cos(q_1 + q_2)] \\
m[rl\cos q_2 + l^2] & ml^2 & [l\cos(q_1 + q_2)] \\
[r\cos q_1 + l\cos(q_1 + q_2)] & [l\cos(q_1 + q_2)] & 0
\end{bmatrix}
\begin{Bmatrix}
\ddot{q}_1 \\
\ddot{q}_1 \\
-R
\end{Bmatrix}
$$

$$
= \begin{Bmatrix}
-m(\dot{q}_2^2 + 2\dot{q}_1\dot{q}_2)rl\sin q_2 \\
mrl\dot{q}_1^2 \sin q_2 \\
0
\end{Bmatrix}
+ \begin{Bmatrix}
-Fr\sin q_2 - mg[r\cos q_1 + l\cos(q_1 + q_2)] \\
-mgl\cos(q_1 + q_2) \\
-r(\dot{q}_1^2)\sin q_1 - l(\dot{q}_1 + \dot{q}_2)^2 \sin(q_1 + q_2)
\end{Bmatrix}
$$

$$(1.105)$$

APPENDIX A: GUIDELINE FOR CHOOSING EFFICIENT MOTION VARIABLES IN KANE'S METHOD

(Developed by Dr Paul Mitiguy, in a PhD thesis at Stanford University)

Guidelines for choosing efficient motion variables for Kane's method are given in Ref. [17].

The underlying rationale is to choose variables that produce the simplest possible expressions for inertial angular velocities of rigid bodies (which figure prominently in expressing other important quantities such as velocities, accelerations, partial angular velocities, partial velocities, angular accelerations, inertia torques, and kinetic energy). This is illustrated by writing the equations of motion for an example system of the space shuttle with its manipulator arm.

Figure 1.6 shows the space shuttle with its manipulator arm, a system made up of the shuttle, body A in the figure, with a revolving body B, and the manipulator links C, D, and E. Bodies B, C, D are mounted on revolute joints, and E has a ball-and-socket

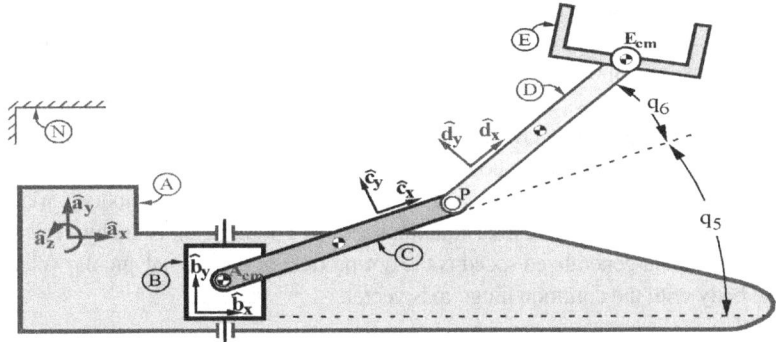

FIGURE 1.6 Space shuttle with its manipulator arm.

joint. While this is really a 12 dof system, in the force-free condition, the constraint of conservation of linear momentum reduces to 3 dof, yielding a 9 dof system.

Below, we show the configuration of the system in terms of the various vector bases, using bold notation, instead of the overbar shown for them in the figure.

Body A rotates freely in inertial frame N, and its orientation in N can be described by three successive right-handed rotations characterized by $q_1\mathbf{a}_x$, $q_2\mathbf{a}_y$, $q_3\mathbf{a}_z$; body B is connected at the center of the mass of A at a revolute joint and turns through the angle $q_4\mathbf{b}_y$; body C is connected at the center of the mass of A at a revolute joint and turns through the angle $q_5\mathbf{c}_z$; body D is connected to body C, at point P by a revolute joint and turns through the angle $q_6\mathbf{d}_z$; body E is connected to body D at a spherical joint subjected to relative right-hand rotations successively about $q_7\mathbf{e}_x$, $q_8\mathbf{e}_y$, $q_9\mathbf{a}_z$. In the

---

Customary

$$\dot{u}_2 = (I_1^B - I_3^B)(s_4 u_1 + c_4 u_3)(s_4 u_3 - c_4 u_1) + c_5(I_1^C(u_5 + s_4 u_1 + c_4 u_3)(s_4 c_5 u_3 - s_5 u_2 - s_5 u_4 - c_4 c_5 u_1) - I_3^C(u_5 + s_4 u_1 +$$
$$c_4 u_3)(s_4 c_5 u_3 - s_5 u_2 - s_5 u_4 - c_4 c_5 u_1) - I_2^C(s_5 u_4(s_4 u_1 + c_4 u_3) - u_5(s_5(u_2 + u_4) - c_5(s_4 u_3 - c_4 u_1)))) + c_56(I_1^D(u_5 + u_6 + s_4 u_1 +$$
$$c_4 u_3)(s_4 c_56 u_3 - s_56 u_2 - s_56 u_4 - c_4 c_56 u_1) - I_3^D(u_5 + u_6 + s_4 u_1 + c_4 u_3)(s_4 c_56 u_3 - s_56 u_2 - s_56 u_4 - c_4 c_56 u_1) - I_2^D(s_6(c_5 u_4(s_4 u_1 +$$
$$c_4 u_3) - u_5(c_5(u_2 + u_4) + s_5(s_4 u_3 - c_4 u_1))) + c_6(s_5 u_4(s_4 u_1 + c_4 u_3) - u_5(s_5(u_2 + u_4) - c_5(s_4 u_3 - c_4 u_1))) - u_6(s_6(c_5 u_2 +$$
$$c_5 u_4 + s_4 s_5 u_3 - s_5 c_4 u_1) - c_6(s_4 c_5 u_3 - s_5 u_2 - s_5 u_4 - c_4 c_5 u_1)))) + c_9(s_9 c_8 s_6 - c_56(c_7 c_9 - s_7 s_8 s_9)(I_1^E(u_9 + c_7 c_8 u_5 + c_7 c_8 u_6 +$$
$$(s_8 s_6 - s_7 c_8 c_56)u_2 + (s_8 s_6 - s_7 c_8 c_56)u_4 + (s_4 c_7 c_8 + s_8 c_4 c_56 + s_7 c_4 c_8 s_56)u_1 + (c_4 c_7 c_8 - s_4 s_8 c_56 - s_4 s_7 c_8 s_56)u_3)(u_7 + (s_7 s_9 -$$
$$s_8 c_7 c_9)u_5 + (s_7 s_9 - s_8 c_7 c_9)u_6 + (c_8 c_9 s_56 + c_56(s_9 c_7 + s_7 s_8 c_9))u_2 + (c_8 c_9 s_56 + c_56(s_9 c_7 + s_7 s_8 c_9))u_4 + (c_4 c_8 c_9 c_56 + s_4(s_7 s_9 -$$
$$s_8 c_7 c_9) - c_4 s_56(s_9 c_7 + s_7 s_8 c_9))u_1 - (s_4 c_8 c_9 c_56 - c_4(s_7 s_9 - s_8 c_7 c_9) - s_4 s_56(s_9 c_7 + s_7 s_8 c_9))u_3) - I_3^E(u_9 + c_7 c_8 u_5 + c_7 c_8 u_6 +$$
$$(s_8 s_6 - s_7 c_8 c_56)u_2 + (s_8 s_6 - s_7 c_8 c_56)u_4 + (s_4 c_7 c_8 + s_8 c_4 c_56 + s_7 c_4 c_8 s_56)u_1 + (c_4 c_7 c_8 - s_4 s_8 c_56 - s_4 s_7 c_8 s_56)u_3)(u_7 + (s_7 s_9 -$$
$$s_8 c_7 c_9)u_5 + (s_7 s_9 - s_8 c_7 c_9)u_6 + (c_8 c_9 s_56 + c_56(s_9 c_7 + s_7 s_8 c_9))u_2 + (c_8 c_9 s_56 + c_56(s_9 c_7 + s_7 s_8 c_9))u_4 + (c_4 c_8 c_9 c_56 + s_4(s_7 s_9 -$$
$$s_8 c_7 c_9) - c_4 s_56(s_9 c_7 + s_7 s_8 c_9))u_1 - (s_4 c_8 c_9 c_56 - c_4(s_7 s_9 - s_8 c_7 c_9) - s_4 s_56(s_9 c_7 + s_7 s_8 c_9))u_3)) - I_2^E((s_7 c_9 + s_8 s_9 c_7)u_4(s_4 u_1 -$$
$$c_4 u_1) - u_7(c_7 c_8(u_5 + u_6 + s_4 u_1 + c_4 u_3) - s_8(s_4 c_56 u_3 - s_56 u_2 - s_56 u_4 - c_4 c_56 u_1) - s_7 c_8(c_56 u_2 + c_56 u_4 + s_4 s_56 u_3 -$$
$$c_4 s_56 u_1)) - u_9(c_8 c_9(s_4 c_56 u_3 - s_56 u_2 - s_56 u_4 - c_4 c_56 u_1) - (s_7 s_9 - s_8 c_7 c_9)(u_5 + u_6 + s_4 u_1 + c_4 u_3) - (s_9 c_7 + s_7 s_8 c_9)(c_56 u_2 +$$
$$c_56 u_4 + s_4 s_56 u_3 - c_4 s_56 u_1)) - s_9 c_8(c_6(c_5 u_4(s_4 u_1 + c_4 u_3) - u_5(c_5(u_2 + u_4) + s_5(s_4 u_3 - c_4 u_1))) - s_6(s_5 u_4(s_4 u_1 + c_4 u_3) -$$
$$u_5(s_5(u_2 + u_4) - c_5(s_4 u_3 - c_4 u_1))) - u_6(c_6(c_5 u_2 + c_5 u_4 + s_4 s_5 u_3 - s_5 c_4 u_1) + s_6(s_4 c_5 u_3 - s_5 u_2 - s_5 u_4 - c_4 c_5 u_1))) -$$
$$(c_7 c_9 - s_7 s_8 s_9)(s_6(c_5 u_4(s_4 u_1 + c_4 u_3) - u_5(c_5(u_2 + u_4) + s_5(s_4 u_3 - c_4 u_1))) + c_6(s_5 u_4(s_4 u_1 + c_4 u_3) - u_5(s_5(u_2 + u_4) -$$
$$c_5(s_4 u_3 - c_4 u_1))) - u_6(s_6(c_5 u_2 + c_5 u_4 + s_4 s_5 u_3 - s_5 c_4 u_1) - c_6(s_4 c_5 u_3 - s_5 u_2 - s_5 u_4 - c_4 c_5 u_1))))) + T_2^A - \cdots$$

New

$$\dot{u}_2 = [\, T_2^A - T^{A/B} \;-\; (I_1^A - I_3^A)\, u_1 u_3 \,] \, / \, I_2^A$$

FIGURE 1.7 Sample dynamical equations in $\dot{\mathbf{u}}_2$ with customary and efficient generalized speeds for the shuttle manipulator system, shown in Ref. [17].

following table, we show three sets of possible choices for generalized speeds, dubbed as simple, customary, and efficient, for this problem. Figure 1.7 lists the kinematical equations corresponding to the customary and efficient choices for generalized speeds. Figure 1.8 shows the dynamical equations corresponding to the generalized speed u_2, for the customary choice and the efficient choice, with $c_4 = \cos q_4$, $s_{56} = \sin (q_5 + q_6)$, $t_4 = \tan (q_4)$, etc. This shows how the equations become simpler with the efficient choice of generalized speeds, and illustrates how the choice of motion variables makes a vast difference in the final equations. For two bodies connected by a revolute joint, the efficient generalized speed is the dot-product of the inertial angular velocity of one body with the common hinge-axis vector.

Simple	Customary	Efficient
$u_1 = \dot{q}_1$	$u_1 = {}^N\omega^A \cdot \mathbf{a}_x$	$u_1 = {}^N\omega^A \cdot \mathbf{a}_x$
$u_2 = \dot{q}_2$	$u_2 = {}^N\omega^A \cdot \mathbf{a}_y$	$u_2 = {}^N\omega^A \cdot \mathbf{a}_y$
$u_3 = \dot{q}_3$	$u_3 = {}^N\omega^A \cdot \mathbf{a}_z$	$u_3 = {}^N\omega^A \cdot \mathbf{a}_z$
$u_4 = \dot{q}_4$	$u_4 = q_4$	$u_4 = {}^N\omega^B \cdot \mathbf{b}_y$
$u_5 = \dot{q}_5$	$u_5 = \dot{q}_5$	$u_5 = {}^N\omega^C \cdot \mathbf{c}_z$
$u_6 = \dot{q}_6$	$u_6 = \dot{q}_6$	$u_6 = {}^N\omega^D \cdot \mathbf{d}_z$
$u_7 = \dot{q}_7$	$u_7 = {}^D\omega^E \cdot \mathbf{e}_x$	$u_7 = {}^N\omega^E \cdot \mathbf{e}_x$
$u_8 = \dot{q}_8$	$u_8 = {}^D\omega^E \cdot \mathbf{e}_x$	$u_8 = {}^N\omega^E \cdot \mathbf{e}_x$
$u_9 = \dot{q}_9$	$u_9 = {}^D\omega^E \cdot \mathbf{e}_z$	$u_9 = {}^N\omega^E \cdot \mathbf{e}_z$

A.1 Customary Choice of Generalized Speeds

$$\dot{q}_1 = (c_3 u_1 - s_3 u_2)/c_2$$

$$\dot{q}_2 = s_3 u_1 + c_3 u_2$$

$$\dot{q}_3 = u_3 + t_2(s_3 u_2 - c_3 u_1)$$

$$\dot{q}_{3+i} = u_{3+i} \quad i = 1, 2, 3 \tag{A.1}$$

$$\dot{q}_7 = (c_9 u_7 - s_9 u_8)/c_8$$

$$\dot{q}_8 = s_9 u_7 + c_9 u_8$$

$$\dot{q}_9 = u_9 + t_8(s_9 u_8 - c_9 u_7)$$

A.2 Efficient Choice of Generalized Speeds

$$\dot{q}_1 = (c_3 u_1 - s_3 u_2)/c_2$$

$$\dot{q}_2 = s_3 u_1 + c_3 u_2$$

$$\dot{q}_3 = u_3 + t_2(s_3 u_2 - c_3 u_1)$$

$$\dot{q}_4 = u_4 - u_2$$

$$\dot{q}_5 = u_5 - (s_4 u_1 + c_3 u_3)$$

$$\dot{q}_6 = u_6 - u_5 \qquad\qquad\qquad\qquad (A.2)$$

$$\dot{q}_7 = (c_7 s_8 u_6 + c_9 u_7 - s_9 u_8)/c_8 + (s_4 c_{56} - s_7 t_8 s_{56})u_3$$

$$\qquad + c_4(s_7 t_8 s_{56} - c_{56})u_1 - (s_7 t_8 c_{56} + s_{56})u_2$$

$$\dot{q}_8 = s_9 u_7 + c_9 u_8 + c_4 c_7 s_{56} u_1 - s_7 u_6 - c_7 c_{56} u_4 - s_4 c_7 s_{56} u_3$$

$$\dot{q}_9 = u_9 + t_8(s_9 u_8 - c_9 u_7) + (s_7 c_{56} u_4$$

$$\qquad + s_4 s_7 s_{56} u_3 - c_7 u_6 - s_7 c_4 s_{56} u_1)/c_8$$

A sample of the dynamical equations obtained by the customary and the efficient variables, shown below is taken from Ref. [17].

APPENDIX B: SLIDING IMPACT WITH FRICTION OF A NOSE CAP ON A PACKAGE OF PARACHUTE

This section deals with the impact dynamics of a parachute package by its enclosing nose cap, which is pulled out to deploy the parachute. The analysis is described in Ref. [18]. Figure 1.8 shows the container nose cap for the parachute package coming off the nose cone of a shuttle solid rocket booster (SRB), before the parachute opens to decelerate the descent of the SRB for its recovery on the ocean, for possible re-use.

Restricting ourselves to the analysis of a planar impact, with generalized coordinates q_r, $r = 1, 2, 3$ representing the centroidal coordinates of the nose cap and its orientation, x, y, θ, respectively, the linear and angular impulses for impact at point Q are obtained from Kane's definition [16] of a generalized impulse:

$$I_r = {}^N\mathbf{v}_r^Q \cdot (N_1 \mathbf{n}_1 + N_2 \mathbf{n}_2) \quad r = 1, 2, 3 \qquad (B.1)$$

$$N_i = \int_{t_1}^{t_2} F_i dt \quad i = 1, 2 \qquad (B.2)$$

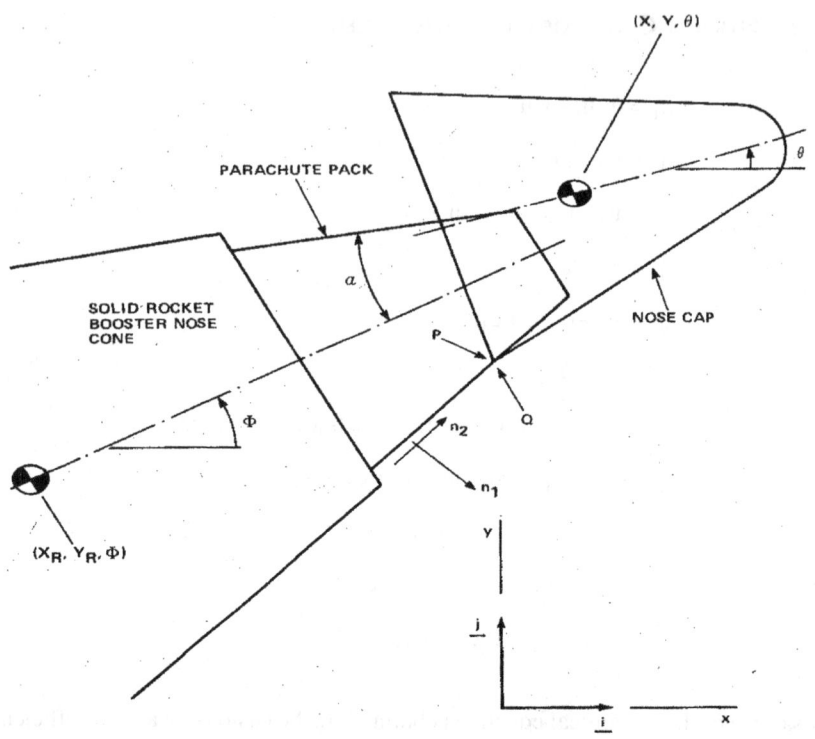

FIGURE 1.8   Nose cap of a parachute package being pushed off a solid rocket booster.

Here, $^N\mathbf{v}_r^Q$ are the r-th partial velocities of Q (see Figure 1.8); N_1, N_2 are the magnitudes of the normal impulse and the tangential impulse; $\mathbf{n}_1$, $\mathbf{n}_2$ are the normal and tangential unit vectors; and t_1, t_2 are two instants of time infinitesimally separated. The basic impulse–momentum relationship governs the difference in generalized momentum at instants *infinitesimally* after and before impact:

$$I_r = p_r(t_+) - p_r(t_-) \quad r = 1, 2, 3 \tag{B.3}$$

where for mass m and the centroidal moment of inertia I:

$$p_r = \frac{\partial K}{\partial \dot{q}_r} = \frac{\partial}{\partial \dot{q}_r}[0.5m(\dot{q}_1^2 + \dot{q}_2^2) + 0.5I\dot{q}_3^2] \quad r = 1, 2, 3 \tag{B.4}$$

Treating the effect of the impact on the massive SRB as negligible, one can look at Eqs. (B.3)–(B.4) as three equations in the five unknowns, $\dot{q}_r, r = 1, 2, 3; N_1, N_2$. Two standard assumptions are now made to complete the solution. The first is the law of restitution, expressed as

$$^N\mathbf{v}^{Q/P}(t_+) \cdot \mathbf{n}_1 = -e\,^N\mathbf{v}^{Q/P}(t_-) \cdot \mathbf{n}_1; \quad ^N\mathbf{v}^{Q/P} = {}^N\mathbf{v}^Q - {}^N\mathbf{v}^P \tag{B.5}$$

Here, superscript P refers to the point P on the parachute pack in Figure 1.8. The second assumption is about the friction behavior of the colliding surfaces, which results in either a normal rebound or a sliding impact. Normal rebound with no slip means:

$$^N\mathbf{v}^{Q/P}(t_+) \cdot \mathbf{n}_2 = 0 \qquad (B.6)$$

and is possible only if

$$|N_2| < \mu N_1 \qquad (B.7)$$

Here, μ is the coefficient of friction. Sliding impact occurs when $|N_2| = \mu N_1$, where N_1, N_2 are the normal and tangential impulse and

$$N_2 = \mu N_1 \qquad (B.8)$$

or

$$N_2 = -\mu N_1 \qquad (B.9)$$

Condition $N_2 > 0$ indicates that the nose cap slips backward on the cone, and $N_2 < 0$ indicates the cap is sliding forward. Both cases are covered by the condition that the tangential component of the separation velocity opposes the sense of the tangential impulse:

$$N_2[^N\mathbf{v}^{Q/P}(t_+).\mathbf{n}_2] < 0 \qquad (B.10)$$

Analysis of an actual impact is started by using Eqs. (B.3), (B.5), and (B.6). If the test in Eq. (B.7) fails, Eq. (B.6) is replaced by Eq. (B.8) and the test, Eq. (B.10), applied. If this too fails, the only possible solution is given by Eqs. (B.5), (B.6), and (B.10).

Impact detection is made by integrating the free flight condition of the two colliding bodies from a non-impact configuration to a situation where the interpenetration of their boundaries is indicated. Noting the penetration depth d of the point Q normal to the surface, a time interval, $\Delta t = t - t_-$, is calculated from the knowledge of relative velocity and relative acceleration:

$$d = -^N\mathbf{v}^{Q/Q'}.\mathbf{n}_1\Delta t - 0.5\,^N\mathbf{a}^{Q/Q'}.\mathbf{n}_1\Delta t^2 \qquad (B.11)$$

The generalized coordinates and their rates are then computed from Δt as follows:

$$q_r(t_-) = q_r(t) - \dot{q}_r(t)\Delta t - 0.5\ddot{q}_r(t)\Delta t^2; \quad \dot{q}_r(t_-) = \dot{q}_r(t) - \ddot{q}_r(t)\Delta t \qquad (B.12)$$

Successive impact configuration values pertaining to the flight condition of the shuttle SRB and its nose cap for the parachute deployment package, are shown

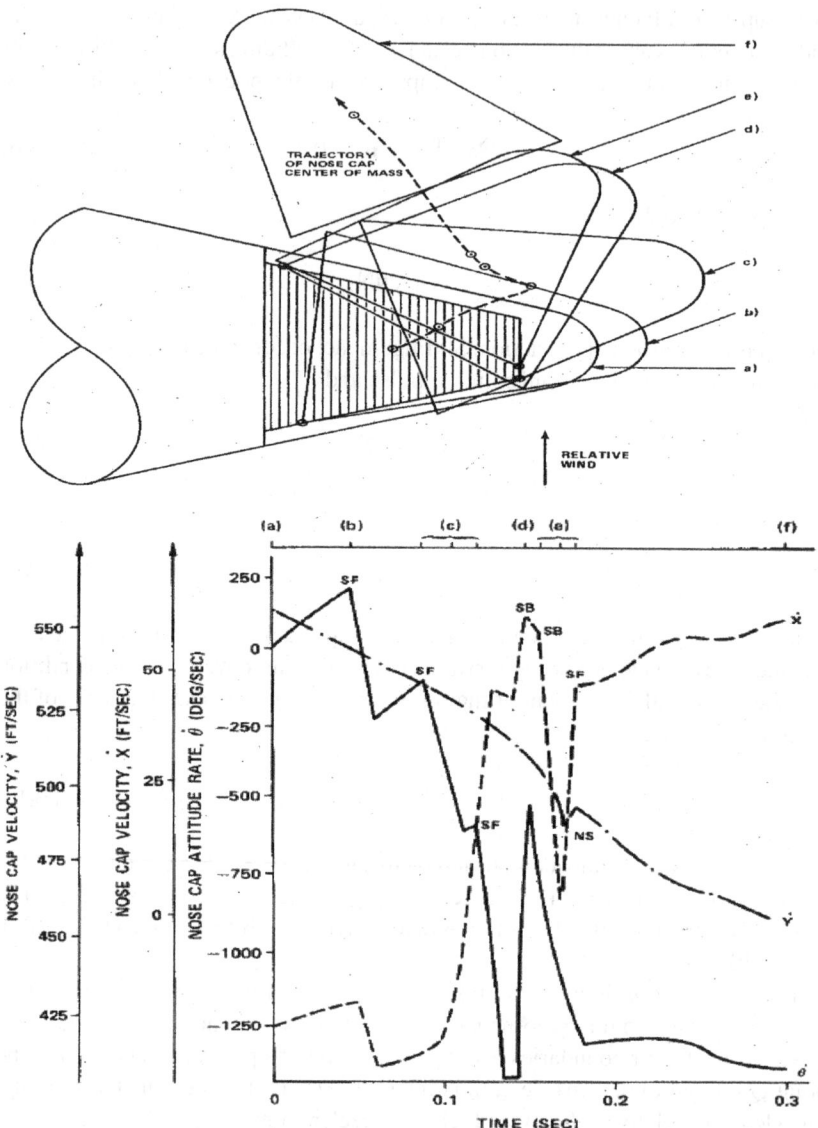

FIGURE 1.9 Successive impact configurations, and slip forward (SF), slip backward (SB), and no slip (NS) instants of the nose cap on the parachute package.

in Figure 1.9 (upper figure), and attendant slip forward, slip backward, and no slip are shown in Figure 1.9 (lower figure). For a three-dimensional impact, the solution procedure is the same but the impact detection process becomes quite difficult [18].

PROBLEM SET 1

1. Consider the double pendulum example worked out in Figure 1.1. Derive the equations of motion of the system for large angles q_1, q_2, in two ways:
   (a) Using Lagrange's equations in the generalized coordinates q_1, q_2.
   (b) Using Kane's equations with the choice of generalized speeds $u_1 = \dot{q}_1; u_2 = \dot{q}_2$.

   Note that the dynamical equations are now more complex than those worked out in Example 1. This exemplifies the simplicity of equations obtainable with the proper choice of generalized speeds.
2. Consider the same double pendulum but placed horizontally, being attached to the rotating disk of Figure 1.5, with torsional springs of stiffness k between the disk and first link and also between the two links. Derive the non-linear dynamical equations by Kane's method.
3. Derive the equations of motion of the double pendulum, linearized in q_1, q_2, using Kane's method of direct linearization, and verify that they agree with the equations obtainable by linearizing the non-linear equations.

REFERENCES

1. Kane, T.R. and Wang, C.F., (1965), "On the Derivation of Equations of Motion", *Journal of the Society of Industrial and Applied Mechanics*, **13**(2), pp. 487–492.
2. Kane, T.R., (1968), *Dynamics*, Holt, Rinehart and Winston, Inc.
3. Kane, T.R. and Levinson, D.A., (1985), *Dynamics*, McGraw-Hill.
4. Kane, T.R., Likins, P.W., and Levinson, D.A., (1983), *Spacecraft Dynamics*, McGraw-Hill.
5. Kane, T.R. and Levinson, D.A., (1980), "Formulation of Equations of Motion for Complex Spacecraft", *Journal of Guidance, Control, and Dynamics*, **3**(2), pp. 99–112.
6. Baruh, H., (1999), Analytical Dynamics, McGraw-Hill.
7. Levinson, D.A. and Banerjee, A.K., (Mar.–Apr., 1990), "Comment on "Efficacy of the Gibbs–Appell Method for Generating Equations for Complex Systems", *Journal of Guidance, Control, and Dynamics*.
8. Banerjee, A.K., (Nov.–Dec., 1987), "Comment on Relationship Between Kane's Equation and the Gibbs–Appell Equations", *Journal of Guidance, Control, and Dynamics*, **10**(6), pp. 596–597.
9. Schaechter, D.B., and Levinson, D.A., (Oct.–Dec., 1988), "Interactive Computerized Symbolic Dynamics for the Dynamicist", **36**(4), pp. 365–388.
10. Ayoubi, M.A., Goodarzi, F.A., and Banerjee, A.K, (Oct.–Dec., 2011), "Attitude Motion of a Spinning Spacecraft with Fuel Sloshing and Nutation Damping", *Journal of the Astronautical Sciences*, pp. 551–568.
11. Shabana, A.A., (2020), *Dynamics of Multibody Systems*, Cambridge University Press.

12. Kane, T.R., Ryan, R.R., and Banerjee, A.K., (1987), "Dynamics of a Cantilever Beam Attached to a Moving Base", *Journal of Guidance, Control, and Dynamics*, **10**(2), pp. 139–151.

13. Banerjee, A.K. and Kane, T.R., (1989), "Dynamics of a Plate in Large Overall Motion", *Journal of Applied Mechanics*, **56**(6), pp. 887–892.

14. Banerjee, A.K and Dickens, J.M., (1990), "Dynamics of an Arbitrary Flexible Body in Large Rotation and Translation", *Journal of Guidance, Control, and Dynamics*, **13**(2), pp. 221–227.

15. Cook, R.D., (1974), *Concepts and Applications of Finite Element Analysis*, John Wiley, p. 282.

16. Wang, J.T. and Huston, E.L., (1987), "Kane's Equations with Undetermined Multipliers: Application to Constrained Multibody Systems", *Journal of Applied Mechanics*, **54**, pp. 424–429.

17. Mitiguy, P.C. and Kane, T.R., (1996), "Motion Variables Leading to Efficient Equations of Motion", *International Journal of Robotics Research*, **15**(5), pp. 522–532.

18. Banerjee, A.K. and Coppey, J.M., (1975), "Post-Ejection Impact of the Space Shuttle Booster Nose Cap", *Journal of Spacecraft and Rockets*, **12**(10), pp. 632–633.

2 Deployment, Station-Keeping, and Retrieval of a Flexible Tether Connecting a Satellite to the Shuttle

In this chapter, we consider the application of Kane's equations of motion, *without any linearization*, to an interesting problem: the dynamics of a long flexible tether connecting a satellite to the space shuttle. The objective of the NASA shuttle-tethered satellite project was to study the earth from a low altitude where a satellite could not stay by itself because of high atmospheric drag, and the idea was to tow the satellite by a 100 km long cable attached to the more stable carrier vehicle, the space shuttle. The dynamics of the system was originally considered in Refs. [1–4]. This chapter considers the same problem in further detail in three aspects, based on the work described in Refs. [5–9]:

(1) deployment of a tethered satellite from the orbiting shuttle, where the satellite is basically pulled by the earth's gravity, with the restraining elastic tether taking up a curved shape; (2) retrieval of the satellite with the elastic tether reeled in by a motor becoming slack, as the satellite comes very close to the shuttle tending to wrap up around it, and thus needing a thruster to apply tension; and (3) the simpler problem of station-keeping of the earth-observing satellite. As previous investigators found, all three processes require control for effectiveness. One basic assumption is that a control system for the shuttle maintains its desired orbit in the presence of disturbance due to mass going out or coming in, represented by the deployment and retrieval of the tether and the satellite. Station-keeping of the satellite after its deployment offers a control problem to keep the tether swing angles small for good earth observation, and is treated here based on Ref. [8]. Finally, we consider the interesting problem of pointing control with tethers as actuators of a space station–supported platform. The Appendix to this chapter treats the problem of the formation flying of tethered satellites as well as orbit boosting by electrodynamic forces on a tethered satellite. We will first analyze the deployment problem of the shuttle-borne tethered satellite.

DOI: 10.1201/9781003231523-3

2.1   EQUATIONS OF MOTION FOR DEPLOYMENT OF A SATELLITE TETHERED TO THE SPACE SHUTTLE

A tether is an elastic cable that can be modeled either as an elastic continuum or as a discrete system of particles connected by springs. The exposition given here for the deployment of a satellite is based on Refs. [5, 6], where a discrete tether model is used to allow possibly large deviations of the tether from a straight line. We will use a continuum tether model for retrieval. Consider the model [6] of a tethered satellite shown in Figure 2.1, where E is the center of the earth, B is the shuttle orbiter, and the tethered satellite system is modeled with a series of spring-connected lumped mass points numbered $P_1, \ldots, P_n$, where P_n is the satellite at the end of the tether.

The (shuttle) orbiter is modeled as a rigid body B, with the mass center at B^*, with orthogonal B basis vectors, $\mathbf{b}_1$, $\mathbf{b}_2$, $\mathbf{b}_3$ (using bold notation in the analysis for vectors, not overbars as shown in the figure). Tether lumped mass points are measured from the tether exit point P_0 on the orbiter B; E is the point mass representing the earth. Generalized speeds, the motion variables used in Kane's method, are chosen to allow a three-dimensional tether notion [6]:

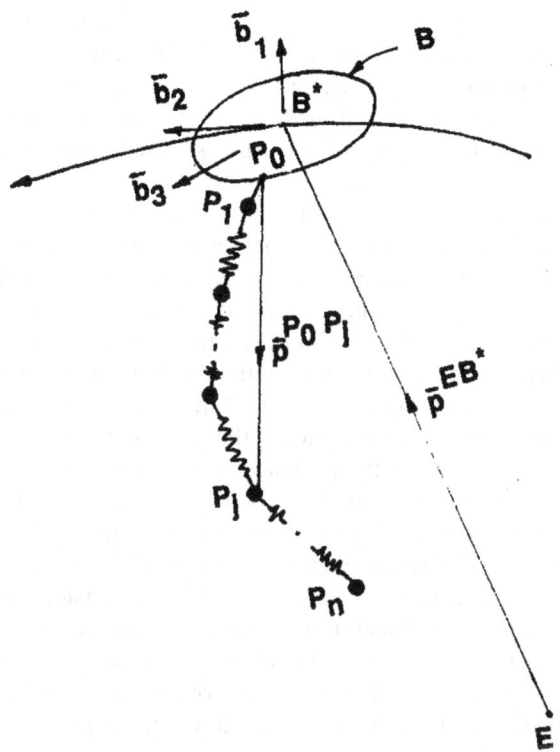

FIGURE 2.1   Discrete model of a tethered satellite deployed out of the orbiter

$$u_i = {}^N\omega^B \cdot \mathbf{b}_i \quad (i = 1, 2, 3)$$

$$u_{3+i} = {}^N\mathbf{v}^{B*} \cdot \mathbf{b}_i \quad (i = 1, 2, 3) \tag{2.1}$$

$$u_{3+3j+i} = {}^N\mathbf{v}^{P_j} \cdot \mathbf{b}_i \quad (i = 1, 2, 3; \ j = 1, \ldots, n)$$

Here, ${}^N\omega^B$, ${}^N\mathbf{v}^{B*}$, ${}^N\mathbf{v}^{P_j}$ represent, respectively, the vectors of the angular velocity of B in a Newtonian reference frame N, the velocity of the mass center of B* in N, and velocity of the j-th lumped mass P_j. The configuration variables for the lumped mass particles of the tether are defined in terms of position vectors from the exit point, P_0, shown in Figure 2.1, rather than by inertial coordinates measured from E, to avoid differencing two nearly close large numbers when measured from E.

$$q_{3j-3+i} = \mathbf{p}^{P_0 P_j} \cdot \mathbf{b}_i \quad (i = 1, 2, 3; \ j = 1, \ldots, n)$$

$$q_{3n+i} = \mathbf{p}^{EB*} \cdot \mathbf{b}_i \quad (i = 1, 2, 3) \tag{2.2}$$

Here, $\mathbf{p}^{P_0 P_j}, \mathbf{p}^{EB*}$ are, respectively, the position vectors from P_0 to P_j $(j = 1, \ldots, n)$ and from E to B*. The tether exits the shuttle from the point P_0 with the location components:

$$p_i = \mathbf{p}^{B* P_0} \cdot \mathbf{b}_i \quad (i = 1, 2, 3) \tag{2.3}$$

2.1.1 KINEMATICAL EQUATIONS

The angular velocity of the shuttle B, the velocity of B*, and the velocity of the lumped mass points P_j $(j = 1, \ldots, n)$ in N follow from Eq. (2.1), and are given by the following three equations:

$$ {}^N\omega^B = \sum_{i=1}^{3} u_i \mathbf{b}_i $$

$$ {}^N\mathbf{v}^{B*} = \sum_{i=1}^{3} u_{3+i} \mathbf{b}_i \tag{2.4}$$

$$ {}^N\mathbf{v}^{P_j} = \sum_{i=1}^{3} u_{3+3j+i} \mathbf{b}_i; \ j = 1, \ldots, n $$

Note that these angular velocity and velocity expressions would have been far more complicated had one used $u_j = \dot{q}_j, j = 1, \ldots, 6 + 3n$, required when deriving equations of motion using Lagrange's method. Substituting Eqs. (2.2) and (2.4) in the equation for the velocity of P_j in N:

$$^{N}\mathbf{v}^{P_j} = {}^{N}\mathbf{v}^{B*} + {}^{N}\boldsymbol{\omega}^{B} \times \left[\mathbf{p}^{B*P_0} + \sum_{i=1}^{3} q_{3j-3+i}\mathbf{b}_i \right] + \sum_{i=1}^{3} \dot{q}_{3j-3+i}\mathbf{b}_i \quad (j=1,\dots n) \quad (2.5)$$

leads to the kinematical equations:

$$\dot{q}_{3j-2} = u_{4+3j} - (u_4 + u_2 z_{3j} - u_3 z_{2j}) \quad (j=1,\dots,n)$$

$$\dot{q}_{3j-1} = u_{5+3j} - (u_5 + u_3 z_{1j} - u_1 z_{3j}) \quad (j=1,\dots,n) \qquad (2.6a)$$

$$\dot{q}_{3j} = u_{6+3j} - (u_6 + u_1 z_{2j} - u_2 z_{1j}) \quad (j=1,\dots,n)$$

Here

$$z_{ij} = p_i + q_{3j-3+i} \quad (i=1,2,3; j=1,\dots,n) \qquad (2.6b)$$

Equating the derivative of the position vector $\mathbf{p}^{EB*}$ with the velocity of B* in N yields

$$\dot{q}_{3n+1} = u_4 - (u_2 q_{3n+3} - u_3 u_{3n+2})$$

$$\dot{q}_{3n+2} = u_5 - (u_3 q_{3n+1} - u_1 u_{3n+3}) \qquad (2.7)$$

$$\dot{q}_{3n+3} = u_6 - (u_1 q_{3n+2} - u_2 u_{3n+1})$$

This completes the kinematical equations.

2.1.2 DYNAMICAL EQUATIONS

Equation (2.4) can be used to calculate the angular acceleration of B and the accelerations of B* and P_j ($j=1,\dots,$ n). The generalized inertia forces can then be written as the following set of equations, as per Kane's method reviewed in Chapter 1:

$$F_1^* = -I_1\dot{u}_1 + (I_2 - I_3)u_2 u_3$$

$$F_2^* = -I_2\dot{u}_2 + (I_3 - I_1)u_3 u_1$$

$$F_3^* = -I_3\dot{u}_3 + (I_1 - I_2)u_1 u_2$$

$$F_4^* = -m_B(\dot{u}_4 + u_2 u_6 - u_3 u_5) \qquad (2.8)$$

$$F_5^* = -m_B(\dot{u}_5 + u_3 u_4 - u_1 u_6)$$

$$F_6^* = -m_B(\dot{u}_6 + u_1 u_5 - u_2 u_4)$$

$$F^*_{4+3j} = -m_j(\dot{u}_{3j+4} + u_2 u_{3j+6} - u_3 u_{3j+5}) \quad (j = 1, \ldots, n)$$

$$F^*_{5+3j} = -m_j(\dot{u}_{3j+5} + u_3 u_{3j+4} - u_1 u_{3j+6}) \quad (j = 1, \ldots, n) \tag{2.9}$$

$$F^*_{6+3j} = -m_j(\dot{u}_{3j+6} + u_1 u_{3j+5} - u_2 u_{3j+4}) \quad (j = 1, \ldots, n)$$

Here I_1, I_2, I_3 are, respectively, the centroidal principal moments of inertia of the shuttle along axes which define the $\mathbf{b}_1$, $\mathbf{b}_2$, $\mathbf{b}_3$ basis vectors of B, and m_B and m_j are the mass of B and P_j. It is *assumed* here that the mass and moment of inertia of the body B do not change to any significant level due to the tether mass deployed or retrieved.

Generalized active forces due to gravity (neglecting gravitational moments) can also be expressed as follows, using Kane's method, as reviewed in Chapter 1:

$$
\left[
\begin{array}{l}
F^g_i = 0 \quad (i = 1, 2, 3) \\[2ex]
F^g_{3+i} = -\mu\, m_B q_{3n+i} \Big/ \left(\displaystyle\sum_{j=1}^{3} q^2_{3n+j} \right)^{3/2} \quad (i = 1, 2, 3) \\[3ex]
z_{3+i,j} = q_{3n+i} + z_{i,j} \quad (i = 1, 2, 3;\ j = 1, \ldots, n) \\[2ex]
z_{7,j} = \mu\, m_j \Big/ \left(\displaystyle\sum_{j=1}^{3} z^2_{3+i,j} \right)^{3/2} \quad (i = 1, 2, 3), (j = 1, \ldots, n) \\[3ex]
F^g_{3+3j+i} = -z_{7,j} z_{3+i,j} \quad (i = 1, 2, 3;\ j = 1, \ldots, n)
\end{array}
\right] \tag{2.10}
$$

Here, μ s the universal gravitational constant multiplied by the mass of the earth. See Eq. (2.6b) for the definitions of $z_{1,1}, \ldots, z_{3,n}$. Generalized active forces due to the springs, in Figure 2.1, representing tether elasticity are formed assuming that the tether cannot sustain compression, and has extensional stiffness k per segment of the tether. Defining intermediate variables:

$$z_{7+i,j} = q_{3j-3+i} - q_{3j-6+i} \quad (i = 1, 2, 3;\ j = 2, \ldots, n)$$

$$z_{11,j} = \left(\sum_{j=1}^{3} z^2_{7+i,j} \right)^{1/2} \quad (j = 2, \ldots, n) \tag{2.11}$$

$$z_{12j} = z_{11,j} - L/n \quad (j = 2, \ldots, n)$$

or

$$z_{12,j} = 0 \quad \text{if} \quad z_{12,j} < 0 \quad (z = 2, \ldots, n)$$

$$z_{13,j} = k z_{12,j} / z_{11,j} \quad (z = 2, \ldots, n) \tag{2.12}$$

one gets the expressions for the generalized active forces due the springs as follows:

$$F_i^s = 0 \quad (i = 1, \ldots, 9; \, n = 1)$$

$$F_i^s = 0 \quad (i = 1, \ldots, 6)$$

$$F_{6+i}^s = z_{13,2} z_{7+i,2} \quad (i = 1, 2, 3; \, n \geq 2) \tag{2.13}$$

$$F_{9+i}^s = -F_{6+i}^s \quad (i = 1, 2, 3; \, n = 2)$$

$$F_{3+3j+i}^s = -z_{13,j} z_{7+i,j} + z_{13,j+1} z_{7+i,j+1} \quad (i = 1, 2, 3; \, j = 2, \ldots, n-1)$$

Force due to the thrust-type interaction needed for continuous deployment (pushing out material) is computed by defining the intermediate variable, τ:

$$\tau = T / \left(\sum_{i=1}^{3} q_i^2 \right)^{1/2} \tag{2.14}$$

Here, T is the tether tension at P_1; q_1, q_2, q_3 being the coordinates of P_1 with respect to the origin P_0, see Eq. (2.2), in the B basis. Then, the generalized force due to this action–reaction thrust on P_0, P_1 (see Figure 2.1) is

$$F_i^t = \left[\tau \sum_{i=1}^{3} q_i \mathbf{b}_i \right] . (\mathbf{v}_i^{P_0} - \mathbf{v}_i^{P_1}) \tag{2.15}$$

Here, $\mathbf{v}_i^{P_0}, \mathbf{v}_i^{P_1}$ are the partial velocity vectors of the points P_0 and P_1 with respect to the i-th generalized speed, which are evaluated from the expression:

$$\mathbf{v}^{P_1} - \mathbf{v}^{P_0} = (u_2 q_3 - u_3 q_2 + \dot{q}_1) \mathbf{b}_1 + (u_3 q_1 - u_1 q_3 + \dot{q}_2) \mathbf{b}_2$$
$$+ (u_1 q_2 - u_2 q_1 + \dot{q}_3) \mathbf{b}_3 \tag{2.16}$$

By using Eq. (2.6) for j = 1, in Eq. (2.16) one finds

$$\mathbf{v}^{P_1} - \mathbf{v}^{P_0} = (u_7 - u_4 - u_2 p_3 + u_3 p_2) \mathbf{b}_1$$
$$+ (u_8 - u_5 - u_3 p_1 + u_1 p_3) \mathbf{b}_2 + (u_9 - u_6 - u_1 p_2 + u_2 p_1) \mathbf{b}_3 \tag{2.17}$$

Now the partial velocity vectors needed in Eq. (2.15) can be obtained by inspection of Eq. (2.17), and the generalized forces expressions due to deployment or releasing of the tether, given in Eq. (2.15), can be evaluated with the following sequence of expressions:

$$y_1 = p_2q_3 - p_3q_2$$

$$y_2 = p_3q_1 - p_1q_3$$

$$y_3 = p_1q_2 - p_2q_1$$

$$F_i^t = y_i \quad (i = 1, 2, 3)$$

$$F_{3+i}^t = q_i \quad (i = 1, 2, 3)$$

$$F_{6+i}^t = -q_i \quad (i = 1, 2, 3)$$

(2.18)

Kane's dynamical equations for tether deployment can now be written as

$$F_i^* + F_i^g + F_i^t + F_i^c = 0 \quad (i = 1, \ldots, 6+n)$$

(2.19)

Here, F_i^c is a generalized force due to motion constraint implied by having to follow a deployment law, which is a statement of how the commanded length of the tether changes as a prescribed function of time, $l(t)$. Note that we have control over how the distance between P_1 and P_0 changes with deployment. This can be stated as a constraint condition on the "error", e, in following the command, defined by

$$y_4 = l(t) - L(n-1)/n$$

(2.20a)

$$e = \sum_{i=1}^{3} q_i^2 - (y_4)^2 = 0$$

(2.20b)

Here, L is the total length of the tether, unstretched. Now Eqs. (2.19) and (2.20b) constitute a system of differential-algebraic equations. We choose to solve them as a system of ordinary differential equations, by differentiating Eq. (2.20b) *twice*:

$$\dot{e} = 2\sum_{i=1}^{3} \{q_i\dot{q}_i - \dot{l}y_4\} = 0$$

$$\ddot{e} = 2\left\{\sum_{i=1}^{3}(q_i\ddot{q}_i + \dot{q}_i^2) - \dot{l}^2 - \ddot{l}y_4\right\} = 0$$

(2.21)

Equation (2.19) and the second derivative equation in Eq. (2.21) can be solved together. However, due to unavoidable approximations in numerical integration, constraint satisfaction will drift with time. To remedy this, we invoke Baumgarte's constraint stabilization procedure [10], which consists of writing the constraint equation involving $\ddot{e}, \dot{e}, e$ as follows, in the manner of a proportional-derivative control of error:

$$\ddot{e} + k_d\dot{e} + k_p e = 0 \quad (k_d > 0. k_p > 0)$$

(2.22)

Use of the kinematical equations of motion, Eq. (2.6a), for the expression for $\ddot{e}$ in Eq. (2.21) renders Eq. (2.22) into the matrix differential equation:

$$\dot{H U}_1 = G \quad \text{where}$$

$$H = [y_1 \ y_2 \ y_3 \ q_1 \ q_2 \ q_3 \ -q_1 \ -q_2 \ -q_3]$$

$$U_1 = [u_1 \ u_2 \ u_3 \ u_4 \ u_5 \ u_6 \ u_7 \ u_8 \ u_9]^T \tag{2.23}$$

$$G = -q_1(u_2\dot{q}_3 - u_3\dot{q}_2) - q_2(u_3\dot{q}_1 - u_1\dot{q}_3)$$

$$-q_3(u_1\dot{q}_2 - u_2\dot{q}_1) - \dot{I}^2 + \sum_{i=1}^{3} \dot{q}_i^2 - \ddot{I}y_4 - k_d\dot{e} - k_p e$$

Looking back on the dynamical equations, Eq. (2.19), these can be rewritten in matrix form, and interpreted as Kane's equations with undetermined multipliers [11]:

$$M_1\dot{U}_1 = -C_1 + F_1 + H^T\tau$$

$$M_2\dot{U}_2 = -C_2 + F_2 \tag{2.24}$$

$$U_2 = [u_{10},\ldots,u_{6+3n}]^T \quad (n > 2)$$

Here M_1, M_2, C_1, C_2, F_1, F_2 are defined implicitly by the negative of the generalized inertia force, followed by the generalized active force:

$$-F^* = \begin{bmatrix} M_1 & 0 \\ 0 & M_2 \end{bmatrix} \begin{Bmatrix} \dot{U}_1 \\ \dot{U}_2 \end{Bmatrix} + \begin{Bmatrix} C_1 \\ C_2 \end{Bmatrix}$$

$$\begin{Bmatrix} F_1 \\ F_2 \end{Bmatrix} = F^g + F^s \tag{2.25}$$

Again F^*, F^g, F^s represent column matrices forming generalized forces due to inertia, gravity, and spring terms in Eqs. (2.9), (2.10), and (2.13). Equations (2.8) and (2.9) show that M_1 and M_2 are diagonal matrices. Equations (2.23) and (2.24) permit the evaluation of τ in Eq. (2.24), and the explicit development of the dynamical equations.

$$\tau = \left\langle G - \sum_{i=1}^{3} [y_i\{[F_1(i) - C_1(i)]/I_i - q_i[F_1(3+i) - C_1(3+i)]/m_B \atop + q_i[F_1(6+i) - C_1(6+i)]/m_1\} \right\rangle / y_5 \tag{2.26}$$

$$y_5 = \sum_{i=1}^{3} y_i^2/I_1 + \sum_{i=1}^{3} q_i^2(m_B + m_1)/(m_B m_1)$$

$$\dot{u}_i = [F_1(i) - C_1(i) + y_i \tau] / I_i \quad (i = 1, 2, 3)$$

$$\dot{u}_{3+i} = [F_1(3+i) - C_1(3+i) + q_i \tau] / m_B \quad (i = 1, 2, 3)$$

$$\dot{u}_{6+i} = [F_1(6+i) - C_1(6+i) + q_i \tau] / m_1 \quad (i = 1, 2, 3)$$

$$\dot{u}_{3+3j+i} = [F_2(3j-6+i) - C_2(3j-6+i)] / m_j \quad (i = 1, 2, 3; j = 2, \ldots, n)$$

(2.27)

Here, $F_1(i)$, $F_2(i)$, $C_1(i)$, $C_2(i)$ are the i-th elements in the column matrices F_1, F_2, C_1, C_2, respectively. Note that the derivatives of the generalized speeds in Eq. (2.26) are uncoupled. This fact is primarily responsible for the speed of the numerical simulations.

2.1.3 SIMULATION RESULTS

Equations (2.6), (2.7), and (2.27) were numerically integrated for the following system parameters: satellite orbit altitude 600 nautical miles, $m_B = 272$ kg, $I_1 = 41$ kg-m^2, $I_2 = I_3 = 542$ kg-m^2, tether attachment offset $p_1 = 1$ m, payload mass 45 kg, tether length $L = 1$ km, tether segment stiffness 373 N/m, Baumgarte constraint gains $k_v = 5.0$, $k_p = 6.32$ in Eq. (2.22) were chosen for good error frequency and damping. The tether mass was divided into five lumped masses from four tether segments when completely deployed, so as to reduce the computational burden, while keeping representative complexity. The method of updating the dynamical states by momentum conservation, as the number of particles increases by one, during deployment, was done following Kallaghan et al. in Ref. [1]. The results [6] given in Figure 2.2 show the plots for a deployed length rate control law, proposed by Baker et al. in Ref. [2], which is as follows:

$$\frac{dl}{dt} = \alpha l \qquad 0 \le l \le l_1$$

$$= \alpha l_1 \qquad l_1 \le l \le l_2 \qquad (2.28)$$

$$= \alpha l_1 - (L - l_2) l_2 \le l \le L$$

where $l_1 = 250$ m and $l_2 = 750$ m.

Figure 2.2 shows the time-behavior of the tether tension, the tether orbit in-plane angle for the line of sight to the payload, and the tether instantaneous deployed length, for a value of $\alpha = 0.7$ in Eq. (2.28). Note the snatch load action as the deployment rates are changed at l_1 l_2 in Eq. (2.28).

The sharp increase in tension in the plot, as the number of particle masses deployed increases by one, appears to be a reflection of the modeling of the lumped mass deployment process. The dark band in the figure is due to plotting at 0.5 sec intervals, the tether tension arising from longitudinal oscillations at 1.36 Hz, added on to the fluctuations due to the in-plane libration period of the tether at 3750 sec. This in-plane angle, shown as "THETA", being positive, implies that the tether leads the orbiting shuttle, see Figure 2.1, during deployment, as has been

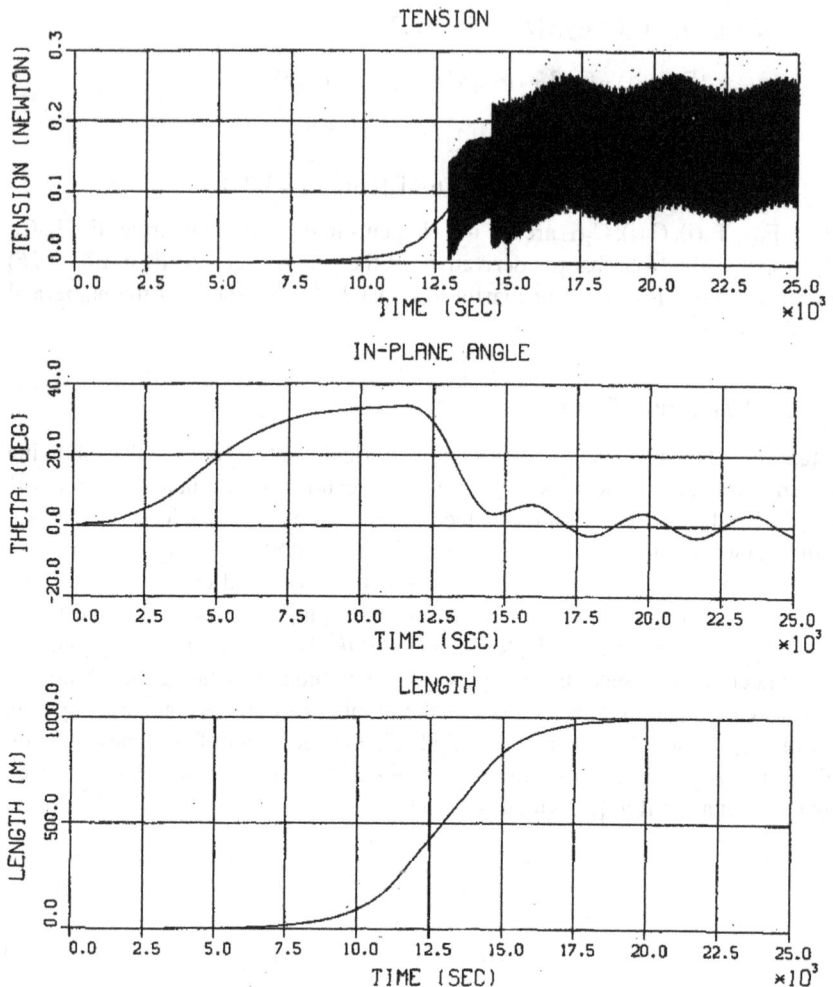

FIGURE 2.2 Numerical simulation results

reported in Ref. [2]. Additional simulations show that the tether becomes slack for a higher rate of deployment, such as $\alpha = 0.8$ in Eq. (2.28). Finally, note that the above tether deployment model can be used for station-keeping by setting i = 0. Station-keeping is basically a control problem, and is presented in Section 2.3.

2.2 THRUSTER AUGMENTED RETRIEVAL OF A TETHERED SATELLITE TO THE ORBITING SHUTTLE

Tethered satellite retrieval is basically an unstable process. This is seen from the consideration of the conservation of angular momentum in the absence of external moments, as follows: ω_i, ω_f, l_i, l_f are the initial and final angular velocities, and the initial and final length, respectively, for a rigid massless tether connecting the shuttle to a satellite of mass m, $ml_i^2\omega_i = ml_f^2\omega_f$, yielding $\omega_f = \omega_i (l_i/l_f)^2$. This means that as the final length is smaller compared to the initial length, the final angular velocity of a rigid tethered satellite system is larger than the initial angular velocity, implying in effect that the tether will tend to "wrap up around the shuttle" for a large angle rate. The problem is exacerbated by aerodynamics which bows out a non-rigid tether and causes it to stretch, and can even make the tether become slack, putting the satellite out of control. Originally, tethered satellite retrieval was studied in Ref. [2] using a tension control law in terms of commanded length, actual length, and its rate. References [2, 3] accounted for tether extensibility and transverse motion flexibility and used an augmented tension control law in feeding back in-plane and out-of-plane tether angles and their rates for improved performance. However, this comes at the cost of a high simulation time; indeed, it appears that the cost of simulating the retrieval of a tether less than 5 km long is prohibitive, which means that information of crucial importance, as the tether gets shorter, cannot be obtained from such a model. With the first mode transverse vibration frequency of a both-ends-fixed string being $(\pi/L)(\sqrt{T/\rho})$, where T is the tension, ρ is the mass per unit length, and L is the length, the frequency increases with decreasing length. In this chapter, we *exclude* transverse flexibility and *retain* extensibility, hoping to make up for what we lose in model fidelity by *usefulness* in simulation, with a tension control law. Tension control has its own drawback in that tension becomes very small for short tether lengths. To augment tension, we propose using a tether-aligned thruster mounted on the satellite connected to the tether, and also a thruster capable of exerting forces transverse to a tensioned tether, to stabilize tether retrieval and speed up the retrieval process. The following exposition is based on the work in Ref. [7].

2.2.1 DYNAMICAL EQUATIONS

Consider Figure 2.3, which keeps the essentials of the system with an orbiter-mounted smooth drum D of radius r and the axial moment of inertia J, an extensible tether T of mass ρ per unit length and having a total length L, and a particle S of mass m, representing the satellite. The orbiter (not shown) is presumed to be earth-pointing and moving in an earth-centered polar circular orbit of radius R.

To describe the orientation of a tether frame, we first introduce a dextral set of mutually perpendicular unit vectors $\mathbf{a}_1$, $\mathbf{a}_2$, $\mathbf{a}_3$ (not using the underbars used for vectors in the figure) with $\mathbf{a}_1$ pointing in the direction of the orbital motion of D*, the center of D, and $\mathbf{a}_3$ pointing from D* to E*, the center of the earth. Next, we let $\mathbf{b}_1$, $\mathbf{b}_2$, $\mathbf{b}_3$ form a similar set of unit vectors, align these with $\mathbf{a}_1$, $\mathbf{a}_2$, $\mathbf{a}_3$, respectively, and then subject them successively to a rotation of amount α about a line parallel to $\mathbf{a}_2$, and a rotation of amount β about a line parallel to $\mathbf{b}_1$. Finally, we take the tether aligned to be parallel to $\mathbf{b}_3$. Thus, α and β are, respectively, the in-plane and out-of-plane tether angle when the tether remains straight, which defines a tether frame. These two tether frame

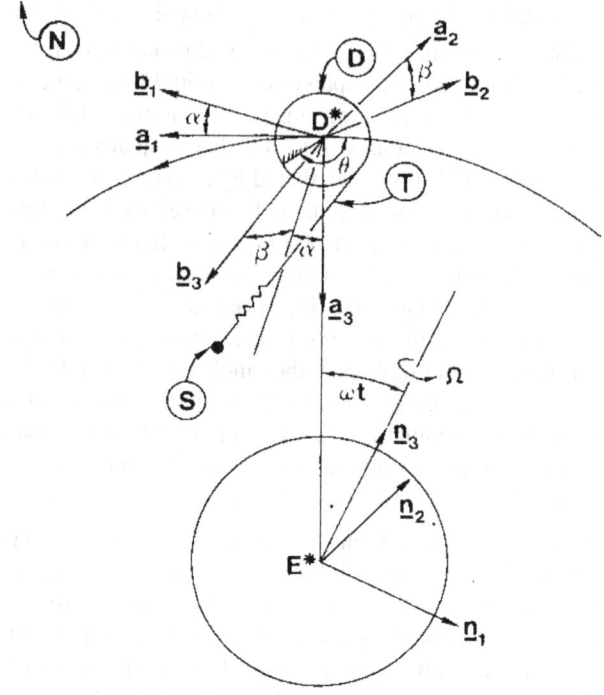

FIGURE 2.3 Representation of a continuum model tether connecting a satellite to a drum mounted on the earth-orbiting shuttle

angles are shown in Figure 2.3. Drum rotations are characterized by the angle θ between a line parallel to $\mathbf{a}_1$ and a line fixed in D. The line joining D* to E* is presumed to move in a Newtonian reference frame N, with a constant orbital rate ω this amounts to assuming that the center of mass of the Shuttle-Tether-Satellite system moves in a circular orbit. Letting $\mathbf{n}_1$, $\mathbf{n}_2$, $\mathbf{n}_3$ be mutually perpendicular unit vectors fixed in the inertial frame N, with $\mathbf{n}_3$ parallel to earth's polar axis, we can thus set the angle between the polar axis and line E*D* equal to ω. Finally, the angular speed of earth in N is denoted by Ω a quantity needed for the evaluation of the aerodynamic drag force.

Let ν be the number of extensional modes of tether vibration to be taken into account, and denoting the associated modal coordinates $q_1,\ldots,q_\nu$, one can introduce $\nu + 3$ generalized speeds as

$$u_1 \quad = r\dot{\theta}$$

$$u_{1+i} = \dot{q}_i \quad (i = 1,\ldots,\nu)$$

$$u_{\nu+2} = \dot{\alpha}$$

$$u_{\nu+3} = \dot{\beta}$$

(2.29)

Now $v + 3$ differential equations can be formulated by Kane's method, considering generalized active forces due to gravity, aerodynamic drag due to a rotating earth, tether elasticity, control torque action on the tether drum D, and thruster action on the satellite S. The key steps in the derivation of the equations of motion are given below. The generalized inertia force, F_i^*, corresponding to the i-th generalized speed is given by

$$F_i^* = - \begin{bmatrix} {}^N\omega_i^D \cdot (\mathbf{J}.{}^N\alpha^D) + \int_0^{L/r} {}^N\mathbf{v}_i^{P'} \cdot {}^N\mathbf{a}^{P'} \rho r d\psi \\ + \int_{L-r\theta}^1 {}^N\mathbf{v}_i^P \cdot {}^N\mathbf{a}^P \rho dx + {}^N\mathbf{v}_i^S \cdot m\mathbf{a}^S \end{bmatrix} \quad (i = 1,\ldots,v+3) \quad (2.30)$$

In Eq. (2.30), ${}^N\omega_i^D, {}^N\mathbf{v}_i^{P'}, {}^N\mathbf{v}_i^P, {}^N\mathbf{v}_i^S$ are, respectively, the i-th partial angular velocity of D; the partial velocity of a particle P' of the tether on D; the partial velocity of a particle P on the deployed part of the tether; and the partial velocity of the particle satellite S. Quantities ${}^N\alpha^D, {}^N\mathbf{a}^{P'}, {}^N\mathbf{a}^P, {}^N\mathbf{a}^S$ are, respectively, the angular acceleration of D, and the accelerations of P', P, and S, P' equally P in the unstretched tether. The required partial velocities and accelerations are all derivable from the corresponding velocity/angular velocity expressions. The inertial angular velocity of the drum D in N is the orbit angular velocity plus the relative angular velocity of the drum.

$$^N\omega^D = (u_1/r - \omega)\mathbf{a}_2 \quad (2.31)$$

The velocity of the point P', the unstretched version of a generic point P for which the tether wrap angle is ψ and the tangent vector $\tau^{P'}$ is

$$^N\mathbf{v}^{P'} = \omega R\mathbf{a}_1 + \left(\frac{u_1}{r} - \omega\right)r\tau^{P'} + \sum_{i=1}^{v}\phi_i\left[\frac{(\psi - \theta)r}{L}\right]u_{1+i}\tau^{P'} \quad (2.32)$$

The velocity of the particle P which, when the tether is unstretched, is at a distance x from the point at which T is attached to the drum D is given in Eq. (2.32), where the stretched location is given by Eq. (2.33):

$$^N\mathbf{v}^P = \omega R\mathbf{a}_1 + \left[(u_{2+v} - \omega)\mathbf{a}_2 + u_{3+v}\mathbf{b}_1\right] \times y\mathbf{b}_3 + \left[u_1 + \sum_{i=1}^{v}\phi_i\left(\frac{x}{L}\right)u_{1+i}\right]\mathbf{b}_3 \quad (2.33)$$

where

$$y = x + \sum_{i=1}^{v}\phi_i\left(\frac{x}{L}\right)q_i(t) - (L - r\theta) \quad (2.34)$$

The velocity of the satellite, $^N\mathbf{v}^S$, is obtained from Eq. (2.32), setting $x = L$ in Eqs. (2.32) and (2.33).

The i-th generalized active force F_i', reflecting contributions from the control torque $T_c\mathbf{a}_2$ on drum D, the gravitational forces on the tether and satellite, and the thrust force $\mathbf{F}$ on the satellite, is given by

$$F_i' = \omega_i^D.(T_c\mathbf{a}_2) + \int_0^{L/r} \mathbf{v}_i^{P'}.(\rho_T g_0 \mathbf{a}_3)r d\psi + \int_{L-r\theta}^L \mathbf{v}_i^P.\mathbf{f}_g(y)dy$$

$$+\mathbf{v}_i^S.[m\mathbf{f}_g(L) + \mathbf{F}] \quad (i = 1,\dots v+3) \tag{2.35a}$$

$$\mathbf{f}_g(y) = \rho_T g_0\{[1 + 3(y/R)\cos\alpha\cos\beta]\mathbf{a}_3 - (y/R)\mathbf{b}_3\} \tag{2.35b}$$

Here, g_0 is the acceleration due to earth's gravity at the orbit, and ρ_T is the linear mass density of the tether. The contribution to the generalized forces associated with rotating earth aerodynamics is given by

$$F_i'' = -0.5C_d f \int_{L-r\theta}^L \mathbf{v}_i^P.\mathbf{v}_w \rho_T dx - 0.5C_D \rho_S A_S \mathbf{v}_i^P.|\mathbf{v}_w|\mathbf{v}_w \quad (i = 1,\dots,v+3) \tag{2.36}$$

Here

$$\mathbf{v}_w = \omega R \mathbf{a}_1 + \Omega R \sin\omega t \, \mathbf{a}_2 \tag{2.37}$$

and the notations C_d, C_D stand for aerodynamic drag coefficients for the tether and the satellite; f is a factor expressed in the details that follow; A_S is the satellite cross-sectional area; ρ_S is the satellite mass density; and ρ_T and A_T are the tether linear mass density and the cross-sectional area. The generalized force due to tether longitudinal viscoelasticity, evaluated as shown in Ref. [4], is obtained by referring to Eq. (2.32) and Eq. (2.33) as

$$F_i''' = -A_T \sum_{i=1}^v (Eq_i + Fu_{1+i})\langle J_1 + J_2 \rangle \quad i = 1,\dots,v+3$$

$$J_1 = \int_{L-r\theta}^L \frac{d}{dx}\left[\phi_i\left(\frac{x}{L}\right)\right]\frac{d}{dx}(\mathbf{b}_3.\,^N\mathbf{v}_i^P)dx \tag{2.38}$$

$$J_1 = \frac{1}{r}\int_0^{L/r} \frac{d}{d\psi}\left[\phi_i\left\{(\psi-\theta)\frac{r}{L}\right\}\right]\frac{d}{d\psi}(^N\mathbf{v}_i^{P'}.\,\tau^{P'}\mathbf{b}_3)d\psi$$

Substituting Eq. (2.2) and Eq. (2.7) through Eq. (2.11) into Kane's dynamical equations:

$$-F_i^* = F_i' + F_i'' + F_i''' \quad (i = 1, \ldots, v+3) \tag{2.39}$$

one arrives at the equations of motion, written as follows.

$$\left(m + \rho L + \frac{J}{r^2} \right) \dot{u}_1 + \sum_{i=1}^{v} \left[m \sin \gamma_i + \frac{\rho L}{\gamma_i} (1 - \cos \gamma_i) \right] \dot{u}_{1+i}$$

$$= I_1 \left[u_{v+3}^2 + (u_{v+2} - \omega)^2 \cos^2 \beta + \omega^2 (3 \cos^2 \alpha \cos^2 \beta - 1) \right] \tag{2.40}$$

$$- 0.5 R (\omega \sin \alpha \cos \beta - \Omega \sin \beta \sin \omega t) \left[f C_d \int_{L-r\theta}^{L} \rho_T dx + h \right] + \frac{T_c}{r}$$

$$\dot{u}_1 \left[m \sin \gamma_i + \frac{\rho L}{\gamma_i} (1 - \cos \gamma_i) \right] + \sum_{j=1}^{v} \dot{u}_{1+j} \left\{ \begin{array}{c} m \sin \gamma_i \sin \gamma_j + \dfrac{\rho L}{2} \\[2mm] \left[\dfrac{\sin(\gamma_i - \gamma_j)}{\gamma_i - \gamma_j} - \dfrac{\sin(\gamma_i + \gamma_j)}{\gamma_i + \gamma_j} \right] \end{array} \right\}$$

$$= I_{3+i} \left[u_{v+3}^2 + (u_{v+2} - \omega)^2 \cos^2 \beta + \omega^2 (3 \cos^2 \alpha \cos^2 \beta - 1) \right]$$

$$- 0.5 R (\omega \sin \alpha \cos \beta - \Omega \sin \beta \sin \omega t) \left[f C_d \int_{L-r\theta}^{L} \rho_T \varphi_i \left(\frac{x}{L} \right) dx + h \phi_i(1) \right]$$

$$- \sum_{j=1}^{v} A_T \left(Eq_j + Fu_{1+j} \right) \frac{\gamma_i \gamma_j}{2L} \left[\frac{\sin(\gamma_i - \gamma_j)}{\gamma_i - \gamma_j} - \frac{\sin(\gamma_i + \gamma_j)}{\gamma_i + \gamma_j} \right]$$

$$+ \tau \phi_i(1) \quad (i = 1, \ldots, v) \tag{2.41}$$

$$\dot{u}_{v+2} = 2(u_{v+2} - \omega) u_{v+3} \tan \beta - 3\omega^2 \sin \alpha \cos \alpha - 2 \frac{I_3}{I_2} (u_{v+3} - \omega)$$

$$- \frac{\omega R \cos \alpha}{2 I_2 \cos \beta} \left\{ f C_d \int_{L-r\theta}^{L} \begin{array}{c} \rho_T \left[x + \displaystyle\sum_{i=1}^{v} \varphi_i \left(\dfrac{x}{L} \right) q_i - L + r\theta \right] dx \\[4mm] + h \left[r\theta + \displaystyle\sum_{i=1}^{v} \varphi_i(1) q_i \right] \end{array} \right\} + F_\alpha$$

$$\tag{2.42}$$

$$\dot{u}_{v+3} = -[(u_{v+2} - \omega)^2 + 3\omega^2 \cos^2 \alpha] \sin \beta \cos \beta - 2 \frac{I_3}{I_2} u_{v+3}$$

$$+ \left[\frac{R(\omega \sin \alpha \sin \beta + \Omega \sin \omega t \cos \beta)}{2I_2} \right] \times Z + F_\beta \qquad (2.43)$$

$$Z = \left\{ f \, C_d \int_{L-r\theta}^{L} \begin{array}{c} \rho_T \left[x + \sum_{i=1}^{v} \phi_i \left(\frac{x}{L} \right) q_i - L + r\theta \right] dx \\ \\ +h \left[r\theta + \sum_{i=1}^{v} \phi_i(1) q_i \right] \end{array} dx \right\}$$

In Eq. (2.42), Eq. (2.43), F_α, F_β are generalized forces due to in-plane and out-of-plane thrust, defined, respectively, by referring to Figure 2.3, with F_α in orbit in-plane, and F_β out-of-plane, along $\mathbf{b}_3$. The *modal integrals* $I_1, \ldots, I_{3+i}$ ($i = 1, \ldots, v$), and f, h, are shown below.

$$I_1 = \int_{L-r\theta}^{L} \rho_T \left[x + \sum_{i=1}^{v} \phi_i \left(\frac{x}{L} \right) q_i - L + r\theta \right] dx + m \left[r\theta + \sum_{i=1}^{v} \phi_i(1) q_i \right] \quad (2.44)$$

$$I_2 = \int_{L-r\theta}^{L} \rho_T \left[x + \sum_{i=1}^{v} \phi_i \left(\frac{x}{L} \right) q_i - L + r\theta \right]^2 dx + m \left[r\theta + \sum_{i=1}^{v} \phi_i(1) q_i \right]^2 \quad (2.45)$$

$$I_3 = \int_{L-r\theta}^{L} \rho_T \left[x + \sum_{i=1}^{v} \phi_i \left(\frac{x}{L} \right) q_i - L + r\theta \right] \left[u_1 + \sum_{j=1}^{v} \phi_i \left(\frac{x}{L} \right) u_{j+1} \right] dx$$

$$+ m \left[r\theta + \sum_{i=1}^{v} \phi_i(1) q_i \right] \left[u_1 + \sum_{j=1}^{v} \phi_i \left(\frac{x}{L} \right) u_{j+1} \right] \qquad (2.46)$$

$$I_{3+v} = \int_{L-r\theta}^{L} \begin{array}{c} \rho_T \left[x + \sum_{j=1}^{v} \phi_j \left(\frac{x}{L} \right) q_j - L + r\theta \right] \phi_i \left(\frac{x}{L} \right) dx \\ \\ + m \left[r\theta + \sum_{i=1}^{v} \phi_i(1) q_i \right] \phi_i(1) \end{array} \qquad (i = 1, \ldots, v) \quad (2.47)$$

$$f = \left\{ \frac{4 A_T R^2}{\pi} \left[\begin{array}{c} \omega^2 (1 - \sin^2 \alpha \cos^2 \beta) + \Omega^2 \cos^2 \beta \sin^2 \omega t \\ + 2\omega\Omega \sin \alpha \sin \beta \cos \beta \sin \omega t \end{array} \right] \right\}^{0.5} \qquad (2.48)$$

$$h = C_d \rho_s A_s R \left\{ \omega^2 + \Omega^2 \sin^2 \omega t \right\}^{0.5} \tag{2.49}$$

All of the above tether *extensional* vibrations are described as in Ref. [5] in terms of modal functions:

$$\varphi_i \left(\frac{x}{L} \right) = \sin \left(\frac{\gamma_i x}{L} \right) \quad (i = 1, \ldots, v) \tag{2.50}$$

where the vibration frequencies satisfy the characteristic equations:

$$\gamma_i L - \tan(\gamma_i L) = \rho L/m \quad (i = 1, \ldots, v) \tag{2.51}$$

In addition to the thrust control, the tension control used by previous investigators is also required. To this end, we use a form of control law proposed in Ref. [2], and a command:

$$\frac{T_c}{r} = -3 \left(m + \frac{\rho r \theta}{2} \right) \omega^2 r \theta + \left(m + \rho L + \frac{J}{r^2} \right) \{ 4\omega^2 (L_c - r\theta)$$

$$-4\omega u_1 - L_c \omega [u_{v+2} + u_{v+3} + \omega(\alpha + \beta)] \} \tag{2.52}$$

Here, the commanded tether length for retrieval is given by

$$L_c = L_0 \exp(-\varepsilon t) \tag{2.53}$$

In Eq. (2.53), describing the commanded length of the tether, L_0, ε are, respectively, the initial length and a retrieval rate control parameter. In the results shown in Figures 2.4–2.8, $\varepsilon = 1.27$, and in Figures 2.9–2.12, $\varepsilon = 2.00$. As regards thrust control, the in-plane thruster, represented by F_α, is turned on when α irst reaches (-0.35) rad and continues thereafter. For out-of-plane control represented by F_β in Eq. (2.43), the thruster is fired whenever β first reaches $|\beta| > 0.35$, and continues thereafter. The threshold values for the angles are set to reduce the burden on the thrusters, and the thrusters go into action only after tension control has removed some of the energy in the system. However, relying only on tension control and delaying the thruster onset time too long can make recovery impossible. The threshold value of ± 0.35 rad represents a compromise that highlights the necessity of doing numerous simulations, as was done here. Expressions for orbit *in-plane* and *out-of- plane thrusters* are chosen as follows:

$$F_\alpha = 2(\delta - \varsigma \omega \sqrt{3}) u_{v+2}$$

$$F_\beta = 2(\delta - 2\varsigma \omega) u_{v+2} \tag{2.54}$$

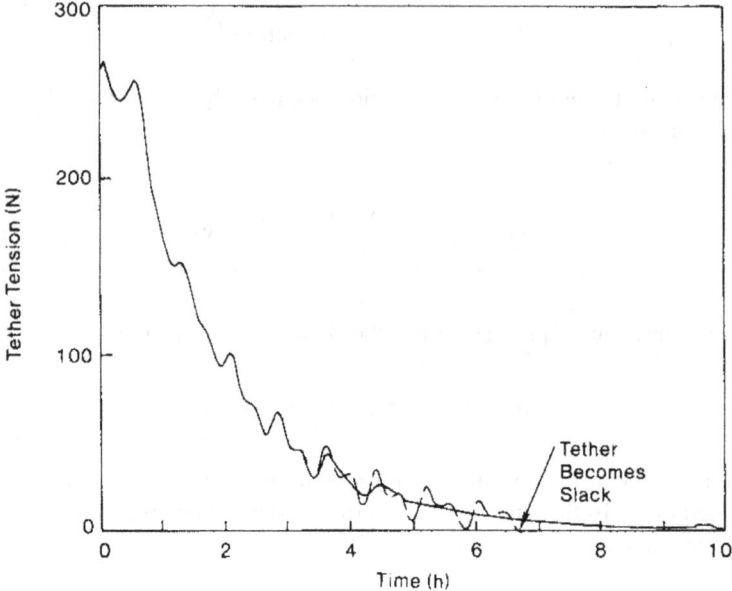

FIGURE 2.4 Tether tension with and without thruster augmentation vs. time; dashed line (tension control only); solid line (tension control + thrust)

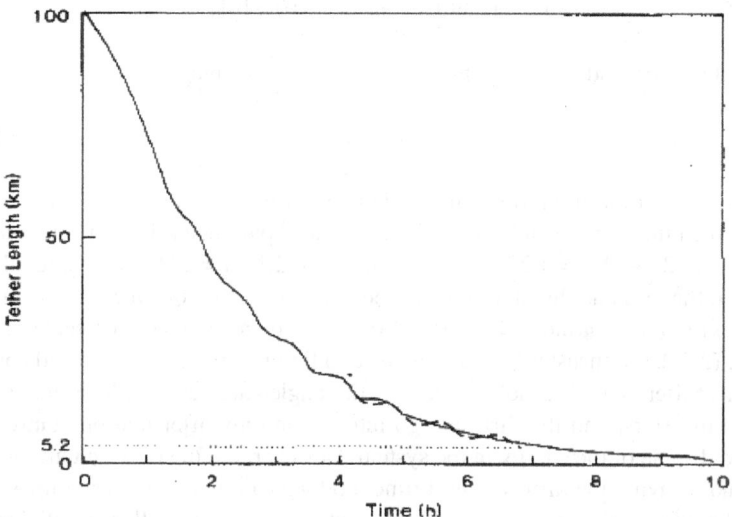

FIGURE 2.5 Tether length with and without thruster, and tension control; tether slack at 6.7 hr

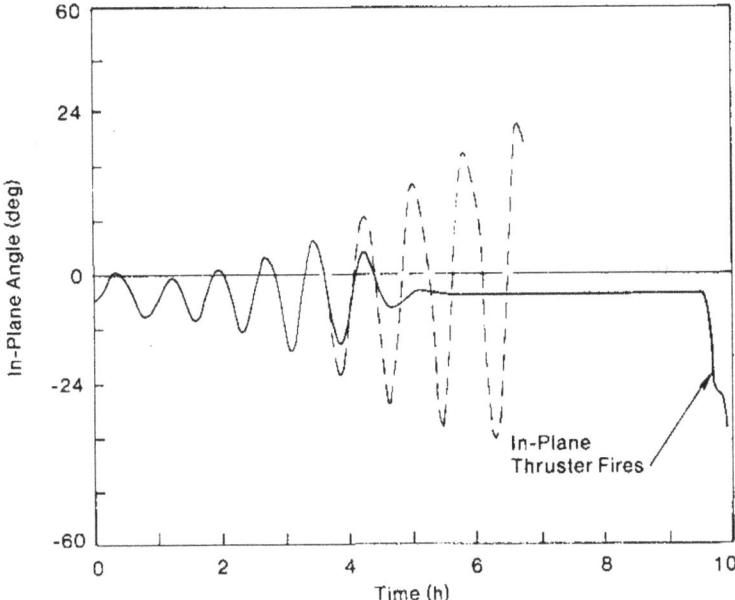

FIGURE 2.6   Tether in-plane angle with and without thrust augmentation of tension control

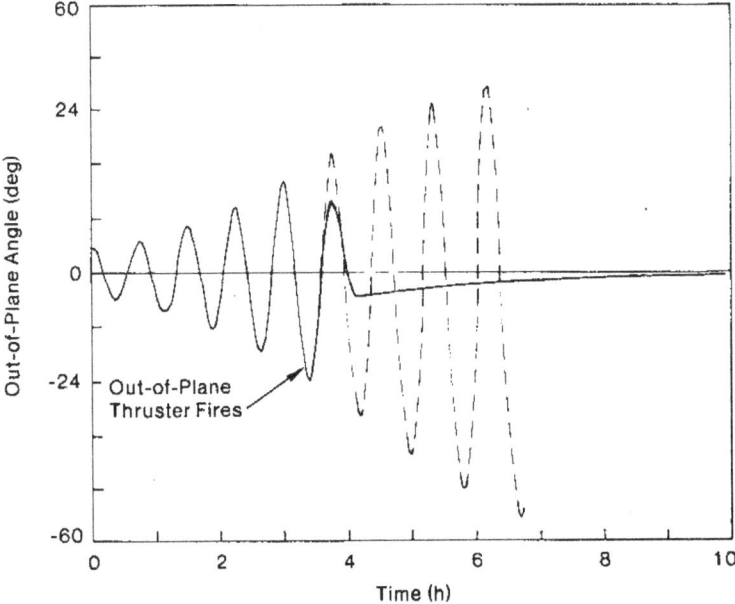

FIGURE 2.7   Tether out-of-plane angle with and without thrust augmentation of tension control

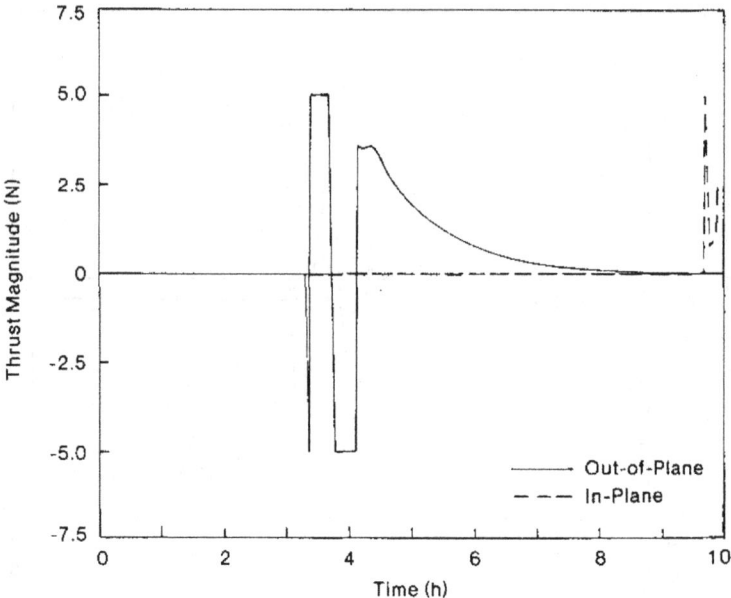

FIGURE 2.8　In-plane and out-of-plane thrust histories for Figures 2.6–2.7 vs. time

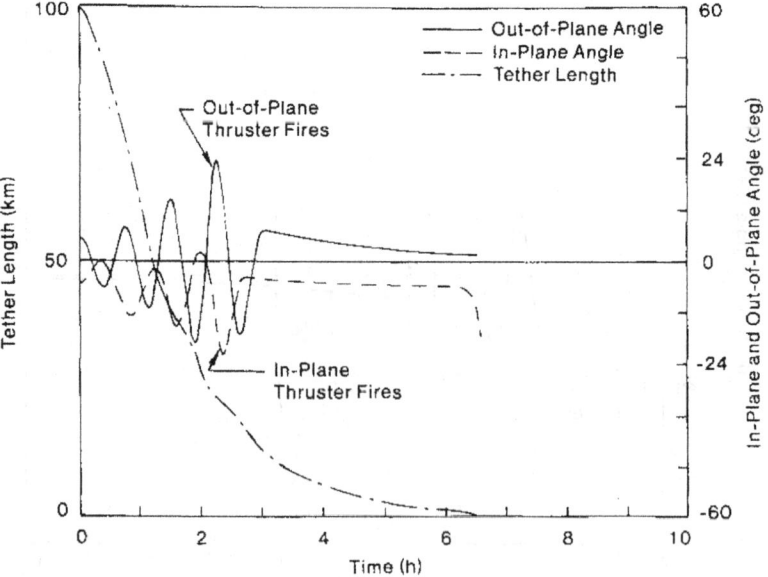

FIGURE 2.9　Fast retrieval with thruster-aided tension control and tension control alone (dashed)

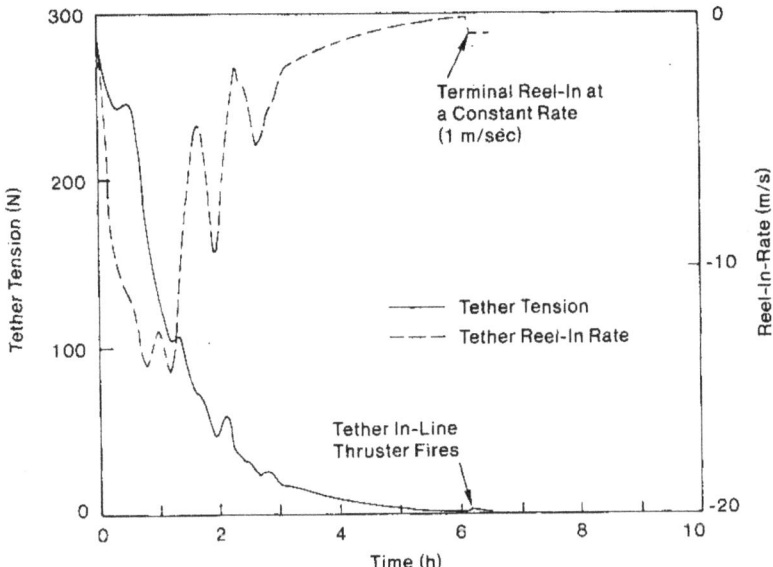

FIGURE 2.10   Fast retrieval tether tension and reel-in rate vs. time

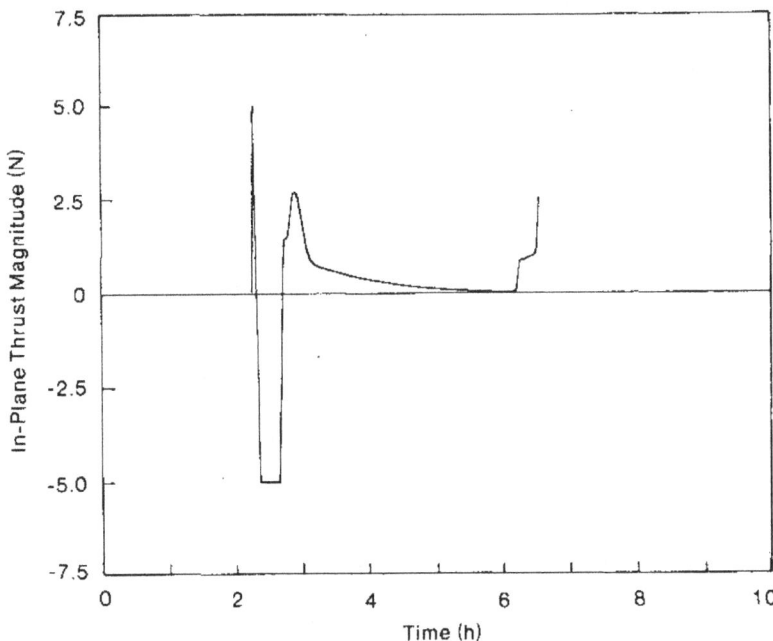

FIGURE 2.11   In-plane thruster firing time history for fast retrieval vs. time

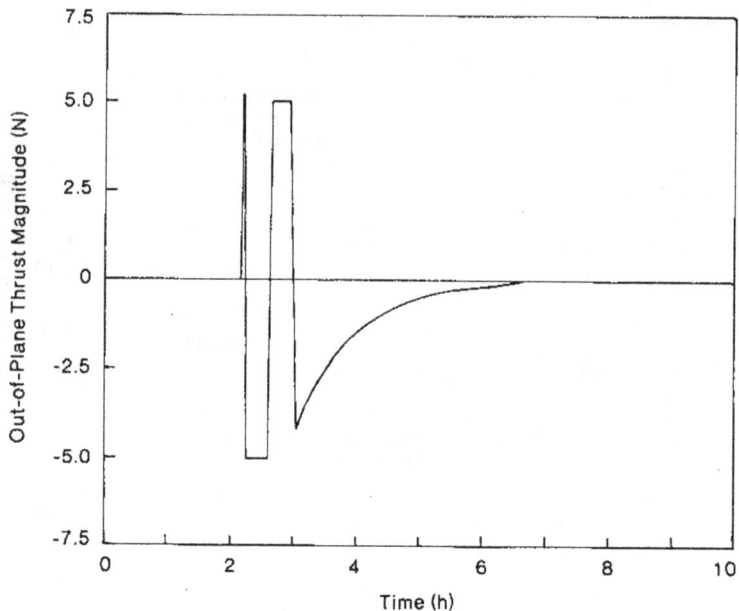

FIGURE 2.12 Out-of-plane thruster time history for fast retrieval vs. time

with $\varsigma = 10.0$ and

$$\delta = (m + 0.5\rho r\theta) r\theta u_1 \Big/ \left\{ mr^2\theta^2 + \frac{\rho L}{3} L^2 \left[1 - \left(1 - \frac{r\theta}{L} \right)^3 \right] - \rho r\theta L^2 \left(1 - \frac{r\theta}{L} \right) \right\} \quad (2.55)$$

Again, Eqs. (2.53) and (2.54) were arrived at after several numerical experiments. The rationale underlying such functional forms is to attempt to annul the $\dot{L}$ terms normally associated with negative damping (coefficients of $\dot{L}$, which is negative for retrieval) in the dynamical equations. Equation (2.54) provides expressions for the generalized active force due to thrust; the corresponding physical thrust has to be of the form:

$$\mathbf{F} = F_1 \mathbf{b}_1 + F_2 \mathbf{b}_2 \quad (2.56)$$

which, in turn, requires the following, with I_2 given by Eq. (2.45):

$$F_1 = F_\alpha I_2 \cos \beta / (r\theta) \quad (2.57)$$

$$F_2 = -F_\beta I_2 / (r\theta) \quad (2.58)$$

Finally, a key terminal step in the control of tethered satellite retrieval is to turn on a tether in-line thruster of magnitude 2 N away from the orbiter and reel in the tether rapidly, while relying on the in-plane thrust to prevent a tether wrap-up. A thrust level of 2 N is chosen because hardware tests reveal that the same level of thrust is needed during the initial deployment of the satellite merely to overcome the friction in the tether pulley system. With the tether tension maintained in this way, the final strategy is then to reel in the satellite at a high rate, such as 1 m/s used in the simulations, while relying on the tether in-plane thruster to prevent a tether wrap-up around the shuttle.

2.2.2 SIMULATION RESULTS

Figure 2.4 shows the results of a tethered satellite retrieval for tether tension with only tension control law (dashed line) and tension control together with tether in-plane and out-of-plane control (solid line). The figure shows that the tether becomes slack after some time, with tension control alone, while thruster augmentation keeps the tether taut. Figure 2.5 shows the corresponding tether lengths.

Figures 2.6 and 2.7 show the in-plane and out-of-plane angles with respect to the local vertical with and without thruster augmentation of the tension control. Again, the dashed line is for tension control alone. Thrust vs. time histories associated with Figures 2.4–2.7 is shown in Figure 2.8.

Figures 2.9 and 2.10 show the results for the tether length and tension for *fast retrieval*; again, the dashed lines mean tension control acting alone. The corresponding thruster firing histories for in-plane and out-of-plane fast retrieval are shown in Figures 2.11 and 2.12.

2.2.3 CONCLUSION

The effectiveness of both in-plane and out-of-plane thrust augmentation control, compared to tension control acting alone has been shown. It has also been shown that thrust augmentation can cut down retrieval time, whereas tension control for fast retrieval can cause a tether wrap-up on the shuttle. Retrieval is basically an unstable process, and thrust control laws can stabilize the system by annulling the negative damping terms inherent in negative rates of length change describing retrieval.

2.3 DYNAMICS AND CONTROL OF STATION-KEEPING OF THE SHUTTLE-TETHERED SATELLITE

The purpose of having a shuttle-borne tethered satellite is, of course, to do station-keeping, with a small satellite to look at the earth from an altitude where the satellite cannot stay by itself due to atmospheric drag. The analysis presented here is based on Ref. [8]. Consider the spherical pendulum shown in Figure 2.13. Let it represent a *rigid rod model of a tether* of length L attached to an orbiting

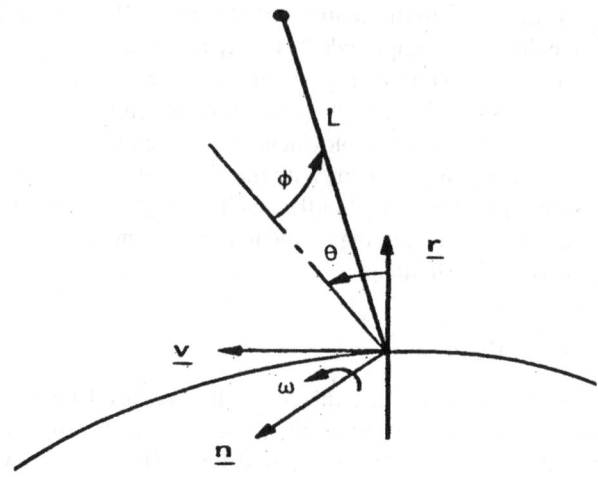

FIGURE 2.13   Spherical pendulum representing a rigid tether attached to an orbiting body

body moving in a circular orbit with angular speed ω. Rotations through an orbit in-plane angle θ followed by an out-of-plane angle ϕ describe the orientation of the tether. To keep the angles θ and ϕ small, we consider a pumping control of the tether, much like a yo-yo toy.

For the prescribed deployment rate $\dot{L}$, the equations of motion of a particle as a lumped end-mass on a massless rigid tether, acted on by gravity gradient force and tether tension, can be written from Newton's law as follows:

$$\ddot{\theta} = -2(\omega + \dot{\theta})[(\dot{L}/L) + \dot{\phi}\tan\phi] - 3\omega^2 \sin\theta\cos\theta \tag{2.59}$$

$$\ddot{\phi} = -2(\dot{L}/L)\dot{\phi} - (\omega + \dot{\theta})^2 \sin\phi\cos\phi - 3\omega^2 \cos^2\theta\sin\phi\cos\phi \tag{2.60}$$

For the tether to always remain taut, it is required that the tension T is always positive, where the tension is evaluated by projecting all forces along the tether, yielding

$$T = m\{L[\dot{\phi}^2 + (\omega + \dot{\theta})^2 \cos^2\phi + \omega^2(3\cos^2\theta\cos^2\phi - 1)] - \ddot{L}\} \tag{2.61}$$

Equation (2.61) places a constraint on the maximum value of $\ddot{L}$, for small angles and angular rates $\dot{\theta} < \omega$, $\dot{\phi} < \omega$, that

$$\ddot{L} < 3\omega^2 L \tag{2.62}$$

The station-keeping objective is to keep θ and φ small. For this, Eqs. (2.59) and (2.60) reduce to

$$\ddot{\theta} = -2(\dot{L}/L)(\omega + \dot{\theta}) - 3\omega^2\theta \qquad (2.63)$$

$$\ddot{\varphi} = -2(\dot{L}/L)\dot{\varphi} - 4\omega^2\varphi \qquad (2.64)$$

Now the yo-yo length control of the tether is visualized as changing the rate of length, $\dot{L}$, to effect reductions in θ and φ. Equations (2.63) and (2.64) show that changing L changes θ more than φ, due to the presence of the forcing term involving ω. As Eq. (2.63) indicates, when φ reaches a maximum, $\dot{\varphi}$ is zero, the tether can be deployed or retracted without building up φ. On the other hand, when the roll angle (or φ) is zero, $\dot{\varphi}$ is maximum and a change in the tether length at this time has the maximum effect on the roll momentum and energy. Since lengthening the tether removes energy by decreasing the roll rate, a roll damping effect can be accomplished by lengthening the tether as it passes through the local vertical and then shortening the tether the same amount at the maximum roll angle.

This action is imitates a child's pumping of a swing; it is shown as the out-of-plane control strategy in Figure 2.14. The situation for in-plane control is similar to the action of a spinning ice-skater. This describes the overall strategy for station-keeping.

If the tether length is changed by an appropriate amount at the time the pitch angle is zero, pitch libration can be completely eliminated. The magnitude and

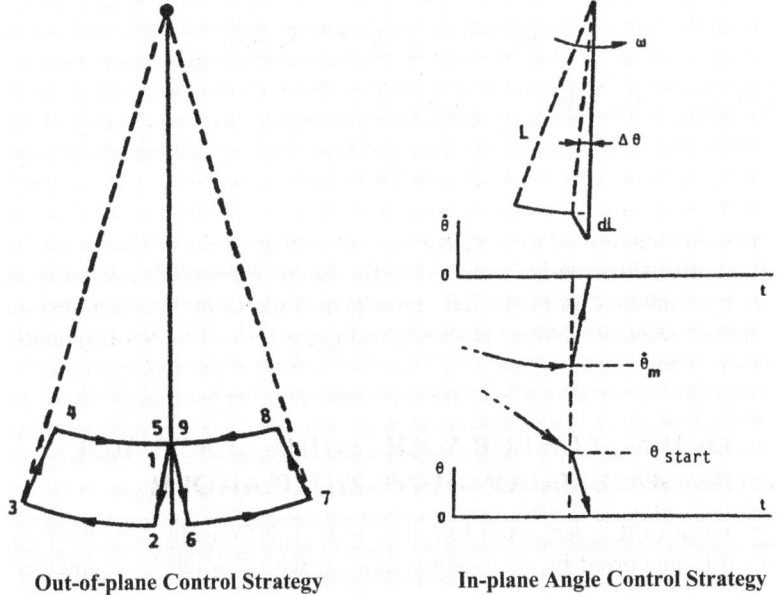

Out-of-plane Control Strategy   **In-plane Angle Control Strategy**

FIGURE 2.14   Out-of-plane and in-plane yo-to control strategy scenarios

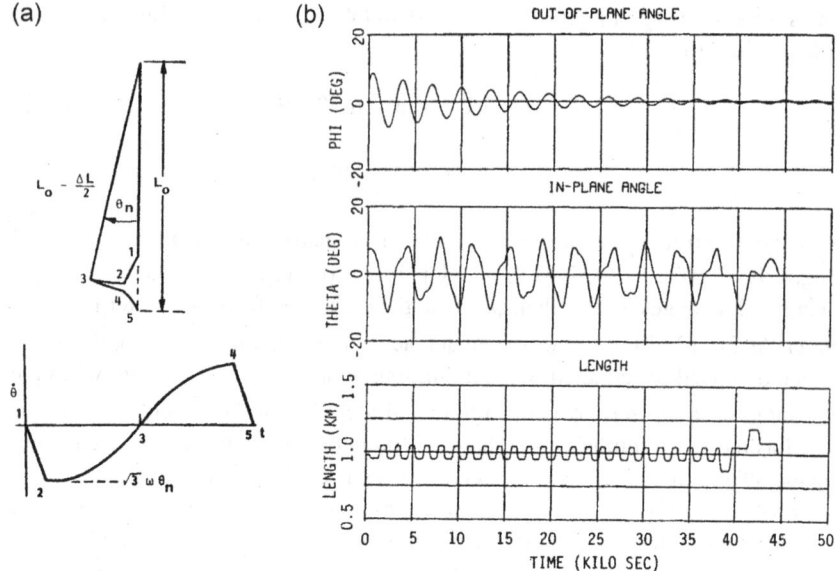

FIGURE 2.15 (a) Yo-yo tether length command control and (b) overall station-keeping control results

direction of the increment in tether length is found by equating the pitch angular momentum before and after length increment:

$$mL^2(\omega + \dot{\theta}) = (mL^2 + 2mLdL)\omega \qquad (2.65)$$

which leads to the result

$$2\omega dL = L\dot{\theta} \qquad (2.66)$$

The sign and magnitude of $\dot{\theta}$ determines the sign and magnitude of dL. This length control situation is described at the top of Figure 2.15, while typical results for combined in-plane and out-of plane control by length adjustment, with digital implementation, is shown in Figure 2.16. Detailed explanations are given in Ref. [8].

2.4 POINTING CONTROL WITH TETHERS AS ACTUATORS OF A SPACE STATION–SUPPORTED PLATFORM

The material in this section is based on Ref. [12]. Consider the system of a platform B supported by the space station A, with two tethers connected at two points D and E of the platform with the common attachment point C on the space station.

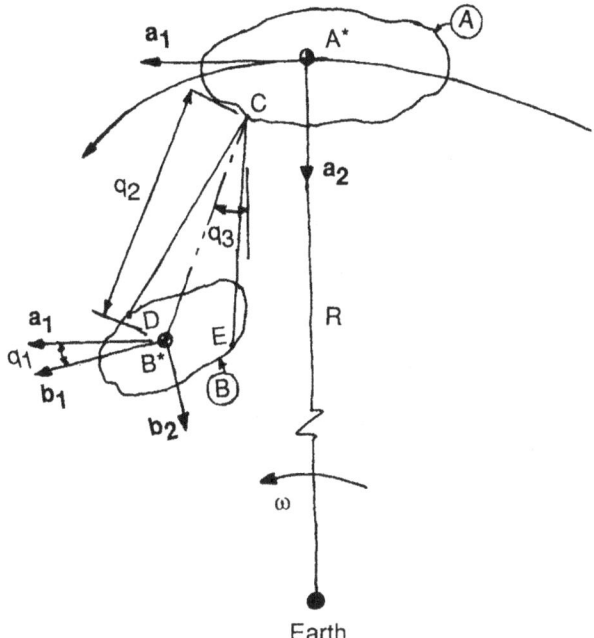

FIGURE 2.16 Platform B supported by the space station A by two tethers, CD and CE

2.4.1 NON-LINEAR EQUATIONS OF MOTION

In Figure 2.16, A*, the mass center of the space station A, is shown moving in a circular orbit of radius R and an angular speed, ω. Two tethers, emerging from A at point C, are attached to a platform B at points D and E and are offset from the mass center B* of B by distances d_1, d_2 and e_1, e_2, respectively. Free to move in the orbital plane, B has 3 degrees of freedom (dof), with q_1 the angle between a unit vector fixed in B and a unit vector fixed in A, q_2 the distance between C and B*, and q_3 the angle between the line CB* and the local vertical and the line of sight from C to B*. Body B has mass m and centroidal moments of inertia I_1, I_2, I_3 about the principal axes inertia parallel to b_1, b_2, b_3. In deriving the equations of motion, tether tensions, T_1, T_2, and gravitational forces and moments due to the earth are taken into account. The equations of motion are given as

$$\ddot{q}_1 = < -(T_D / z_D)q_2[d_1 \cos(q_1 + q_3) - d_2 \sin(q_1 + q_3)]$$

$$-(T_E / z_E)q_2[e_1 \cos(q_1 + q_3) - e_2 \sin(q_1 + q_3)]$$

$$-3\omega^2(I_1 - I_2)[\sin q_1 \cos q_1 - (q_2/R)\sin(2q_1 + q_3)$$

$$+5\sin q_1 \cos q_1 \cos q_3] > /I_3$$

(2.67)

$$\ddot{q}_2 = < -(T_D / z_D)[q_2 + d_1 \sin(q_1 + q_3) + d_2 \cos(q_1 + q_3)]$$
$$-(T_E / z_E)[q_2 + e_1 \sin(q_1 + q_3) + e_2 \sin(q_1 + q_3)] > /m \qquad (2.68)$$
$$+3\omega^2(q_2 \cos q_3 + c_2)\cos q_3 + q_2\dot{q}_3(\dot{q}_3 - 2\omega)$$

$$\ddot{q}_3 = < (T_D / z_D)[d_1\cos(q_1 + q_3) - d_2 \sin(q_1 + q_3)]$$
$$-(T_E / z_E)[e_1 \cos(q_1 + q_3) - e_2 \sin(q_1 + q_3)]/m \qquad (2.69)$$
$$+[2(\omega - \dot{q}_3)\dot{q}_2 - 3\omega^2(q_2 \cos q_3 + c_2)\sin q_3] > /q_2$$

In simplifying the form of the equations of motion given above, we have introduced two intermediate variables, z_D, z_E, defined as follows:

$$z_D = [q_2^2 + d_1^2 + d_2^2 + 2q_2\{d_1 \sin(q_1 + q_3) + d_2 \cos(q_1 + q_3)\}]^{0.5}$$
$$z_E = [q_2^2 + e_1^2 + e_2^2 + 2q_2\{e_1 \sin(q_1 + q_3) + e_2 \cos(q_1 + q_3)\}]^{0.5} \qquad (2.70)$$

2.4.2 LINEARIZED EQUATIONS

An equilibrium state is found from Eqs. (2.67)–(2.69) by setting

$$\dot{q}_i = \ddot{q}_i = 0 \quad (i = 1, 2, 3) \qquad (2.71)$$

This yields three equations in the five unknown variables, $\overline{T}_D, \overline{T}_D, \overline{q}_1, \overline{q}_2, \overline{q}_3$, where an overbar indicates the equilibrium values of the corresponding entity. Thus, we are free to specify two of the barred quantities, which we do so by setting

$$\overline{q}_1 = 0, \quad \overline{q}_2 = L \qquad (2.72)$$

Here, L is an arbitrarily selected reference length. The resulting equations can only be solved numerically for $\overline{T}_D, \overline{T}_E, \overline{q}_3$, unless special values are assigned to c_i, d_i, e_i $(i = 1, 2)$ and $\overline{q}_3$. For example, with

$$\overline{q}_3 = 0$$
$$c_1 = 0, \quad c_2 = a$$
$$d_1 = b, \quad d_2 = 0 \qquad (2.73)$$
$$e_1 = -b, \quad e_2 = 0$$

a simple closed-form solution is obtained for $\overline{T}_D, \overline{T}_E$, namely,

$$\overline{T}_D = \overline{T}_E = 3m\omega^2(L+a)\sqrt{L^2+b^2}/(2L) \tag{2.74}$$

We now take as a reference motion the ones corresponding to Eqs. (2.72)–(2.74), and we introduce perturbations, $T_D^*, T_E^*, q_1^*, q_2^*, q_3^*$, such that

$$T_D = \overline{T}_D + T_D^*$$

$$T_E = \overline{T}_E + T_E^*$$

$$q_1 = 0 + q_1^* \tag{2.75}$$

$$q_2 = L + q_2^*$$

$$q_3 = 0 + q_3^*$$

Substitution from Eqs. (2.67)–(2.70) and linearization in the starred quantities give rise to two sets of terms in each of the resulting equations. The first set involves the equilibrium equations and sums to zero, and the set produces the equations linearized in the perturbations. When used with the control algorithms to be described later, these linear equations are found to be numerically ill-conditioned. To surmount this difficulty, we introduce scaling by defining the following non-dimensional variables:

$$\tau = \omega t$$

$$\delta = b/L; \quad \gamma = b/L$$

$$x_1 = q_1^*; \quad x_2 = q_2^*/L; \quad x_3 = q_3^* \tag{2.76}$$

$$u_1 = T_D^*; \quad u_2 = T_E^*$$

In preparation for the development of the state space suitable for control law development, we define the following state and control vectors x and u as

$$x = [x_1, x_2, x_3, \dot{x}_1, \dot{x}_2, \dot{x}_3]^T \tag{2.77}$$

$$u = [u_1, u_2]^T$$

The state space equations can now be written as

$$\dot{x} = Ax + Bu \tag{2.78}$$

Here, the non-zero elements of the 6×6 linear dynamics matrix A and the 6×2 matrix B are as follows:

$$A(1,4) = 1.0$$

$$A(2,5) = 1.0$$

$$A(3,6) = 1.0$$

$$A(4,1) = \frac{\delta^2}{(1+\delta^2)^{1.5}} \frac{(\bar{T}_D + \bar{T}_E)L}{\omega^2 I_3} + 3\left(\frac{I_2 - I_1}{I_3}\right)\left(1 + 3\frac{L}{R}\right)$$

$$A(4,3) = \frac{\delta^2}{(1+\delta^2)^{1.5}} \frac{(\bar{T}_D + \bar{T}_E)L}{\omega^2 I_3} - 3\left(\frac{I_2 - I_1}{I_3}\right)\frac{L}{R}$$

$$A(5,2) = -\frac{(\bar{T}_D + \bar{T}_E)}{m\omega^2 L} \frac{\delta^2}{(1+\delta^2)^{1.5}} + 3$$

$$A(5,6) = -2.0$$

$$A(6,1) = \frac{(\bar{T}_D + \bar{T}_E)}{m\omega^2 L} \frac{\delta^2}{(1+\delta^2)^{1.5}}$$

(2.79)

$$A(6,3) = A(6,1) - 3(1 + \gamma)$$

$$A(6,5) = 2.0$$

$$B(4,1) = -\frac{\delta}{(1+\delta^2)^{1.5}} \frac{\bar{T}_D}{\omega^2} \frac{L}{I_3}$$

$$B(4,2) = -B(4,1)$$

$$B(5,1) = -\frac{1}{(1+\delta^2)^{0.5}} \frac{\bar{T}_D}{m\omega^2 L}$$

$$B(5,2) = B(5,1)$$

$$B(6,1) = -\frac{\delta}{(1+\delta^2)^{0.5}} \frac{\bar{T}_D}{m\omega^2 L}$$

$$B(6,2) = -B(6,1)$$

The analytical form of the linear equations may be exploited in a parametric examination of control design.

2.4.3 CONTROL LAW DESIGN

The objective of control design is to bring initial deviations from the equilibrium state to zero in a reasonable amount of time by controlling the tether tensions in such a way that they never become negative, a requirement that is fundamental to the use of a tether as an actuator. The system parameters are taken to be m = 1000 kg; I_1 = 70,000 kg-m^2; I_2 = 85,000 kg-m^2;

I_3 = 100,000 kg-m^2; L = 100 m;

a = 2 m; b = 20 m; R = 6.878×10^6 m (corresponding to an orbit

altitude of 500 km and ω = 1.1068×10^{-3} rad/sec)

For these parameters, the open-loop plant equations described by Eqs. (2.78) and (2.79) are found to be unstable with open loop poles at (±7.707, ±j2.398, ±j1.242). The system is, however, controllable, as may be checked by the controllability condition [13] for Eqs. (12) and (13). A full-state feedback controller was generated by using the linear quadratic regulator (LQR) theory [13]. The cost function J was taken as

$$J = \int_0^\infty e^{2\alpha\tau}(x^T Q x + u^T R u)d\tau \tag{2.80}$$

Here, the parameter α was used to select the bandwidth [13], with the consequence that the linear dynmics matrix A in Eq. (12) had to be replaced by [A + αI], where I is the 6×6 identity matrix. The state and control weights, Q and R, were chosen following Bryson's rule [13], with assumed values of the maximum state and control magnitudes. The feedback control as computed by the LQR algorithm was implemented by scaling back to the variables in Eq. (2.76). The total control effort consists of LQR gains, [k_{ij}; i = 1,2; j = 1,...,6] and the feed-forward control for nominal motion, $\bar{T}_D, \bar{T}_E$ given by Eq. (2.74):

$$T_D = \bar{T}_D - k_{11}q_1 - k_{12}[(q_2 / L) - 1] - k_{13}q_3$$
$$- k_{14}(\dot{q}_1 / \omega) - k_{15}\dot{q}_2 / (\omega L) - k_{16}\dot{q}_3 / \omega$$
$$T_E = \bar{T}_E - k_{21}q_1 - k_{22}[(q_2 / L) - 1] - k_{23}q_3 \tag{2.81}$$
$$- k_{24}(\dot{q}_1 / \omega) - k_{25}\dot{q}_2 / (\omega L) - k_{26}\dot{q}_3 / \omega$$

A simulation of the system motion was performed by substituting from Eq. (2.81) in Eq. (2.67), Eq. (2.68) and Eq. (2.69). For α= 1, Q=diag(25,25,25,1,1,1), R=diag(10,10), (where "diag" stands for a diagonal matrix with arguments in the diagonal) the dashed curve in the three plots in Figure 2.17 represent q_1, q_2, q_3, starting with the initial values, $q_1(0)$=6 deg; $q_2(0)$= 101 m, $q_3(0)$=6 deg; the nominal values for q_1, q_2, q_3 are 0 deg, 100 m, and 0 deg, respectively. As can be seen, the settling times of the perturbations are of the order of the orbit period. The corresponding tension vs. time plots of the tethers CD and CE are shown in Figure 2.18, which shows that the tensions never become negative. It must be stated that the LQR solution does not guarantee that the tension will never be negative, and one must adjust the parameters α, Q, R in the cost function, Eq. (2.80), to get an acceptable solution. Ref. [12] gives details, not presented here, of adding a

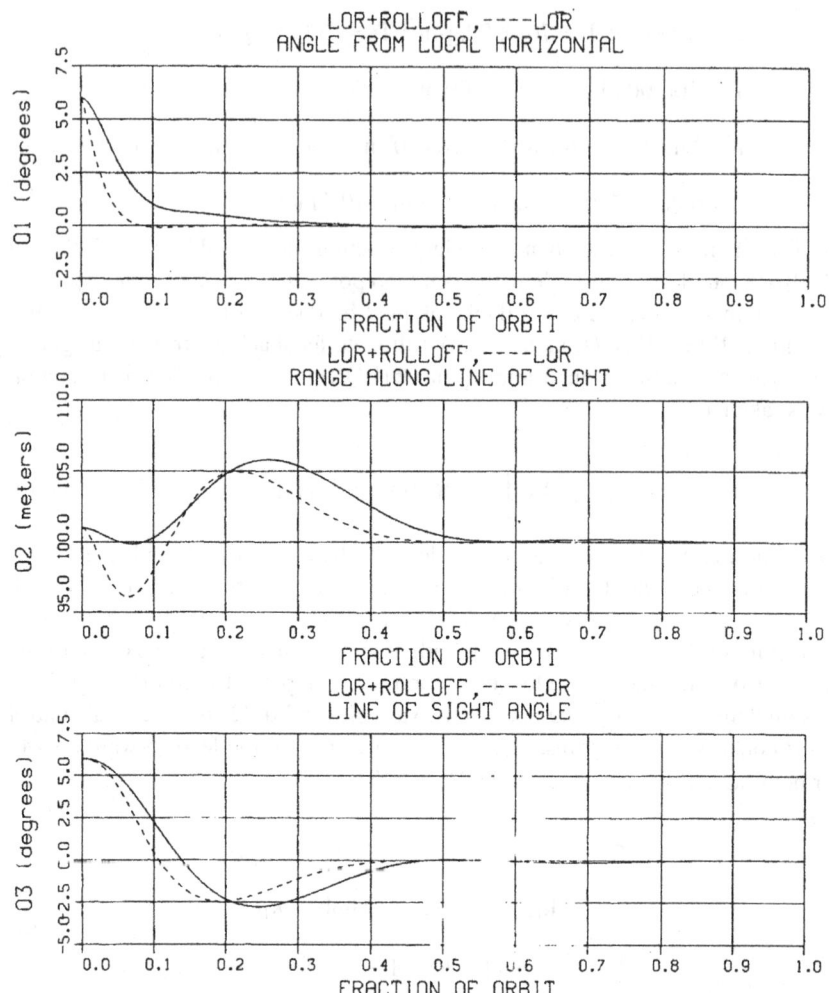

FIGURE 2.17   Response to initial disturbance with LQR, and LQR plus a roll-off

roll-off filter to the LQR control to guarantee that the tether tension will never be negative.

APPENDIX 2.A FORMATION FLYING WITH MULTIPLE TETHERED SATELLITES

Tethered satellite dynamics has reached a level of maturity where an entire book [14] has now been written. We conclude this chapter by pointing to a very interesting body of work on formation flying in space, with tethers deployed from a central body, deploying four or more sub-satellites. The principal reference is Ref. [15]. We explain the idea of formation flying below.

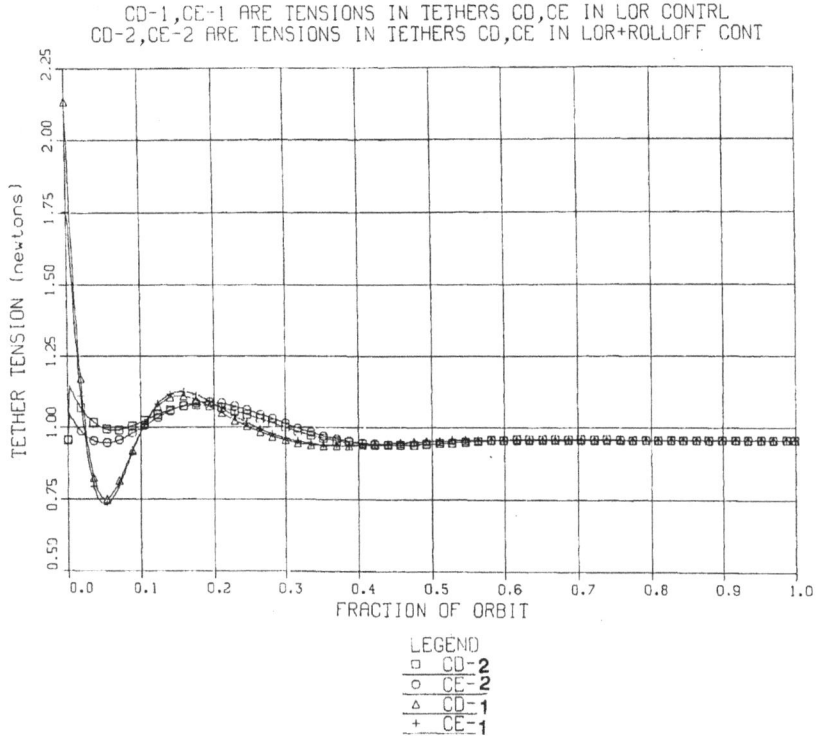

FIGURE 2.18   Tension in tethers 1 and 2 with LQR, and LQR plus a roll-off

Two examples of formation flying, as shown in Figure 2.A.1, are considered in Ref. [12], with N satellites connected to a main hub, the $(N+1)$ satellite, with the closed hub version forming an outer loop. Each particle satellite has 3 dof, because all tethers are considered stretchable. Another possible scenario is a seven-body double pyramid configuration shown in Figure 2.A.2.

With the motion of the primary body prescribed, the gravitational potential energy of the system is obtained by expanding the standard denominator term of the gravitational potential in a binomial series and neglecting terms beyond the third power, as follows:

$$V_g = GM_e \sum_{i=1}^{N+1} \frac{m_i}{R_p} - \frac{GM_e}{R_P^2} \sum_{i=1}^{N+1} m_i (\mathbf{O}_1 \cdot \mathbf{r}_i) + \frac{GM_e}{2R_P^3} m_i [\mathbf{r}_i \cdot \mathbf{r}_i - 3(\mathbf{O}_1 \cdot \mathbf{r}_i)^2] \qquad (A2.1)$$

where $\mathbf{O}_1$ is an orbit radius unit vector going from the earth center to the primary satellite, and $\mathbf{r}_i$ is the three-dimensional position vector from the primary to the i-th satellite. Potential energy stored in the springs is given in terms of the spring stiffness k_i, the instantaneous and nominal length l_i, l_{ni} for the i-th tether (see Figure 2.20) as

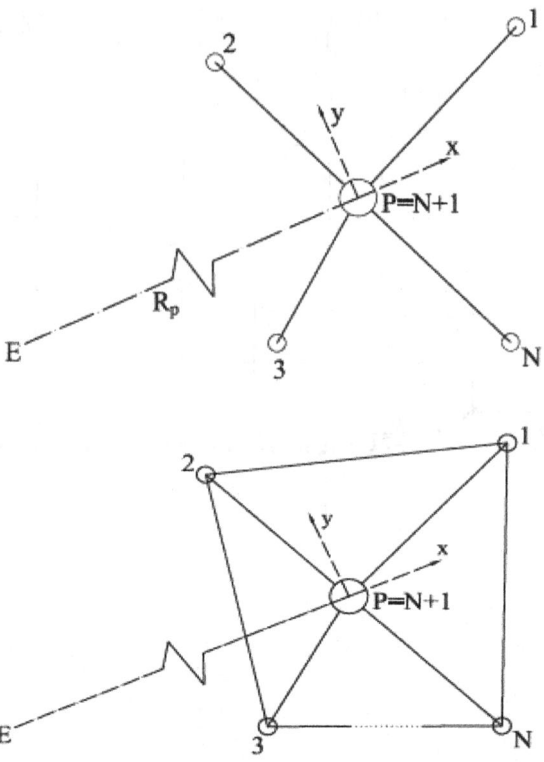

FIGURE 2.A.1   Open hub-and-spoke and closed hub-and-spoke formation flying

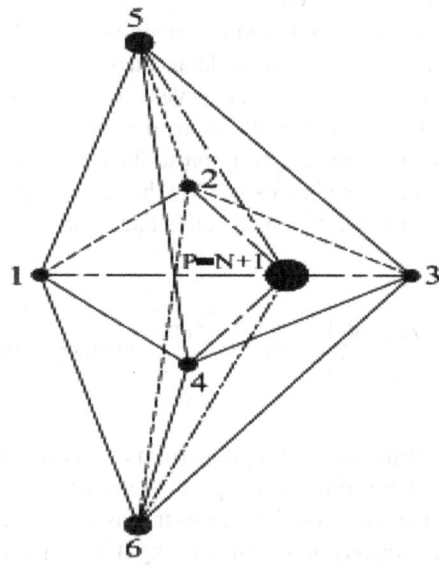

FIGURE 2.A.2   A seven-body double pyramid configuration

$$V_e = \sum_{i=1}^{N} 0.5 k_i (l_i - l_{ni})^2 \qquad (A2.2)$$

The generalized force for damping in the i-th tether is stated via the Rayleigh dissipation function:

$$Q_{d,j} = \frac{\partial}{\partial \dot{q}_j} \left[0.5 \sum_{i=1}^{N} c_i (\dot{l})_i^2 \right] \qquad (A2.3)$$

Finally, Lagrange's equation of motion, used in Ref. [15], is written as

$$\frac{d}{dt} \frac{\partial K}{\partial \dot{q}_j} - \frac{\partial K}{\partial q_j} = -\frac{\partial}{\partial q_j} (V_g + V_e) - Q_{d,j} \quad (j = 1, \ldots, n) \qquad (A2.4)$$

Ref. [15] gives explicit equations of motion, based on Eqs. (A2.1)–(A2.4), for the case when the parent body is in a circular orbit. Figure 2.A.3 shows that α_i, β_i are angles of the projection of the i-th tether on the y–z plane and the angle from this projection to the actual tether.

Figure 2.A.3 shows the time response of one of the alignment angles for bodies 1 and 3 in the closed hub-and-spoke four-body system of Figure 2.A.1. Figure 2.A.4 shows the other angle needed for describing the tether alignment, for bodies 1–4 in the seven-body pyramid configuration of Figure 2.A.2. The stability of the configurations is investigated by simulation in Ref. [15]. It was noted that, for planar formations, the closed configuration is stable, but there is no guarantee that the open one will be stable. For the three-dimensional case, only the double pyramid configuration was found to be stable.

APPENDIX 2.B ORBIT BOOSTING OF TETHERED SATELLITE SYSTEMS BY ELECTRODYNAMIC FORCES

Use of the tether in a shuttle-supported tethered satellite system as a current-carrying conductor moving in the earth's magnetic field to generate a propulsive force has been investigated by several authors, an example being Ref. [16]. The process is executed by turning the current in the tether for a time to generate the force-time impulse needed for orbit boosting. A mathematical basis for the idea, and the means to modulate the current pulse shaped to suppress librations of the tethered satellite system, and to steady down the orbit, are given below [17].

Basically, the tether is discretized into several rigid rods, as shown in Section 2.1, and forces due to gravity and aerodynamic drag are computed as before. A schematic of the electrodynamic tethered satellite system is shown in Figure 2.B.1. The electrodynamic force on a tether of length L moving in the earth's magnetic field vector **B** with a current I flowing along the segment direction $\mathbf{u}_j$ is given by

$$\mathbf{F}_e^{P_j} = I L \mathbf{u}_j \times \mathbf{B} \qquad (A2.5)$$

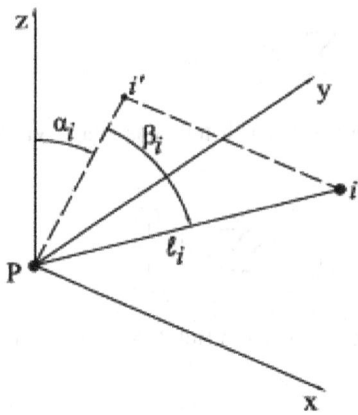

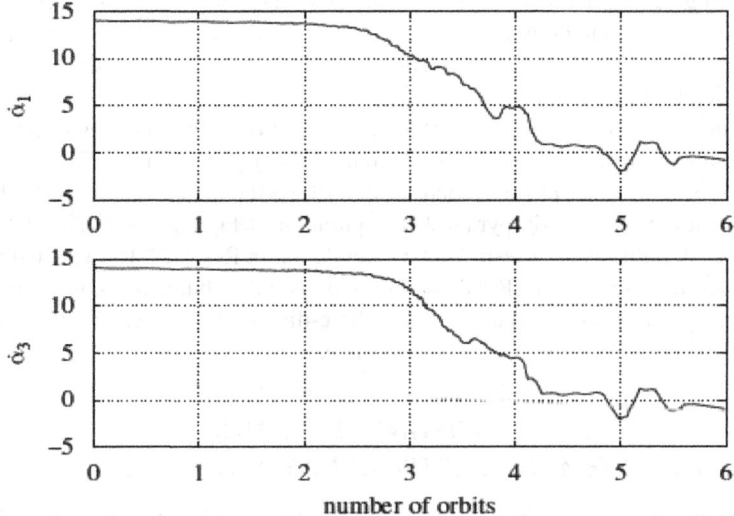

FIGURE 2.A.3 (a) Generalized coordinates α_i, β_i, l_i for the i-th satellite from the hub; and (b) in-plane angle rate $\dot{\alpha}_1, \dot{\alpha}_3$ for bodies 1 and 3 in the closed hub-and-spoke four-body system

The magnetic field of the earth can be evaluated at the midpoint R_j of the tether segment as due to a magnetic dipole, and written as

$$\mathbf{B} = \mu_e R_e^3 [\cos\lambda \mathbf{u}_n + 2\sin\lambda \mathbf{u}_r]/(\mathbf{p}^{ER_j} \cdot \mathbf{p}^{ER_j})^{1.5}$$

$$\sin\lambda = \mathbf{u}_m \cdot \mathbf{u}_r$$

$$\mathbf{u}_n = \mathbf{u}_r \times (\mathbf{u}_r \times \mathbf{u}_m)/|\mathbf{u}_r \times \mathbf{u}_m| \qquad (A2.6)$$

$$\mathbf{u}_m = \sin\lambda(\cos\omega_e t \mathbf{n}_1 + \sin\omega_e t \mathbf{n}_2) + \cos\lambda \mathbf{n}_3$$

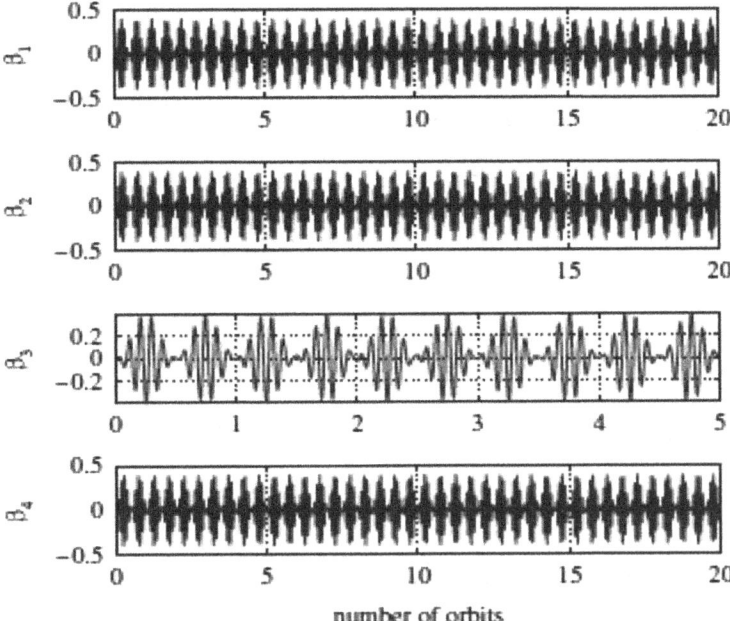

number of orbits

FIGURE 2.A.4 Out-of-plane angle β history for bodies 1–4 in the seven-body double pyramid of Figure 2.A.3 given in Pizarro-Chong and Misra [15]

Here, μ_e, $\mathbf{u}_r$, $\mathbf{u}_m$ are, respectively, the earth's magnetic field constant, 3.5e-5 Wb/m², the unit vector from the earth center E to R_j, and the unit vector along the earth's magnetic axis; δ is the angle between the earth's magnetic axis and its spin axis. Once the electrodynamic force is calculated, using Eq. (A2.5) and Eq. (A2.6), generalized active forces can be evaluated by dot-multiplying the force vector given by Eq. (A2.1) with the partial velocity of point R_j for all j, and added to the equations of motion.

The application of the tether electrodynamic force for orbit boosting is considered here. For higher energy orbits, the system mass center in Figure 2.B.1, the required change in angular momentum is equated to the time-integral of the component of the electrodynamic force in the orbit tangent vector $\mathbf{b}_2$, see Figure 2.B.2. Approximating this force to be constant, the duration of the pulse to raise the orbit from R_1 to R_2 for a *total system mass* M at R*, becomes

$$\Delta t = [\sqrt{\mu R_2} - \sqrt{\mu R_1}] / [F(R_1 + R_2) / (2M)]$$

$$F = \mathbf{F}^{R^*} \cdot \mathbf{b}_2$$

(A2.7)

In the following, we show the results for imparting a pulse of force level F for duration Δt without pre-shaping, and pre-shaping by convolution of the pulse with a three-impulse sequence [15] for each mode of vibrations in the tethered model of Figure 2.B.1. The following data were used for generating the results in Figure

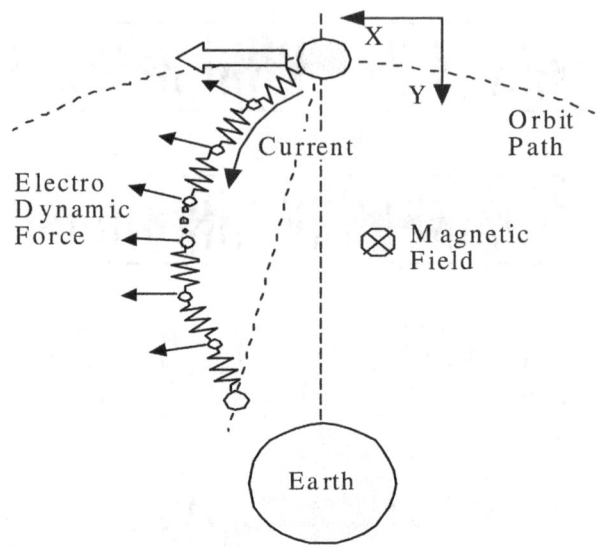

FIGURE 2.B.1   Schematic of the satellite model with electrodynamic forces on the tether

2.B.2–2.B.5: mass of satellite 300 kg; tether mass 20 kg; end mass 5 kg; initial circular earth orbit altitude 500 km; total tether length 20 km; length of tether segment conducting current, upper 10 km; length of tether with aerodynamic drag, lower 10 km; orbit altitude pulsation period 5817 sec; tether in-plane libration period 3398 sec; tether out-of-plane libration period 2828 sec; tether string vibration period 652 sec.

PROBLEM SET 2

1. Derive the equations of motion of a rigid tether with a lumped mass as in Eqs. (2.59)–(2.61), for a 3 degrees of freedom system.
2. Derive the equations of motion of deployment and retrieval of a satellite tethered to the space shuttle going on a circular orbit, by considering the *tether to be a rigid massless rod* with the only mass being that of the satellite, i.e., with the tether instantaneous length, in-plane and out-of-plane angles, as the 3 degrees of freedom.
3. Derive the equations of motion for one main satellite deploying four sub-satellites in the orbital plane of the main satellite, as in the closed hub-and-spoke configuration. Refer to Ref. [15] for details of the derivation.
4. Consider the station-keeping problem of a 20 km long tether of mass 1 kg/km connected to a small 200 kg satellite and a large spacecraft orbiting the earth with an orbit radius of 6800 km. Examine small longitudinal elastic oscillations of the tether, assuming that the transverse vibrations are negligible. Derive linearized dynamical equations of the

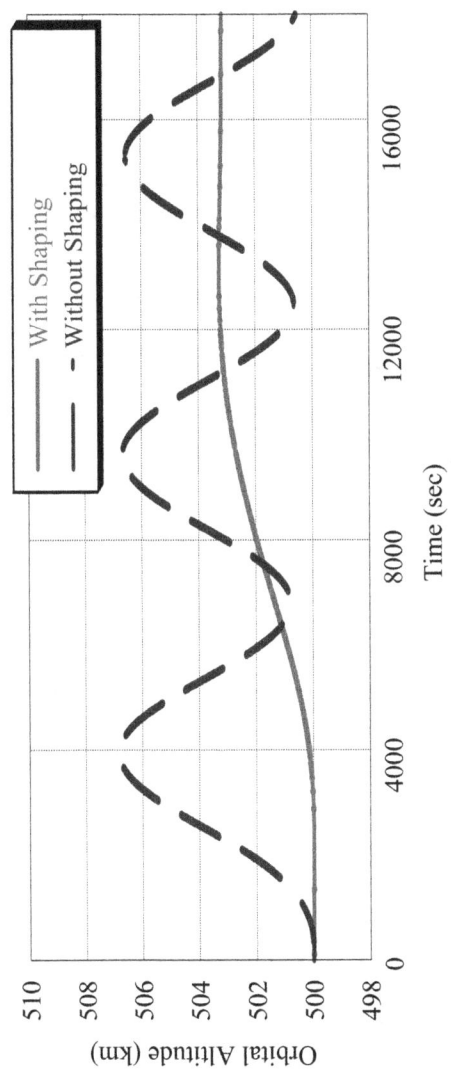

FIGURE 2.B.2   Orbit altitude vs. time with and without current shaping

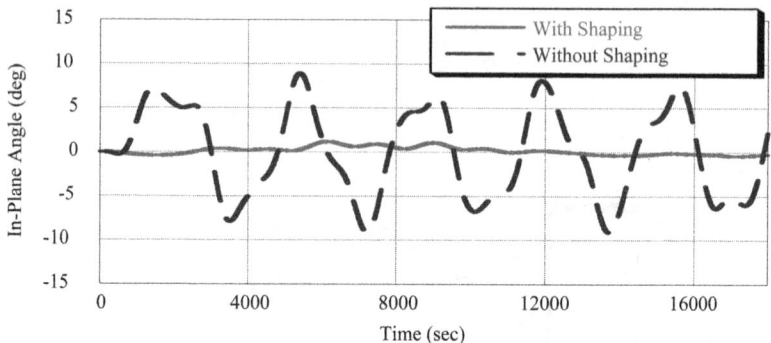

FIGURE 2.B.3 Tether in-plane angle with and without current shaping

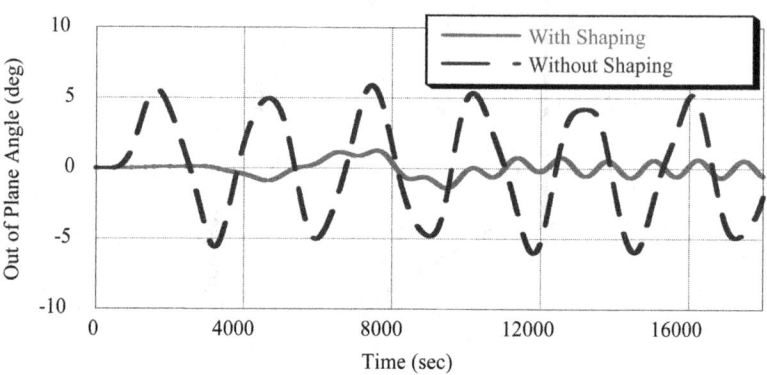

FIGURE 2.B.4 Tether out-of-plane angle with and without current shaping

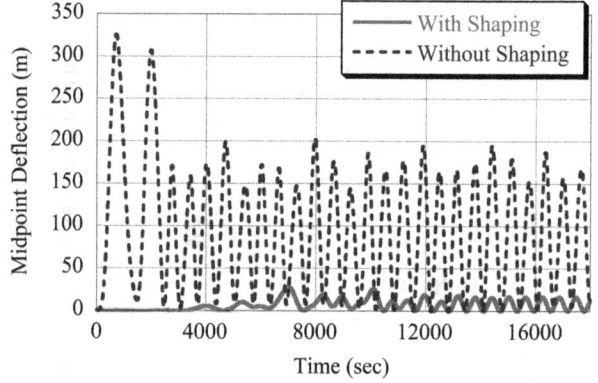

FIGURE 2.B.5 Tether mid-point deflection in the discretized tether model

system by Kane's method, and calculate the lowest four longitudinal natural frequencies of the system from the associated eigenvalue problem.

5. Consider the same problem as above and assume that the tether has a steady longitudinal strain but no longitudinal oscillations. Assuming that the tether executes small transverse vibrations, develop a dynamics model to describe this motion, and obtain the first four transverse vibration frequencies.

REFERENCES

1. Baker, W.P., Dunkin, J.A., Galaboff, Z.J., Johnston, K.D., Kissel, R.R., Rheinfurth, M.H., and Seibel, M.P., (1976), *Tethered Subsatellite Study*, Marshall Space Flight Center, NASA TM X-73314.
2. Kallaghn, P.N., Arnold, D.A., Colombo, G., Grossi, M.D., Kirschner, L. R., and Orringer, O., (1978), "Study of the Dynamics of a Tethered Satellite System (Skyhook)", Smithsonian Astrophysical Observatory, Cambridge, Mass., Final Report.
3. Misra, A.K. and Modi, V.J., (1979), "A General Dynamical Model for the Space Shuttle Based Tethered Satellite System", *Advances in the Astronautical Sciences*, **40**, pp. 537–557.
4. Misra, A.K. and Modi, V.J., (1982), "Deployment and Retrieval of a Subsatellite Connected by a Tether to the Space-Shuttle", *Journal of Guidance, Control, and Dynamics*, **5**(May), pp. 278–285.
5. Banerjee, A.K. and Kane, T.R., (1982), "Tether Deployment Dynamics", *Journal of the Astronautical Sciences*, **30**(Oct.–Dec.), pp. 347–366.
6. Banerjee, A.K., (1990), "Dynamics of Tethered Payloads with Deployment Rate Control", *Journal of Guidance, Control, and Dynamics*, July–August, pp. 759–762.
7. Banerjee, A.K. and Kane, T.R., (1984), "Tethered Satellite Retrieval with Thruster Augmented Control", *Journal of Guidance, Control, and Dynamics*, **7**(1), pp. 45–50.
8. Davis, W.R. and Banerjee, A.K., (1990), "Libration Damping of a Tethered Satellite by Yo-Yo Control with Angle Measurement", *Journal of Guidance, Control, and Dynamics*, **13**(2), pp. 370–374.
9. Kane, T.R. and Levinson, D.A., (1985), *Dynamics: Theory and Application*, McGraw-Hill, pp. 225–241.
10. Baumgarte, J., (1972), "Stabilization of Constraints and Integrals of Motion", *Computer Methods in Applied Mechanics and Engineering*, **1**, pp. 1–16.
11. Wang, J.T. and Huston, E.L., (1987), "Kane's Equations with Undetermined Multipliers: Application to Constrained Multibody Systems", *Journal of Applied Mechanics*, **54**, pp. 424–429.
12. Banerjee, A.K. and Kane, T.R., (1992), "Pointing Control, with Tethered as Actuators, of a Space-Station Supported Platform", *Journal of Guidance, Control, and Dynamics*, **16**(2), pp. 396–399.
13. Bryson, A.E. and Ho, Y.C., (1969), *Applied Optimal Control*, Blaisdell Publishing Co.
14. Alpatov, A.P., Beletsky, V.V., Dranovskii, V.I., Khoroshilov, V.S., Pirozhenko, A.V., Troger, H., and Zakrzhevskii, A.E., (2010), *Dynamics of Tethered Space Systems*, CRC Press, Taylor and Francis Group, pp. 1–241.
15. Pizarro-Chong, A. and Misra, A.K., (2008), "Dynamics of Multi-Tethered Satellite Formations Containing a Parent Body", *Acta Astronautica*, **63**, pp. 1188–1202.

16. Vas, I.E., Kelley, T.J., and Scarl, E.A., (2000), "Space Station Reboost with Electrodynamic Tethers", *Journal of Spacecraft and Rockets*, March–April, pp. 154–164.

17. Banerjee, A., Singhose, W., and Blackburn, D., (2005), "Orbit Boosting of an Electrodynamic Tethered Satellite with Input-Shaped Current", AAS/AIAA Astrodynamics Specialist Conference, Lake Tahoe, CA.

3 Kane's Method of Linearization Applied to the Dynamics of a Beam in Large Overall Motion

This chapter is based on a paper [1] that deals with the issue of linearization with respect to modal coordinates associated with vibration mode shapes, for a beam executing small vibrations in a reference frame that is itself undergoing large motion. Study of the behavior of flexible bodies attached to moving supports has been vigorously pursued for over 60 years in diverse disciplines, such as machine design, robotics, helicopter dynamics, and spacecraft dynamics. In particular, beams attached to moving bases have received attention in hundreds of technical papers dealing variously with elastic linkages, rotating machinery, robotic manipulator arms, aircraft propellers, helicopter rotor blades, and spinning satellites with flexible appendages. Indeed, the existing literature is so voluminous as to preclude a comprehensive review even within the confines of a chapter. A 1974 review by Modi [2] of the literature on rigid bodies with flexible appendages contains more than 200 references, and we will add to it with relatively later pertinent publications, as the discussion proceeds. Here, we will derive equations of motion of beams in large overall motion, by Kane's method [3], for small deformations about a frame undergoing large motion. The method forms linearized equations, without deriving the non-linear equations, but requires keeping non-linear terms up to a certain point in the analysis.

3.1 NON-LINEAR BEAM KINEMATICS WITH NEUTRAL AXIS STRETCH, SHEAR, AND TORSION

The system to be described, shown in Figure 3.1, consists of a cantilever beam B built into a rigid body A whose motion in a Newtonian frame N is prescribed as a function of time. The beam is characterized by material properties, $E_0(x)$, $G(x)$, $\rho(x)$, and cross-sectional properties $A_0(x)$, $I_2(x)$, $I_3(x)$, $\alpha_2(x)$, $\alpha_3(x)$, $\kappa(x)$, $\Gamma(x)$, $e_2(x)$, $e_3(x)$ defined as follows. Let x be the distance from a point O, located at the root of B, to the plane of a generic cross section of B, when B is undeformed. Then, $E_0(x)$, $G(x)$, $\rho(x)$ are the modulus of elasticity, the shear modulus, and the mass per unit length of the beam, respectively, as functions of x. The area of the cross section located at a distance x from O is denoted as $A(x)$, and the Saint Venant torsion

DOI: 10.1201/9781003231523-4

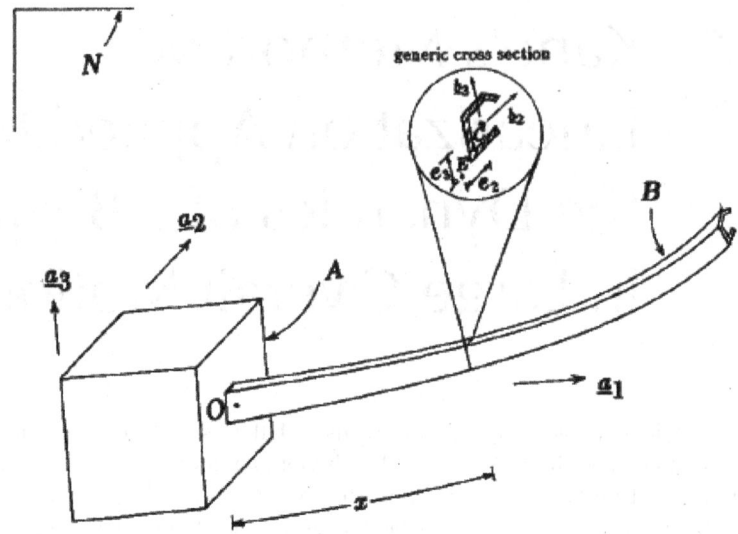

FIGURE 3.1   Beam attached to a moving rigid base

factor and warping factor are denoted by κ(x), Γ(x). Symbols $I_2(x)$, $I_3(x)$, $\alpha_2(x)$, $\alpha_3(x)$, $e_2(x)$, $e_3(x)$ can be defined by introducing a dextral set of unit vectors, $\mathbf{b}_1$, $\mathbf{b}_2$, $\mathbf{b}_3$ (again using bold symbols instead of underbars shown in the figure), fixed in the plane of the cross section located at a distance x from point O, and oriented such that $\mathbf{b}_1$ is parallel to the centroidal axis of B, while $\mathbf{b}_2$ and $\mathbf{b}_3$ are parallel to the central principal axes of the cross section. Additionally, a similar set of unit vectors, $\mathbf{a}_1$, $\mathbf{a}_2$, $\mathbf{a}_3$, is introduced, fixed in A and parallel, respectively, to $\mathbf{b}_1$, $\mathbf{b}_2$, $\mathbf{b}_3$, when B is undeformed. The symbols $I_2(x)$, $I_3(x)$ represent, respectively, the central principal second moments of inertia of the cross section about $\mathbf{b}_2$, $\mathbf{b}_3$, respectively, and $\alpha_2(x)$, $\alpha_3(x)$ are the shear correction factors; $e_2(x)$, $e_3(x)$ are components of the eccentricity vector from the elastic center to the centroid of the area.

The goal of the following analysis is to produce equations governing the extension, transverse bending deflection, shear, and torsion of the beam. The motion of A in N is characterized by the six scalars v_1, v_2, v_3, ω_1, ω_2, ω_3, defined as follows:

$$v_i = {}^N\mathbf{v}^O \cdot \mathbf{a}_i \quad i = 1,2,3$$

$$\omega_i = {}^N\boldsymbol{\omega}^A \cdot \mathbf{a}_i \quad i = 1,2,3$$

(3.1)

Equation (3.1) implies that the velocity of O in N and the angular velocity of A in N, in Figure 3.2 are

$$^N\mathbf{v}^O = v_1\mathbf{a}_1 + v_2\mathbf{a}_2 + v_3\mathbf{a}_3$$

$$^N\boldsymbol{\omega}^A = \omega_1\mathbf{a}_1 + \omega_2\mathbf{a}_2 + \omega_3\mathbf{a}_3$$

(3.2)

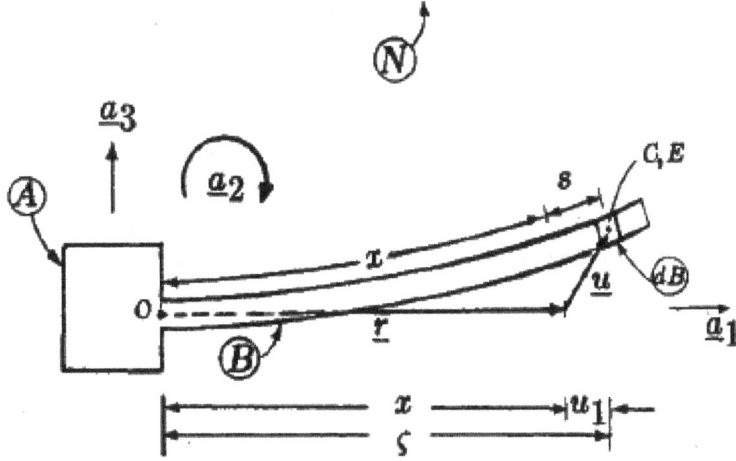

FIGURE 3.2   Deformed beam with differential element

To describe the motion of B in N a rigid differential element dB of B is introduced, with centroid C and elastic center E located at a distance x from point O when B is undeformed. The differential element dB can be brought into a general orientation in A by aligning $\mathbf{b}_1$, $\mathbf{b}_2$, $\mathbf{b}_3$, with $\mathbf{a}_1$, $\mathbf{a}_2$, $\mathbf{a}_3$, respectively, and subjecting dB to successive dextral rotations in A of amounts θ_1, θ_2, θ_3 about lines parallel to $\mathbf{a}_1$, $\mathbf{a}_2$, $\mathbf{a}_3$, in turn. Similarly, point E of dB can be brought into a general position by subjecting E to successive translations $u_1\mathbf{a}_1$, $u_2\mathbf{a}_2$, $u_3\mathbf{a}_3$, which allow us to define two vectors:

$$\mathbf{u} = u_1\mathbf{a}_1 + u_2\mathbf{a}_2 + u_3\mathbf{a}_3$$

$$\mathbf{r} = x\mathbf{a}_1 \tag{3.3}$$

so that the position vector from O to E is

$$\mathbf{p}^{OE} = \mathbf{r} + \mathbf{u} \tag{3.4}$$

The position vector from E to C is given by $\mathbf{p}^{EC} = e_2\mathbf{b}_2 + e_3\mathbf{b}_3$, so that

$$\mathbf{p}^{OC} = (x + u_1)\mathbf{a}_1 + u_2\mathbf{a}_2 + u_3\mathbf{a}_3 + e_2\mathbf{b}_2 + e_3\mathbf{b}_3 \tag{3.5}$$

Orientation of dB in A is described by the space 3 orientation matrix $^A C^{db}$ such that its elements are [4]

$$^A C_{ij}^{dB} = \mathbf{a}_i \cdot \mathbf{b}_j \quad i, j = 1, 2, 3$$

$$^A C^{dB} = \begin{bmatrix} c_2 c_3 & s_1 s_2 c_3 - s_3 c_1 & c_1 s_2 c_3 + s_3 s_1 \\ c_2 s_3 & s_1 s_2 s_3 + c_3 c_1 & c_1 s_2 s_3 - c_3 s_1 \\ -s_2 & s_1 c_2 & c_1 c_2 \end{bmatrix} \tag{3.6}$$

Here, $s_i = \sin\theta_i$, $c_i = \cos\theta_i$, $i = 1,2,3$. The velocity of the centroid C of the cross section in N is given by

$$^N\mathbf{v}^C = {}^N\mathbf{v}^O + {}^N\boldsymbol{\omega}^A \times \mathbf{p}^{OC} + {}^A\mathbf{v}^C \tag{3.7}$$

where $^A\mathbf{v}^C$, the velocity of C in A, can be found by differentiating $\mathbf{p}^{OC}$ with respect to time in frame A. Substitution of the resulting expression and using Eqs. (3.2) and (3.5) in Eq. (3.7) yields

$$
^N\mathbf{v}^C = \mathbf{a}_1 \left\{
\begin{aligned}
&v_1 + \omega_2\left[u_3 + e_2 s_1 c_2 + e_3 c_1 c_2\right] - \omega_3\left[\begin{aligned}&u_2 + e_2(s_1 s_2 s_3 + c_3 c_1)\\&+e_3(c_1 s_2 s_3 - c_3 s_1)\end{aligned}\right]\\
&+\dot{u}_1 + e_2\left[(c_1 s_2 c_3 + s_3 s_1)\dot{\theta}_1 + s_1 c_2 c_3 \dot{\theta}_2 - (s_1 s_2 s_3 + c_3 c_1)\dot{\theta}_3\right]\\
&+e_3\left[(-s_1 s_2 c_3 + s_3 c_1)\dot{\theta}_1 + c_1 c_2 c_3 \dot{\theta}_2 - (c_1 s_2 s_3 - c_3 s_1)\dot{\theta}_3\right]
\end{aligned}
\right\}
$$

$$
+\mathbf{a}_2 \left\{
\begin{aligned}
&v_2 + \omega_3\left[x + u_1 + e_2\left(s_1 s_2 c_3 - s_3 c_1\right) + e_3\left(c_1 s_2 c_3 + s_3 s_1\right)\right] + \dot{u}_2\\
&-\omega_1\left[u_3 + e_2 s_1 c_2 + e_3 c_1 c_2\right] + e_2\left[\begin{aligned}&\left(c_1 s_2 s_3 - c_3 s_1\right)\dot{\theta}_1 + s_1 c_2 s_3 \dot{\theta}_2\\&+\left(s_1 s_2 c_3 - s_3 c_1\right)\dot{\theta}_3\end{aligned}\right]\\
&+e_3\left[-\left(s_1 s_2 s_3 + c_3 c_1\right)\dot{\theta}_1 + c_1 c_2 s_3 \dot{\theta}_2 + \left(c_1 s_2 c_3 + s_3 s_1\right)\dot{\theta}_3\right]
\end{aligned}
\right\}
$$

$$
+\mathbf{a}_3 \left\{
\begin{aligned}
&v_3 + \omega_1\left[u_2 + e_2\left(s_1 s_2 s_3 + c_3 c_1\right) + e_3\left(c_1 s_2 s_3 - c_3 s_1\right)\right] - e_3\left[s_1 c_2 \dot{\theta}_1 + c_1 s_2 \dot{\theta}_2\right]\\
&-\omega_2\left[x + u_1 + e_2\left(s_1 s_2 c_3 - s_3 c_1\right) + e_3\left(c_1 s_2 c_3 + s_3 s_1\right)\right] + \dot{u}_3\\
&+e_2\left[c_1 c_2 \dot{\theta}_1 - s_1 s_2 \dot{\theta}_2\right]
\end{aligned}
\right\}
\tag{3.8}
$$

The rotation sequence $\theta_1\,\theta_2\,\theta_3$, consistent with the direction cosine matrix of Eq. (3.6), gives the angular velocity expression for dB in A, as shown in Ref. [3]:

$$^A\boldsymbol{\omega}^{dB} = (\dot{\theta}_1 - \dot{\theta}_3 s_2)\mathbf{b}_1 + (\dot{\theta}_2 c_1 + \dot{\theta}_3 s_1 c_2)\mathbf{b}_2 + (-\dot{\theta}_2 s_1 + \dot{\theta}_3 c_1 c_2)\mathbf{b}_3 \tag{3.9}$$

The angular velocity of dB in N is found by the angular velocity addition theorem as

$$^N\omega^{dB} = {}^N\omega^A + {}^A\omega^{dB}$$

$$
\begin{aligned}
= \mathbf{b}_1 &\left[\omega_1 c_2 c_3 + \omega_2 c_2 s_3 - \omega_3 s_2 + \dot\theta_1 - \dot\theta_3 s_2 \right] \\
+ \mathbf{b}_2 &\left[\omega_1 (s_1 s_2 c_3 - s_3 c_1) + \omega_2 (s_1 s_2 s_3 + c_3 c_1) + (\omega_3 + \dot\theta_3) s_1 c_2 + \dot\theta_2 c_1 \right] \\
+ \mathbf{b}_3 &\left[\omega_1 (c_1 s_2 c_3 + s_3 c_1) + \omega_2 (c_1 s_2 s_3 - c_3 s_1) + (\omega_3 + \dot\theta_3) c_1 c_2 - \dot\theta_2 s_1 \right]
\end{aligned}
\tag{3.10}
$$

The variables $\dot u_1, \dot u_2, \dot u_3, \dot\theta_1, \dot\theta_2, \dot\theta_3$ in Eqs. (3.8) and (3.10) have to be expressed in terms of system generalized coordinates. To that end, we introduce the variable $\zeta = \mathbf{p}^{OE} \cdot \mathbf{a}_1$, see Figure 3.2, and let $s(x,t)$ denote the *stretch in the beam along the elastic axis*; then, the distance along the deformed elastic axis from O to E, the point of which is at a distance x from O when B is undeformed, is

$$
x + s(x,t) = \int_0^\zeta \left\{ 1 + \left[\frac{\partial u_2(\sigma,t)}{\partial\sigma} \right]^2 + \left[\frac{\partial u_3(\sigma,t)}{\partial\sigma} \right]^2 \right\}^{0.5} d\sigma
\tag{3.11}
$$

where σ is a dummy variable along a_1. Denoting for convenience:

$$
J(\sigma,t) \equiv 1 + \left[\frac{\partial u_2(\sigma,t)}{\partial\sigma} \right]^2 + \left[\frac{\partial u_3(\sigma,t)}{\partial\sigma} \right]^2
\tag{3.12}
$$

and differentiating under the integral sign in Eq. (3.11) with respect to time, one obtains

$$
\dot s(x,t) = 0.5 \int_0^\zeta [J(\sigma,t)]^{-0.5} \frac{\partial J(\sigma,t)}{\partial t} d\sigma + \dot\zeta [J(\zeta,t)]^{0.5}
\tag{3.13}
$$

which produces the solution for the stretch rate:

$$
\dot\zeta = \frac{-0.5 \int_0^\zeta [J(\sigma,t)]^{-0.5} [\partial J(\sigma,t)/\partial t] d\sigma + \dot s(x,t)}{[J(\zeta,t)]^{0.5}}
\tag{3.14}
$$

From Eqs. (3.5) to (3.7) it is apparent that $\zeta = x + u_1$. Therefore,

$$
\dot u_1 = \dot\zeta
\tag{3.15}
$$

because x is the unstretched length. At this stage, we express the stretch and the transverse displacements and rotations, as follows:

$$s(x,t) = \sum_{i=1}^{v} \varphi_{1i}(x)q_i(t)$$

$$u_j(x,t) = \sum_{i=1}^{v} \varphi_{ji}(x)q_i(t) \qquad (j=2,3) \tag{3.16}$$

$$\theta_j(x,t) = \sum_{i=1}^{v} \varphi_{j+3,i}(x)q_i(t) \qquad (j=1,2,3)$$

Here, ϕ_{ij} ($i=1,6$; $j=1,...,v$) are as yet totally unrestricted spatial functions; q_i ($i=1,...,v$) are the generalized coordinates; and v is an arbitrary integer. Based on Eq. (3.16), one can differentiate Eq. (3.12) with respect to t to form

$$\frac{\partial J(\sigma,t)}{\partial t} = 2\sum_{i=1}^{v}\sum_{j=1}^{v}\left[\varphi_{2i}'(\sigma)\varphi_{2j}'(\sigma)+\varphi_{3i}'(\sigma)\varphi_{3j}'(\sigma)\right]\dot{q}_i(t)q_j(t) \tag{3.17}$$

In Eq. (3.17) a prime denotes differentiation with respect to the dummy variable, σ. Using this relation in Eq. (3.14) and Eq. (3.15), yields Eq. (3.18), with the other derivatives as in Eq. (3.19):

$$\dot{u}_1 = [J(\zeta,t)]^{-0.5}\sum_{i=1}^{v}\left\{ \begin{matrix} \varphi_{1i}(x) - \sum_{j=1}^{v} q_j(t)\int_0^{\zeta} [J(\sigma,t)]^{-0.5}[\varphi_{2i}'(\sigma)\varphi_{2j}'(\sigma) \\ +\varphi_{3i}'(\sigma)\varphi_{3j}'(\sigma)]d\sigma \end{matrix} \right\}\dot{q}_i(t) \tag{3.18}$$

$$\dot{u}_j(x,t) = \sum_{i=1}^{v} \varphi_{ji}(x)\dot{q}_i(t) \qquad (j=2,3)$$

$$\tag{3.19}$$

$$\dot{\theta}_j(x,t) = \sum_{i=1}^{v} \varphi_{j+3,i}(x)\dot{q}_i(t) \qquad (j=1,2,3)$$

Substituting $\dot{u}_1, \dot{u}_2, \dot{u}_3, \dot{\theta}_1, \dot{\theta}_2, \dot{\theta}_3$ in Eq. (3.8) results in the velocity expression, $^N v^C$, Eq. (3.20), the angular velocity $^N\omega^{dB}$, in Eq. (3.21), shown below. Non-linear partial velocities, needed for correctly deriving linearized equations, are the coefficients of the generalized speeds, u_i, $i = 1,...,v$, in these equations, and are given in Eqs. (3.22) and (3.23).

$$^N\mathbf{v}^C = \mathbf{a}_1 \left\langle v_1 + \omega_2(u_3 + e_2s_1c_2 + e_3c_1c_2) - \omega_3\left[u_2 + e_2(s_1s_2s_3 + c_3c_1) + e_3(c_1s_2s_3 - c_3s_1)\right]\right.$$

$$+ [J(\zeta,t)]^{-1/2}\sum_{i=1}^{v}\left\{\varphi_{1i}(x) - \sum_{j=1}^{v}q_j(t)\int_0^\zeta [J(\sigma,t)]^{-1/2}[\varphi'_{2i}(\sigma)\varphi'_{2j}(\sigma) + \varphi'_{3i}(\sigma)\varphi'_{3j}(\sigma)]d\sigma\right\}\dot{q}_i(t)$$

$$+ e_2\sum_{i=1}^{v}\left\{[c_1s_2s_3 + s_3s_1]\varphi_{4i}(x) + s_1c_2c_3\varphi_{5i}(x) + [s_1s_2s_3 + c_3c_1]\varphi_{6i}(x)\right\}\dot{q}_i(t)$$

$$+ e_3\sum_{i=1}^{v}\left\{[-s_1s_2c_3 + s_3c_1]\varphi_{4i}(x) + c_1c_2c_3\varphi_{5i}(x) + [-c_1s_2s_3 + c_3s_1]\varphi_{6i}(x)\right\}\dot{q}_i(t)\right\rangle$$

$$+ \mathbf{a}_2\left\langle v_2 + \omega_3\left[x + u_1 + e_2(s_1s_2c_3 - s_3c_1) + e_3(c_1s_2c_3 + s_3s_1)\right] - \omega_1(u_3 + e_2s_1c_2 + e_3c_1c_2)\right.$$

$$+ \sum_{i=1}^{v}\varphi_{2i}(x)\dot{q}_i(t) + e_2\sum_{i=1}^{v}\left\{(c_1s_2s_3 - c_3s_1)\varphi_{4i}(x) + s_1c_2s_3\varphi_{5i}(x) + (s_1s_2c_3 - s_3c_1)\varphi_{6i}(x)\right\}\dot{q}_i(t)$$

$$+ \left\{e_3\sum_{i=1}^{v} -\left[(s_1s_2s_3 + c_3c_1)\varphi_{4i}(x) + c_1c_2s_3\varphi_{5i}(x) + (c_1s_2c_3 + s_3s_1)\varphi_{6i}(x)\right]\dot{q}_i(t)\right\}\right\rangle$$

$$+ \mathbf{a}_3\left\langle v_3 + \omega_1\left[u_2 + e_2(s_1s_2s_3 + c_3c_1) + e_3(c_1s_2s_3 - c_3s_1)\right]\right.$$

$$- \omega_2\left[x + u_1 + e_2(s_1s_2c_3 - s_3c_1) + e_3(c_1s_2c_3 + s_3s_1)\right] + \sum_{i=1}^{v}\varphi_{3i}(x)\dot{q}_i(t)$$

$$+ e_2\sum_{i=1}^{v}\left[c_1c_2\varphi_{4i}(x) - s_1s_2\varphi_{5i}(x)\right]\dot{q}_i(t) - e_3\sum_{i=1}^{v}\left\{s_1c_2\varphi_{4i}(x) + c_1s_2\varphi_{5i}(x)\right\}\dot{q}_i(t)\right\rangle$$

$$\text{(3.20)}$$

$$^N\omega^{dB} = \mathbf{b}_1\left\{\omega_1c_2c_3 + \omega_2c_2s_3 - \omega_3s_2 + \sum_{i=1}^{v}\left[\varphi_{4i}(x) - s_2\varphi_{6i}(x)\right]\dot{q}_i\right\}$$

$$+ \mathbf{b}_2\left\{\begin{bmatrix}\omega_1(s_1s_2c_3 - s_3c_1) + \omega_2(s_1s_2s_3 + c_3c_1) + \omega_3s_1c_2 \\ + \sum_{i=1}^{v}[\varphi_{5i}(x)c_1 + \varphi_{6i}(x)s_1c_2]\dot{q}_i\end{bmatrix}\right\} \qquad \text{(3.21)}$$

$$+ \mathbf{b}_3\left\{\begin{matrix}\omega_1(c_1s_2c_3 + s_3s_1) + \omega_2(c_1s_2s_3 - c_3s_1) + \omega_3c_1c_2 \\ + \sum_{i=1}^{v}[-\varphi_{5i}(x)s_1 + \varphi_{6i}(x)c_1c_2]\dot{q}_i\end{matrix}\right\}$$

3.2 NON-LINEAR PARTIAL VELOCITIES AND PARTIAL ANGULAR VELOCITIES, AND THEIR LINEARIZATION

$$
{}^{N}\mathbf{v}_{i}^{C} = \mathbf{a}_{1}\left\langle
\begin{array}{l}
[J(\zeta,t)]^{-1/2}\left\{
\begin{array}{l}
\varphi_{1i}(x) - \displaystyle\sum_{j=1}^{v}\int_{0}^{\zeta}\ [J(\sigma,t)]^{-1/2}[\varphi_{2i}'(\sigma)\varphi_{2j}'(\sigma)\\[2mm]
+\varphi_{3i}'(\sigma)\varphi_{3j}'(\sigma)]d\sigma q_{j}
\end{array}\right\}\\[6mm]
+e_{2}\left\{
\begin{array}{l}
[c_{1}s_{2}c_{3}+s_{3}s_{1}]\varphi_{4i}(x)+s_{1}c_{2}c_{3}\varphi_{5i}(x)\\[1mm]
-[s_{1}s_{2}s_{3}+c_{3}c_{1}]\varphi_{6i}(x)
\end{array}\right\}\\[6mm]
+e_{3}\left\{
\begin{array}{l}
[-s_{1}s_{2}c_{3}+s_{3}c_{1}]\varphi_{4i}(x)+c_{1}c_{2}c_{3}\varphi_{5i}(x)\\[1mm]
+[-c_{1}s_{2}s_{3}+c_{3}s_{1}]\varphi_{6i}(x)
\end{array}\right\}
\end{array}\right\rangle
$$

$$
+\mathbf{a}_{2}\left\langle
\begin{array}{l}
\varphi_{2i}(x)+e_{2}\{[c_{1}s_{2}s_{3}-c_{3}s_{1}]\varphi_{4i}(x)+s_{1}c_{2}s_{3}\varphi_{5i}(x)+[s_{1}s_{2}c_{3}-s_{3}c_{1}]\varphi_{6i}(x)\}\\[1mm]
+e_{3}\{-[s_{1}s_{2}s_{3}+c_{3}c_{1}]\varphi_{4i}(x)+c_{1}c_{2}s_{3}\varphi_{5i}(x)+[c_{1}s_{2}c_{3}+s_{3}s_{1}]\varphi_{6i}(x)\}
\end{array}\right\rangle
$$

$$
+\mathbf{a}_{3}\left\langle
\begin{array}{l}
\varphi_{3i}(x)+e_{2}\{c_{1}c_{2}\varphi_{4i}(x)-s_{1}s_{2}\varphi_{5i}(x)-e_{3}\{s_{1}c_{2}\varphi_{4i}(x)\\[1mm]
+c_{1}s_{2}\varphi_{5i}(x)\}
\end{array}\right\rangle \qquad i=(1,...,v))
$$

$$\tag{3.22}$$

$$
{}^{N}\omega_{i}^{dB} = \mathbf{b}_{1}\left[\varphi_{4i}(x)-s_{2}\varphi_{6i}(x)\right]+\mathbf{b}_{2}\left[c_{1}\varphi_{5i}(x)+s_{1}c_{2}\varphi_{6i}(x)\right]
$$

$$\tag{3.23}$$

$$
+\mathbf{b}_{3}\left[-s_{1}\varphi_{5i}(x)+c_{1}c_{2}\varphi_{6i}(x)\right] \quad (i=1,\cdots v)
$$

To produce linear equations in $q_{1},...,q_{v},\dot{q}_{1},...,\dot{q}_{v}$ by the direct linearization method [3], to describe small deformations, we can now linearize the non-linear expressions of partial velocity and partial angular velocity in Eqs. (3.22) and (3.23), respectively. Noting from Eqs. (3.12) and (3.16) that

$$
J(x,t) = 1 + \sum_{i=1}^{v}\sum_{j=1}^{v}[\varphi_{2i}'(x)\varphi_{2j}'(x)+\varphi_{3i}'(x)\varphi_{3j}'(x)]q_{i}q_{j} \tag{3.24}
$$

the linearized version of Eq. (3.24) in modal coordinates is

$$
\tilde{J}(x,t) = 1, \qquad \tilde{J}(\zeta,t) = 1 \tag{3.25}
$$

Here and in the following expressions, a tilde over a symbol denotes its linearized version. Defining two new symbols:

$$\beta_{ij} = \int_0^\zeta \varphi'_{2i}(\sigma)\,\varphi'_{2j}(\sigma)\,d\sigma$$

$$\gamma_{ij} = \int_0^\zeta \varphi'_{3i}(\sigma)\varphi'_{3j}(\sigma)\,d\sigma \qquad (i, j = 1,...., v)$$

$$(3.26)$$

we form the linearized partial velocity expression from Eq. (3.22) as follows.

$$
{}^N\tilde{\mathbf{v}}_i^C = \mathbf{a}_1 \left\langle
\begin{array}{l}
\varphi_{1i}(x) - \displaystyle\sum_{j=1}^{v}(\beta_{ij}+\gamma_{ij})q_j + e_2 \displaystyle\sum_{j=1}^{v}\left\{\begin{array}{l}[\varphi_{4i}(x)+\varphi_{5j}(x)+\varphi_{5i}(x)\varphi_{4j}(x)]q_j \\ -\varphi_{6i}(x)\end{array}\right. \\[4mm]
+e_3\left\{\displaystyle\sum_{j=1}^{v}[\varphi_{4i}(x)\varphi_{6j}(x)+\varphi_{6i}(x)\varphi_{4j}(x)]q_j + \varphi_{5i}(x)\right\}
\end{array}
\right\rangle
$$

$$
+\mathbf{a}_2 \left\langle
\begin{array}{l}
\left\{\varphi_{2i}(x)-e_2\displaystyle\sum_{j=1}^{v}[\varphi_{4i}(x)\varphi_{4j}(x)+\varphi_{6i}(x)\varphi_{6j}(x)]\right\}q_j \\[4mm]
+e_3\left\{\displaystyle\sum_{j=1}^{v}[\varphi_{5i}(x)\varphi_{6j}(x)+\varphi_{6i}(x)\varphi_{5j}(x)]q_j - \varphi_{4i}(x)\right\}
\end{array}
\right\rangle
$$

$$
+\mathbf{a}_3 \left\langle \varphi_{3i}(x)+e_2\varphi_{4i}(x)-e_3\displaystyle\sum_{j=1}^{v}\left\{\begin{array}{l}[\varphi_{4i}(x)\varphi_{4j}(x) \\ +\varphi_{5i}(x)\varphi_{5j}(x)]q_j\end{array}\right\}\right\rangle \qquad (i = 1,...,v)
$$

$$(3.27)$$

The linearized partial angular velocity of dB in N follows from Eq. (3.23):

$$
{}^N\omega_i^{dB} = \mathbf{b}_1\left[\phi_{4i}(x)-\sum_{j=1}^{v}\varphi_{5j}(x)\varphi_{5i}(x)q_j\right] + \mathbf{b}_2\left[\phi_{5i}(x)+\sum_{j=1}^{v}\varphi_{4j}(x)\varphi_{6i}(x)q_j\right]
$$

$$
+\mathbf{b}_3[\phi_{6i}(x)-\sum_{j=1}^{v}\varphi_{4j}(x)\varphi_{5i}(x)q_j] \qquad (i = 1,.....,v)
$$

$$(3.28)$$

3.3   USE OF KANE'S METHOD FOR DIRECT DERIVATION OF LINEARIZED DYNAMICAL EQUATIONS

As per Kane's method of direct linearization [3], without deriving the non-linear equations, the velocity and the angular velocity expressions can *now* be linearized, because the partial velocity and partial angular velocity expressions have been extracted from their non-linear counterparts. The linearized velocity of C follows from Eq. (3.20) as

$$
\begin{aligned}
{}^N\tilde{\mathbf{v}}^C = \mathbf{a}_1 &\left\langle
\begin{array}{l}
v_1 + \omega_2 e_3 - \omega_3 e_2 + \displaystyle\sum_{k=1}^{v}
\left\{ \begin{array}{l} \omega_2\left[\varphi_{3k}(x) + e_2\varphi_{4k}(x)\right] \\ -\omega_3\left[\varphi_{2k}(x) - e_3\varphi_{4k}(x)\right]\end{array}\right\} q_k \\[4mm]
+ \displaystyle\sum_{k=1}^{v}\left\{\varphi_{1k}(x) - e_2\varphi_{6k}(x) + e_3\varphi_{5k}(x)\right\}\dot{q}_k
\end{array}
\right\rangle \\[6mm]
+\mathbf{a}_2 &\left\langle
\begin{array}{l}
v_2 + \omega_3 x - \omega_1 e_3 + \displaystyle\sum_{k=1}^{v}
\left\{ \begin{array}{l} \omega_3\left[\varphi_{1k}(x) - e_2\varphi_{6k}(x) + e_3\varphi_{5k}(x)\right] \\ -\omega_1\left[\varphi_{3k}(x) + e_2\varphi_{4k}(x)\right]\end{array}\right\} q_k \\[4mm]
+ \displaystyle\sum_{k=1}^{v}\left\{\varphi_{2k}(x) - e_3\varphi_{4k}(x)\right\}\dot{q}_k
\end{array}
\right\rangle \\[6mm]
+\mathbf{a}_3 &\left\langle
\begin{array}{l}
v_3 - \omega_2 x + \omega_1 e_2 + \displaystyle\sum_{k=1}^{v}
\left\{ \begin{array}{l} \omega_1\left[\varphi_{2k}(x) - e_3\varphi_{4k}(x)\right] \\ -\omega_2\left[\varphi_{1k}(x) - e_2\varphi_{6k}(x) + e_3\varphi_{5k}(x)\right]\end{array}\right\} q_k \\[4mm]
+ \displaystyle\sum_{k=1}^{v}\left\{\varphi_{3k}(x) + e_2\varphi_{4k}(x)\right\}\dot{q}_k
\end{array}
\right\rangle
\end{aligned} \tag{3.29}
$$

The linearized angular velocity of the differential element dB in N follows from Eq. (3.21):

$$
\begin{aligned}
{}^N\omega^{dB} = \mathbf{b}_1 &\left\langle \omega_1 + \sum_{k=1}^{v}\left[\omega_2 f_{6k}(x) - \omega_3 f_{5k}(x)\right]q_k + \sum_{k=1}^{v} f_{4k}(x)\dot{q}_k \right\rangle \\[4mm]
+\mathbf{b}_2 &\left\langle \omega_2 + \sum_{k=1}^{v}\left[\omega_3 f_{4k}(x) - \omega_1 f_{6k}(x)\right]q_k + \sum_{k=1}^{v} f_{5k}(x)\dot{q}_k \right\rangle \tag{3.30} \\[4mm]
+\mathbf{b}_3 &\left\langle \omega_3 + \sum_{k=1}^{v}\left[\omega_1 f_{5k}(x) - \omega_2 f_{4k}(x)\right]q_k + \sum_{k=1}^{v} f_{6k}(x)\dot{q}_k \right\rangle
\end{aligned}
$$

Now the acceleration of the generic point C and the angular acceleration of the elemental body dB, both in N, can be expressed by differentiating Eqs. (3.29) and (3.30), and the resulting expressions substituted in the definition of the system generalized inertia forces:

$$
\tilde{F}_i^* = -\int_0^L \rho \, {}^N\tilde{v}_i^C \cdot {}^N\tilde{a}^C dx
$$

$$
-\int_0^L {}^N\omega^{dB} \cdot \left[\mathbf{I} \cdot {}^N\alpha^{dB} + {}^N\omega^{dB} \times (\mathbf{I} \cdot {}^N\omega^{dB}) \right] dx \qquad (i = 1,...,v)
$$

$$(3.31)$$

Here, ρdx, Idx are the mass and central inertia dyadic of dB, ${}^N\tilde{a}^C$, ${}^N\tilde{\alpha}^{dB}$ are, respectively, the linearized acceleration of C and the angular acceleration of dB in N, and the element central inertia dyadic can be defined as

$$
Idx = \rho(I_{11}\mathbf{b}_1\mathbf{b}_1 + I_{22}\mathbf{b}_2\mathbf{b}_2 + I_{33}\mathbf{b}_3\mathbf{b}_3)dx \qquad (3.32)
$$

where I_{11}, I_{22}, I_{33}, expressed in terms of the central principal moments of inertia, are

$$
I_{22} = \frac{I_2}{A_0}, \quad I_{33} = \frac{I_3}{A_0}, \quad I_{11} = \frac{I_2 + I_3}{A_0} = I_{22} + I_{33} \qquad (3.33)
$$

Regarding the generalized active force, since only the elastic restoring forces contribute, these can be derived from a strain energy function U as

$$
F_i = -\frac{\partial U}{\partial q_i} \qquad (i = 1,...,v) \qquad (3.34)
$$

where U is given by

$$
U = \int_0^L \frac{(P_1)^2}{2E_0A_0} dx + \int_0^L \frac{\alpha_2(V_2)^2}{2G_0A_0} dx + \int_0^L \frac{\alpha_3(V_3)^2}{2G_0A_0} dx
$$

$$
+ \int_0^L \frac{(T_1)^2}{2G\kappa'} dx + \int_0^L \frac{(M_2)^2}{2E_0I_2} dx + \int_0^L \frac{(M_3)^2}{2E_0I_3} dx
$$

$$(3.35)$$

Here, the set of forces acting on the cross section of the beam is $P_1\mathbf{a}_1 + V_1\mathbf{a}_2 + V_1\mathbf{a}_3$, and the torque acting on the section is $T_1\mathbf{a}_1 + M_1\mathbf{a}_2 + M_1\mathbf{a}_3$; further, α_2, α_3 are Timoshenko's shear coefficients [5], and κ' is an effective torsional constant if one uses the first-order relation, rather than the exact third-order equation for torsion, such that

$$
T_1 = G\kappa' \frac{d\theta_1}{dx} \qquad (3.36)
$$

When the effects of shear are included in the analysis, θ_2, θ_3 are related to u_3, u_2 by the following relations:

$$\frac{\partial u_3(x,t)}{\partial x} = -[\theta_2(x,t) + \psi_2(x,t)]$$

$$\frac{\partial u_2(x,t)}{\partial x} = [\theta_3(x,t) + \psi_3(x,t)]$$

(3.37)

where ψ_2, ψ_3 denote angles of shear distortion of a cross section measured at the elastic axis for the directions of a_2, a_3, respectively. With these relationships, V_2, V_3 in Eq. (3.35) are

$$V_2 = \frac{A_0 G}{\alpha_2} \psi_3 = \frac{A_0 G}{\alpha_2} \left[\frac{\partial u_2(x,t)}{\partial x} - \theta_3(x,t) \right]$$

$$V_3 = -\frac{A_0 G}{\alpha_3} \psi_2 = -\frac{A_0 G}{\alpha_3} \left[\frac{\partial u_3(x,t)}{\partial x} + \theta_2(x,t) \right]$$

(3.38)

The other force and moment measure numbers appearing in Eq. (3.35) are expressed as

$$P_1 = E_0 A_0 \frac{\partial s(x,t)}{\partial x}, \qquad T_1 = G\kappa' \frac{\partial \theta_1(x,t)}{\partial x}$$

$$M_2 = E_0 I_2 \frac{\partial \theta_2(x,t)}{\partial x}, \qquad M_3 - E_0 I_3 \frac{\partial \theta_3(x,t)}{\partial x}$$

(3.39)

Substitution of these expressions for P_1, V_2, V_3, T_1, M_2, M_3 in Eq. (3.35) yields

$$U = (1/2) \int_0^L \left\{ \begin{array}{l} E_0 A_0 \left[\dfrac{\partial s(x,t)}{\partial x} \right]^2 + \dfrac{GA_0}{\alpha_2} \left[\dfrac{\partial u_2(x,t)}{\partial x} - \theta_3(x,t) \right]^2 \\[2ex] + \dfrac{GA_0}{\alpha_3} \left[\dfrac{\partial u_3(x,t)}{\partial x} + \theta_2(x,t) \right]^2 \end{array} \right\} dx$$

$$+ (1/2) \int_0^L \left\{ G\kappa' \left[\frac{\partial \theta_1(x,t)}{\partial x} \right]^2 + E_0 I_2 \left[\frac{\partial \theta_2(x,t)}{\partial x} \right]^2 + E_0 I_3 \left[\frac{\partial \theta_3(x,t)}{\partial x} \right]^2 \right\} dx$$

(3.40)

Substitution for s_1, u_2, u_3, θ_1, θ_2, θ_3 from Eq. (3.16) in Eq. (3.40), and invoking Eq. (3.34) yields

$$F_i = -\sum_{j=1}^{v}\left\{\int_0^L E_0\, A_0\varphi'_{1i}(x)\varphi'_{1j}(x)\,dx + \frac{GA_0}{\alpha_3}[\varphi_{5i}(x)\varphi_{5j}(x) + \varphi_{5i}(x)\varphi'_{3j}(x)]\right\}q_j\,dx$$

$$-\sum_{j=1}^{v}\left\{\int_0^L \frac{GA_0}{\alpha_3}[\varphi_{5j}(x)\varphi'_{3i}(x) + \varphi'_{3i}(x)\varphi'_{3j}(x)] + \frac{GA_0}{\alpha_2}[\varphi_{6i}(x)\varphi_{6j}(x) - \varphi_{6i}(x)\varphi'_{2j}(x)]\right\}q_j\,dx$$

$$-\sum_{j=1}^{v}\frac{GA_0}{\alpha_2}\int_0^L [\varphi_{6j}(x)\varphi'_{2i}(x) + \varphi'_{2i}(x)\varphi'_{2j}(x)]q_j\,dx$$

$$-\sum_{j=1}^{v}\left\{\begin{array}{l}\int_0^L [G\kappa'\varphi'_{4i}(x)\varphi'_{4j}(x) + E_0[I_2\varphi'_{5i}(x)\varphi'_{5j}(x)\\ +I_3\varphi'_{6i}(x)\varphi'_{6j}(x)]\end{array}\right\}q_j\,dx \qquad (i=1,...,v)$$

$$(3.41)$$

The term within the brace above can be compactly represented for later reference as

$$\tilde{F}_i = -\sum_{j=1}^{v}H_{ij}q_j \qquad (i=1,...,v) \qquad (3.42)$$

Now, if one writes Kane's dynamical equations:

$$F_i + F_i^* = 0 \qquad (i=1,...,v) \qquad (3.43)$$

using Eqs. (3.41) and (3.31), and re-casts them in matrix form, one gets the equations of motion in the matrix form:

$$M\,\ddot{q} + G\,\dot{q} + K\,q = F \qquad (3.44)$$

Note that damping can be included by adding a matrix $C\dot{q}$ to the term, $G\dot{q}$.

Here, the elements of the matrices M, G, K, F are quite lengthy expressions that are best defined by defining certain intermediate terms, called *modal integrals*.

$$\breve{W}_{ki} = \int_0^L \rho\varphi_{ki}(x)\,dx \qquad (i=1,...,v;\quad k=1,...,6) \qquad (3.45)$$

$$W_{klij} = \int_0^L \rho\varphi_{ki}(x)\varphi_{lj}(x)\,dx \qquad (i,j=1,...,v;\quad k,l=1,...,6) \qquad (3.46)$$

$$\breve{X}_{ki} = \int_0^L \rho\,x\,\varphi_{ki}(x)\,dx \qquad (i=1,...,v;\quad k=1,...,6) \qquad (3.47)$$

$$X_{klij} = \int_0^L \rho\,x\,\varphi_{ki}(x)\varphi_{lj}(x)\,dx \qquad (i,j=1,...,v;\quad k,l=1,...,6) \qquad (3.48)$$

$$\breve{Y}_{ki} = \int_0^L \rho I_{22} \varphi_{ki}(x) dx \qquad (i = 1,...,v; \quad k = 4,5,6) \qquad (3.49)$$

$$Y_{klij} = \int_0^L \rho I_{22} \varphi_{ki}(x) \varphi_{lj}(x) dx \qquad (i,j = 1,...,v; \quad k,l = 4,5,6) \qquad (3.50)$$

$$\breve{Z}_{ki} = \int_0^L \rho I_{33} \varphi_{ki}(x) dx \qquad (i = 1,...,v; \quad k = 4,5,6) \qquad (3.51)$$

$$Z_{klij} = \int_0^L \rho I_{33} \varphi_{ki}(x) \varphi_{lj}(x) dx \qquad (i,j = 1,...,v; \quad k,l = 4,5,6) \qquad (3.52)$$

$$\mu_{ij} = \int_0^L \rho(\beta_{ij} + \gamma_{ij}) dx \qquad (i,j = 1,...,v) \qquad (3.53)$$

$$\eta_{ij} = \int_0^L \rho x (\beta_{ij} + \gamma_{ij}) dx \qquad (i,j = 1,...,v) \qquad (3.54)$$

$$H_{ij} \equiv \int_0^L \{ E_0 A_0 \varphi'_{1i}(x) \varphi'_{1j}(x) + \frac{GA_0}{\alpha_3}[\varphi_{5i}(x)\varphi_{5j}(x) + \varphi_{5i}(x)\varphi'_{3j}(x) + \varphi'_{3i}(x)\varphi_{5j}(x) +$$

$$\varphi'_{3i}(x)\varphi'_{3j}(x)] + \frac{GA_0}{\alpha_2}[\varphi_{6i}(x)\varphi_{6j}(x) - \varphi'_{2j}(x)\varphi_{6i}(x) - \varphi'_{2i}(x)\varphi_{6j}(x) + \varphi'_{2i}(x)\varphi'_{2j}(x)] +$$

$$G\kappa'\varphi'_{4i}(x)\varphi'_{4j}(x) + E_0 I_2 \varphi'_{5i}(x)\varphi'_{5j}(x) + E_0 I_3 \varphi'_{6i}(x)\varphi'_{6j}(x) \} dx \qquad (i,j = 1,...,v)$$

$$(3.55)$$

The equations given so far can be simplified for special cases, as shown in Table 3.1.

Note that setting $\phi_{ni} = 0$ ($n = 1,..., 6$) means the following simplifications in Eqs. (3.45)–(3.55): $W_{klij} = X_{klij} = Y_{klij} = Z_{klij} = \breve{W}_{ki} = \breve{X}_{ki} = \breve{Y}_{ki} = \breve{Z}_{ki} = 0$ (k or l = n; j = 1,...,v).

Now we are ready to express the elements of the M, G, K, F matrices in Eq. (3.44). In what follows certain terms are underlined to facilitate a later discussion on discrepancies between the present formulation and those underlying many public domain software. Briefly, only the underlined double-barred terms are present in a conventional theory embedded in public domain software, but not the massive set of the rest of the terms.

$$M_{ij} = \{ [\underline{W_{11ij}} + W_{22ij} + W_{33ij} + Y_{44ij} + Y_{55ij} + Z_{44ij} + Z_{66ij}]$$

$$+ e_2[W_{44ij} + W_{66ij} - W_{16ij} + W_{34ij} + W_{43ij} - W_{61ij}]$$

$$+ e_3[W_{15ij} - W_{24ij} - W_{42ij} + W_{51ij} + W_{44ij} + W_{55ij}] - e_2 e_3[W_{56ij} + W_{65ij}] \}$$

$$(3.56)$$

$$(i,j = 1,...,v)$$

TABLE 3.1
Simplification of General Equations

To neglect	Set
Extension	$\phi_{1i}=0$, $i=1,\ldots,\nu$
Bending in a_2	$\Phi_{2i}=0$, $i=1,\ldots,\nu$
Bending in a_3	$\Phi_{3i}=0$, $i=1,\ldots,\nu$
Torsion	$\Phi_{4i}=0$, $i=1,\ldots,\nu$
Rotatory inertia	$I_{22}=I_{33}=0$
Shear	$\alpha_2=\alpha_2=0$
Warping restraint	$\Gamma=0$
Non-uniform torsion	$\kappa'=\kappa$

$$G_{ij} = 2\{\omega_1(W_{32ij}-W_{23ij}) + \omega_2[W_{13ij}-W_{31ij}+Z_{46ij}-Z_{64ij}] + \omega_3[W_{21ij}-W_{12ij}+Y_{54ij}$$

$$-Y_{45ij}] + e_2[\omega_1(W_{42ij}-W_{24ij}) + \omega_2(W_{14ij}-W_{41ij}+W_{36ij}-W_{63ij}) + \omega_3(W_{62ij}-W_{26ij})$$

$$+e_2\omega_2(W_{46ij}-W_{64ij})] + e_3[\omega_1(W_{43ij}-W_{34ij}) + \omega_2(W_{53ij}-W_{35ij})$$

$$+\omega_3(W_{14ij}-W_{41ij}+W_{25ij}-W_{52ij}) + e_3\omega_3(W_{54ij}-W_{45ij})]$$

$$-e_2e_3[\omega_2(W_{45ij}-W_{54ij}) + \omega_3(W_{64ij}-W_{46ij})]\} \qquad (i,j=1,\ldots,\nu)$$

$$(3.57)$$

$$F_i = -\{(\dot{v}_1+\omega_2v_3-\omega_3v_2)\breve{W}_{1i} + (\dot{v}_2+\omega_3v_1-\omega_1v_3)\breve{W}_{2i} + (\dot{v}_3+\omega_1v_2-\omega_2v_1)\breve{W}_{3i}$$

$$+\omega_2\omega_3(\breve{Z}_{4i}-\breve{Y}_{4i}) + \dot{\omega}_1(\breve{Y}_{4i}+\breve{Z}_{4i}) + (\dot{\omega}_2+\omega_3\omega_1)\breve{Y}_{5i} + (\dot{\omega}_3-\omega_1\omega_2)\breve{Z}_{6i}$$

$$-(\omega_2^2+\omega_3^2)\breve{X}_{1i} + (\dot{\omega}_3+\omega_1\omega_2)\breve{X}_{2i} - (\dot{\omega}_2-\omega_3\omega_1)\breve{X}_{3i} + e_2[-(\dot{\omega}_3-\omega_1\omega_2)\breve{W}_{1i} - (\omega_1^2+$$

$$\omega_3^2)\breve{W}_{2i} + (\dot{\omega}_1+\omega_2\omega_3)\breve{W}_{3i} + (\dot{v}_3+\omega_1v_2-\omega_2v_1+\omega_2\omega_3e_2+\dot{\omega}_1e_2)\breve{W}_{4i} - (\dot{\omega}_2-$$

$$\omega_3\omega_1)\breve{X}_{4i} + (\dot{v}_3+\omega_1v_2-\omega_2v_1+\omega_2\omega_3e_2+\dot{\omega}_1e_2)\breve{W}_{4i} - (\dot{\omega}_2-\omega_1\omega_3)\breve{X}_{4i}$$

$$-(\dot{v}_1+\omega_2v_3-\omega_3v_2+\omega_2\omega_3e_2+\dot{\omega}_1e_2)\breve{W}_{6i} + (\omega_2^2+\omega_3^2)\breve{X}_{6i}]$$

$$e_3[(\dot{\omega}_2+\omega_3\omega_1)\breve{W}_{1i} - (\dot{\omega}_1-\omega_2\omega_3)\breve{W}_{2i} - (\omega_1^2+\omega_2^2)\breve{W}_{3i}$$

$$-(\dot{v}_2+\omega_3v_1-\omega_1v_3+\omega_2\omega_3e_3-\dot{\omega}_1e_3)\breve{W}_{4i}$$

$$-(\dot{\omega}_3+\omega_1\omega_2)\breve{X}_{4i} + (\dot{v}_1+\omega_2v_3-\omega_3v_2+\omega_1\omega_3e_3+\dot{\omega}_2e_3)\breve{W}_{5i}$$

$$-(\omega_2^2+\omega_3^2)\breve{X}_{5i}] + e_2e_3[(\omega_3^2-\omega_2^2)\breve{W}_{4i} - (\dot{\omega}_3-\omega_1\omega_2)\breve{W}_{5i} - (\dot{\omega}_2+\omega_3\omega_1)\breve{W}_{6i}]\}$$

$$(i=1,\ldots,\nu)$$

$$(3.58)$$

Next we show the expression for K_{ij}, which is quite lengthy:

$$K_{ij} = \Big\{ -(\omega_2^2 + \omega_3^2)W_{11ij} + \omega_1\omega_2(W_{12ij} + W_{21ij}) - (\omega_1^2 + \omega_3^2)W_{22ij} + \omega_1\omega_3(W_{13ij} + W_{31ij})$$

$$-(\omega_1^2 + \omega_2^2)W_{33ij} + \omega_2\omega_3(W_{23ij} + W_{32ij}) - (\omega_2^2 - \omega_3^2)(Z_{44ij} - Y_{44ij})$$

$$+ \omega_1\omega_2(Z_{45ij} + Z_{54ij} - Y_{45ij} - Y_{54ij}) - \omega_1\omega_3(Z_{46ij} + Z_{64ij} - Y_{46ij} - Y_{64ij})$$

$$-(\omega_3^2 - \omega_1^2)Y_{55ij} + \omega_2\omega_3(Y_{56ij} + Y_{65ij}) - (\omega_2^2 - \omega_1^2)Z_{66ij}$$

$$- \dot{\omega}_1(W_{23ij} - W_{32ij} + Y_{56ij} + Y_{65ij}) + \dot{\omega}_2(W_{13ij} - W_{31ij} + Y_{46ij} + Y_{64ij} + Z_{46ij} - Z_{64ij})$$

$$- \dot{\omega}_3(W_{12ij} - W_{21ij} + Y_{45ij} - Y_{54ij} + Z_{45ij} + Z_{54ij}) - (\dot{v}_1 + \omega_2 v_3 - \omega_3 v_2)\mu_{ij}$$

$$+ (\omega_2^2 + \omega_3^2)\eta_{ij} + e_2[\omega_1\omega_3(W_{14ij} + W_{41ij}) + (\omega_2^2 + \omega_3^2)(W_{16ij} + W_{61ij})$$

$$+ \omega_2\omega_3(W_{24ij} + W_{42ij}) - \omega_1\omega_2(W_{26ij} + W_{62ij}) - (\omega_1^2 + \omega_2^2)(W_{34ij} + W_{43ij})$$

$$- \omega_1\omega_3(W_{36ij} + W_{63ij}) - (\dot{v}_2 + \omega_3 v_1 - \omega_1 v_3 - e_2\omega_3^2 + e_2\omega_2^2)W_{44ij}$$

$$+ (\dot{v}_1 + \omega_2 v_3 - \omega_3 v_2 + \omega_1\omega_2 e_2)(W_{45ij} + W_{54ij}) - e_2\omega_1\omega_3(W_{46ij} + W_{64ij})$$

$$- (\dot{v}_2 + \omega_3 v_1 - \omega_1 v_3 - e_2\omega_1^2 + e_2\omega_2^2)W_{66ij} + \dot{\omega}_2(W_{14ij} - W_{41ij})$$

$$- \dot{\omega}_1(W_{24ij} - W_{42ij}) - \dot{\omega}_3(W_{26ij} - W_{62ij}) + \dot{\omega}_2(W_{36ij} - W_{63ij})$$

$$- \dot{\omega}_3 e_2(W_{45ij} + W_{54ij}) + \dot{\omega}_2 e_2(W_{46ij} - W_{64ij}) + (\dot{\omega}_3 - \omega_1\omega_2)\mu_{ij} - (\dot{\omega}_3 + \omega_1\omega_2)X_{44ij}$$

$$- (\omega_2^2 + \omega_3^2)(X_{45ij} + X_{54ij}) - (\dot{\omega}_3 + \omega_1\omega_2)X_{66ij}] + e_3[-\omega_1\omega_2(W_{14ij} + W_{41ij})$$

$$- (\omega_2^2 + \omega_3^2)(W_{15ij} + W_{51ij}) + (\omega_1^2 + \omega_3^2)(W_{24ij} + W_{42ij}) + \omega_1\omega_2(W_{25ij} + W_{52ij})$$

$$- \omega_2\omega_3(W_{34ij} + W_{43ij}) + \omega_1\omega_3(W_{35ij} + W_{53ij})$$

$$- (\dot{v}_3 + \omega_1 v_2 - \omega_2 v_1 - e_3\omega_2^2 + e_3\omega_3^2)W_{44ij} - e_3\omega_1\omega_2(W_{45ij} + W_{54ij})$$

$$- (\dot{v}_3 + \omega_1 v_2 - \omega_2 v_1 - e_3\omega_1^2 + e_3\omega_3^2)W_{55ij} + (\dot{v}_2 + \omega_3 v_1 - \omega_1 v_3$$

$$+ \omega_1\omega_3 e_3(W_{56ij} + W_{65ij}) + (\dot{v}_1 + \omega_2 v_3 - \omega_3 v_2 + \omega_1\omega_2 e_2)(W_{46ij} + W_{64ij})$$

$$+ \dot{\omega}_3(W_{14ij} - W_{41ij} + W_{25ij} - W_{52ij}) - \dot{\omega}_1(W_{34ij} - W_{43ij}) - \dot{\omega}_2(W_{35ij} - W_{53ij})$$

$$- e_3\dot{\omega}_3(W_{45ij} - W_{54ij}) + e_3\dot{\omega}_2(W_{46ij} + W_{64ij}) - e_3\dot{\omega}_1(W_{56ij} + W_{65ij}) - (\dot{\omega}_2 + \omega_1\omega_3)\mu_{ij}$$

$$+ (\dot{\omega}_2 - \omega_1\omega_3)(X_{44ij} + X_{55ij}) - (\omega_2^2 + \omega_3^2)(X_{46ij} + X_{64ij}) + (\dot{\omega}_3 + \omega_1\omega_2)(X_{56ij} + X_{65ij})$$

$$+ e_2 e_3[-4\omega_2\omega_3 W_{44ij} + 2\omega_1\omega_3(W_{45ij} + W_{54ij}) + 2\omega_1\omega_2(W_{46ij} + W_{64ij}) - \omega_2\omega_3 W_{55ij}$$

$$+ (\omega_2^2 - \omega_1^2)(W_{56ij} + W_{65ij}) - \omega_2\omega_3 W_{66ij} - \dot{\omega}_1(W_{55ij} - W_{66ij}) + 2\dot{\omega}_2 W_{54ij} - 2\dot{\omega}_3 W_{64ij}]$$

$$+ \underline{H}_{ij} \Big\} \qquad (i, j = 1, \ldots, v)$$

$$(3.59)$$

3.4 SIMULATION RESULTS FOR A SPACE-BASED ROBOTIC MANIPULATOR

One representative problem involving a beam subjected to general base motion is that of simulating the behavior of a space-based robotic manipulator, such as the one shown in Figure 3.3, which consists of three links, L_1, L_2, L_3, connected by revolute joints. The outboard link L_3 consists of a base A and two distinct segments B_1, B_2, of which B_1 is 2.667 m long and has a symmetric box cross section, while B_2 is a 5.333 m long channel. Both segments are made of a material for which $E = 6.895 \times 10^{10}$ Nm^2, $G = 2.652 \times 10^{10}$ Nm^2, $\rho/A_0 = 2766.67$ kg/m^3. The section properties for B_1 are

$$A_0 = 3.84 \times 10^{-4} \text{ m}^2; \quad I_2 = 1.50 \times 10^{-7} \text{ m}^4; \quad I_3 = I_2; \quad \alpha_2 = 2.09; \quad \alpha_3 = \alpha_2;$$

$$\kappa = 2.2 \times 10^{-7} \text{ m}^4; \quad \Gamma = 0; \quad e_2 = e_3 = 0$$

The properties of B_2 are

$$A_0 = 7.3 \times 10^{-5} \text{ m}^2; \quad I_2 = 14.8746 \times 10^{-9} \text{ m}^4; \quad I_3 = 8.2181 \times 10^{-9} \text{ m}^4;$$

$$\alpha_2 = 3.174; \quad \alpha_3 = 1.52;$$

$$\kappa = 2.433 \times 10^{-11} \text{ m}^4; \quad \Gamma = 5.0156 \times 10^{-13} \text{ m}^6; \quad e_2 = 0; \quad e_3 = 0.01875 \text{ m}$$

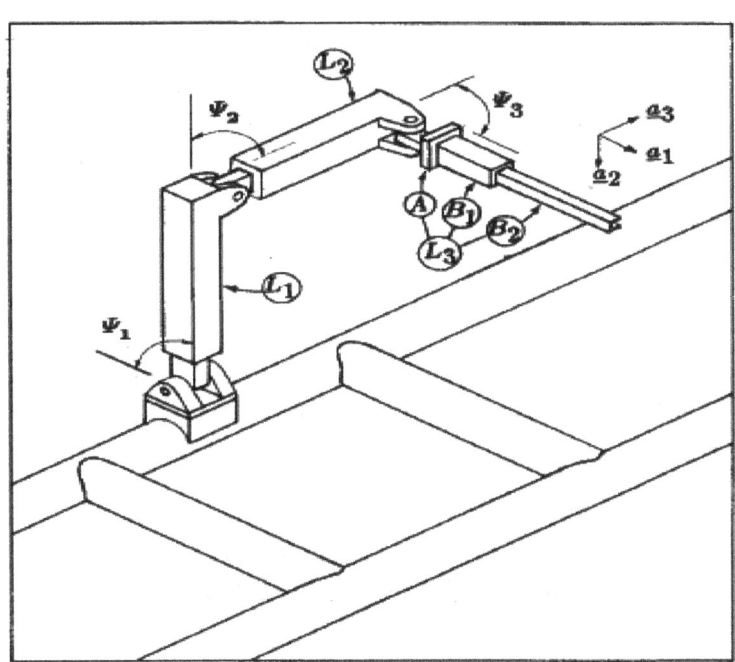

FIGURE 3.3   Space-based robotic manipulator

To demonstrate the simulation algorithm, we examine the behavior of L_3 during deployment of the manipulator from a stowed configuration to a fully operational configuration, evolving as follows. The deployment process is presumed to last for 15 sec ($T=15$), during which time the values of the angles Ψ_1, Ψ_2, Ψ_3, shown in Figure 3.3, change from 180 to 90 deg, 180 to 45 deg, and 180 to 90 deg, respectively. If the two inboard links (beams) L_1, L_2, each of length d, are treated as rigid, this maneuver results in a prescribed motion of the base A of the outboard link, a motion characterized by the following six temporal functions:

$$v_1 = d\,(1+\bar{c}_2)\bar{s}_3\dot{\psi}_1; \quad v_2 = d\dot{\psi}_2; \quad v_3 = -d\,(1+\bar{c}_2)c_3\dot{\psi}_1 \qquad (3.60)$$

$$\omega_1 = \dot{\psi}_1\bar{s}_2\,\bar{c}_3 - \dot{\psi}_2\bar{s}_3\,; \quad \omega_2 = \dot{\psi}_1\bar{c}_2 + \dot{\psi}_3\,; \quad \omega_3 = \dot{\psi}_1\bar{s}_2\,\bar{s}_3 + \dot{\psi}_2\bar{c}_3 \qquad (3.61)$$

Here

$$\bar{c}_i = \cos\psi_i\,; \quad \bar{s}_i = \sin\psi_i \quad (i=1,2,3) \qquad (3.62)$$

We take d = 8 m and let

$$\psi_1(t) = \pi - \frac{\pi}{2T}\left(t - \frac{T}{2\pi}\sin\frac{2\pi t}{T}\right)\text{rad} \qquad 0 < t < T$$

$$= \frac{\pi}{2}\,\text{rad} \qquad\qquad\qquad\qquad\qquad t > T \tag{3.63}$$

$$\psi_2(t) = \pi - \frac{3\pi}{4T}\left(t - \frac{T}{2\pi}\sin\frac{2\pi t}{T}\right)\text{rad} \qquad 0 < t < T$$

$$= \frac{\pi}{4}\,\text{rad} \qquad\qquad\qquad\qquad\qquad t > T \tag{3.64}$$

$$\psi_3(t) = \pi - \frac{\pi}{T}\left(t - \frac{T}{2\pi}\sin\frac{2\pi t}{T}\right)\text{rad} \qquad 0 < t < T$$

$$= 0\,\text{rad} \qquad\qquad\qquad\qquad\qquad t > T \tag{3.65}$$

Figure 3.4 shows, for the tip of the beam B_2, the transverse displacements and twist, u_2, u_3, θ_1, respectively, during a time interval of 30 sec, which is twice the deployment time. The first 10 vibration modes of each beam were used to obtain the solution.

3.5   ERRONEOUS RESULTS OBTAINED USING VIBRATION MODES WITH LARGE ROTATION IN CONVENTIONAL ANALYSIS

In passing, we note that the conventional solutions used in representative public domain software such as in Refs. [6, 7] differ from the preceding analysis in one

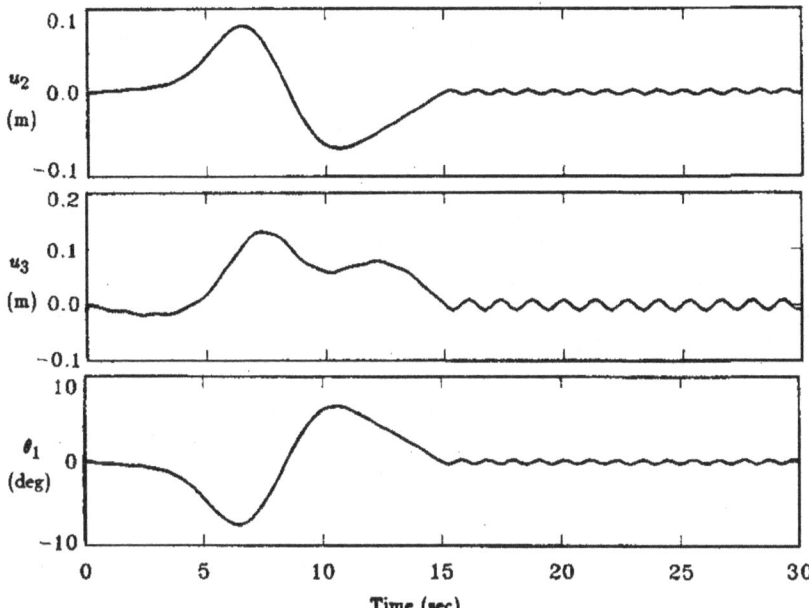

FIGURE 3.4 Manipulator deployment simulation results

fundamental issue: they *assume that the elastic deformations in transverse and axial directions are independent* spatial functions. That is, the elastic deformation, and the associated velocity of a generic point C in N (see Figure 3.5), are expressed in the conventional approach in terms of vibration modes, which come from a linear theory of structural dynamics, as

$$\mathbf{u} = \sum_{j=1}^{\nu} \{\varphi_{1j}(x_1, x_2, x_3)\mathbf{b}_1 + \varphi_{2j}(x_1, x_2, x_3)\mathbf{b}_2 + \varphi_{3j}(x_1, x_2, x_3)\mathbf{b}_3\}q_j(t) \quad (3.66)$$

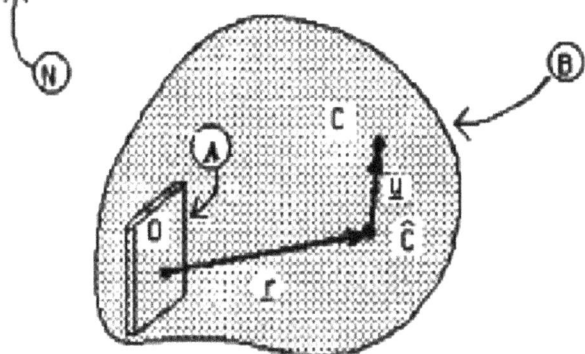

FIGURE 3.5 Generic body in a multibody system

$$
{}^{N}\mathbf{v}^{C} = {}^{N}\mathbf{v}^{O} + {}^{N}\boldsymbol{\omega}^{A}\mathbf{x}(\mathbf{r} + \mathbf{u}) + \sum_{j=1}^{v}\left\{\varphi_{1j}(x_1, x_2, x_3)\mathbf{b}_1 + \varphi_{2j}(x_1, x_2, x_3)\mathbf{b}_2\right.
$$

$$
\left. + \varphi_{3j}(x_1, x_2, x_3)\mathbf{b}_3\right\}\dot{q}_j(t)
$$

(3.67)

This arrangement of using independent spatial functions in the three orthogonal directions is in direct contrast to Eq. (3.11) where the elastic axis stretch is tied to the transverse deformations in two directions. This conventional approach misses the crucial effect of dynamic stiffness, associated with the terms involving μ_{ij}, η_{ij} in the present theory, because of the assumption of independent orthogonal deformations. This is a consequence of premature linearization. This is depicted in Figure 3.6, showing disastrous deflections in transverse directions and axial rotations with time, for a transverse axis spin-up maneuver. Also, when beams or plates are considered, rotations of the elements must be taken into account, which is not done in conventional theories; the difference between the present and the conventional theories for beam section axial rotations is also shown in Figure 3.6, for example in axial rotations. Other features of the present theory are in the nature of refinements. These include shear and warping using the concepts of shear area ratio and an effective torsional factor calculated with the aid of a warping factor, and coupled bending and torsion made possible by the introduction of an eccentricity vector from the shear center to the centroid of a cross section. These effects too are not considered in conventional formulations. Finally, one may note that the present chapter also does not consider flexural-torsional coupled vibrations of a beam spinning about the beam axis [8], but rigid body motions of the support in such cases would still require to be treated by the method presented here. Spinning beam vibrations for composite materials have been discussed by dynamic stiffness methods in Ref. [9].

PROBLEM SET 3

1. Derive the equations of motion of a rotating cantilever beam, of length L, considering only in-plane bending, neglecting rotatory inertia and shear deformation, and using the vibration modes of a non-rotating cantilever beam [8]:

$$
\varphi_i(x) = (\sin\beta_i L - \sinh\beta_i L)(\sin\beta_i x - \sinh\beta_i x) + (\cos\beta_i L + \cosh\beta_i L)(\cos\beta_i x - \cosh\beta_i x);
$$

$$
\beta_i = (m\omega_i^2 / EI)^{(1/4)} \; i = 1,...,n; \; \beta_1 L = 1.875; \; \beta_2 L = 4.694; \; \beta_3 L = 7.855;...
$$

(3.68)

where m, E, I, ω are the mass per unit length, modulus of elasticity, section inertia, and vibration frequency. Work out the modal integrals for your choice of beam parameters, and do a simulation using only two modes, i.e., $v = 2$.

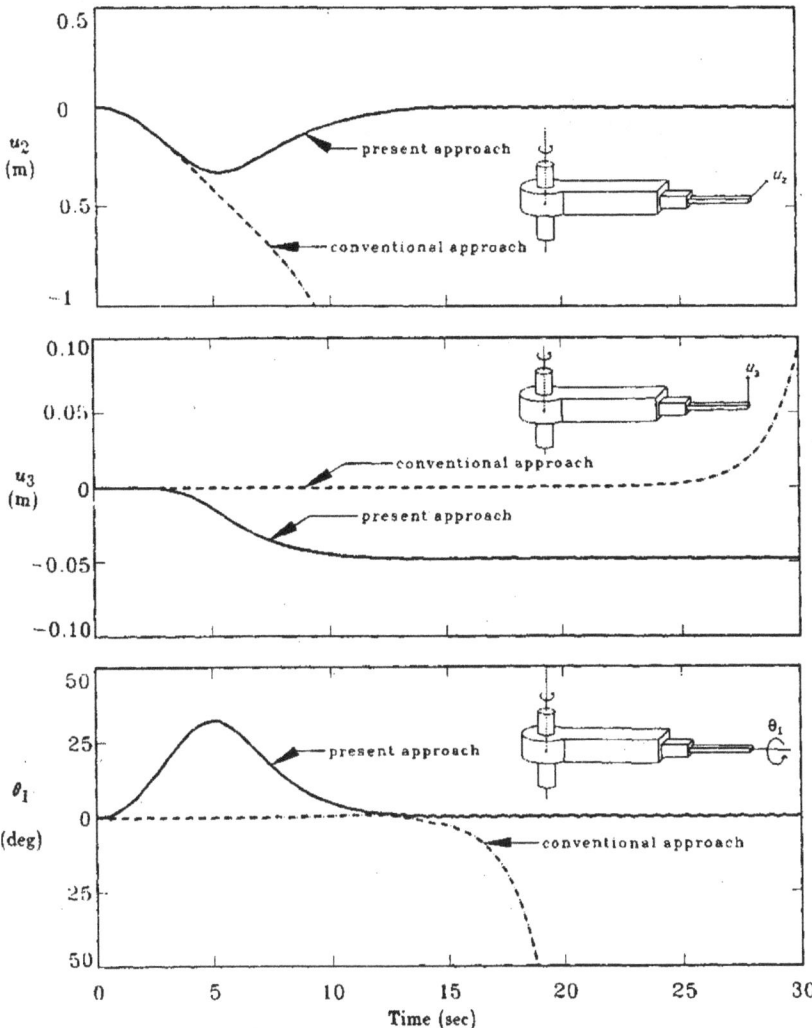

FIGURE 3.6 Spin-up maneuver results for u_2, u_3, θ_1 for transverse displacement and axial rotation for a channel section beam

2. Consider the spin-up function where the spin-up steady speed is $\omega > \omega_1$, ω_1 being the first natural frequency of the beam:

$$\omega = \frac{2}{5}\left[t - \frac{7.5}{\pi}\sin\left(\frac{\pi t}{7.5}\right) \right] \qquad 0 \le t \le 15$$

$$= 6 \qquad\qquad\qquad 15 \le t \le 30$$

(3.69)

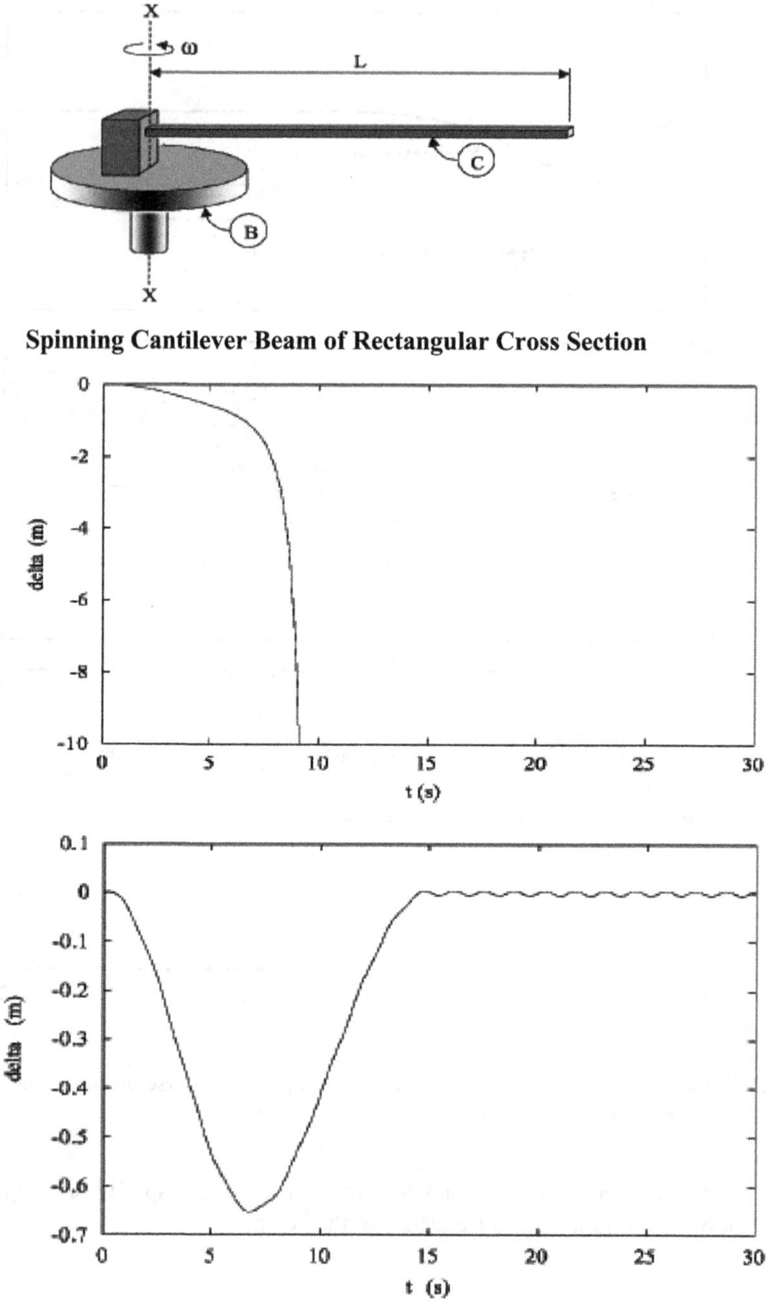

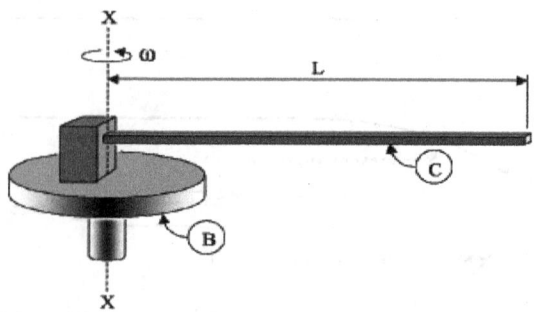

Spinning Cantilever Beam of Rectangular Cross Section

FIGURE 3.7 Conventional analysis with premature linearization (top figure), using only transverse vibration modes, and proper linearization given in this chapter, with spin frequency exceeding the first mode bending frequency

Derive the equations of motion using the vibration modes (take one mode) and show that one gets the first curve in Figure 3.7, whereas doing proper linearization one gets the correct result given by the lower curve, see Ref. [10].

3. Consider a cantilever beam of length L of uniform mass m per length and flexural rigidity EI. The beam can be treated as a Bernoulli–Euler beam, which neglects rotary inertia and shear deformation. Refer to Figure 3.1 and assume that the base of the beam undergoes planar motion with no rotation and only $^N\mathbf{v}^O = v_1\mathbf{a}_1 + v_2\mathbf{a}_2$ and base acceleration $\dot{v}_1\mathbf{a}_1 + \dot{v}_2\mathbf{a}_2$. Derive the equations of motion of the beam using non-linear kinematics, and examine the effects of proper and improper linearization using vibration modes.

REFERENCES

1. Kane, T.R., Ryan, R.R., and Banerjee, A.K., (1987), "Dynamics of a Cantilever Beam Attached to a Moving Base", *Journal of Guidance, Control, and Dynamics*, **10**(2), pp. 139–151.
2. Modi, V.J., (1974), "Attitude Dynamics of Satellites with Flexible Appendages", *Journal of Spacecraft and Rockets*, **11**, pp. 743–751.
3. Kane, T.R. and Levinson, D.A., (1985), *Dynamics: Theory and Application*, McGraw-Hill.
4. Kane, T.R., Likins, P.W., and Levinson, D.A., (1983), *Spacecraft Dynamics*, McGraw-Hill.
5. Timoshenko, S., Young, D.H., and Weaver, W., Jr., (1974), *Vibration Problems in Engineering*, Fourth Edition, Wiley.
6. Bodley, C.S., Devers, A.D., Park, A.C., and Frisch, H.P., (1978), *A Digital Computer Program for the Dynamic Interaction Simulation of Controls and Structures (DISCOS)*, vol. I, NASA Technical Paper 1219.
7. Singh, R.P., van der Voort, R.J., and Likins, P.W., (1984), "Dynamics of Flexible Bodies in Tree Topology: A Computer Oriented Approach", *Journal of Guidance, Control, and Dynamics*, **8**(5), pp. 584–590.
8. Meirovitch, L., (1967), *Analytical Methods in Vibrations*, Macmillan Company.
9. Banerjee, J.R. and Su, H., (2006), "Dynamic Stiffness Formulation and Free Vibration Analysis of a Spinning Composite Beam", *Computers & Structures*, **84**, pp. 1208–1214.
10. Banerjee, A.K., (2003), "Contributions of Multibody Dynamics to Space Flight: A Brief Review", *Journal of Guidance, Control, and Dynamics*, **26**(3), pp. 385–394.

4 Dynamics of a Plate in Large Overall Motion

This chapter is concerned with the formulation of equations of motion of a thin elastic plate that is executing small vibrations relative to a reference frame undergoing large rigid body motion (three-dimensional translation and translation) in a Newtonian reference frame. We use Kane's method of direct linearization with respect to modal coordinates associated with vibration modes defining the small deformation, as discussed generally in Chapter 1, and used in the preceding chapter for the case of beams undergoing large overall motion. As an illustrative example, we consider the spin-up maneuver for a simply supported rectangular plate, and using the vibration modes of such a plate, show that the theory given here captures the phenomenon of dynamic stiffening. The analysis and results presented here are taken from Ref. [1].

4.1 BEGINNING AT THE END: SIMULATION RESULTS

Figure 4.1 shows a rectangular plate moving with its reference frame R undergoing large rotation and translation in the Newtonian frame N, while the plate executes small vibrations in R. Orthogonal unit vectors $\mathbf{r}_1$, $\mathbf{r}_2$, $\mathbf{r}_3$ (again using bold symbols instead of underbars for vectors shown in the figure) are fixed on R. Point O designates a particle of the plate fixed in R. Our objective is to produce an algorithm that can be used to simulate the motions of the plate when the motion of R in N is given. Referring to Figure 4.1, consider R as a rigid body, and suppose that a 72×48 in. rectangular plate with 0.1 in. thickness is *simply supported* by R along all four edges of the plate, the 72 in. long edge coinciding with X_1, and the 48 in. long edge coinciding with X_2; note that X_1, X_2 align, respectively, with $\mathbf{r}_1$, $\mathbf{r}_2$. Let R undergo a *spin-up motion about the edge* such that X_2 remains fixed in N, and $\omega_2(t)$, the $\mathbf{r}_2$ measure number of the angular velocity of R in N, measured in radians per second, is prescribed as

$$\omega_2(t) = \frac{2\pi n^*}{60 t^*}\left(t - \frac{t^*}{2\pi}\sin\frac{2\pi t}{t^*}\right), \quad t \le t^*;$$

$$= \frac{2\pi n^*}{60} \quad t > t^* \tag{4.1}$$

Here, n^* is the final spin speed, in revolutions per minute, and t^* is the spin-up time, in seconds. Under these circumstances, Q, the center point of the plate, must experience an elastic deflection, say u_3^Q, parallel to the X_3-axis, and one can predict this deflection in qualitative terms on physical grounds. Initially, u_3^Q is equal to zero. As R begins to rotate, Q must begin to move in the direction of $\mathbf{r}_3$ so that

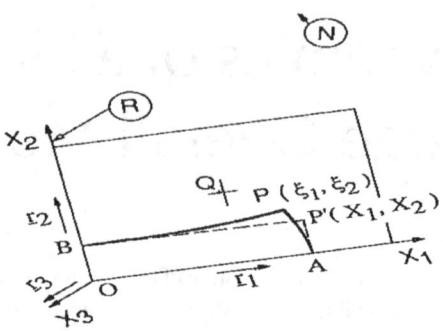

FIGURE 4.1 Plate showing small elastic deformation of line elements in a reference frame R that undergoes large rotation and translation in a Newtonian frame N.

u_3^Q takes on positive values, because, initially the plate cannot keep up with the motion of the frame R. But soon, restoring forces arising from the elasticity of the plate make themselves felt, and eventually become sufficiently large not only to arrest the growth of u_3^Q but also to cause u_3^Q to decrease. Indeed, sooner or later, Q should not only catch up with R, that is, u_3^Q should become equal to zero, but Q should also overshoot its original position in R, in which event u_3^Q takes on negative values; and after the motion of R has settled down, that is, once ω_2 has reached its maximum value, from which time onward R is rotating with constant angular speed, it is to be expected that a residual "ringing" will take place, that is, that u_3^Q will have an oscillatory character. Thus, a plot of u_3^Q versus time should have the general appearance of the curve in Figure 4.2, which was generated by the theory set forth in the sequel, with $n^* = 12$ revolutions per minute. Figure 4.3 deals with the effect of spin on a steady-state response frequency; that is, the ratio of ω_d, the dynamic response frequency when spin-up has been completed, to ω_1, the first natural frequency of the plate, is plotted versus the ratio of ω_s, the spin speed, to ω_1. As can be seen, the theory predicts *dynamic stiffening*, that is, an increase in ω_d with increasing ω_s, and this for *all* spin speeds.

In what follows, we set forth the underlying theory behind these simulation results in detail, with conclusions, and an Appendix showing the relationship between certain "modal integrals" arising in this theory, and geometric stiffness coefficients associated with the inertia forces of the plate due to the large motion of its reference frame.

4.2 APPLICATION OF KANE'S METHODOLOGY FOR LINEARIZATION WITHOUT DERIVING THE NON-LINEAR EQUATIONS

Refer to Figure 4.1, and let x_1, x_2 be the coordinates along orthogonal axes parallel to $\mathbf{r}_1$, $\mathbf{r}_2$ (vectors replacing underbars in the figure) of P', a generic material point

P of the plate when the plate is undeformed. Then, the displacement of P after deformation can be expressed in terms of measure numbers u_1, u_2, u_3, as

$$\mathbf{u} = u_1(x_1, x_2, t)\mathbf{r}_1 + u_2(x_1, x_2, t)\mathbf{r}_2 + u_3(x_1, x_2, t)\mathbf{r}_3 \tag{4.2}$$

Now consider the line elements in Figure 4.1, AP′, BP′, which are parallel to x_1, x_2, respectively, when the plate is undeformed, with A, B being two points fixed on the edge of the plate (Figures 4.2 and 4.3).

If we restrict our analysis to a plate whose middle surface is inextensible, then the length of these line elements remains constant during deformation of the plate, when the coordinates of P become (ξ_1, ξ_2, t), that is

$$x_1 = \int_0^{\xi_1} \left\{ 1 + \left[\frac{\partial u_2(\sigma, x_2, t)}{\partial \sigma} \right]^2 + \left[\frac{\partial u_3(\sigma, x_2, t)}{\partial \sigma} \right]^2 \right\}^{1/2} d\sigma \tag{4.3}$$

$$x_2 = \int_0^{\xi_2} \left\{ 1 + \left[\frac{\partial u_1(x_1, \sigma, t)}{\partial \sigma} \right]^2 + \left[\frac{\partial u_3(x_1, \sigma, t)}{\partial \sigma} \right]^2 \right\}^{1/2} d\sigma \tag{4.4}$$

To save labor in writing, we introduce the symbols J_1 and J_2 as

$$J_1(\sigma, x_2, t) = 1 + \left[\frac{\partial u_2(\sigma, x_2, t)}{\partial \sigma} \right]^2 + \left[\frac{\partial u_3(\sigma, x_2, t)}{\partial \sigma} \right]^2 \tag{4.5}$$

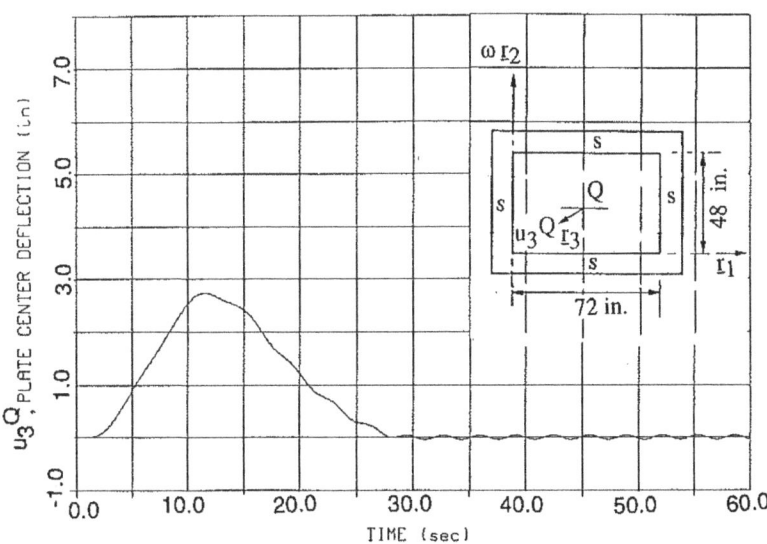

FIGURE 4.2 Transient deflection in inches of the center point u_3^Q of a simply supported plate (supports indicated in the inset by s) spun-up from rest about an edge along $\mathbf{r}_2$ to a maximum spin rate of 12 rpm.

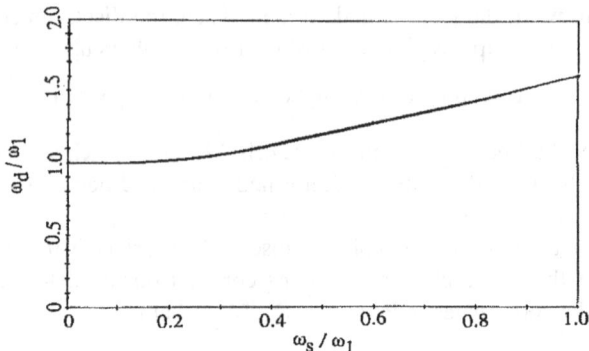

FIGURE 4.3 Ratio of dynamic frequency due to centrifugal stiffening, to first natural frequency, ω_d/ω_1, versus ratio of spin frequency to first natural frequency, ω_s/ω_1, of the plate.

$$J_2(x_1,\sigma,t) = 1 + \left[\frac{\partial u_1(x_1,\sigma,t)}{\partial\sigma}\right]^2 + \left[\frac{\partial u_3(x_1,\sigma,t)}{\partial\sigma}\right]^2 \tag{4.6}$$

Differentiating under the integral sign in Eqs. (4.3) and (4.4) with respect to t, we get

$$0 = \frac{1}{2}\int_0^{\xi_1}\left[J_1(\sigma,x_2,t)\right]^{-1/2}\frac{\partial J_1(\sigma,x_2,t)}{\partial t}d\sigma + \dot{\xi}_1\left[J_1(\xi_1,x_2,t)\right]^{1/2} \tag{4.7}$$

$$0 = \frac{1}{2}\int_0^{\xi_2}\left[J_2(x_1,\sigma,t)\right]^{-1/2}\frac{\partial J_2(x_1,\sigma,t)}{\partial t}d\sigma + \dot{\xi}_2\left[J_2(x_1,\xi_2,t)\right]^{1/2} \tag{4.8}$$

The *instantaneous coordinates* of the point P are related to the coordinates of P′ by

$$\xi_1 = x_1 + u_1(x_1,x_2,t) \tag{4.9}$$

$$\xi_2 = x_2 + u_2(x_1,x_2,t) \tag{4.10}$$

Differentiating Eqs. (4.9) and (4.10) with respect to t, and using the results in Eqs. (4.7) and (4.8):

$$\dot{u}_1 = -\frac{1}{2}[J_1(\xi_1,x_2,t)]^{-1/2}\int_0^{\xi_1}[J_1(\sigma,x_2,t)]^{-1/2}\frac{\partial J_1(\sigma,x_2,t)}{\partial t}d\sigma \tag{4.11}$$

$$\dot{u}_2 = -\frac{1}{2}\left[J_2(x_1,\xi_2,t)\right]^{-1/2} \int_0^{\xi_2} \left[J_2(x_1,\sigma,t)\right]^{-1/2} \frac{\partial J_2(x_1,\sigma,t)}{\partial t}\, d\sigma \qquad (4.12)$$

Considering points on the line element BP', we express u_2 and u_3 in series expansions:

$$u_2(\sigma,x_2,t) = \sum_{j=1}^{n} \phi_{2j}(\sigma,x_2)q_j(t) \qquad (4.13)$$

$$u_3(\sigma,x_2,t) = \sum_{j=1}^{n} \varphi_{3j}(\sigma,x_2)q_j(t) \qquad (4.14)$$

Similarly, for the line element AP', we express u_1 and u_3 as

$$u_1(x_1,\sigma,t) = \sum_{j=1}^{n} \varphi_{1j}(x_1,\sigma)q_j(t) \qquad (4.15)$$

$$u_3(x_1,\sigma,t) = \sum_{j=1}^{n} \phi_{3j}(x_1,\sigma)q_j(t) \qquad (4.16)$$

The spatial functions $\phi_{1j},\phi_{2j},\phi_{3j}$ $j=1,\ldots,n$ introduced in Eqs. (4.13)–(4.16) are as yet unrestricted; ultimately, we will take them to be the vibration mode shapes for a plate with appropriate boundary conditions. The temporal functions $q_j(t)$ ($j=1,\ldots,n$) play the role of generalized coordinates in the sense of Lagrange, and n is simply any integer indicating how many functions or modes are kept in the series. Differentiating Eqs. (4.5) and (4.6) with respect to t, and using Eqs. (4.13)–(4.16), we get the following expressions:

$$\frac{\partial J_1(\sigma,x_2,t)}{\partial t} = 2\sum_{i=1}^{n}\sum_{j=1}^{n}\left[\phi'_{2i}(\sigma,x_2)\phi'_{2j}(\sigma,x_2)\right.$$
$$\left. +\phi'_{3i}(\sigma,x_2)\phi'_{3j}(\sigma,x_2)\right]\dot{q}_i(t)q_j(t) \qquad (4.17)$$

$$\frac{\partial J_2(x_1,\sigma,t)}{\partial t} = 2\sum_{i=1}^{n}\sum_{j=1}^{n}\left[\phi'_{1i}(x_1,\sigma)\phi'_{1j}(x_1,\sigma)\right.$$
$$\left. +\phi'_{3i}(x_1,\sigma)\phi'_{3j}(x_1,\sigma)\right]\dot{q}_i(t)q_j(t) \qquad (4.18)$$

where primes denote the partial differentiation with respect to the dummy space variable σ. Now, substituting Eqs. (4.17) and (4.18) into Eqs. (4.11) and (4.12) gives

$$\dot{u}_1 = -\left[J_1(\xi_1, x_2, t)\right]^{-(1/2)} \int_0^{\xi_1} \left\{ \begin{array}{l} [J_1(\sigma, x_2, t)]^{-(1/2)} \times \\ \sum_{i=1}^n \sum_{j=1}^n \left[\begin{array}{l} \phi_{2i}'(\sigma, x_2)\phi_{2j}'(\sigma, x_2) \\ +\phi_{3i}'(\sigma, x_2)\phi_{3j}'(\sigma, x_2) \end{array} \right] \dot{q}_i q_j \end{array} \right\} d\sigma \qquad (4.19)$$

$$\dot{u}_2 = -\left[J_2(x_1, \xi_2, t)\right]^{-(1/2)} \int_0^{\xi_2} \left\{ \begin{array}{l} [J_2(x_1, \sigma, t)]^{-(1/2)} \times \\ \sum_{i=1}^n \sum_{j=1}^n \left[\begin{array}{l} \phi_{1i}'(x_1, \sigma)\phi_{1j}'(x_1, \sigma) \\ +\phi_{3i}'(x_1, \sigma)\phi_{3j}'(x_1, \sigma) \end{array} \right] \dot{q}_i q_j \end{array} \right\} d\sigma \qquad (4.20)$$

In addition, differentiation of Eq. (4.14) or (4.16) with respect to t yields

$$\dot{u}_3 = \sum_{i=1}^n \phi_{3i}(x_1, x_2)\dot{q}_i \qquad (4.21)$$

The velocity of point P in the Newtonian frame N can now be formally written as

$$^N\mathbf{v}^P = {}^N\mathbf{v}^O + {}^N\boldsymbol{\omega}^R \times \left\{ \begin{array}{l} \left[x_1 + \sum_{j=1}^n \phi_{1j}(x_1, x_2)q_j(t) \right]\mathbf{r}_1 \\ + \left[x_2 + \sum_{j=1}^n \phi_{2j}(x_1, x_2)q_j(t) \right]\mathbf{r}_2 \\ + \sum_{j=1}^n \phi_{3j}(x_1, x_2)q_j(t)\mathbf{r}_3 \end{array} \right\} \qquad (4.22)$$

$$+ \dot{u}_1\mathbf{r}_1 + \dot{u}_2\mathbf{r}_2 + \dot{u}_3\mathbf{r}_3$$

where $^N\mathbf{v}^O$, $^N\boldsymbol{\omega}^R$, respectively, are the velocity of the point O in N, and the angular velocity of the frame R in N, which can be expressed in terms of the *specified time functions* $v_i(t)$ and $\omega_i(t)$ (i = 1,2,3) as follows:

$$^N\mathbf{v}^O = v_1(t)\mathbf{r}_1 + v_2(t)\mathbf{r}_2 + v_3(t)\mathbf{r}_3 \qquad (4.23)$$

$$^N\boldsymbol{\omega}^R = \omega_1(t)\mathbf{r}_1 + \omega_2(t)\mathbf{r}_2 + \omega_3(t)\mathbf{r}_3 \qquad (4.24)$$

Using $\dot{q}_1, \ldots, \dot{q}_n$ as generalized speeds, one can now form the i-th non-linear partial velocity of P in N, by inspection of Eqs. (4.19)–(4.22), as

$$^N\mathbf{v}_i^P = -\mathbf{r}_1 \left\langle \left[J_1(\xi_1,x_2,t)\right]^{-(1/2)} \int_0^{\xi_1} \left\{ \begin{aligned} &[J_1(\sigma,x_2,t)]^{-(1/2)} \times \\ &\sum_{j=1}^{n}\left[\begin{aligned} &\phi_{2i}'(\sigma,x_2)\phi_{2j}'(\sigma,x_2) \\ &+\phi_{3i}'(\sigma,x_2)\phi_{3j}'(\sigma,x_2)\end{aligned}\right]q_j \end{aligned}\right\}d\sigma \right\rangle$$

$$-\mathbf{r}_2 \left\langle \left[J_2(x_1,\xi_2,t)\right]^{-(1/2)} \int_0^{\xi_2} \left\{ \begin{aligned} &[J_2(x_1,\sigma,t)]^{-(1/2)} \times \\ &\sum_{j=1}^{n}\left[\begin{aligned} &\phi_{1i}'(x_1,\sigma)\phi_{1j}'(x_1,\sigma) \\ &+\phi_{3i}'(x_1,\sigma)\phi_{3j}'(x_1,\sigma)\end{aligned}\right]q_j \end{aligned}\right\}d\sigma \right\rangle \qquad (4.25)$$

$$+\mathbf{r}_3\,\phi_{3i}(x_1,x_2) \qquad (i=1,\dots,n)$$

So far, no linearization has been performed; but now that non-linear expressions for partial velocities are in hand, Kane's method of direct realization [2], without linearizing the non-linear equations, can be used, i.e., the partial velocities can be linearized, along with ignoring the non-linear terms in expressions for velocities and accelerations everywhere in $q_j, \dot{q}_j$ ($j=1,\dots,n$), to produce the correct equations for small motions of the plate relative to R. In the sequel, we perform such linearization and place a tilde over a quantity linearized in $q_j, \dot{q}_j$ ($j=1,\dots,n$). Thus, using Eqs. (4.13) and (4.14) in Eq. (4.5) leads to

$$\tilde{J}_1(\sigma,x_2,t)=1; \quad \tilde{J}_1(\zeta_1,x_2,t)=1 \qquad (4.26)$$

Similarly, from Eqs. (4.5) and (4.15) in Eq. (4.16), we have

$$\tilde{J}_2(x_1,\sigma,t)=1; \quad \tilde{J}_2(x_1,\zeta_2,t)=1 \qquad (4.27)$$

Again, to save labor in writing, we introduce the notations:

$$\alpha_{ij}(x_1,x_2) = \int_0^{x_1}[\phi_{2i}'(\sigma,x_2)\phi_{2j}'(\sigma,x_2)$$
$$+\phi_{3i}'(\sigma,x_2)\phi_{3j}'(\sigma,x_2)]d\sigma \quad (i,j=1,\dots,n) \qquad (4.28)$$

$$\beta_{ij}(x_1,x_2) = \int_0^{x_2}[\phi_{1i}'(x_1,\sigma)\phi_{1j}'(x_1,\sigma)$$
$$+\phi_{3i}'(x_1,\sigma)\phi_{3j}'(x_1,\sigma)]d\sigma \quad (i,j=1,\dots,n) \qquad (4.29)$$

The correctly linearized partial velocity of P in N is now written from Eq. (4.25) as follows:

$$^N\tilde{\mathbf{v}}_i^P = -\mathbf{r}_1\sum_{j=1}^{n}\alpha_{ij}(x_1,x_2)q_j(t) - \mathbf{r}_2\sum_{j=1}^{n}\beta_{ij}(x_1,x_2)q_j(t)$$
$$+\mathbf{r}_3\varphi_{3i}(x_1,x_2) \quad (i=1,\dots,n) \qquad (4.30)$$

The linearized velocity of P in N follows from Eqs. (4.22)–(4.24) as

$$
\begin{aligned}
{}^{N}\tilde{\mathbf{v}}^{P} = \mathbf{r}_{1} &\left\{ v_{1} + \omega_{2} \sum_{j=1}^{n} \phi_{3j}(x_{1},x_{2})q_{j}(t) - \omega_{3}\left[x_{2} + \sum_{j=1}^{n} \phi_{2j}(x_{1},x_{2})q_{j}(t) \right] \right\} \\
+ \mathbf{r}_{2} &\left\{ v_{2} + \omega_{3}\left[x_{1} + \sum_{j=1}^{n} \phi_{1j}(x_{1},x_{2})q_{j}(t) \right] - \omega_{1} \sum_{j=1}^{n} \phi_{3j}(x_{1},x_{2})q_{j}(t) \right\} \\
+ \mathbf{r}_{3} &\left\{ v_{3} + \omega_{1}\left[x_{2} + \sum_{j=1}^{n} \phi_{2j}(x_{1},x_{2})q_{j}(t) \right] \right. \\
&\left. - \omega_{2}\left[x_{1} + \sum_{j=1}^{n} \phi_{1j}(x_{1},x_{2})q_{j}(t) \right] + \sum_{j=1}^{n} \phi_{3j}(x_{1},x_{2})\dot{q}_{j}(t) \right\}
\end{aligned}
\tag{4.31}
$$

The acceleration of P in N, linearized in $q_{1},\ldots q_{n},\dot{q}_{1},\ldots \dot{q}_{n}$ is obtained by differentiating ${}^{N}\tilde{\mathbf{v}}^{P}$ in N, and discarding non-linear terms:

$$
{}^{N}\tilde{\mathbf{a}}^{P} = \frac{{}^{N}d}{dt}({}^{N}\tilde{\mathbf{v}}^{P}) = \frac{{}^{R}d}{dt}({}^{N}\tilde{\mathbf{v}}^{P}) + {}^{N}\boldsymbol{\omega}^{R} \times {}^{N}\tilde{\mathbf{v}}^{P}
\tag{4.32}
$$

The generalized inertia force [2] corresponding to the i-th generalized speed is given by

$$
\tilde{F}_{i}^{*} = -\int_{S} {}^{N}\tilde{\mathbf{v}}_{i}^{P}\cdot{}^{N}\tilde{\mathbf{a}}^{P}\rho dx_{1}dx_{2} \quad (i = 1,\ldots,n)
\tag{4.33}
$$

Here, the integration is carried out over S, the area domain of the plate, and ρ is the mass density per unit area. In expanding Eq. (4.33), it is convenient to define the following *modal integrals*:

$$
A_{i} = \int_{S} \phi_{3i}\rho dx_{1}dx_{2} \quad (i = 1,\ldots,n)
\tag{4.34}
$$

$$
B_{ki} = \int_{S} x_{k}\phi_{3i}\rho dx_{1}dx_{2} \quad (k = 1,2; \ i = 1,\ldots,n)
\tag{4.35}
$$

$$
C_{ij} = \int_{S} \alpha_{ij}\rho dx_{1}dx_{2} \quad (i,j = 1,\ldots,n)
\tag{4.36}
$$

$$
C_{kij} = \int_{S} x_{k}\alpha_{ij}\rho dx_{1}dx_{2} \quad (k = 1,2; \ i,j = 1,\ldots,n)
\tag{4.37}
$$

$$D_{ij} = \int_S \beta_{ij}\rho dx_1 dx_2 \quad (i,j = 1,\ldots,n) \tag{4.38}$$

$$D_{kij} = \int_S x_k \beta_{ij}\rho dx_1 dx_2 \quad (k = 1,2; \; i,j = 1,\ldots,n) \tag{4.39}$$

$$E_{kij} = \int_S \phi_{ki}\phi_{kj}\rho dx_1 dx_2 \quad (k = 1,2,3; \; i,j = 1,\ldots,n) \tag{4.40}$$

Using Eqs. (4.34)–(4.40) in Eq. (4.33) and neglecting all non-linear terms in the generalized coordinates and their derivatives yield

$$\tilde{F}_i^* = -\sum_{j=1}^n E_{3ij}\ddot{q}_j - \sum_{j=1}^n (\omega_1 E_{2ij} - \omega_2 F_{1ij})\dot{q}_j - \sum_{j=1}^n [-(\dot{v}_1 + \omega_2 v_3 - \omega_3 v_2)C_{ij}$$

$$- (\omega_1\omega_2 - \dot{\omega}_3)C_{2ij} + (\omega_2^2 + \omega_3^2)C_{1ij} - (\dot{v}_2 + \omega_3 v_1 - \omega_1 v_3)D_{ij}$$

$$- (\dot{\omega}_3 + \omega_1\omega_2)D_{1ij} - (\omega_1^2 + \omega_3^2)D_{2ij} + (\dot{\omega}_1 + \omega_2\omega_3)E_{2ij} \tag{4.41}$$

$$+ (\omega_1\omega_3 - \dot{\omega}_2)E_{1ij} - (\omega_1^2 + \omega_2^2)E_{3ij}]q_j - (\dot{v}_3 + \omega_1 v_2 - \omega_2 v_1)A_i$$

$$- (\dot{\omega}_1 + \omega_2\omega_3)B_{2i} - (\omega_1\omega_3 - \dot{\omega}_2)B_{1i} \quad (i = 1,\ldots,n)$$

Each coefficient of $q_j(t)$ $(j = 1,\ldots, n)$ in Eq. (4.41) can be called a *dynamic stiffness coefficient*, and terms involving C_{ij}, C_{1ij}, C_{2ij}, D_{ij}, D_{1ij}, D_{2ij} can be related to the geometric stiffness associated with translational and rotational inertia loads, as pointed out by Dickens in Ref. [3]. This topic is discussed further in the Appendix to this chapter. The generalized active forces due to the material elasticity of the plate are given by

$$\tilde{F}_i = -\sum_{j=1}^n \lambda_{ij}q_j \quad (i = 1,\ldots n) \tag{4.42}$$

Here, λ_{ij} is a generic element of the modal stiffness matrix, $\phi^T K \phi$, in the standard notation of structural dynamics, which may or may not be diagonal, depending on the type of component modes (Ref. [3], pp. 467–492) used to represent the elastic deformation of the plate. Finally, the linearized equations of the motion of the plate that is executing small elastic motions in a reference frame undergoing large rigid body motion are obtained by substituting Eqs. (4.41) and (4.42) into Kane's dynamical equations:

$$\tilde{F}_i + \tilde{F}_i^* = 0, \quad (i = 1,\ldots,n) \tag{4.43}$$

4.3 SIMULATION ALGORITHM

A step-by-step procedure for simulating large overall motions of the plate follows.

(a) Specify the base motion temporal functions v_1, v_2, v_3, ω_1, ω_2, ω_3, in Eqs. (4.23) and (4.24).

(b) Specify the plate geometry and material constants such as ρ, the mass per unit of area; E, the modulus of elasticity; and ν, Poisson's ratio.

(c) Specify the number n of modes to be used.

(d) Determine the vibration mode shapes and frequencies of the plate, for example, with boundaries supported and fixed in a Newtonian frame. This is normally done with a finite element code and yields the mode shapes $\phi_{ij}(x_1, x_2)$, $(i = 1, 2, 3; j = 1, \ldots, n)$ (see Eqs. (4.13) and (4.15), and the elements of the modal stiffness matrix $\lambda_{ij}(i, j = 1, \ldots, n)$ in Eq. (4.42).

(e) Select m node points; let $x_1^k, x_2^k (k = 1, \ldots, m)$ denote their coordinates; specify initial displacements $d(x_1^k, x_2^k)$ and initial velocities $v(x_1^k, x_2^k)$ $(k = 1, \ldots, m)$ by assigning numerical values to $d_i(x_1^k, x_2^k)$ and $v_i(x_1^k, x_2^k)$ $(i = 1, 2, 3; k = 1, \ldots, m)$ with

$$\mathbf{d} = d_1(x_1^k, x_2^k)\mathbf{r}_1 + d_2(x_1^k, x_2^k)\mathbf{r}_2 + d_3(x_1^k, x_2^k)\mathbf{r}_3 \tag{4.44}$$

$$\mathbf{v} = v_1(x_1^k, x_2^k)\mathbf{r}_1 + v_2(x_1^k, x_2^k)\mathbf{r}_2 + v_3(x_1^k, x_2^k)\mathbf{r}_3 \tag{4.45}$$

To find the associated initial values of $q_i, \dot{q}_i$ $(i = 1, \ldots, n)$, denoted by $q_i(0), \dot{q}_i(0), i = 1, \ldots, n$, introduce the matrices d, v, Φ, q(0), $\dot{q}$(0) as follows:

$$d = \begin{bmatrix} d_1(x_1^1, x_2^1) \\ d_2(x_1^1, x_2^1) \\ d_3(x_1^1, x_2^1) \\ \vdots \\ \vdots \\ \vdots \\ d_1(x_1^m, x_2^m) \\ d_2(x_1^m, x_2^m) \\ d_3(x_1^m, x_2^m) \end{bmatrix}, \qquad v = \begin{bmatrix} v_1(x_1^1, x_2^1) \\ v_2(x_1^1, x_2^1) \\ v_3(x_1^1, x_2^1) \\ \vdots \\ \vdots \\ \vdots \\ v_1(x_1^m, x_2^m) \\ v_2(x_1^m, x_2^m) \\ v_3(x_1^m, x_2^m) \end{bmatrix} \tag{4.46}$$

$$\phi = \begin{bmatrix} \phi_{11}(x_1^1,x_2^1) & . & . & . & \phi_{1n}(x_1^1,x_2^1) \\ \phi_{21}(x_1^1,x_2^1) & . & . & . & \phi_{2n}(x_1^1,x_2^1) \\ \phi_{31}(x_1^1,x_2^1) & . & . & . & \phi_{3n}(x_1^1,x_2^1) \\ . & . & . & . & . \\ . & . & . & . & . \\ . & . & . & . & . \\ \phi_{11}(x_1^m,x_2^m) & . & . & . & \phi_{1n}(x_1^m,x_2^m) \\ \phi_{21}(x_1^m,x_2^m) & . & . & . & \phi_{2n}(x_1^m,x_2^m) \\ \phi_{31}(x_1^m,x_2^m) & . & . & . & \phi_{3n}(x_1^m,x_2^m) \end{bmatrix} \tag{4.47}$$

$$q(0) = \begin{Bmatrix} q_1(0) \\ . \\ . \\ . \\ q_n(0) \end{Bmatrix}, \qquad \dot{q}(0) = \begin{Bmatrix} \dot{q}_1(0) \\ . \\ . \\ . \\ \dot{q}_n(0) \end{Bmatrix} \tag{4.48}$$

Then $q(0)$ and $\dot{q}(0)$ must satisfy, see Eq. (4.2) and Eqs. (4.13)–(4.15), the equations:

$$\phi q(0) = d, \quad \phi \dot{q}(0) = v \tag{4.49}$$

Hence, solving Eq. (4.49) for *least square error* by the Moore–Penrose pseudo-inverse, evaluate $q(0)$ and $\dot{q}(0)$:

$$q(0) = [\phi^T \phi]^{-1} \phi^T d, \quad \dot{q}(0) = [\phi^T \phi]^{-1} \phi^T v \tag{4.50}$$

(f) Evaluate the modal integrals defined in Eqs. (4.34)–(4.40), employing integration by parts wherever possible (see the Appendix to this chapter), using finite element interpolations (Ref. [5], pp. 234–235).

(g) Substitute the modal integrals from Step (f) into Eq. (4.41), and the modal stiffness matrix elements from Step (d) into Eq. (4.41).

(h) Make Eq. (4.43) in the form:

$$M\ddot{q} + G\dot{q} + Kq = F \tag{4.51}$$

where M, G, K are $(n \times n)$ mass, gyroscopic damping, and overall stiffness matrices, and F is an $(n \times 1)$ vector, and integrate Eq. (4.51) numerically. Evaluate the displacement d at (x_1, x_2) at time t as

$$\mathbf{d} = \sum_{j=1}^{n} [\phi_{1j}(x_1,x_2)\mathbf{r}_1 + \phi_{2j}(x_1,x_2)\mathbf{r}_2 + \phi_{3j}(x_1,x_2)\mathbf{r}_3] q_j(t) \tag{4.52}$$

We note that all of the numerical results reported in this chapter are based on the exact mode shapes and frequencies for natural vibrations of a *simply supported plate*, that is [8]

$$\phi_{1j}(x_1, x_2) = \phi_{2j}(x_1, x_2) = 0 \quad (j = 1, \ldots, n) \tag{4.53}$$

$$\phi_{3j}(x_1, x_2) = f_j \sin(\alpha_j x_1 / a) \sin(\beta x_2 / b) \quad (j = 1, \ldots, n) \tag{4.54}$$

where f_j is a normalizing factor and α_j, β_j are tabulated in Blevins (Ref. [8], p. 262) for values of (a/b) for a rectangular plate, where a is the length and b is the width.

4.4 CONCLUSION

It has been shown that by retaining non-linearities up to the partial velocity expression and then linearizing it and the velocity and acceleration expressions, rather than linearizing the fully non-linear equations, one can arrive at a theory that captures the phenomenon of dynamic stiffening. The theory assumes that *the middle surface of a plate does not stretch*. If one started with a linearized velocity, this constraint is missed, as is customarily done in all public domain software, and one obtains dynamic softening instead, for a rotating plate. The underlying cause is premature linearization, implicit with the use of mode shapes. The present theory does suffer from one limitation, namely, it is essentially based on the strip method version of plate theory (see Urugal [6], p. 80). Recently, improvements in plate vibration theory have been developed [7]. The incorporation of a plate theory that accounts for section rotation and also shear, membrane, and large deformation effects could yield even greater dynamic stiffening, provided the fundamental constraint of the mid-surface deformation or consistent modification of it is taken into consideration.

APPENDIX 4 SPECIALIZED MODAL INTEGRALS

The modal integrals in Eqs. (4.34), (4.35), and (4.40) are standard features in flexible body dynamics; see, for example, in Refs. [9, 10] which do not account for dynamic stiffening. Additional integrals arise in the present theory and are given by Eqs. (4.36)–(4.39). Their evaluation is facilitated by employing integration by parts, and the results are listed as follows for a rectangular plate of sides length a along the X_1-axis and b along the X_2-axis; note that $i = 1, \ldots, n$, $j = 1, \ldots, n$

$$C_{ij} = \rho \int_0^b \left\{ \int_0^a (a - x_1) \left[\frac{\partial \phi_{2i}}{\partial x_1} \frac{\partial \phi_{2j}}{\partial x_1} + \frac{\partial \phi_{3i}}{\partial x_1} \frac{\partial \phi_{3j}}{\partial x_1} \right] dx_1 \right\} dx_2 \tag{4.55}$$

$$C_{1ij} = 0.5\rho \int_0^b \left\{ \int_0^a (a^2 - x_1^2) \left[\frac{\partial \phi_{2i}}{\partial x_1} \frac{\partial \phi_{2j}}{\partial x_1} + \frac{\partial \phi_{3i}}{\partial x_1} \frac{\partial \phi_{3j}}{\partial x_1} \right] dx_1 \right\} dx_2 \tag{4.56}$$

$$C_{2ij} = \rho \int_0^b x_2 \left\{ \int_0^a (a - x_1) \left[\frac{\partial \phi_{2i}}{\partial x_1} \frac{\partial \phi_{2j}}{\partial x_1} + \frac{\partial \phi_{3i}}{\partial x_1} \frac{\partial \phi_{3j}}{\partial x_1} \right] dx_1 \right\} dx_2 \qquad (4.57)$$

$$D_{ij} = \rho \int_0^a \left\{ \int_0^b (b - x_2) \left[\frac{\partial \phi_{1i}}{\partial x_2} \frac{\partial \phi_{1j}}{\partial x_2} + \frac{\partial \phi_{3i}}{\partial x_2} \frac{\partial \phi_{3j}}{\partial x_2} \right] dx_2 \right\} dx_1 \qquad (4.58)$$

$$D_{1ij} = 0.5\rho \int_0^a \left\{ \int_0^b (b^2 - x_2^2) \left[\frac{\partial \phi_{1i}}{\partial x_2} \frac{\partial \phi_{1j}}{\partial x_2} + \frac{\partial \phi_{3i}}{\partial x_2} \frac{\partial \phi_{3j}}{\partial x_2} \right] dx_2 \right\} dx_1 \qquad (4.59)$$

$$D_{2ij} = \rho \int_0^a x_1 \left\{ \int_0^b (b - x_2) \left[\frac{\partial \phi_{1i}}{\partial x_2} \frac{\partial \phi_{1j}}{\partial x_2} + \frac{\partial \phi_{3i}}{\partial x_2} \frac{\partial \phi_{3j}}{\partial x_2} \right] dx_2 \right\} dx_1 \qquad (4.60)$$

The terms involving C_{ij}, C_{1ij}, C_{2ij}, D_{ij}, D_{1ij}, and D_{2ij} in Eq. (4.41) can be related to geometric stiffness due to inertia loading. This is seen as follows. Consider the quantity Q_i defined as

$$Q_i = \sum_{j=1}^n (\dot{v}_1 + \omega_2 v_3 - \omega_3 v_2) C_{ij} q_j \qquad (4.61)$$

This is one of the terms in Eq. (4.41). Using Eq. (4.55) in Eq. (4.61):

$$Q_i = \rho(\dot{v}_1 + \omega_2 v_3 - \omega_3 v_2) \int_0^b \left\{ \int_0^a (a - x_1) \sum_{j=1}^n \left[\frac{\partial \phi_{2i}}{\partial x_1} \frac{\partial \phi_{2j}}{\partial x_1} + \frac{\partial \phi_{3i}}{\partial x_1} \frac{\partial \phi_{3j}}{\partial x_1} \right] q_j dx_1 \right\} dx_2 \qquad (4.62)$$

or equivalently

$$Q_i = \frac{\partial}{\partial q_i} \left\langle \frac{1}{2} \rho(\dot{v}_1 + \omega_2 v_3 - \omega_3 v_2) \int_0^b \int_0^a (a - x_1) \sum_{i=1}^n \sum_{j=1}^n \left[\frac{\partial \phi_{2i}}{\partial x_1} \frac{\partial \phi_{2j}}{\partial x_1} + \frac{\partial \phi_{3i}}{\partial x_1} \frac{\partial \phi_{3j}}{\partial x_1} \right] q_i q_j dx_1 \, dx_2 \right\rangle$$

$$(4.63)$$

Now, in view of Eqs. (4.13) and (4.14), with σ_1 defined as

$$\sigma_1(x_1, x_2) = -\rho(\dot{v}_1 + \omega_2 v_3 - \omega_3 v_2)(a - x_1) \qquad (4.64)$$

one can replace Eq. (4.63) with

$$Q_i = -\frac{\partial}{\partial q_i} \left\{ \int_0^b \int_0^a \frac{\sigma_1}{2} \left[\left(\frac{\partial u_2}{\partial x_1} \right)^2 + \left(\frac{\partial u_3}{\partial x_1} \right)^2 \right] dx_1 dx_2 \right\} \qquad (4.65)$$

and the term within the brace in this equation can be recognized as the potential energy due to geometric stiffness from a stressed state due to σ_1 (Ref. [5], p. 281), provided the contribution of $\left(\dfrac{\partial u_1}{\partial x_1}\right)^2$ to this geometric stiffness is neglected, which is the case for a plate that cannot stretch. Moreover, the form of Eq. (4.65) shows that Q_i can be thought of as a generalized force derivable from a potential function associated with strain energy. Finally, σ_1 as defined in Eq. (4.64) also happens to be the solution of the differential equation, with the boundary condition:

$$\frac{\partial \sigma_1}{\partial x_1} = \rho(\dot{v}_1 + \omega_2 v_3 - \omega_3 v_2) \tag{4.66}$$

$$\sigma_1(a, x_2) = 0 \tag{4.67}$$

Eq. (4.66) has the character of a stress equation of motion, for the quantity $\dot{v}_1 + \omega_2 v_3 - \omega_3 v_2$ is a portion of the $\mathbf{r}_1$ measure number of the inertial acceleration of point P' in Figure 4.1. Thus, σ_1 defined in Eq. (4.64) plays the role of an acceleration-induced stress or *stress due to inertia load*; and, proceeding similarly, one finds that C_{1ij}, C_{2ij}, D_{ij}, D_{1ij}, D_{2ij} are also associated with geometric stiffness due to inertia loading.

PROBLEM SET 4

1. Fill in the steps of forming the generalized inertia force, uncovering the formation of the modal integrals given here. Finally, code and simulate the motion of a rectangular plate with your choice of prescribed frame motion.
2. Consider a panel of a wind turbine spinning up as in Eq. (4.1) with t* = 10, for your choice of blade size. Assume that the plate is mounted on rigid frames moving with the shaft, and use modes for a simply supported plate. Take two modes [6]:

$$u_3(x, y, t) = q_{mn}(t)\sin\frac{m\pi x}{a}\sin\frac{n\pi y}{b} \quad m, n = 1, 2, \ldots$$

$$\lambda_{mn} = \pi^4\left[\left(\frac{m}{a}\right)^2 + \left(\frac{n}{b}\right)^2\right]^2 \frac{Eh^3}{12\rho\left(1 - v^2\right)} \tag{4.68}$$

Here, E is the modulus of elasticity; ρ, v are the mass density and Poisson's ratio for the plate, respectively, and a, b are its lengths along the x- and y-axes; h is the plate thickness. Assume a wind pressure as follows:

$$p = p_0 \cos \omega t \quad 0 \le t \le \pi/2; \quad 3\pi/2 \le t \le 2\pi$$

$$p = 0 \quad \pi/2 \le t \le 3\pi/2$$

$$(4.69)$$

Simulate the response of the wind turbine panel.

3. Derive the equations for a "free-free" rectangular plate, with the plane of the plate along the x- and y-axes being reoriented by a bang-bang torque of magnitude T about the y-axis. Keep only two modes of the plate, so that the generalized speeds are $u_1 = \dot{\theta}, u_2 = \dot{q}_1, u_3 = \dot{q}_2$, where θ, q_1, q_2 are, respectively, the rigid body rotation angle, and the two modal coordinates for the vibration modes of the plate. The transverse vibration modes for a square plate of side length a, with all edges free can be taken from Ref. [6]:

$$u_3(x,y,t) = \sum_{i=1}^{2} \left\{ \left[\alpha_i \cos(\beta_i x/a) + \cosh(\beta_i x/a) \right] + \sin(\beta_i x/a) + \sinh(\beta_i x/a) \right\}$$

$$\times \left\{ \left[\alpha_i \cos(\beta_i y/a) + \cosh(\beta_i y/a) \right] + \sin(\beta_i y/a) \right.$$

$$\left. + \sinh(\beta_i y/a) \right\} q_i(t) \, \alpha_i \qquad (4.70)$$

$$= \left[\sin(\beta_i a) - \sinh(\beta_i a) \right] / \left[\cos(\beta_i a) - \cosh(\beta_i a) \right]$$

$$i = 1,2 \; \beta_1 = 4.730; \; \beta_2 = 7.853$$

4. Show that if one starts with a linear velocity expression with assumed modes, and uses Kane's (or Lagrange's) method, as in

$$^N\mathbf{v}^P = {}^N\mathbf{v}^O + {}^N\boldsymbol{\omega}^R \times \{ [x_1 + \sum_{j=1}^{n} \phi_{1j}(x_1,x_2)q_j(t)]\mathbf{r}_1 + [x_2$$

$$+ \sum_{j=1}^{n} \phi_{2j}(x_1,x_2)q_j(t)]\mathbf{r}_2 + \sum_{j=1}^{n} \phi_{3j}(x_1,x_2)q_j(t)\mathbf{r}_3 \}$$

$$+ \sum_{j=1}^{n} \phi_{1j}(x_1,x_2)\dot{q}_j\mathbf{r}_1 + \sum_{j=1}^{n} \phi_{2j}(x_1,x_2)\dot{q}_j\mathbf{r}_2$$

$$+ \sum_{j=1}^{n} \phi_{3j}(x_1,x_2)\dot{q}_j\mathbf{r}_3$$

$$(4.71)$$

one would obtain an incorrect solution due to premature linearization that manifests itself as dynamic softening for the spin-up Eq. (4.1). This feature was unfortunately embedded in many public domain software such as Refs [9, 10].

REFERENCES

1. Banerjee, A.K. and Kane, T.R., (1989), "Dynamics of a Plate in Large Overall Motion", *Journal of Applied Mechanics*, **56**, pp. 887–892.
2. Kane, T.R. and Levinson, D.A., (1985), *Dynamics: Theory and Applications*, McGraw-Hill.
3. Craig, Jr., R.R., (1981), *Structural Dynamics*, Wiley.
4. Banerjee, A.K. and Dickens, J.M., (1990), "Dynamics of an Arbitrary Flexible Body in Large Rotation and Translation", *Journal of Guidance, Control, and Dynamics*, **13**(2), pp. 221–227.
5. Cook, R.D., (1981), *Concepts and Applications in Finite Element Analysis*, Second Edition, Wiley.
6. Urugal, A.C., (1981), *Stresses in Plates and Shells*, McGraw-Hill.
7. Boscolo, M. and Banerjee, J.R., (2012), "Dynamic Stiffness Formulation for Composite Mindlin Plates for Exact Modal Analysis of Structures. Part I: Theory", *Computers & Structures*, **96–97**, pp. 61–73.
8. Blevins, R.D., (1979), *Formulas for Natural Frequency and Mode Shape*, van Nostrand Reinhold Publishing.
9. Bodley, C.S., Devers, A.D., Park, A.C., and Frisch, H.P., (1978), *A Digital Computer Program for the Dynamic Interaction Simulation of Controls and Structures (DISCOS)*, vol. I, NASA Technical Paper 1219.
10. Singh, R.P., van der Voort, R.J., and Likins, P.W., (1985), "Dynamics of Flexible Bodies in Tree Topology: A Computer Oriented Approach", *AIAA Journal of Guidance, Control, and Dynamics*, **8**(5), pp. 584–590.

5 Dynamics of an Arbitrary Flexible Body in Large Overall Motion

It has been shown in the preceding chapters on large overall motions of beams and plates, that the *customary way* of describing small elastic deformation in terms of vibration modes *produces incorrect results*, when the reference frame with respect which vibration occurs is undergoing large rotation and translation. To be more explicit, one obtains *dynamic softening* for a rotating beam or plate, where dynamic stiffening is expected. It is pointed out that this error can be traced to *premature linearization*. In terms of Kane's method, this happens when partial velocities and partial angular velocities are extracted from velocity, and angular velocity expressions written using vibration modes that comes from a linear theory. To correct the error, one must *obtain the nonlinear partial velocity and partial angular velocity expressions* from the nonlinear expressions for velocity and angular velocity, and *then linearize these and all other kinematical quantities*. It was shown that this process, *strictly in that sequence*, recovers dynamic stiffening for rotating beams and plates. Unfortunately, for arbitrarily general elastic bodies undergoing large frame motion, such a procedure is not practicable because one must start with non-linear deformation theories. Non-linear theories for large overall motion of continua, including large elastic deformation, were developed by Simo and Vu-Quoc [3], but their numerical *solution* procedure, being based on the non-linear finite element method, is very time-consuming. In practice, for arbitrary elastic bodies in large overall motion premature linearization is unavoidable, and one can only compensate for this error after the fact, by ad hoc addition of geometric stiffness due to inertia loads. We show the effectiveness of this process with a simple example in Section 1.5. A consummate process will be outlined for arbitrary elastic bodies in this chapter. The material is based on the papers by Banerjee and Dickens [1] and Banerjee and Lemak [2]. The representation of dynamic stiffening through direct use of geometric stiffness has been made for general structures by Modi and Ibrahim [4] and for beams by Amirouche and Ider [5]. Again, the implementation of geometric stiffness, in these formulations based on the *instantaneous* nodal displacements, is also computationally intensive. The formulation given here for arbitrarily general structures avoids this burden, with pre-computed geometric stiffness by the instantaneous inertia forces. Numerical results show that the general theory reduces to special case theories for beams and plates.

DOI: 10.1201/9781003231523-6

5.1   DYNAMICAL EQUATIONS FOR GENERAL STRUCTURES WITH THE USE OF VIBRATION MODES

To make the chapter self-contained, we will review some of the basic material on kinematics, generalized inertia force, and generalized active force [6] for an arbitrary elastic body going through large overall motion in an inertial frame. Consider the body of Figure 5.1; its elastic motion, assumed "small", is measured with respect to a flying reference frame or "body frame" B with basis vectors $\mathbf{b}_1$, $\mathbf{b}_2$, $\mathbf{b}_3$ (denoted with underbars in the figure) attached in the undeformed state of the flexible body at the origin of the coordinates O; the frame B has "large" rigid rotations and translations in a Newtonian frame N. It is assumed that the flexible body is discretized as a system of "nodal" rigid bodies connected by springs, as in finite element modeling. Let G* be the mass center of a small nodal rigid body G. Let the elastic displacement vector at G from its location $\hat{G}$ in the undeformed body, see Figure 5.1, be given by a superposition of ν number of modes and modal coordinates:

$$\delta = \sum_{j=1}^{\nu} \varphi_j q_j \qquad (5.1)$$

where φ_j, q_j are the j-th vibration mode at $\hat{G}$, and the j-th modal coordinate, respectively.

Then, in terms of a vector basis $\mathbf{b}_1$, $\mathbf{b}_2$, $\mathbf{b}_3$, fixed in B, one can characterize the overall motion of the nodal rigid body at G by introducing generalized speeds [6] for the frame rotation and translation, and a number ν of modal coordinate rates for vibration of the arbitrary body:

$$u_i = {}^N\mathbf{v}^O \cdot \mathbf{b}_i \quad (i = 1, 2, 3) \qquad (5.2)$$

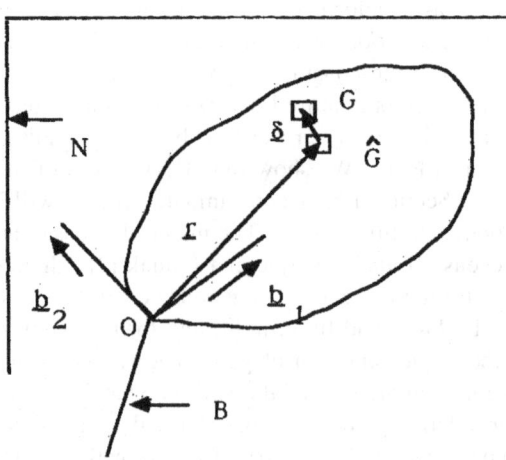

FIGURE 5.1   Arbitrary flexible body with "small" elastic displacements in a flying reference frame B undergoing "large" rotation and translation in a Newtonian frame N.

$$u_{3+i} = {}^N\boldsymbol{\omega}^B \cdot \mathbf{b}_i \quad (i = 1,2,3) \tag{5.3}$$

$$u_{6+i} = \dot{q}_i \quad (i = 1,\ldots,\nu) \tag{5.4}$$

Here, ${}^N\mathbf{v}^O$ is the velocity of the point O in N, and ${}^N\boldsymbol{\omega}^B$ is the angular velocity of B in N. The velocity of G in N is then written as

$$^N\mathbf{v}^{G*} = \sum_{i=1}^{3} u_i\mathbf{b}_i + \sum_{i=1}^{3} u_{3+i}\,\mathbf{b}_i \times (\mathbf{r}+\boldsymbol{\delta}) + \sum_{j=1}^{\nu} \boldsymbol{\varphi}_j\,u_{6+j} \tag{5.5}$$

where $\mathbf{r}$ is the position vector from O to $\hat{G}$, the location of G in the undeformed configuration. The small elastic rotation vector of the infinitesimally small nodal rigid body at G of B is given by a sum of the j-th space-dependent modal rotation vector, multiplied by the generalized coordinates for ν modes:

$$\boldsymbol{\theta} = \sum_{j=1}^{\nu} \boldsymbol{\Theta}_j q_j \tag{5.6}$$

Here, $\boldsymbol{\Theta}_j$ is the j-th modal vector representing small elastic rotation. One can now write the angular velocity of G in N by invoking the angular velocity addition theorem [6] and doing the time-differentiation of Eq. (5.6) using Eq. (5.4) for small rotation:

$$^N\boldsymbol{\omega}^G = {}^N\boldsymbol{\omega}^B + {}^B\boldsymbol{\omega}^G = \sum_{i=1}^{3} u_{3+i}\,\mathbf{b}_i + \sum_{i=1}^{\nu} \boldsymbol{\Theta}_i\,u_{6+i} \tag{5.7}$$

Note that Eqs. (5.5) and (5.7) are already linear in the modal generalized speeds, u_{6+1} (i = 1,...,ν), associated with small deformation. As shown in Chapter 1, this is an act of premature linearization, leading to error manifested by a loss of stiffness that can be compensated only by an ad hoc addition of geometric stiffness due to inertia loads. To start the process, we first note that the i-th partial velocity vector of G* in N and the i-th partial angular velocity vector of G in N associated with the i-th generalized speed can be seen by inspection of Eqs. (5.5) and (5.7), as shown in Table 5.1.

The acceleration of G* is developed by differentiating the vectors in Eq. (5.5) in N, by using the theorem of a particle moving on a moving frame:

$$\mathbf{a}^{G*} = \sum_{j=1}^{3} \left[\dot{u}_j\mathbf{b}_j + \dot{u}_{3+j}\mathbf{b}_j \times (\mathbf{r}+\boldsymbol{\delta}) \right] + \sum_{j=1}^{\nu} \boldsymbol{\varphi}_j\dot{u}_{6+j}$$

$$+ \sum_{j=1}^{3} u_{3+j}\mathbf{b}_j \times \left\{ \sum_{j=1}^{3} \left[u_j\mathbf{b}_j + u_{3+j}\mathbf{b}_j \times (\mathbf{r}+\boldsymbol{\delta}) \right] + 2\sum_{j=1}^{\nu} \boldsymbol{\varphi}_j u_{6+j} \right\} \tag{5.8}$$

TABLE 5.1

Partial Velocities of G and Partial

Angular Velocities of G of Figure 5.1

i	$^N\mathbf{v}_i^G$	$^N\boldsymbol{\omega}_i^G$
1,2,3	$\mathbf{b}_i$	0
4,5,6	$\mathbf{b}_i \times (\mathbf{r}+\boldsymbol{\delta})$	$\mathbf{b}_i$
7,...,6+ν	$\boldsymbol{\varphi}_i$	$\boldsymbol{\Theta}_i$

Similarly, the angular acceleration of G follows from Eq. (5.7):

$$\boldsymbol{\alpha}^G = \sum_{j=1}^{3}\left(\dot{u}_{3+j}\mathbf{b}_j + u_{3+j}\mathbf{b}_j \times \sum_{j=1}^{\nu}\boldsymbol{\Theta}_j u_{6+j} \right) + \sum_{j=1}^{\nu}\boldsymbol{\Theta}_j \dot{u}_{6+j} \tag{5.9}$$

The generalized inertia force corresponding to the i-th generalized speed, with contributions from all differential elements in the entire body, is obtained by integrating over the whole body:

$$F_i^* = -\int_D {}^N\mathbf{v}_i^{G*} \cdot {}^N\mathbf{a}^{G*}\, dm$$

$$-\int_D {}^N\boldsymbol{\omega}_i^G \cdot \left[d\mathbf{I}^G \cdot {}^N\boldsymbol{\alpha}^G + {}^N\boldsymbol{\omega}^G \times (d\mathbf{I}^G \cdot {}^N\boldsymbol{\omega}^G) \right] \quad (i=1,...,6+v) \tag{5.10}$$

Here, dm is the mass and $d\mathbf{I}^G$ is the central inertia dyadic for the differential element G, respectively. The integrations in Eq. (5.10), performed over the domain D occupied by the flexible body, are straightforward but extremely tedious. This requires doing the integrations in such a way that the integrands do not contain any time-varying quantities. The process of extracting time-invariant groups of terms from the integrand in Eq. (5.10) is performed by invoking certain vector-dyadic identities, to produce all the modal integrals usually reported [13, 14] with some additional ones arising because of the inclusion of the rotatory inertia of the nodal rigid bodies. The details of writing Eq. (5.10) in terms of modal integrals are reported in Ref. [2], with the integrals listed in Appendix A at the end of this book. These details are not central to the issue at hand. It is important, however, to realize that Eq. (5.10) will be of the form:

$$-F_i^* = \sum_{j=1}^{6+\nu} M_{ij}\,\dot{u}_j + C_i \quad (i=1,...,6+\nu) \tag{5.11}$$

where elements M_{ij} of the so-called "mass matrix" M, with $(i,j=1,...,6+v)$, are functions of the modal coordinates q_j $(i=1,...,\nu)$, and C_i, $(i=1,...,6+v)$ are functions of the modal coordinates as well as the generalized speeds u_j, $(i=1,...,6+v)$.

5.2 COMPENSATING FOR PREMATURE LINEARIZATION BY ADDING GEOMETRIC STIFFNESS DUE TO INERTIA LOADS

In this section, we consider generalized active force due to two kinds of internal forces. The first one is associated with standard, structural stiffness. The second one, associated with geometric stiffness due to inertia forces, compensates for errors due to premature linearization, and will be discussed in detail here. To deal with the structural stiffness first, we note that generalized active force due to nominal structural elasticity is commonly written, on the basis of the mass normalization of the vibration modes as yielding the elastic vibration frequencies, ω_{i-6} ($i=7,\ldots, 6+v$).

$$F_i = 0 \qquad\qquad (i = 1,\ldots,6)$$
$$= -\omega_{i-6}^2\, q_{i-6} \qquad (i = 7,\ldots,6+v)$$

(5.12)

Motion-induced stiffness is a special case of geometric stiffness that is treated in detail by the non-linear finite element theory. Cook [7] has shown that *geometric stiffness accounts for the effect of existing forces on bending stiffness*, and that it only depends on the element's geometry, together with the displacement field, and the state of existing stress components, σ_{xx0}, σ_{xy0}, σ_{xz0}, σ_{yy0}, σ_{yz0}, σ_{zz0}, and the ($3 \times$ ndof) matrix of interpolation functions $[N(x,y,z)]$, where ndof stands for the number of element degrees of freedom describing the displacement at a point (x,y,z):

$$\{\delta\} = \left[N(x,y,z)\right]\{d\} \tag{5.13}$$

Here, $\{\delta\}$ is a column matrix of three components of the elastic displacement at a point within the element and $\{d\}$ is the column matrix of ndof element nodal displacements. Then, the element geometric stiffness matrix is derived from the strain energy induced, as the existing stresses work through the strains representing the non-linear parts of the Lagrangian strain components. This element geometric stiffness matrix is given in Refs. [7, 8] as follows:

$$k_g^e = \int_e \begin{bmatrix} N_{,x}^T & N_{,y}^T & N_{,z}^T \end{bmatrix} \begin{bmatrix} \sigma_{xx0}U_3 & \sigma_{xy0}U_3 & \sigma_{xz0}U_3 \\ \sigma_{xy0}U_3 & \sigma_{yy0}U_3 & \sigma_{yz0}U_3 \\ \sigma_{xz0}U_3 & \sigma_{yz0}U_3 & \sigma_{zz0}U_3 \end{bmatrix} \begin{Bmatrix} N_{,x} \\ N_{,y} \\ N_{,z} \end{Bmatrix} dv \tag{5.14}$$

Here, a variable in the subscripts for the interpolation function $N(x, y z)$ preceded by a comma indicates differentiation with respect to the variables x,y,z, and U_3 is the (3×3) unity matrix. In a finite element structural analysis code such as NASTRAN, computation of geometric stiffness proceeds as follows: First the distributed loading on the structure is prescribed, and the corresponding linear static equilibrium problem is solved to determine the element stresses; then, the element geometric stiffnesses are evaluated following Eq. (5.14), and the standard assembly procedure of the finite element method is used to construct the geometric stiffness matrix, K_g, of the whole structure. Now the following question arises: If geometric stiffness depends on existing loads, and is to explain the phenomenon of dynamic stiffening with motion, then where is the load coming from? The natural answer is, *the loading in the initial state of stress due to the rigid body motion itself,* namely, the stresses due to *inertia forces* and *inertia torques* acting throughout the body. This is the reason we refer to the associated geometric stiffness as motion-induced stiffness. We will now proceed to compute the distributed inertia force and inertia torque loading. The inertia force on a nodal rigid body G in the "existing state" is due to the *rigid body acceleration* of G, multiplied by the nodal mass dm:

$$\mathbf{f}_{rigid}^* = -dm\,^N\mathbf{a}^G = -dm\left\{\sum_{j=1}^{3}\left[\dot{u}_j\mathbf{b}_j + \dot{u}_{3+j}\mathbf{b}_j \times \mathbf{r}\right] + \sum_{j=1}^{3}u_{3+j}\mathbf{b}_j \times \sum_{j=1}^{3}\left[u_j\mathbf{b}_j + u_{3+j}\mathbf{b}_j \times \mathbf{r}\right]\right\}$$

$$(5.15)$$

Expanding the terms of $^N\mathbf{a}^G$ from Eq. (5.15) in the reference *rigid body* configuration of the body $\mathbf{b}_1$, $\mathbf{b}_2$, $\mathbf{b}_3$ basis, the components of $\mathbf{f}_{rigid}^*$ can be written in the *matrix notation* as the inertia load, depending on the global coordinates of the generic nodal mass dm as follows:

$$\begin{Bmatrix} f_1^* \\ f_2^* \\ f_3^* \end{Bmatrix} = -dm\begin{bmatrix} U_3 & x_1U_3 & x_2U_3 & x_3U_3 \end{bmatrix}\begin{Bmatrix} A_1 \\ : \\ : \\ : \\ A_{12} \end{Bmatrix} \qquad (5.16)$$

Here, U_3 is a 3×3 unity matrix; x_1, x_2, x_3 are coordinates of the element mass dm; and $A_1,\ldots, A_{12}$ are 12 acceleration terms in the *existing or undeformed (rigid body) state*, expressed as follows:

$$A_1 = \dot{u}_1 + z_1$$

$$A_2 = \dot{u}_2 + z_2$$

$$A_3 = \dot{u}_3 + z_3$$

$$A_4 = -(u_5^2 + u_6^2)$$

$$A_5 = \dot{u}_6 + z_4$$

$$A_6 = -\dot{u}_5 + z_5$$

$$A_7 = -\dot{u}_6 + z_4$$

$$A_8 = -(u_6^2 + u_4^2)$$

$$A_9 = \dot{u}_4 + z_6$$

$$A_{10} = \dot{u}_5 + z_5$$

$$A_{11} = -\dot{u}_4 + z_6$$

$$A_{12} = -(u_5^2 + u_4^2)$$

$$(5.17)$$

The above terms have been defined using the following *intermediate variables*:

$$z_1 = u_5 u_3 - u_6 u_2$$
$$z_2 = u_6 u_1 - u_4 u_3$$
$$z_3 = u_4 u_2 - u_5 u_1$$
$$z_4 = u_4 u_5 \tag{5.18}$$
$$z_5 = u_6 u_4$$
$$z_6 = u_5 u_6$$

Considering now the inertia torque vector on the nodal rigid body G, and neglecting the deformation-related terms, from an *existing* state of rigid body rotation, we see that this is obtained by using the angular velocity and angular acceleration vectors given by Eqs. (5.7) and (5.9) and the nodal rigid body inertia dyadic:

$$t^* = -\left[d\mathbf{I}^G \cdot \sum_{j=1}^{3} \dot{u}_{3+j}\mathbf{b}_j + \sum_{j=1}^{3} u_{3+j}\mathbf{b}_j \times \left(d\mathbf{I}^G \cdot \sum_{j=1}^{4} u_{3+j}\mathbf{b}_j \right) \right] \qquad (i = 1,..,6+v) \tag{5.19a}$$

This is written in the vector basis of the flying reference B, using element inertia dyadics:

$$t^* = -[(I_{11}\mathbf{b}_1\mathbf{b}_1 + I_{12}\mathbf{b}_1\mathbf{b}_2 + I_{13}\mathbf{b}_1\mathbf{b}_3 + I_{21}\mathbf{b}_2\mathbf{b}_1 + I_{22}\mathbf{b}_2\mathbf{b}_2 + I_{23}\mathbf{b}_2\mathbf{b}_3 + I_{31}\mathbf{b}_3\mathbf{b}_1 +$$
$$I_{32}\mathbf{b}_3\mathbf{b}_2 + I_{33}\mathbf{b}_3\mathbf{b}_3) \cdot (\dot{u}_4\mathbf{b}_1 + \dot{u}_5\mathbf{b}_2 + \dot{u}_6\mathbf{b}_3) + (u_4\mathbf{b}_1 + u_5\mathbf{b}_2 + u_6\mathbf{b}_3) \times [(I_{11}\mathbf{b}_1\mathbf{b}_1 + I_{12}\mathbf{b}_1\mathbf{b}_2$$
$$+ I_{13}\mathbf{b}_1\mathbf{b}_3 + I_{21}\mathbf{b}_2\mathbf{b}_1 + I_{22}\mathbf{b}_2\mathbf{b}_2 + I_{23}\mathbf{b}_2\mathbf{b}_3 + I_{31}\mathbf{b}_3\mathbf{b}_1 + I_{32}\mathbf{b}_3\mathbf{b}_2 + I_{33}\mathbf{b}_3\mathbf{b}_3) \cdot (u_4\mathbf{b}_1 + u_5\mathbf{b}_2 + u_6\mathbf{b}_3)]$$

$$\tag{5.19b}$$

Equation (5.19b) in the $\mathbf{b}_1$, $\mathbf{b}_2$, $\mathbf{b}_3$ basis leads to the following (3×9) matrix relation expressing the inertia torque of the nodal rigid body G, in terms of the nodal rotational inertia elements and the angular acceleration components:

$$\begin{Bmatrix} t_1^* \\ t_2^* \\ t_3^* \end{Bmatrix} = - \begin{bmatrix} I_{11} & I_{12} & I_{13} & I_{13} & I_{33}\text{-}I_{22} & \text{-}I_{12} & 0 & I_{23} & 0 \\ I_{12} & I_{22} & I_{23} & \text{-}I_{23} & I_{12} & I_{11}\text{-}I_{33} & 0 & 0 & I_{13} \\ I_{13} & I_{23} & I_{33} & I_{22}\text{-}I_{11} & \text{-}I_{13} & I_{23} & I_{12} & 0 & 0 \end{bmatrix} \begin{Bmatrix} A_{13} \\ \vdots \\ \vdots \\ A_{21} \end{Bmatrix}$$

$$\tag{5.20}$$

Here, the angular acceleration terms $A_{13}, \ldots, A_{21}$ stand for the following:

$$A_{13} = \dot{u}_4$$

$$A_{14} = \dot{u}_5$$

$$A_{15} = \dot{u}_6$$

$$A_{16} = z_4$$

$$A_{17} = z_6 \qquad\qquad (5.21)$$

$$A_{18} = z_5$$

$$A_{19} = u_4^2 - u_5^2$$

$$A_{20} = u_5^2 - u_6^2$$

$$A_{21} = u_6^2 - u_4^2$$

With the force and torque loadings defined by Eqs. (5.16) and (5.20), respectively, a *non-linear finite element code* can now be used to generate 12 geometric matrices if nodal rotation is ignored, or 21 geometric stiffness matrices, $K_g^{(i)}$ ($i = 1, \ldots, 21$) for unit values of each of the acceleration or angular acceleration variables A_i ($i = 1, \ldots, 21$). Motion-induced stiffness due to the actual values of translational inertia and rotational inertia corresponding to large translations and rotations of a nodal rigid body can then be obtained by simple multiplication with the current states of these angular accelerations. Generalized active force due to these motion-induced stiffness components can be written for ν number of modes in either 12 (or 21, if nodal rotations are considered loads) as:

$$\left\{ \begin{array}{c} F_7^g \\ \vdots \\ \vdots \\ F_{6+\nu}^g \end{array} \right\} = -\sum_{i=1}^{NL} S^{(i)} A_i q, \quad NL = 12 \text{ or } 21 \qquad (5.22)$$

Here, $S^{(i)}$ denotes the $\nu \times \nu$ generalized geometric stiffness matrices due to unit values of the i-th inertia loading, A_i ($i = 1, \ldots, NL$), $NL = 12$ or 21, and is given for each i, by

$$S^{(i)} = \Phi^T K_g^{(i)} \Phi \qquad\qquad (5.23)$$

In Eq. (5.23), Φ is the matrix of mode shapes and q is the ($\nu \times 1$) matrix of modal generalized coordinates. With expressions for the generalized inertia forces and generalized actives forces at hand, one can express *Kane's equations of motion* in matrix form:

$$\sum_{j=1}^{6+v} M_{ij}\dot{u}_j + C_i = 0 \quad (i = 1,\ldots,6) \tag{5.24}$$

$$\sum_{j=1}^{6+v} M_{i+6,j}\dot{u}_j + C_{i+6} + \omega_i^2 q_i + \sum_{j=1}^{v}\left\{ S_{ij}^{(1)}\dot{u}_1 + S_{ij}^{(2)}\dot{u}_2 + S_{ij}^{(3)}\dot{u}_3 + [S_{ij}^{(13)} - S_{ij}^{(9)} - S_{ij}^{(11)}]\dot{u}_4 \right\} q_j +$$

$$\sum_{j=1}^{v}\left[\begin{aligned} &[S_{ij}^{(14)} + S_{ij}^{(10)} - S_{ij}^{(6)}]\dot{u}_5 + [S_{ij}^{(16)} + S_{ij}^{(5)} - S_{ij}^{(7)}]\dot{u}_6 + S_{ij}^{(1)}z_1 + S_{ij}^{(2)}z_2 + S_{ij}^{(3)}z_3 \\ &+[S_{ij}^{(5)} + S_{ij}^{(7)}]z_4 \end{aligned} \right\} q_j$$

$$+\sum_{j=1}^{v}\left\{ [S_{ij}^{(6)} + S_{ij}^{(10)}]z_5 + [S_{ij}^{(9)} + S_{ij}^{(11)}]z_6 + S_{ij}^{(4)}A_4 + S_{ij}^{(8)}A_8 + S_{ij}^{(12)}A_{12} \right\} q_j$$

$$+\sum_{k=13}^{21} S_{ij}^{(k)}A_k q_j = 0$$

$$(i = 1,\ldots,v)$$

$$\tag{5.25}$$

Note that here we have *not considered* any generalized active force due to *external loads*; if these are present, their contribution to the generalized active forces has to be added to the right-hand sides of Eqs. (5.24) and (5.25), as described in previous chapters. Equations (5.24) and (5.25) can be written in compact form as the matrix differential equation:

$$M\dot{U} = R \tag{5.26}$$

where U is the column matrix of generalized speeds $u_1,\ldots,u_{6+v}$. Note from Eqs. (5.24) and (5.25) that the coefficient matrix M is time-varying and asymmetric. Note also that if the rotatory inertia effects of the nodes of a structure can be neglected, which is the usual case, then the second integral in Eq. (5.10) vanishes and the last term in Eq. (5.25) becomes zero, i.e.,

$$S_{ij}^{(k)} = 0 \quad (k = 13,\ldots,21; \, i, j = 1,\ldots,v) \tag{5.27}$$

with the simplification that only 12, not 21, of the motion-induced geometric stiffness matrices in Eq. (5.25) are computed. This completes the formulation of the dynamical equations, with geometric stiffness added to compensate for the error of premature linearization inherent with the use of vibration modes given by linear structural dynamics.

Before leaving this presentation of the dynamical equations, one has to ask again, *when is the error of premature linearization acceptable?* The answer is: whenever the rotational speed of the frame with respect to which the body vibrates is far smaller than the first vibration frequency of the body.

5.2.1 RIGID BODY KINEMATICAL EQUATIONS

To complete the equations of motion, we now consider the kinematical equations for the inertial position coordinates x,y,z of the point O of B in Fig. 5.1 and the attitude of body B, expressed in terms of the Euler parameters (quaternions), $\varepsilon_1, \varepsilon_2, \varepsilon_3, \varepsilon_4$ [9]:

$$\begin{Bmatrix} \dot{x} \\ \dot{y} \\ \dot{z} \end{Bmatrix} = \begin{bmatrix} 1-2(\varepsilon_2^2+\varepsilon_3^2) & 2(\varepsilon_1\varepsilon_2-\varepsilon_3\varepsilon_4) & 2(\varepsilon_3\varepsilon_1+\varepsilon_2\varepsilon_4) \\ 2(\varepsilon_1\varepsilon_2+\varepsilon_3\varepsilon_4) & 1-2(\varepsilon_3^2+\varepsilon_1^2) & 2(\varepsilon_2\varepsilon_3-\varepsilon_1\varepsilon_4) \\ 2(\varepsilon_3\varepsilon_1-\varepsilon_2\varepsilon_4) & 2(\varepsilon_2\varepsilon_3+\varepsilon_1\varepsilon_4) & 1-2(\varepsilon_1^2+\varepsilon_2^2) \end{bmatrix} \begin{Bmatrix} u_1 \\ u_2 \\ u_3 \end{Bmatrix}$$

$$(5.28)$$

$$\text{or} \begin{Bmatrix} \dot{x} \\ \dot{y} \\ \dot{z} \end{Bmatrix} = \begin{bmatrix} {}^N C^B \end{bmatrix} \begin{Bmatrix} u_1 \\ u_2 \\ u_3 \end{Bmatrix}$$

The attitude parameters $\varepsilon_1, \varepsilon_2, \varepsilon_3, \varepsilon_4$ change with time as given by the kinematical equations [9]:

$$\begin{Bmatrix} \dot{\varepsilon}_1 \\ \dot{\varepsilon}_2 \\ \dot{\varepsilon}_3 \\ \dot{\varepsilon}_4 \end{Bmatrix} = \frac{1}{2} \begin{bmatrix} \varepsilon_4 & -\varepsilon_3 & \varepsilon_2 \\ \varepsilon_3 & \varepsilon_4 & -\varepsilon_1 \\ -\varepsilon_2 & \varepsilon_1 & \varepsilon_4 \\ -\varepsilon_1 & -\varepsilon_2 & -\varepsilon_3 \end{bmatrix} \begin{Bmatrix} u_4 \\ u_5 \\ u_6 \end{Bmatrix} \qquad (5.29)$$

One can determine the body 3, 1-2-3 sequence [9] angles in terms of the quaternions by equating the direction cosine matrix in Eq. (5.28) to the direction matrix:

$$\begin{bmatrix} {}^N C^B \end{bmatrix} = \begin{bmatrix} c_2 c_3 & -c_2 s_3 & s_2 \\ s_1 s_2 c_3 + s_3 c_1 & -s_1 s_2 s_3 + c_3 c_1 & -s_1 c_2 \\ -c_1 s_2 c_3 + s_3 s_1 & c_1 s_2 s_3 + c_3 s_1 & c_1 c_2 \end{bmatrix} \qquad (5.30)$$

Thus, the middle rotation (pitch angle) is given by $\theta_2 = \sin^{-1}[2(\varepsilon_3\varepsilon_1+\varepsilon_2\varepsilon_3)]$.

5.3 SUMMARY OF THE ALGORITHM

To implement the theory given above, the following step-by-step procedure must be used in sequence.

1. Using a finite element code such as NASTRAN or equivalent [12], construct the finite element model of the structure with the origin of coordinates constrained to have zero rigid body degrees of freedom; generate a

number v of column matrices representing v number of vibration modes and frequencies of the structure.

2. On each node of the finite element model, set up 12 loads defined in Eq. (5.16) (in terms of the x-, y-, z-coordinates of the node) for unit values of $A_1,...,A_{12}$. If rotatory inertia is deemed significant, then impose nine torque loads as per Eq. (5.20). Then, for each of these 12 or 21 loads distributed throughout the body, compute the geometric stiffness matrix due to unit loads, and convert the latter into generalized force due to motion-induced stiffness, using Eqs. (5.22) and (5.23).

3. All of the above are inputs to a flexible multibody dynamics code. Now undertake the major work of building the large motion dynamics code on the basis of Eqs. (5.24) and (5.25). This involves the work of constructing the M_{ij} and C_i matrix elements in these two equations. That work, in turn, depends on developing the modal integrals, described in detail in the Appendix at the end of this book, and formed out of the modes generated in Step 1.

4. Append equations for $\dot{q}_i, i = 1,...,(6+v)$ related to $u_i, i = 1,...,(6+v)$, from Eqs. (5.2)–(5.4), to those given by Eqs. (5.28) and (5.29). This involves the relationship between the position coordinates of O, the orientation of B, and the modal coordinates of the flexible body to their corresponding generalized speeds for this large overall motion problem.

5. Assign initial values to q_i, u_i $(i = 1,...,6+v)$. If the initial deformations and their rates are given, the generalized coordinates and generalized speed can be obtained as a best approximation fit by taking a pseudo-inverse, see Eq. (4.50).

6. Integrate numerically the dynamical and kinematical equations with these initial conditions.

5.4   CRUCIAL TEST AND VALIDATION OF THE THEORY IN APPLICATION

To validate the foregoing theory, which is applicable to any arbitrary flexible body undergoing large overall motion, we compare the motion predicted by this general theory to predictions already available from special purpose theories for beams [10] and plates [11] attached to a moving base. For a crucial test, we use the spin-up function:

$$\omega_1 = \frac{\Omega}{T}\left(t - \frac{T}{2\pi}\sin\frac{2\pi t}{T}\right) \quad t \le T$$

$$= \Omega \qquad\qquad\qquad t > T \tag{5.31}$$

where ω_1, Ω, T are, respectively, the $\mathbf{b}_1$ component of the spin angular velocity of body B in N, the final spin-up speed, both in radians per second and time in seconds to reach the final spin. Figure 5.2 compares the results for the *spin-up* of

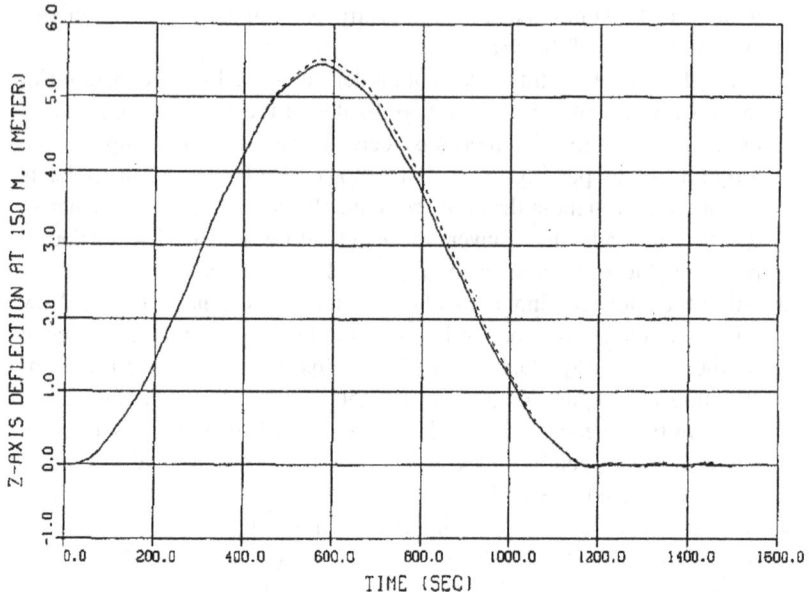

FIGURE 5.2 Cantilever beam tip deflection at 150 m during spin-up, given by the present general theory (solid line) and the special beam theory of Ref. [11]; beam first bending frequency 0.546 rad/sec, spin-up speed 0.06 rad/sec, spin-up time 1200 sec.

a 150 m long WISP antenna attached to the space shuttle, with the antenna tube having a diameter of 0.0635 m and a wall thickness of 0.00254 m, with the first bending mode frequency of $\omega_1 = 0.0546$ rad/sec. In each of the two simulations, with the present general theory and the beam theory, only the first four vibration modes and frequencies are used. Base spin-up is described by Eq. (5.31) with T = 1200 sec and $\Omega = 0.06$ rad/sec; in other words, *maximum spin frequency is higher than the first mode bending frequency*. With the steady spin rate higher than the first bending frequency, premature linearization through the use of vibration modes would produce disastrously wrong results. Compensatory geometric stiffness needed in the present theory is computed by using the theory in Ref. [10] for beam elements with existing axial force due to spin and does not include the effect of transverse shear, and this may explain the slight discrepancy between the two curves in Figure 5.2. The overall agreement between the two curves in Figure 5.2 provides evidence of the adequacy of the present general theory in reproducing dynamic stiffening of a beam given by a beam-specific theory, during and following spin-up.

The results of comparing the present theory with a special purpose rotating plate theory [11] for a cantilever plate, again during *spin-up about an edge*, are shown in Figure 5.3. The plate is 72 in long, 48 in wide, and 0.1 in thick, and has a first natural frequency of $\omega_1 = 0.75$ rad/sec. The spin-up is described by Eq. (5.31) with $\Omega = 1.25$ rad/sec, T = 30 sec. In this case, the first six vibration modes and

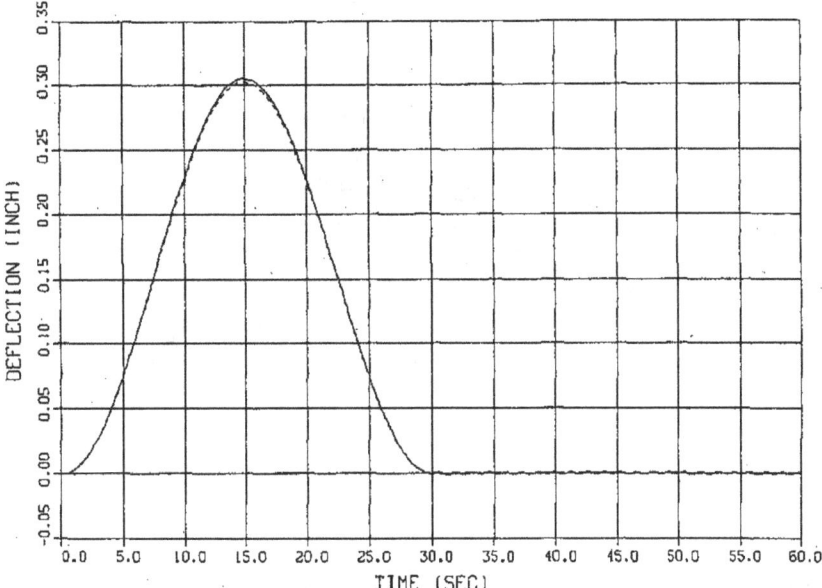

FIGURE 5.3 Cantilever plate corner deflection in inches during spin-up given by the present general theory (solid line) and the special plate theory of Ref. [12]; plate first bending frequency 0.75 rad/sec, spin-up speed 1.25 rad/sec, spin-up time 30 sec.

frequencies of the plate are used in each simulation. The difference between the results of the general theory and the plate theory is hardly discernible. When one recalls that the special purpose theories for beam and plate, in Figures 5.2 and 5.3, are obtained by taking the steps for correct linearization, the reproduction of those results by the general theory, that compensates for premature linearization due to the use of vibration modes, by the addition of geometric stiffness due to inertia loading, testifies to the correctness of the present theory for any arbitrary flexible body.

Now we apply this general theory to the large motion dynamics of two structures, one a rotating paraboloidal shell structure like the Galileo spacecraft, and the other a truss in spin or in translational acceleration. Figure 5.4 shows a dual-spin spacecraft. The paraboloidal, flexible antenna rotates with respect to a rigid base, and we apply the spin-up check with $\Omega=0.5$ Hz, $T=30$ sec in Eq. (5.31) and the first natural frequency of 0.433 Hz, lower than the spin-up speed. Figure 5.5 shows what happens when one uses the present theory that incorporates motion-induced stiffness, compared to the predictions of conventional theories [13, 14] using only vibration modes and no geometric stiffness; the latter case is re-created here by deleting the motion-induced stiffness terms. A theory with no motion-induced stiffness, predicting unbounded motion at a particular node arbitrarily chosen (node number 572 in the finite element model) is clearly erroneous. For the structure made up of two trusses rigidly attached to a base, in Figure 5.6, we again do a ramp up to a steady spin or a

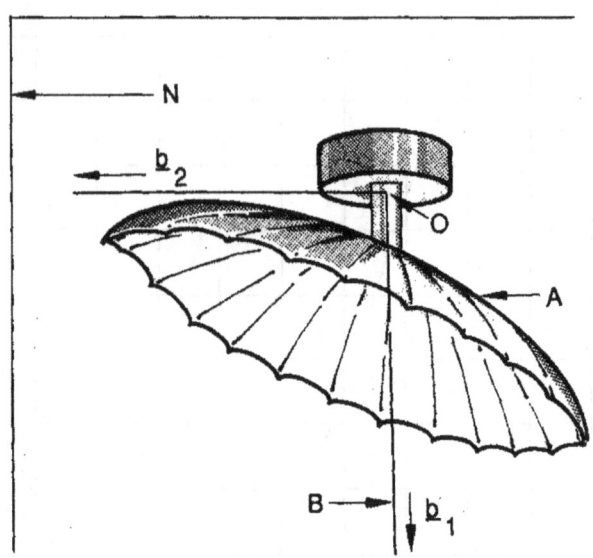

FIGURE 5.4   Dual-spin spacecraft with flexible, spinning offset paraboloidal antenna.

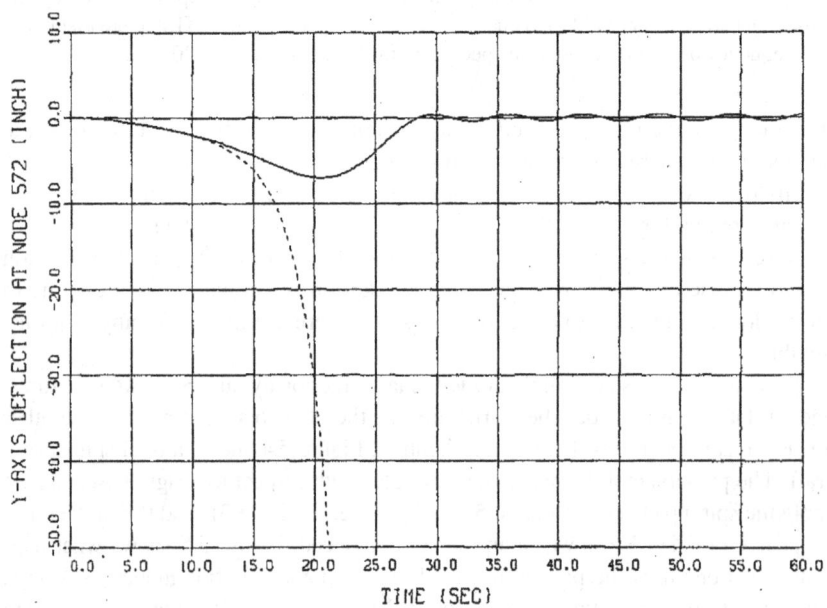

FIGURE 5.5   Elastic displacement along the y-axis at finite element node 572, of a spinning offset paraboloidal antenna for spin-up given by the present general theory (solid line) and a conventional theory with no geometric stiffening (dashed line); steady-state spin frequency (0.5 Hz) is greater than the first natural vibration frequency (0.433); spin-up time is 30 sec.

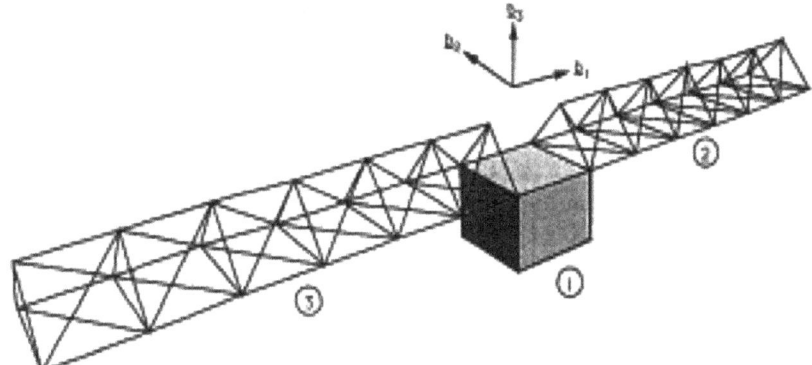

FIGURE 5.6 A single "flexible body spacecraft" made up of a rigid body fixed to two identical trusses in spin-up about b_3 or translational acceleration along b_1.

steady translational speed. Again the vibration modes were obtained using the finite element code [12], with a first mode frequency of 3.72 rad/sec; the *translation* speed time history is again given by Eq. (5.31), with ω now standing for the translational speed and Ω the maximum translation speed attained. For this "spin-up" test, the maximum value of the translational speed was carefully chosen at 4 rad/sec to be higher than the first mode frequency to accentuate the effect of motion-induced stiffness (softness or buckling in this situation due to inertia force being compressive). Figure 5.7 shows the tip deflection of a corner of the truss in spin-up, with and without

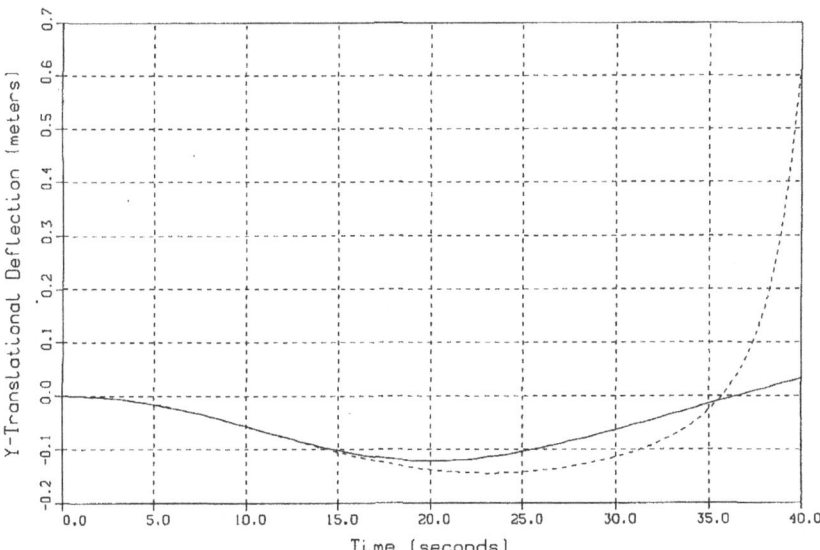

FIGURE 5.7 Transverse tip deflection of a corner of the truss in translation motion (solid line, with geometric stiffness; dashed line, no geometric stiffness).

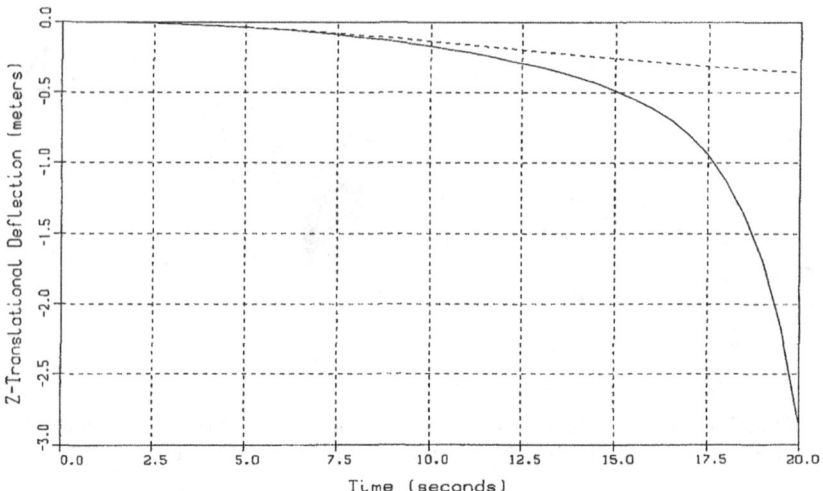

FIGURE 5.8 Dynamic buckling of truss #2 due to compression under translational acceleration (solid line, with geometric stiffness; dashed line, no geometric stiffness).

geometric stiffness consideration. Next, we do a translation test, using Eq. (5.31) as the speed-up function. The maximum acceleration is then $2\ \Omega/T$. The critical *buckling* acceleration for a uniform loading on the truss was calculated by the finite element code [12] to be 53.6 m/s^2. For translation acceleration along $\mathbf{b}_1$ in Figure 5.6, the truss at the right end of the rigid bus is in dynamic compression due to inertia forces, and the truss on the left is in tension. Figure 5.8 shows $\mathbf{b}_3$ direction tip deflection of the truss under compression due to inertia load. Here, we see the evolution of buckling when the maximum acceleration corresponds to the critical acceleration. The solid line, for inclusion of geometric stiffness (softness in this case), shows buckling as it should happen, while the dashed line for no geometric stiffness fails to predict buckling. Figures 5.9 and 5.10 show the results of the dynamic softening and stiffening of the trusses when the system is accelerated to 75% of the critical buckling acceleration. Finally, Figure 5.9 shows a tendency to dynamic buckling, i.e., loss of stiffness that is expected to occur with translational acceleration, as in the case of rockets; it is correctly reproduced with the present theory, whereas a theory with no motion-induced stiffness (softness in this case) fails to predict this catastrophe. Figure 5.10 shows the softening of truss marked body #3 in Figure 5.6 which is under tension, when no motion-induced stiffness is considered, while the present theory with motion-induced stiffness produces an intuitively correct result.

5.5 CONCLUSION

We have produced a general theory for an arbitrary flexible body in large overall motion.

The theory compensates for the error of premature linearization, inherent with the use of vibration modes representing small elastic motion along with large

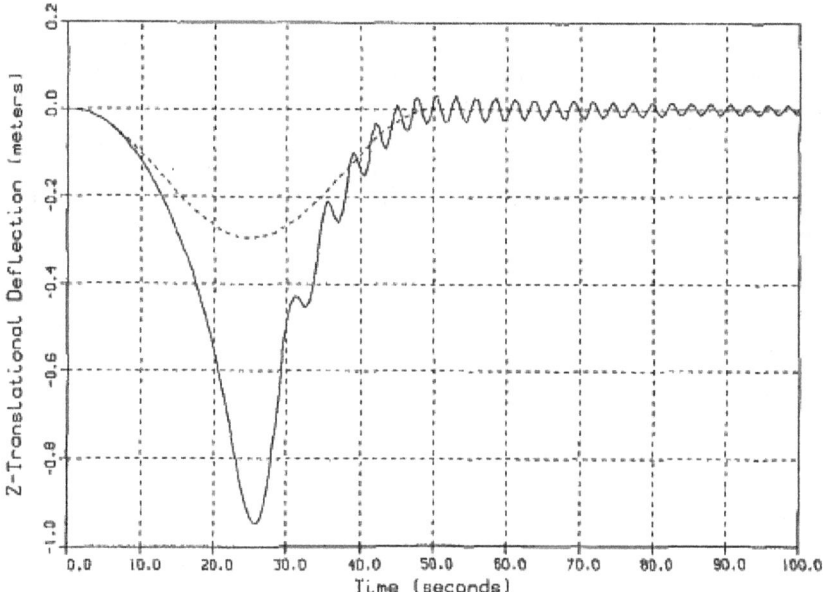

FIGURE 5.9 Buckling tendency of truss #2 under compression due to inertia load in sub-critical translational acceleration along $\mathbf{b}_1$ (solid line with geometric stiffness; dashed line, no geometric stiffness).

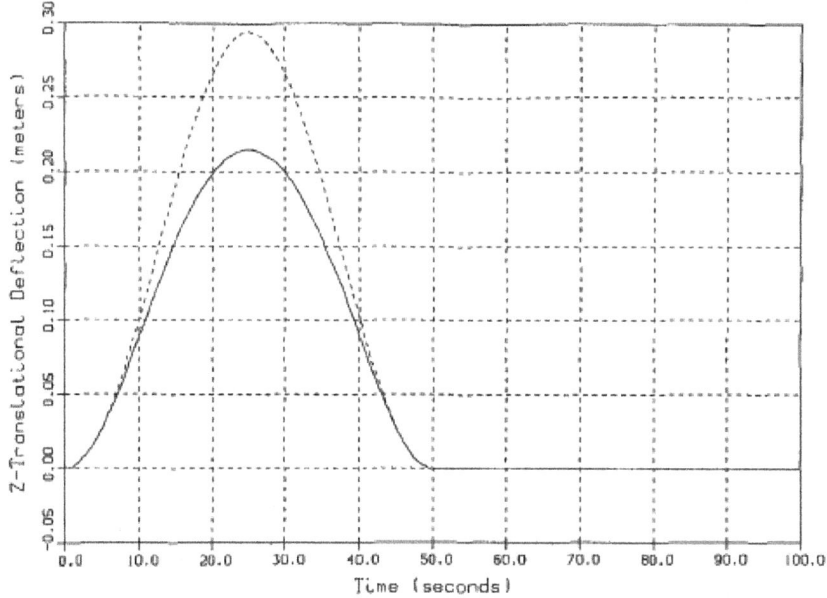

FIGURE 5.10 Dynamic stiffening of truss #3, in tension with $\mathbf{b}_1$ direction translational acceleration (solid line with geometric stiffness; dashed line, no geometric stiffness).

frame rotation and translation, by augmenting the natural stiffness of the structure by geometric stiffness due to inertia forces for translation and inertia torques for rotation. Details of the geometric stiffness due to 12 inertia forces and 9 inertia torques distributed throughout the body are quantified. This general theory reproduces the results of the dynamic stiffening of special purpose theories for beams and plates. It is shown that for structures with components, parts of which undergo dynamic stiffening and parts that must show dynamic softening, like buckling, this general theory of motion-induced stiffness correctly brings out the expected results.

PROBLEM SET 5

1. Code the general theory for an arbitrary elastic solid in large overall motion with motion-induced stiffness. Verify that the results of the theory reduce to those for special purpose rotating beam or plate formulations given in the preceding chapters.

2. Use the code to simulate the spin-up motion of a helicopter rotor blade, modeled by a cantilever mode [16]:

$$\varphi(x) = (\sin\beta_1 L - \sinh\beta_1 L)(\sin\beta_1 x - \sinh\beta_1 x)$$

$$+(\cos\beta_1 L + \cosh\beta_1 L)(\cos\beta_1 x - \cosh\beta_1 x).$$

3. Derive the equations of motion of a free-flying pre-tensioned large solar sail in the sky, of sides a and b, tension S per unit length of edge, γ mass per unit area, with natural frequency and mode shapes as follow [16], neglecting geometric stiffness.

$$\omega_{ij} = \frac{\lambda_{ij}}{2}\left(\frac{S}{\gamma ab}\right)^{1/2} \quad i = 1,2,3,\ldots; \ j = 1,2,3,\ldots; \quad \lambda_{ij} = \left(i^2\frac{b}{a} + j^2\frac{a}{b}\right)^{1/2}$$

$$\varphi_{ij} = \sin\left(\frac{i\pi x}{a}\right)\sin\left(\frac{j\pi y}{b}\right)$$

REFERENCES

1. Banerjee, A.K and Dickens, J.M., (1990), "Dynamics of an Arbitrary Flexible Body in Large Rotation and Translation", *Journal of Guidance, Control, and Dynamics*, 13(2), pp. 221–227.
2. Banerjee, A.K. and Lemak, M.E., (1991), "Multi-Flexible Body Dynamics Capturing Motion-Induced Stiffness", *Journal of Applied Mechanics*, 58, pp. 766–775.
3. Simo, J.C. and Vu-Quoc, L., (1986), "On the Dynamics of Flexible Bodies Under Large Overall Motions: The Plane Case: Parts I & II", *Journal of Applied Mechanics*, 53, pp. 849–863.
4. Modi, V.J. and Ibrahim, A.M., (1988), "On the Dynamics of Flexible Orbiting Structures", in *Large Space Structures: Dynamics and Control*, Edited by S.N. Atluri and A.K. Amos, Springer, pp. 93–114.

5. Amirouche, F.M.L. and Ider, S.K., (1989), "The Influence of Geometric Nonlinearities in the Dynamics of Flexible Tree-Like Structures", *Journal of Guidance, Control, and Dynamics*, **12**(6), pp. 830–837.

6. Kane, T.R and Levinson, D.A., (1985), *Dynamics: Theory and Application*, McGraw-Hill.

7. Cook, R.D., (1985), *Concepts and Applications of Finite Element Analysis*, McGraw-Hill. pp. 331–341.

8. Zienkiewicz, O.C., (1977), *The Finite Element Method*, McGraw-Hill, pp. 517–519.

9. Kane, T.R., Likins, P.W., and Levinson, D.A., (1983), *Spacecraft Dynamics*, McGraw-Hill, pp. 12–13.

10. Kane, T.R., Ryan, R.R., and Banerjee, A.K., (1987), "Dynamics of a Cantilever Beam Attached to a Moving Base", *Journal of Guidance, Control, and Dynamics*, **10**(2), pp. 139–151.

11. Banerjee, A.K and Kane, T.R., (1989), "Dynamics of a Plate in Large Overall Motion", *Journal of Applied Mechanics*, **56**(6), pp. 887–892.

12. Wheatstone, W.D., (1983), *EISI-EAL Engineering Analysis Language Reference Manual*, Engineering Transformations Systems, Inc., July.

13. Bodley, C.S., Devers, A.D., Park, A.C., and Frisch, H.P., (1978), *A Digital Computer Program for the Dynamic Interaction Simulation of Controls and Structures (DISCOS)*, vol. I & II, NASA TP-1219.

14. Singh, R.P., van der Voort, R.J., and Likins, P.W., (1985), "Dynamics of Flexible Bodies in Tree Topology: A Computer Oriented Approach", *Journal of Guidance, Control, and Dynamics*, **8**(5), pp. 584–590.

15. Banerjee, A.K. and Lemak, M.E., (2008), "Dynamics of a Flexible Body with Rapid Mass Loss", in Proceedings of the International Congress on Theoretical and Applied Mechanics, August, Adelaide, Australia.

16. Blevins, R.D., (2001), *Formulas for Natural Frequency and Mode Shape*, Krieger Publishing Company, p. 226.

6 Flexible Multibody Dynamics
Dense Matrix Formulation

6.1 FLEXIBLE BODY SYSTEM IN A TREE TOPOLOGY

In this chapter, we analyze the large overall motion of a multibody system of hinge-connected rigid and elastic bodies, and in the process adapt the theory of capturing motion-induced stiffness to such systems. The material is based principally on Ref. [1]. The interconnection of the bodies in the system is described by a "topological tree" shown in Figure 6.1, where the bodies are numbered arbitrarily as 1, 2, 3,…, n, with the body numbered 0 being the inertial frame of reference.

Following Huston and Passerello [2], a topological array for the system in Figure 6.1 is defined such that body j has an inboard connecting body c(j) along the path going from body j to body 0. Thus, for the system in Figure 6.1 6.1, we have the following topological tree array, with all bodies connected by rotational hinges, except body 7 which is assumed to be connected to body 6 by a relative translational joint, like a piston in a cylinder:

$$\begin{bmatrix} j & 1 & 2 & 3 & 4 & 5 & 6 & 7 & 8 & 9 \\ c(j) & 0 & 1 & 1 & 3 & 4 & 1 & 6 & 6 & 8 \end{bmatrix}$$

6.2 KINEMATICS OF A JOINT IN A FLEXIBLE MULTIBODY SYSTEM

Figure 6.2 shows two adjacent flexible bodies, B_j, $B_{c(j)}$, connected at a hinge allowing both rotation and translation between them, with Q_j a hinge point between B_j, and P_j, the corresponding hinge point on $B_{c(j)}$. Reference frames j and P_j are fixed to Q_j and P_j, respectively.

Following Kane and Levinson [3], we introduce generalized speeds as motion variables, as many in number as there are total degrees of freedom (dof) in the system. At the point Q_j, let there be R_j number of degrees of freedom along rotational hinge axis unit vectors $\mathbf{h}_i^j, (i = 1,…, R_j)$, and T_j number of degrees of freedom along translational direction unit vectors $\mathbf{t}_i^j, (i = 1,…, T_j)$; note that an underscore is used in Figure 6.2 to indicate a basis vector, which we denote with a bold sign in the text. If M_j number of vibration modal coordinates are used to describe the small elastic deformation of body B_j in its body frame basis ($\mathbf{b}_1^j, \mathbf{b}_2^j, \mathbf{b}_3^j$ in Figure 6.2), then the following generalized speeds are defined in terms of $\mathbf{h}_i^j, (i = 1,…, R_j)$ and $\mathbf{t}_i^j, (i = 1,…, T_j)$ for a system of n number of bodies:

DOI: 10.1201/9781003231523-7

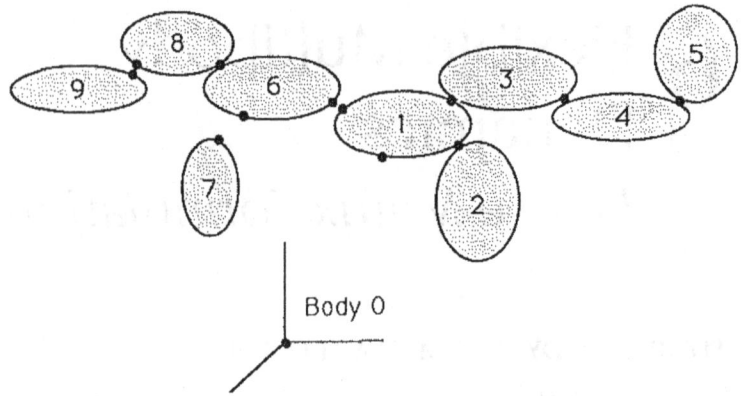

FIGURE 6.1   A system of hinge-connected rigid and flexible bodies in a tree topology.

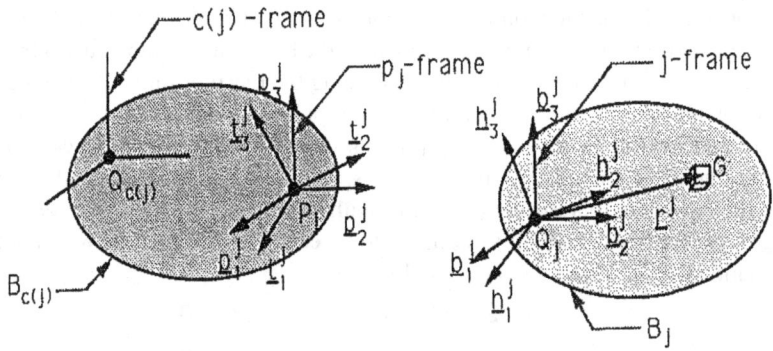

FIGURE 6.2   Two hinge-connected bodies in a tree topology.

$$u_i = {}^{p_j}\boldsymbol{\omega}^j \cdot \mathbf{h}_i^j \quad (j=1,...,n; i=1,..,R_j)$$

$$u_{R_j+i} = {}^{p_j}\mathbf{v}^{Q_j} \cdot \mathbf{t}_i^j \quad (j=1,...,n; i=1,..,T_j) \qquad (6.1)$$

$$u_{R_j+T_j+i} = \dot{\eta}_i^j \quad (j=1,...,n; i=1,..,M_j)$$

Here, ${}^{p_j}\boldsymbol{\omega}^j$ is the angular velocity of frame j with respect frame p_j mounted on body c(j) at its point P_j; ${}^{p_j}\mathbf{v}^{Q_j}$ is the translational velocity of the point Q_j with respect to point P_j; and η_i^j is the i-th modal coordinate of body B_j. Note that the total degree of freedom of the NB body system, with Eq. (6.1) defined for a generic j-th body, is ndof, where

$$\text{ndof} = \sum_{j=1}^{NB}(R_j + T_j + M_j) \qquad (6.2)$$

6.3 KINEMATICS AND GENERALIZED INERTIA FORCES FOR A SINGLE BODY

Let the elastic deformation vector at a generic point G of body B_j in Figure 6.2 be assumed in terms of the i-th mode shape φ_i^j at G and modal coordinates η_i^j of body B_j as

$$d^j = \sum_{i=1}^{M_j} \varphi_i^j \eta_i^j \tag{6.3}$$

The velocity of G, located by $\mathbf{r}^j$ from Q_j, in the undeformed state in the inertial frame N is

$$^N\mathbf{v}^G = {}^N\mathbf{v}^{Q_j} + {}^N\omega^j \times (\mathbf{r}^j + \mathbf{d}^j) + \sum_{i=1}^{M_j} \varphi_i^j u_{R_j+T_j+i}^j \tag{6.4}$$

Note here that $^N\mathbf{v}^G$, $^N\omega^j$ are the *inertial* frame velocity of G and the angular velocity of frame j, respectively. The inertial angular velocity of the nodal rigid body G is expressed by the i-th modal small rotation vector ψ_i^j at G by the angular velocity addition theorem.

$$^N\omega^G = {}^N\omega^j + \sum_{i=1}^{M_j} \psi_i^j u_{R_j+T_j+i}^j \tag{6.5}$$

At this stage, note that Eqs. (6.4) and (6.5) are already linear in the elastic motion modal coordinates, η_i^j, and this is an act of premature linearization. The partial velocity of point G and the partial angular velocity of the nodal body at G follow from Eqs. (6.4) and (6.5):

$$^N\mathbf{v}_i^G = {}^N\mathbf{v}_i^{Q_j} + {}^N\omega_i^j \times (\mathbf{r}^j + \mathbf{d}^j) + \delta_{ik}\,\varphi_k^j \qquad (i = 1,\ldots,\text{ndof}) \tag{6.6}$$

$$^N\omega_i^G = {}^N\omega_i^j + \delta_{ik}\,\psi_k^j \qquad (i = 1,\ldots,\text{ndof}) \tag{6.7}$$

Note that the Kronecker delta δ_{ik} has the value 1 when the i-th generalized speed of the *system* coincides with the general speed corresponding to the k-th modal coordinate rate of body B_j in Eq. (6.1) and is zero otherwise. The inertial acceleration of point G and the inertial angular acceleration of the infinitesimal nodal rigid body at G are, respectively, as follows:

$$^N\mathbf{a}^G = {}^N\mathbf{a}^{Q_j} + {}^N\alpha^j \times (\mathbf{r}^j + \mathbf{d}^j) + \sum_{i=1}^{M_j} \varphi_i^j \ddot{\eta}_i^j + {}^N\omega^j \times \left[{}^N\omega^j \times (\mathbf{r}^j + \mathbf{d}^j) + 2\sum_{i=1}^{M_j} \varphi_i^j \dot{\eta}_i^j \right] \tag{6.8}$$

$$^N\alpha^G = {}^N\alpha^j + \sum_{i=1}^{M_j} \psi_i^j \ddot{\eta}_i^j + {}^N\omega^j \times \sum_{i=1}^{M_j} \psi_i^j \dot{\eta}_i^j \tag{6.9}$$

The i-th generalized inertia force of a single body B_j is summed up over the system as

$$F_i^{j*} = -\int_{B_j} {}^N v_i^G \cdot {}^N a^G dm - \int_{B_j} {}^N \omega_i^G \cdot \left[d\mathbf{I}^G \cdot {}^N \alpha^G + {}^N \omega^G \times \left(d\mathbf{I}^G \cdot {}^N \omega^G \right) \right] \qquad (6.10)$$

$(j = 1, \ldots, NB; \; i = 1, \ldots, ndof)$

With Eqs. (6.6)–(6.9) available, the operations needed in Eq. (6.10) are straightforward but massive. As said before, what one must do is to perform all the integrals over the spatial domain in such a way the integrands do not contain any time-varying quantities. This is done by invoking certain vector-dyadic identities to form "modal integrals", collected in Appendix A at the end of the book. The generalized inertia force due to the j-th body for the i-th generalized speed is then summed up over all bodies, NB, as shown below, where all non-linear terms in the modal coordinates and their derivatives have been dropped.

$$F_i^* = -\sum_{j=1}^{NB} {}^N v_i^{Q_j} \cdot \left\langle m^j \, {}^N a^{Q_j} - s^j \times {}^N \alpha^j + \sum_{k=1}^{M_j} b_k^j \ddot{\eta}_k^j + {}^N \omega^j \times \left[{}^N \omega^j \times s^j + 2 \sum_{k=1}^{M_j} b_k^j \dot{\eta}_k^j \right] \right\rangle$$

$$- \sum_{j=1}^{NB} {}^N \omega_i^j \cdot \left\langle s^j \times {}^N a^{Q_j} + \mathbf{I}^j \cdot {}^N \alpha^j + \sum_{k=1}^{M_j} c_k^j \ddot{\eta}_k^j + {}^N \omega^j \times \left(\mathbf{I}^j \cdot {}^N \omega^j \right) + 2 \sum_{k=1}^{M_j} N_k^{j*} \dot{\eta}_k^j \cdot {}^N \omega^j \right\rangle$$

$$- \sum_{j=1}^{NB} \delta_{ik} \left\langle b_k^j \cdot {}^N a^{Q_j} + g_k^j \cdot {}^N \alpha^j + \sum_{k=1}^{M_j} e_{ik}^j \ddot{\eta}_k^j - {}^N \omega^j \cdot D_k^j \cdot {}^N \omega^j + 2 \sum_{k=1}^{M_j} d_{ik}^j \dot{\eta}_k^j \cdot {}^N \omega^j \right\rangle$$

- -

$$- \sum_{j=1}^{NB} {}^N \omega_i^j \cdot \left\langle \left(\tilde{W}_1^j + \sum_{k=1}^{M_j} \tilde{W}_{2k}^j \eta_k^j \right) \cdot {}^N \alpha^j + {}^N \omega^j \cdot \sum_{k=1}^{M_j} \tilde{W}_{3k}^j \dot{\eta}_k^j + \sum_{k=1}^{M_j} W_{4k}^j \ddot{\eta}_k^j \right\rangle$$

$$- \sum_{j-1}^{NB} {}^N \omega_i^j \cdot \left\langle {}^N \omega^j \times \left[\left(\tilde{W}_1^j + \sum_{k=1}^{M_j} \tilde{W}_{2k}^j \eta_k^j \right) \cdot {}^N \omega^j + \sum_{k=1}^{M_j} W_{4k}^j \dot{\eta}_k^j \right] - {}^N \omega^j \cdot \sum_{k=1}^{M_j} \tilde{W}_{5k}^j \dot{\eta}_k^j \right\rangle$$

$$- \sum_{j=1}^{NB} \delta_{ik} \left\langle \left(W_{6k}^j + \sum_{n=1}^{M_j} W_{7nk}^j \eta_k^j \right) \cdot {}^N \alpha^j + {}^N \omega^j \cdot \sum_{n=1}^{M_j} W_{8nk}^j \dot{\eta}_k^j + \sum_{n=1}^{M_j} W_{9nk}^j \ddot{\eta}_k^j \right\rangle$$

$$- \sum_{j=1}^{NB} {}^N \omega^j \cdot \left[\tilde{W}_{3k}^j \cdot {}^N \omega^j + \sum_{n=1}^{M_j} \tilde{W}_{10nk}^j \eta_k^j \cdot {}^N \omega^j - \sum_{n=1}^{M_j} \left(W_{11nk}^j + W_{12nk}^j \right) \dot{\eta}_k^j \right]$$

$i = 1, \ldots, ndof, \; j = 1, \ldots, NB$

$$(6.11)$$

In Eq. (6.11) the group of terms below the dashed line relates to rotational inertia of the of the finite element nodal bodies and can be ignored if such rotations are not of concern. The term $g_k^j \cdot {}^N \alpha^j$ in Eq. (6.11) ultimately gives rise to a mass matrix that is non-symmetric. Replacing this term with $c_k^j \cdot {}^N \alpha^j$ is an acceptable approximation, to make the mass matrix symmetric. In Eq. (6.11), I^j, N_k^{j*}, D_k^j are dyadics, with * denoting a transpose. All terms with overhead tilde in the lower half of Eq. (6.11) are dyadics.

6.4 KINEMATICAL RECURRENCE RELATIONS PERTAINING TO A BODY AND ITS INBOARD BODY

In this section, we develop the interrelationships between ${}^N \omega^j, {}^N v^{Q_j}, {}^N \alpha^j, {}^N a^{Q_j}$, of body j in Eq. (6.11), in terms of those variables for the inboard body c(j), $B_{c(j)}$ in Figure 6.2, that is, ${}^N \omega^{c(j)}, {}^N v^{Q_{c(j)}}, {}^N \alpha^{c(j)}, {}^N a^{Q_{c(j)}}$ Introducing an orthogonal triad of basis vectors p_i^j fixed to the frame p_j, and assuming R_j number of rotations about possibly non-orthogonal axes, the angular velocity of frame j is given by the angular velocity addition theorem:

$$
{}^N \omega^j = {}^N \omega^{c(j)} + \sum_{i=1}^{M_{c(j)}} \psi_i^{c(j)}(P_j) \dot{\eta}_i^{c(j)} + \sum_{i=1}^{3} \sum_{k=1}^{R_j} p_i^j G_{ik}^j \dot{\theta}_k^j \tag{6.12}
$$

Here, the first two terms give the angular velocity of a frame mounted at the terminal point P_j of the elastic body c(j), and the third term refers to the relative rotations at the hinge for the frame of body j with respect to the frame at P_j. We also have $G_{ik}^j = p_i^j \cdot h_k^j$, where h_k^j is the hinge k-th axis vector, and $\psi_i^{c(j)}(P_j), \dot{\theta}_k^j$ are, respectively, the i-th modal rotation at P_j and the k-th relative rotation rate of the hinge at Q_j. The velocity of point Q_j can be written by assuming T_j number of translational degrees of freedom from P_j along possibly non-orthogonal axes as follows, where τ_k^j is the k-th translation vector at the hinge Q_j, with τ_k^j its magnitude:

$$
v^{Q_j} = {}^N v^{Q_{c(j)}} + {}^N \omega^{c(j)} \times \left[r^{Q_{c(j)}P_j} + \sum_{k=1}^{M_{c(j)}} \varphi_k^{c(j)}(P_j) \eta_k^{c(j)} \right] + \sum_{k=1}^{M_{c(j)}} \varphi_k^{c(j)}(P_j) \dot{\eta}_k^{c(j)}
$$

$$
+ \left[{}^N \omega^{c(j)} + \sum_{k=1}^{M_{c(j)}} \psi_k^{c(j)}(P_j) \dot{\eta}_k^{c(j)} \right] \times \sum_{i=1}^{3} \sum_{k=1}^{T_j} p_i^j L_{ik}^j \tau_k^j \tag{6.13}
$$

$$
+ \sum_{i=1}^{3} \sum_{k=1}^{T_j} p_i^j \left[\dot{L}_{ik}^j \tau_k^j + L_{ik}^j \dot{\tau}_k^j \right]
$$

$$
L_{ik}^j = p_i^j \cdot \tau_k^j
$$

Eqs. (6.12) and (6.13) provide the mechanism for developing the expressions for the inertial (or N-frame) velocities of a hinge point Q_j and the angular velocity of frame j of body B_j for all values of $j = 1,\ldots,$ NB, in the tree topology of Figure 6.1, starting at body 1. The partial velocities and partial angular velocities needed in Eq. (6.11) are obtained from their defining relations (assuming no prescribed motion) as

$$^N\mathbf{v}^{Q_j} = \sum_{i=1}^{ndof} {}^N\mathbf{v}_i^{Q_j} u_i \tag{6.14}$$

$$^N\boldsymbol{\omega}^j = \sum_{i=1}^{ndof} {}^N\boldsymbol{\omega}_i^j u_i \tag{6.15}$$

The acceleration and angular accelerations of Q_j and frame j are expressed from Eqs. (6.14) and (6.15) in terms of the derivatives of the generalized speeds and remainder acceleration terms:

$$^N\mathbf{a}^{Q_j} = \sum_{i=1}^{3} \mathbf{v}_i^{Q_j} \dot{u}_i + \mathbf{h}^j$$

$$^N\boldsymbol{\alpha}^j = \sum_{i=1}^{3} \boldsymbol{\omega}_i^j \dot{u}_i + \mathbf{f}^j \tag{6.16}$$

6.5   GENERALIZED ACTIVE FORCES DUE TO NOMINAL AND MOTION-INDUCED STIFFNESS

This section represents a multibody application of the work in Chapter 5. Here, we will express generalized active forces due to nominal structural stiffness and motion-induced geometric stiffness, and we will do so in matrix form because both these quantities are typically developed by finite element methods, which use matrix representations. Generalized active force due to nominal elasticity is customarily represented on the basis of component modes [4] for the j-th body in the matrix form:

$$F_n^j = -[\Phi^j]^T K_n^j \Phi^j \eta^j = -\Lambda_n^j \eta^j \tag{6.17}$$

Here, Φ^j, K_n^j, η^j are, respectively, the mode shape matrix (with superscript T denoting matrix transpose), the nominal structural stiffness matrix, and the column matrix of modal coordinates for body j. The matrix Λ_n^j is diagonal if mass-normalized vibration modes are used for body j, and non-diagonal if constraint modes are used in addition to the normal modes, as per the Craig–Bampton method [4]. It should be noted that modes here reflect deformations about the initial state of stress of the body, and motion-induced stiffness is due to a special case of initial state of stress [5] caused by inertia loading. It has its origin in the strain energy term:

$$P_{NL} = \int_B \varepsilon_{NL}^T \sigma_0 dv \tag{6.18}$$

where ε_{NL} is the column matrix of the Lagrangian strain, and σ_0 initial stress column matrix.

$$\sigma_0 = \left(\sigma_{110}, \sigma_{120}, \sigma_{130}, \sigma_{220}, \sigma_{230}, \sigma_{330}\right)^{\mathrm{T}}$$

$$\varepsilon_{\mathrm{NL}} = \left\{\begin{array}{c} 0.5(w_{1,1}^2 + w_{2,1}^2 + w_{3,1}^2) \\ 0.5(w_{1,2}^2 + w_{2,2}^2 + w_{3,2}^2) \\ 0.5(w_{1,3}^2 + w_{2,3}^2 + w_{3,3}^2) \\ w_{1,3}w_{1,2} + w_{2,3}w_{2,2} + w_{3,3}w_{3,2} \\ w_{1,3}w_{1,1} + w_{2,3}w_{2,1} + w_{3,3}w_{3,1} \\ w_{1,2}w_{1,1} + w_{2,2}w_{2,1} + w_{3,2}w_{3,1} \end{array}\right\} \tag{6.19}$$

Here w_1, w_2, w_3 are 1-, 2-, and 3-orthogonal components of the elastic displacement, and differentiation with respect to a component is indicated by a subscript comma. In the standard procedure of the finite element theory (Zienkiewicz [6]), one assumes interpolation functions $N(x,y,z)$ between the nodal displacements d for the displacement of a point within an element, and derives elemental stiffness matrices based on the potential energy function. Gradient of the potential energy function of Eq. (6.18) provides the geometric or initial stress stiffness matrix K_g^e as before:

$$\{w\} = \left[N(x_1, x_2, x_3)\right]\{d\}$$
$$\tag{6.20a}$$
$$P_{\mathrm{NL}} = \{d\}^{\mathrm{t}}[K_g^e]\{d\}$$

$$K_g^e = \int_e \begin{bmatrix} N_{,x}^{\mathrm{T}} & N_{,y}^{\mathrm{T}} & N_{,z}^{\mathrm{T}} \end{bmatrix} \begin{bmatrix} \sigma_{xx0}I_3 & \sigma_{xy0}I_3 & \sigma_{xz0}I_3 \\ \sigma_{xy0}I_3 & \sigma_{yy0}I_3 & \sigma_{yz0}I_3 \\ \sigma_{xz0}I_3 & \sigma_{yz0}I_3 & \sigma_{zz0}I_3 \end{bmatrix} \begin{Bmatrix} N_{,x} \\ N_{,y} \\ N_{,z} \end{Bmatrix} dv \tag{6.20b}$$

Here, I_3 is a 3×3 unity matrix. In a finite element code such as NASTRAN or EISI-EAL, with the latter [7] used to generate the results in this chapter, the reference stress stiffness matrix is computed by first evaluating the stresses due to a distributed loading, and computing the element geometric stiffness matrices as per Eq. (6.20b), and finally assembling the latter into a global geometric stiffness matrix for the structure. For structures in large overall motion, the distributed loading is a system of inertia forces and torques corresponding to the motion. To produce a theory that is fully linear in the modal coordinates, consistent with the small vibration assumption, we consider inertia force and torques that involve only zeroth-order terms in the modal coordinates. In other words, we neglect the terms related to elastic displacement in Eqs. (6.6)–(6.8) to form expressions for the rigid body inertia force and the inertia torque on a nodal rigid body G in body j, which produce the "existing" state [5], incremental deformations with respect to which give rise to geometric stiffness, that is, we write

$$\mathbf{f}^{jG*} = -dm\left[{}^N\mathbf{a}^{Qj} + {}^N\boldsymbol{\alpha}^j \times \mathbf{r}^j + {}^N\boldsymbol{\omega}^j \times \left({}^N\boldsymbol{\omega}^j \times \mathbf{r}^j\right)\right]$$

$$\mathbf{t}^{jG*} = -\left[d\mathbf{I}^G \cdot {}^N\boldsymbol{\alpha}^j + {}^N\boldsymbol{\omega}^j \times \left(d\mathbf{I}^G \cdot {}^N\boldsymbol{\omega}^j\right)\right] \tag{6.21}$$

Equation (6.21) is now written in matrix form in terms of the j-basis vector elements, and direction cosine elements between body 1 and body j.

$$[\mathbf{b}^j]^T = [\mathbf{b}_1^j \, \mathbf{b}_2^j \, \mathbf{b}_3^j] \quad (j = 1, \ldots, NB)$$

$$C_{1j}(i, k) = \mathbf{b}_i^1 \cdot \mathbf{b}_k^j$$

(6.22)

We now define a $(n \times 1)$ column matrix U formed by stacking the generalized speeds given by Eq. (6.1) for all the bodies in the system. Then, Eq. (6.15) is rewritten in matrix form in terms of the partial velocity for point Q_j and the partial angular velocity matrix from frame-j, expressed in body 1 basis.

$$^N\omega^j = [\mathbf{b}^1]^T \omega_u^j U$$

$$^N v^{Q_j} = [\mathbf{b}^1]^T v_u^{Q_j} U$$

(6.23)

Equation (6.16) is then rewritten in terms of matrices as

$$^N a^{Q_j} = [\mathbf{b}^1]^T C_{1j}^T (v_u^{Q_j} \dot{U} + h^j)$$

$$^N \alpha^j = [\mathbf{b}^1]^T C_{1j}^T (\omega_u^j \dot{U} + f^j)$$

(6.24)

Now we introduce a set of intermediate scalar variables, z_i^j, $i = 1, \ldots, 15$; $j = 1, \ldots, NB$, to facilitate writing the acceleration and angular velocity expressions in matrix forms.

$$^N\omega^j = z_1^j \mathbf{b}_1^j + z_2^j \mathbf{b}_2^j + z_3^j \mathbf{b}_3^j$$

$$\tilde{\omega}^j = \begin{bmatrix} 0 & -z_3^j & z_2^j \\ z_3^j & 0 & -z_1^j \\ -z_2^j & z_1^j & 0 \end{bmatrix}$$

$$\begin{bmatrix} z_4^j & z_5^j & z_6^j \\ z_5^j & z_7^j & z_8^j \\ z_6^j & z_8^j & z_9^j \end{bmatrix} = C_{1j}^T \tilde{\omega}^j \tilde{\omega}^j C_{1j}$$

$$z_{10}^j = z_1^j z_2^j$$

(6.25)

$$z_{11}^j = z_2^j z_3^j$$

$$z_{12}^j = z_3^j z_1^j$$

$$z_{13}^j = (z_1^j)^2 - (z_2^j)^2$$

$$z_{14}^j = (z_2^j)^2 - (z_3^j)^2$$

$$z_{15}^j = (z_3^j)^2 - (z_1^j)^2$$

$$\mathbf{r}^j = x_1^j \mathbf{b}_1^j + x_2^j \mathbf{b}_2^j + x_3^j \mathbf{b}_3^j$$

In terms of the above-defined scalars, the matrix form of the inertia force at the nodal body G of mass dm, whose coordinates are x_1^j, x_2^j, x_3^j, in Eq. (6.21) can be written as representing the following set of 12 acceleration loadings, just as used in Chapter 5:

$$\begin{Bmatrix} f_1^{j*} \\ f_2^{j*} \\ f_3^{j*} \end{Bmatrix}_{inertia} = -dm \begin{bmatrix} I_3 & x_1^j I_3 & x_2^j I_3 & x_3^j I_3 \end{bmatrix} \begin{Bmatrix} A_1^j \\ \vdots \\ \vdots \\ A_{12}^j \end{Bmatrix} \tag{6.26}$$

where I_3 is the 3×3 unity matrix and A_i^j ($i = 1, \ldots, 12$) are acceleration terms expressed in terms of previously defined variables, with e_i as the i-th row of the unity matrix I_3:

$$\begin{Bmatrix} A_1^j \\ A_2^j \\ A_3^j \end{Bmatrix} = C_{1j}^T \{ v_u^{Qj} \dot{U} + h^j \}$$

$$A_4^j = z_4^j$$

$$\begin{Bmatrix} A_5^j \\ A_6^j \\ A_7^j \end{Bmatrix} = \begin{Bmatrix} z_5^j + e_3 C_{1j}^T (v_u^{Qj} \dot{U} + f^j) \\ z_6^j - e_2 C_{1j}^T (v_u^{Qj} \dot{U} + f^j) \\ z_5^j - e_3 C_{1j}^T (v_u^{Qj} \dot{U} + f^j) \end{Bmatrix} \tag{6.27}$$

$$A_8^j = z_7^j$$

$$\begin{Bmatrix} A_9^j \\ A_{10}^j \\ A_{11}^j \end{Bmatrix} = \begin{Bmatrix} z_8^j + e_1 C_{1j}^T (\omega_u^j \dot{U} + f^j) \\ z_6^j + e_2 C_{1j}^T (\omega_u^j \dot{U} + f^j) \\ z_8^j - e_1 C_{1j}^T (\omega_u^j \dot{U} + f^j) \end{Bmatrix}$$

$$A_{12}^j = z_9^j$$

The inertia torque on a nodal rigid body in body j, given in Eq. (6.21), can be similarly rewritten in terms of the moment of inertia components of the nodal body in frame j, as constituting the set of nine torque loadings distributed over the body:

$$\begin{Bmatrix} t_1^* \\ t_2^* \\ t_3^* \end{Bmatrix} = - \begin{bmatrix} I_{11} & I_{12} & I_{13} & I_{13} & I_{33}-I_{22} & -I_{12} & 0 & I_{23} & 0 \\ I_{12} & I_{22} & I_{23} & -I_{23} & I_{12} & I_{11}-I_{33} & 0 & 0 & I_{13} \\ I_{13} & I_{23} & I_{33} & I_{22}-I_{11} & -I_{13} & I_{23} & I_{12} & 0 & 0 \end{bmatrix} \begin{Bmatrix} A_{13}^j \\ \vdots \\ \vdots \\ A_{21}^j \end{Bmatrix}$$

$$\tag{6.28}$$

Here, for the j-th body in the multibody case, the scalars $A_i^j, (i = 13, \ldots, 21)$ are

$$\begin{Bmatrix} A_{13}^j \\ A_{14}^j \\ A_{15}^j \end{Bmatrix} = -C_{1j}^T (\omega_u^j \dot{U} + h^j) \tag{6.29}$$

$$A_{15+i}^j = z_{9+i}^j \quad (i = 1, \ldots, 6)$$

With the force and torque loadings at the nodes now defined by Eqs. (6.26) and (6.28), these provide inputs to a finite element code, to generate the corresponding geometric stiffness matrices $K_g^{(i)}, i = 1, \ldots 21$ for unit values of the time-varying quantities A_i $i = 1, \ldots, 21$. These geometric stiffness matrices are post-multiplied by the global modal matrix Φ and the result are pre-multiplied by Φ^{j^T}, and the instantaneous values of A_i give generalized active force for motion-induced geometric stiffness. The time-varying nature of A_i is input from the multibody computer program. Dealing with a "generic j-th body" for the i-th inertia loading at a time, the generalized geometric stiffness and corresponding generalized active force contributions are

$$S^{j(i)} = \Phi^{j^T} K_g^{j(i)} \Phi^j \quad i = 1, \ldots 21$$

$$F_g^j = -\sum_{i=1}^{21} A_i^j S^{j(i)} \eta^j \tag{6.30}$$

Combining contributions from the nominal and geometric stiffness for the j-th body in the i-th generalized speed, using Eqs. (6.17) and (6.30), we have the scalar form of the generalized active force due to stiffness:

$$F_i^j = -\delta_{ik} \sum_{k=1}^{M_j} \left[\Lambda_k^{j(i)} + \sum_{i=1}^{21} A_i^j S^{j(i)} \right] \eta_k^j \quad (j = 1, \ldots, NB; i = 1, \ldots, ndof) \tag{6.31}$$

If rotatory inertia of the nodal bodies can be ignored, then the second summation index, i, in Eq. (6.31) only goes up to 12. Generalized forces due to external forces and torques are now added to update F_i^j. System general active forces are obtained by summing over all N bodies j:

$$F_i = \sum_{j=1}^{NB} F_i^j \quad (i = 1, \ldots, ndof) \tag{6.32}$$

Finally, to complete the *system* equations of motion, we invoke Kane's dynamical equation [3]:

$$F_i^* + F_i = 0 \quad (i = 1, \ldots, ndof) \tag{6.33}$$

Considering the forms of Eq. (6.15) embedded in Eq. (6.10), and of Eq. (6.26) and (6.28) in Eq. (6.29), it can be shown that the dynamical equation, Eq. (6.33), takes the matrix form:

$$M\dot{U} + E = 0 \tag{6.34}$$

The coefficient matrix M in Eq. (6.34) is asymmetric because of acceleration-dependent stiffness, and is a function of the generalized coordinates, while E is a function of both generalized coordinates and generalized speeds.

6.6 TREATMENT OF PRESCRIBED MOTION AND INTERNAL FORCES

Here, we consider two types of problems that can be treated alike. In one, relative motion between two contiguous bodies may be prescribed and the interaction force is to be determined to realize this motion; in another, the non-working internal forces at a joint in a structure may be of interest. If some of the generalized coordinates are prescribed, which amounts to prescribing generalized speeds and their time derivatives, then Eq. (6.34) can be written in partitioned matrix form as follows, with a new term added to represent the generalized active force F representing the interaction forces needed to realize the prescribed motion.

$$\begin{bmatrix} M_{ff} & M_{fp} \\ M_{pf} & M_{pp} \end{bmatrix} \begin{Bmatrix} \dot{U}_f \\ \dot{U}_p \end{Bmatrix} = -\begin{Bmatrix} E_{fp} \\ E_{pf} \end{Bmatrix} - \begin{Bmatrix} 0 \\ F \end{Bmatrix} \tag{6.35}$$

Here, subscripts f and p denote free and prescribed variables. The solution is

$$M_{ff}\dot{U}_f = -M_{ff}\dot{U}_p - E_{fp}$$

$$F = -\begin{bmatrix} M_{pf} & M_{pp} \end{bmatrix} \begin{Bmatrix} \dot{U}_f \\ \dot{U}_p \end{Bmatrix} - E_{pf} \tag{6.36}$$

For our choice of generalized speeds, which are time derivatives of relative angles and distances, see Eq. (6.1), the elements of the column matrices F are directly the forces and torques that do work over the prescribed motion. In a problem like antenna shape control prescribing the elastic displacement, $F = \Phi^t f$, the physical force f being the column matrix of the required actuator force distribution at the corresponding finite element nodes, can be determined in the least-square sense by using a pseudo-inverse.

If the internal force at a hinge is of interest, the hinge is prescribed extra degrees of freedom and corresponding interaction forces and torques are exposed such that these forces and torques do work when motions involving those degrees of freedom take place. This prescribed motion is subsequently set equal to zero. The interaction forces are then obtained from Eq. (6.36).

6.7 "RUTHLESS LINEARIZATION" FOR VERY SLOWLY MOVING ARTICULATING FLEXIBLE STRUCTURES

When the rotation rate or translation rate is an order of magnitude less than the first natural frequency of a system, then certain drastic simplifications can be made that make the computer simulations run faster. This is the case of flexible robotic manipulators that go through a large angle change between component members, but at a relatively very slow rotation rate compared to the first modal frequency of the system. In these cases, all geometric stiffness terms due to motion-induced inertia loads can be ignored. What is more, terms in Eq. (6.11) involving modal coordinates in the "mass matrix", products of the modal coordinates and their rates, and products of the rigid body rotation/translation rates with the modal coordinates or their rates are also neglected. This is the approximation for slow motion used by Sincarsin and Hughes [8], and employed for the simulation of the Canada Arm space manipulator. Such a process was called "ruthless linearization" by Padilla and von Flotow [9]. Later, Sharf and Damaren [10] showed that retaining terms containing products of rigid body rotation rates and derivatives of modal coordinates to the ruthlessly linearized equations removed the deficiency in many cases, and by keeping modal coordinate–dependent terms in the mass matrix produced accurate results for a wide range of maneuvers and joint rotational speeds. Ghosh [11] implemented online control with the dynamics of a ruthlessly linearized model for a very slowly rotating system with articulating appendages, and showed that simulation results match well with measurements. It is noteworthy, however, that geometric stiffness effects due to interbody forces on a component may not be ignorable for a very slowly moving, articulated system, and they should be incorporated.

6.8 SIMULATION RESULTS

Having established the effects of including or excluding geometric stiffness on the spin-up or translational acceleration response of a composite truss-type structure in Chapter 5, we present here some new results on the slewing maneuver of an articulated system of flexible bodies. Figure 6.3 shows a space crane consisting of a rigid body with two hinge-connected articulated trusses. We assume that the rigid body undergoes a simple rotation through an angle q_1 in inertial space about the axis shown, while the trusses rotate through the relative angles q_2 and q_3 as shown. In a slewing maneuver, the commanded values of these angles are q_{1c}, q_{2c}, and q_{3c}, respectively, and the following joint torques are applied at the hinges to realize the desired step commands:

$$T_i = -k_i(q_{ic} - q_i) - c_i\dot{q}_i \quad (i = 1, 2, 3) \tag{6.37}$$

The control gains in Eq. (6.37) are chosen for the worst case, or the largest overall outlying body inertia to turn at the respective revolute joints, for a bandwidth of 0.1 Hz and a damping factor of 0.707. This controller bandwidth is below the first natural frequency of the combined system in the initial configuration. Each truss is 12 m

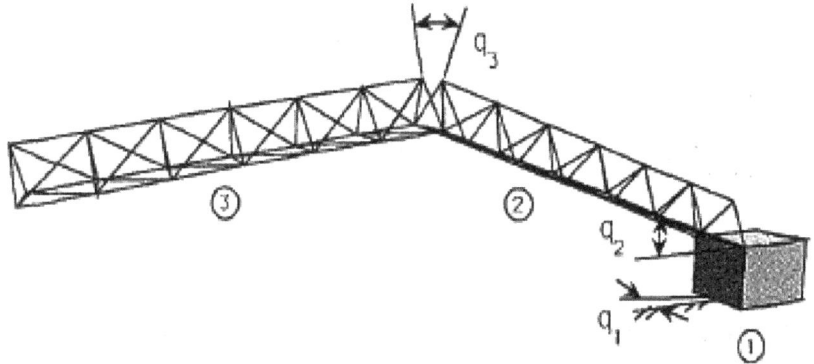

FIGURE 6.3 A space crane of two hinged trusses with three rigid body degrees of freedom in a commanded slewing maneuver.

long and consists of tubular members interconnected as shown at 21 joints; the cross section of the truss is a right isosceles triangle, the members at the right angle being 2 m long. Young's modulus for all members is 10^9 N/m^2, the cross-sectional area of each member is 8.46×10^{-4}, and each member has a mass density 10^4 kg/m^3. The rigid base has a mass of 10^4 kg, and a rotation-axis moment of inertia of 1200 kg/m^2. While both trusses are essentially the same, a change is made to the inner truss by assigning a lumped stiffness at the free end to model a possible gimbal fixture that attaches to all three nodes of the truss tip, so that all three nodes are in one plane. The system is modeled by the finite element code [7], and the modal representation for the inner truss includes the first 10 fixed-fixed vibration modes and six constraint modes [4] representing unit displacements for the end planes of the truss, as per the Craig–Bampton component modes. With 3 rigid body degrees of freedom and 10 modes for the outboard truss, the system in Figure 6.3 has 29 degrees of freedom.

Figure 6.4 shows the elastic rotation due to bending at the free end of the second truss, body number 3, during the slewing maneuver. Figure 6.5 displays the angle of twist at the end of the inner truss at the elbow. Plots such as these can obviously be used to assess control–structure interaction corresponding to various control law designs for meeting a specified control objective. Finally, an animation of the motion of the crane during the slewing maneuver is shown in Figure 6.6, where the structure, initially in the plane of the maneuver, comes out of the plane during the maneuver. Before leaving this chapter, we note that flexible multibody dynamics has been the subject of several public domain software, notably [12] and [13], that failed to include geometric stiffness and hence are of limited utility. These software would thus predict dynamic softening when stiffening is physically expected, and vice versa.

6.9 CONCLUSION

A formulation for the dynamics of a system of hinge-connected flexible bodies is given, with a dense mass matrix M in the equations of motion in the matrix form:

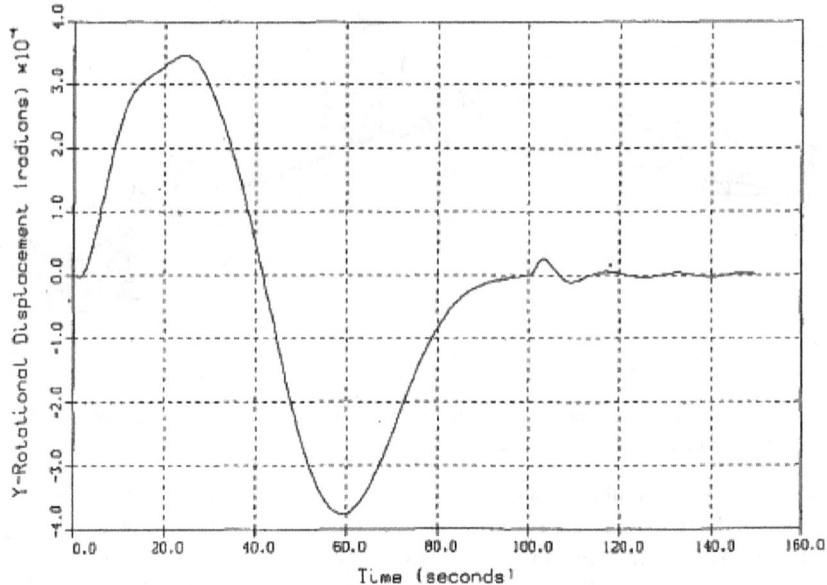

FIGURE 6.4   Elastic rotation due to bending at the tip of the end truss during controlled slewing.

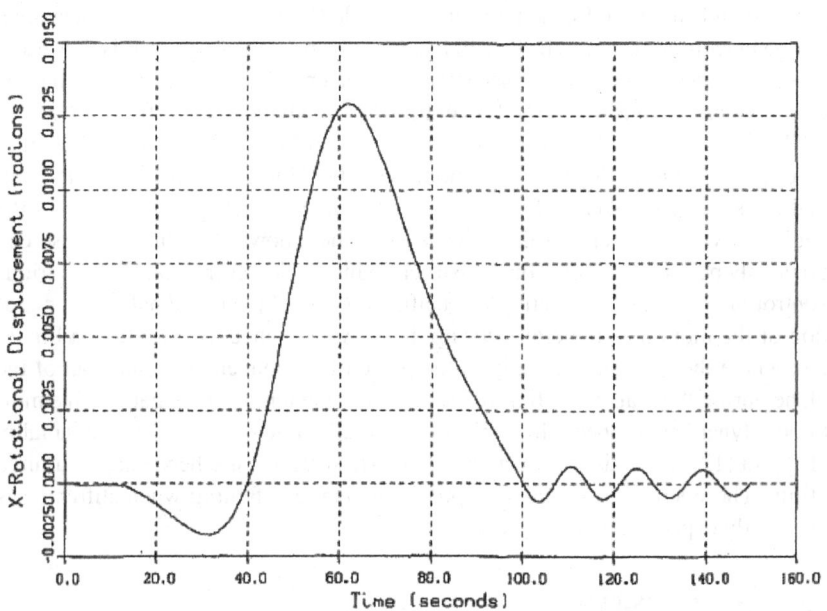

FIGURE 6.5   Angle of twist at the end of the inner truss during the slewing maneuver.

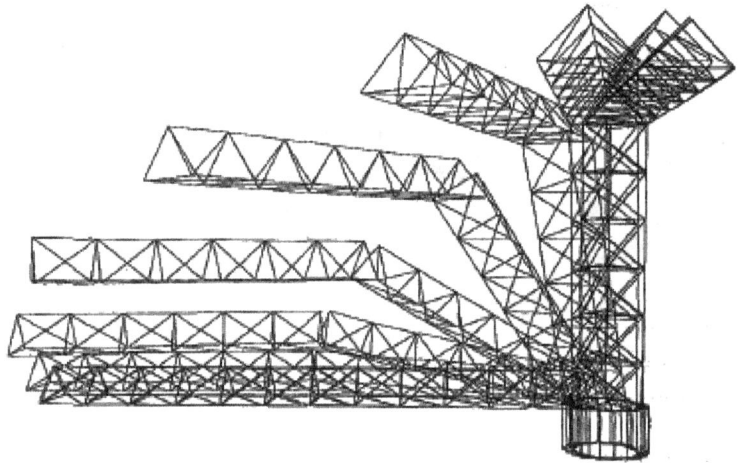

FIGURE 6.6 Animation of the slewing maneuver of the space crane initially in the plane of the maneuver, coming out of the plane during the maneuver.

[M(q)] {U} = {C(q, U)}. The mass matrix M is not symmetric because of the consideration of geometric stiffness due to inertia forces. The formulation spells out the components of the generalized inertia force expression in terms of the modal integrals of a flexible body. Some of the special features of the formulation include kinematic recurrence relation pertaining to a body and its inboard body, treatment of prescribed motion, and allowable simplifications for very slowly moving structures. While several complex problems are solved showing physically meaningful results, there is no reason to believe that further improvements on the formulation are out of the question. For example, consideration of geometric stiffness effects due to interbody forces, and more significantly, a matrix formulation that produces a mass matrix that is not dense, but block-diagonal in nature (with sub-matrices corresponding to component bodies in the system), are highly desirable. These features are taken up later in Chapter 8.

PROBLEM SET 6

1. Derive the multibody equations for two initially co-linear and overlying beams connected by massive revolute joints, with the system tumbling at a high altitude under gravity, and an initial small angle, 5 degrees, between the beams increasing to 90 degrees, due to aerodynamic force. Use free-free bending modes for one, and free-hinged modes for the other [14].

   Free-free:

$$\phi_i^{(1)}(x) = \cosh\left(\frac{\lambda_i x}{L}\right) + \cos\left(\frac{\lambda_i x}{L}\right) - \sigma_i\left[\sinh\left(\frac{\lambda_i x}{L}\right) + \sin\left(\frac{\lambda_i x}{L}\right)\right]$$

$\sigma_1 = 0.9825; \lambda_1 = 4.7300; \sigma_2 = 1.0008; \lambda_2 = 7.8352; \sigma_3 = 0.9999; \lambda_3 = 10.9956$

FIGURE 6.7   A six degrees of freedom robotic manipulator.

$$\phi_i^{(1)}(x) = \cosh\left(\frac{\lambda_i x}{L}\right) + \cos\left(\frac{\lambda_i x}{L}\right) - \sigma_i\left[\sinh\left(\frac{\lambda_i x}{L}\right) + \sin\left(\frac{\lambda_i x}{L}\right)\right]$$

$\sigma_1 = 1.0008;\ \lambda_1 = 3.9266;\ \sigma_2 = 1.0000;\ \lambda_2 = 7.0685;\ \sigma_3 = 1.0000;\ \lambda_3 = 10.210256$

2. Code the equations and use numerical values of your choice for a simulation of a uniform pressure load on both beams, and plot the free end deflections from 5 degrees to 90 degrees for the two beams.

3. Figure 6.7 shows the space station Canada Arm manipulator [11].

Do this exercise departing from the details of the picture in Figure 6.7. Assume that one end of the manipulator is attached to a heavy object on the ground. The first link is short and rotates in "roll" with respect to the attaching base; the second link is also short and rotates in "yaw" with respect to its inboard body. These two links, of the same length, can be modeled as rigid. Links 3 and 4 are both flexible beams of the same length, and they both rotate in "pitch". Link 5 is short and the same length as link 1, modeled as rigid, and rotates in yaw. Finally, link 6 is a long beam of the same length as link 3; it rotates in roll, and is attached to a rigid body called the "end-effector". Derive the equations of motion of this 6 dof robotic manipulator, ignoring all geometric stiffness terms. Use parameter values of your choice to simulate the motion, keeping one mode per beam.

REFERENCES

1. Banerjee, A.K. and Lemak, M.E., (1991), "Multi-Flexible Body Dynamics Capturing Motion-Induced Stiffness", *Journal of Applied Mechanics*, **58**(September), pp. 766–775.

2. Huston, R.L. and Passerello, C.E., (1980), "Multibody Structural Dynamics Including Translation Between Bodies", *Computers and Structures*, **12**, pp. 713–720.
3. Kane, T.R. and Levinson. D.A., (1985), *Dynamics: Theory and Application*, McGraw-Hill.
4. Craig, R.R., Jr., (1981), "Structural Dynamics", in *An Introduction to Computer Methods*, Wiley.
5. Cook, R.D., (1974), *Concepts and Applications of Finite Element Analysis*, Wiley.
6. Zienkiewicz, O.C., (1977), *The Finite Element Method*, McGraw-Hill.
7. Whetstone, W.D., (1983), *EISI-EAL, Engineering Analysis Language Reference Manual*, Engineering Information Systems, Inc.
8. Sincarsin, G.B. and Hughes, P.C., (1989), "Dynamics of an Elastic Multibody Chain: Part A—Body Motion Equations", *Dynamics and Stability of Systems*, **3**, pp. 209–226.
9. Padilla, C.E. and von Flotow, A.H., (1992), "Nonlinear Strain-Displacement Relations and Flexible Multibody Dynamics", *Journal of Guidance, Control, and Dynamics*, **15**(1), pp. 128–136.
10. Sharf, I. and Damaren, C., (May, 1992), "Simulation of Flexible-Link Manipulators: Basis Functions and Nonlinear Terms in the Motion Equations", in Proceedings of the 1992 IEEE International Conference on Robotics and Automation, Nice, France.
11. Ghosh, T.K., (2013), *Correspondence on Simulation of Space Robotic Manipulators at the NASA Johnson Space Center* [private communication of an unpublished Memo]
12. Bodley, C.S., Devers, A.D., Park, A.C., and Frisch, H.P., (1978), *A Digital Computer Program for the Dynamic Interaction Simulation of Controls and Structures (DISCOS)*, vols. I and II, NASA TP-1219.
13. Singh, R.P., van der Voort, R.J., and Likins, P.W., (1985), "Dynamics of Flexible Bodies in Tree Topology: A Computer Oriented Approach", *Journal of Guidance, Control, and Dynamics*, **8**(5), pp. 584–590.
14. Blevins, R.D., (2001), *Formulas for Natural Frequency and Mode Shape*, Krieger Publishing.

7 Component Mode Selection and Model Reduction
A Review

In this chapter, we review the selection of modes and model reduction for a component of a flexible multibody system, to represent the deformations of the component with good fidelity, using as few modes as possible. We must note that for a multibody system with hinge degrees of freedom (dof) locked or driven, vibration modes of the combined *system* change with the relative change in configuration between components, and thus we can only speak of the modes for a *component*. While a high number of modes always model deformations more accurately, it also makes the system model more expensive in simulation or for use in on-line control. Model reduction by modal truncation (MT) is thus a necessary step for time-efficient simulation. This, of course, produces errors in representing deformation, and we will show a way of compensating for errors due to modal truncation. Our main reference for component mode selection is the book by Craig [1], which is complemented by the book by Wijker [2]. We will first discuss two widely used methods of model representation, namely, the Craig–Bampton method and the method of static condensation. In model reduction, we will discuss Guyan reduction [3] and modal effective mass methods. Model reduction by frequency filtering [4] is described next along with component mode selection for multibody systems [5]. Modal truncation vectors (MTV) will be presented [6, 7] as a "catch-all" tool of compensation for modal truncation. Next, we will review a synthesis of component modes for a multibody system. Further model reduction, after a system model is synthesized, will be discussed in terms of singular value decomposition (SVD), which we will illustrate with an example.

7.1 CRAIG–BAMPTON COMPONENT MODES FOR CONSTRAINED FLEXIBLE BODIES

A physical component is typically modeled with many nodes, each node having 6 physical degrees of freedom. Using the notation u for elastic displacement, *not generalized speeds*, as in previous chapters, the n physical degrees of freedom of a component can always be divided into a set of *interior dof*, u_i, and a set of juncture or *boundary dof*, u_b, when the component is part of a multibody system. The equations of free vibration of a component are written

DOI: 10.1201/9781003231523-8

in terms of its mass and stiffness matrices obtained from a finite element model, in the form:

$$\begin{bmatrix} m_{ii} & m_{ib} \\ m_{ib}^t & m_{bb} \end{bmatrix} \begin{Bmatrix} \ddot{u}_i \\ \ddot{u}_b \end{Bmatrix} + \begin{bmatrix} k_{ii} & k_{ib} \\ k_{ib}^t & k_{bb} \end{bmatrix} \begin{Bmatrix} u_i \\ u_b \end{Bmatrix} = \begin{Bmatrix} 0 \\ 0 \end{Bmatrix} \quad \text{or} \quad M\,\ddot{u} + K\,u = 0 \quad (7.1)$$

These n physical dof u are reduced by a *Rayleigh–Ritz approximation* by m number of modes as

$$u = \psi\, q \qquad\qquad (7.2)$$

where ψ is an (n × m) matrix of m "component modes", and q is an (m × 1) matrix of "modal coordinates". Component modes can be of various kinds, such as fixed-fixed or fixed-free vibration modes, and the so-called "constraint modes" and "attachment modes", described below [1]. First, however, we review the definition of *vibration modes* of the component represented by Eq. (7.1) in terms of an *eigenvalue* problem for the form set up from Eq. (7.1) in the following operational sequence of steps:

$$u = \varphi\eta$$

$$M\varphi\ddot{\eta} + K\varphi\eta = 0$$

$$\varphi^t M\varphi\ddot{\eta} + \varphi^t K\varphi\eta = 0$$

$$\text{Let } \hat{M} = \varphi^t M\varphi; \quad \hat{K} = \varphi^t K\varphi \qquad\qquad (7.3a)$$

$$\hat{M}\ddot{\eta} + \hat{K}\eta = 0$$

$$\text{Let } \eta = \hat{\varphi}\sin\omega t$$

$$[\hat{K} - \omega^2\,\hat{M}]\hat{\varphi}\sin\omega t = 0$$

The last statement defines the following algebraic eigenvalue problem:

$$\left[\hat{K} - \omega^2\hat{M}\right]\hat{\varphi} = 0 \qquad\qquad (7.3b)$$

The eigenvalues are the squares of the natural frequencies, ω_i^2, $i = 1,\ldots,n$, and are obtained from the above by solving the determinant equation:

$$\det(\hat{K} - \omega^2\hat{M}) = 0 \qquad\qquad (7.3c)$$

for non-zero values of the modes; the modes are then normalized, or scaled, to the element masses m_i, $i = 1,\ldots, n$, as follows:

$$\hat{\varphi}^t M\hat{\varphi} = \hat{M} = \text{diag}[m_1, m_2, \ldots., m_n] \quad ; \quad \hat{\varphi}^t K\hat{\varphi} = \hat{K} = \text{diag}[\omega_1^2, \omega_2^2, \ldots., \omega_n^2] \quad (7.4)$$

Usually, one keeps only a few number, k, of the modes. In most flexible multibody dynamics representations, component modes consist of a mixture of fixed-interface vibration modes together with a set of *constraint modes*. The latter, also called in the industry *boundary node functions*, are defined as follows [1]:

Let the elastic degrees of freedom u of the component be partitioned into a set B, usually the boundary degrees of freedom, in terms of which constraint modes are to be defined, and let I be the *complement* of the set B, usually the interior dof. A constraint mode or boundary node function is defined as the static deformation shape that results from imposing unit displacement on one boundary coordinate of the B set at a time, while holding the remaining coordinates of the B set fixed to zero displacement. From this definition, the set of constraint modes satisfies the matrix equation:

$$\begin{bmatrix} k_{ii} & k_{ib} \\ k_{ib}^t & k_{bb} \end{bmatrix} \begin{bmatrix} \Psi_{ib} \\ I_{bb} \end{bmatrix} = \begin{bmatrix} 0_{ib} \\ R_{bb} \end{bmatrix} \tag{7.5}$$

Here, I_{bb} is a b ×b unity matrix, where b is the number of elements in the set B, and R_{bb} is the set of reaction forces, at the dof of the B set, brought into play by prescribing unit displacements in the dofs of the B set, in the manner described above. Solving from the top row:

$$\Psi_{ib} = - k_{ii}^{-1} k_{ib} \tag{7.6}$$

The set of "constraint modes" is then completely defined as

$$\Psi_c = \begin{bmatrix} \Psi_{ib} \\ I_{bb} \end{bmatrix} = \begin{bmatrix} - k_{ii}^{-1} k_{ib} \\ I_{bb} \end{bmatrix} \tag{7.7}$$

Most flexible multibody dynamics component mode representations use a set of normal modes augmented by a set of constraint modes. When the component is an intermediate body between two outlying bodies, the normal modes are for the boundary condition with both ends fixed, and the constraint modes use six unit displacements for *each end*. For a terminal body, cantilever modes for interior points, with respect to the attaching body are augmented by constrained modes generated by six constraint modes from the attaching body. Craig–Bampton modes for the component can now be formed by augmenting a truncated set of fixed interface normal vibration modes with the constraint modes, or boundary node functions, associated with unit boundary displacements. Thus, the physical degrees of freedom of the component, described in this manner, are called *Craig–Bampton modes*, and are represented as

$$\begin{Bmatrix} u_i \\ u_b \end{Bmatrix} = \begin{bmatrix} \varphi_k & -k_{ii}^{-1}k_{ib} \\ 0 & I_{bb} \end{bmatrix} \begin{Bmatrix} \eta_k \\ u_b \end{Bmatrix} \tag{7.8}$$

An alternative to the use of the Craig–Bampton modes is *attachment modes*, which are described in Ref. [1]. Two other methods, component modes by Guyan reduction and modal effective mass, are discussed next. Most major aerospace organizations, using flexible multibody dynamics codes, choose the Craig–Bampton component modes, because of the inherent accounting of component boundary conditions.

7.2   COMPONENT MODES BY GUYAN REDUCTION

Guyan reduction [3], also known as *static condensation*, is a method of obtaining component modes when the inertia effects of certain degrees of freedom can be ignored. Consider a model with a given loading:

$$\begin{bmatrix} m_{aa} & m_{ae} \\ m_{ea} & m_{ee} \end{bmatrix} \begin{Bmatrix} \ddot{u}_a \\ \ddot{u}_e \end{Bmatrix} + \begin{bmatrix} k_{aa} & k_{ae} \\ k_{ea} & k_{ee} \end{bmatrix} \begin{Bmatrix} u_a \\ u_e \end{Bmatrix} = \begin{Bmatrix} F_a \\ F_e \end{Bmatrix} \tag{7.9}$$

Here, subscript a refers to "active" dof, and subscript e refers to the to-be-eliminated degrees of freedom that may be assumed as inactive, motion-wise, in that they do not respond significantly to the loading. The choice of u_a, u_e is then solely based on the assumption that the inertia loads $m_{aa}\ddot{u}_a$ are significantly larger than the other inertia loads, and the load F_e in Eq. (7.9) is small relative to F_a. The implied approximation to Eq. (7.9) is then

$$\begin{bmatrix} m_{aa} & m_{ae} \\ 0 & 0 \end{bmatrix} \begin{Bmatrix} \ddot{u}_a \\ \ddot{u}_e \end{Bmatrix} + \begin{bmatrix} k_{aa} & k_{ae} \\ k_{ea} & k_{ee} \end{bmatrix} \begin{Bmatrix} u_a \\ u_e \end{Bmatrix} = \begin{Bmatrix} F_a \\ 0 \end{Bmatrix} \tag{7.10}$$

From the bottom row, we have

$$u_e = -k_{ee}^{-1} k_{ea} u_a \tag{7.11}$$

This yields the reduction transformation for the overall degrees of freedom:

$$\begin{Bmatrix} u_a \\ u_e \end{Bmatrix} = \begin{bmatrix} I \\ -k_{ee}^{-1} k_{ea} \end{bmatrix} \{ u_a \} \equiv T u_a \tag{7.12}$$

Substituting Eq. (7.12) in the force-free vibration equation corresponding to Eq. (7.9), written as $M\ddot{u} + Ku = 0$, and pre-multiplying by T^T, where the transformation matrix T is defined in Eq. (7.12), yields

$$T^t M T \ddot{u}_a + T^t K T u_a = 0 \tag{7.13}$$

One can compute eigenvalues and eigenvectors corresponding to Eq. (7.13) with

$$u_a = \varphi_a \sin \omega t$$

$$[T^T M T - \omega_a^2 T^T K T] \varphi_a = 0 \tag{7.14}$$

Here, ω_a, ϕ_a are, respectively, the natural frequency and the mode shape for the *active dof*. From Eqs. (7.12) and (7.14) then, we have the component mode shapes from static condensation:

$$\begin{bmatrix} \varphi_a \\ \varphi_e \end{bmatrix} = \begin{bmatrix} I \\ -k_{ee}^{-1}k_{ea} \end{bmatrix} \begin{bmatrix} \varphi_a \end{bmatrix} \tag{7.15}$$

or in more compact notation for all degrees of freedom with $u=\psi q$:

$$\psi = T\varphi_a \tag{7.16}$$

7.3 MODAL EFFECTIVE MASS

Modal effective mass [2] is a measure to assign importance to individual modes from the viewpoint of the excitability of structures with base acceleration. A mode with a high modal effective mass corresponds to a high transmitted force at the base. This then becomes a basis for reducing the number of modes. Any component structure can be thought of as having two sets of dof, a set of boundary dof and the internal elastic dof. For the purpose of computing the modal effective mass, the boundary dofs are given rigid body motion, one at a time for a maximum of six in number, and internal elastic deformations are obtained by keeping all other juncture dofs set to zero. With u_j, u_e denoting juncture and elastic dof:

$$u = \varphi_r u_j + \varphi_e u_e \tag{7.17}$$

Note that the dimension of each of ϕ_r, ϕ_e is $n \times 1$.

For rigid body modes:

$$\varphi_r^t M \varphi_r = M_{rr} \tag{7.18}$$

Here, M is the $n \times n$ *nodal* mass matrix for the component, and M_{rr} is the 6×6 rigid body matrix, of mass m, mass moment s, a 3×1 matrix (with $\tilde{s}$ representing a skew-symmetric matrix formed out of its three components), and the mass moment of inertia I, a 3×3 matrix, both about the boundary dof:

$$M_{rr} = \begin{bmatrix} mU & -\tilde{s} \\ \tilde{s} & I \end{bmatrix} \tag{7.19}$$

On the other hand, if the k-th elastic mode is *orthogonalized by mass*, then

$$\varphi_k^t M \varphi_k = m_k \tag{7.20}$$

Now define the 1×6 row matrix, L_k, the rigid-elastic mode coupling vector, for the k-th mode ϕ_k and any one of the six rigid body modes, see Eq. (7.17), taken one at a time:

$$L_k = \varphi_k^t M \varphi_r \qquad (7.21)$$

The modal effective mass for the k-th elastic mode is then defined as the following 6×6 matrix:

$$M_{eff,k} = \frac{L_k^t L_k}{m_k} \qquad (7.22)$$

The goal of model reduction, keeping p number of modes, with $p < n$, is that the 6×6 matrix:

$$M_{rr} \cong \sum_{k=1}^{p} M_{eff,k} \qquad (7.23)$$

This is true exactly if p is the number n of *all* possible modes, because from Eqs. (7.20)–(7.22):

$$M_{eff} = [\varphi_r^t M \varphi_k][\varphi_k^t M \varphi_k]^{-1}[\varphi_k^t M \varphi_r] = [\varphi_r^t M \varphi_k][\varphi_k^{-1} M^{-1} \varphi_k^{-t} \varphi_k^t M \varphi_r = \varphi_r^t M \varphi_r = M_{rr}$$

$$(7.24)$$

A mode with a high modal effective mass is strongly excited by the base motion, giving rise to a high reaction force at the base. This is because, if $\ddot{u}_b$ represents a 6×1 column matrix of unit base acceleration in the 6 dof base motion, then $(-M_{rr}\ddot{u}_b)$ represents the reaction force transmitted to the base providing the base acceleration. Thus, any mode that contributes strongly to the sum in Eq. (7.23) contributes strongly to the transmitted force, and has to be kept in a reduced order model, while a mode with low modal effective mass can be eliminated for a multibody system.

7.4 COMPONENT MODEL REDUCTION BY FREQUENCY FILTERING

Another technique of model reduction [4] is based on the highest frequency of a component that is consistent with a desirable time step for stable numerical integration. As a rule of thumb, 10 integration steps are used to track a harmonic of period T. With $T = 2\pi/\omega$, this allows a highest frequency of vibration for a period of numerical integration step-size, Δt, as

$$\omega_{max} = (2\pi / \Delta t)/10 \qquad (7.25)$$

For a forced vibration model described by the sequence of steps:

$$[M]\{\ddot{u}\} + [K]\{u\} = [F]\{g(t)\}$$

$$\{u\} = [\Phi]\{\eta\}$$

$$[\Phi]^t[M][\Phi]\{\ddot{\eta}\} + [\Phi]^t[K][\Phi]\{\eta\} = [\Phi]^t[F]\{g(t)\}$$

$$\{\ddot{\eta}\} + [\Omega^2]\{\eta\} = [\Phi]^t[F]\{g(t)\}$$

(7.26)

the i-th component mode equation follows, after keeping n modes and adding viscous damping, as

$$\ddot{\eta}_i + 2\xi_i\omega_i\dot{\eta}_i + \omega_i^2\eta_i = \varphi_i^t[F]\{g(t)\} \quad i = 1,\ldots,n \qquad (7.27)$$

If $\omega_i \leq \omega_{max}$, then Eq. (7.27) is used. If $\omega_i \geq \omega_{max}$, then the frequency filtering technique consists in using, instead of Eq. (7.27), the following:

$$\ddot{\eta}_i + 2\xi_i\omega_{max}\dot{\eta}_i + \omega_{max}^2\eta_i = \left(\frac{\omega_{max}^2}{\omega_i^2}\right)\varphi_i^t[F]\{g(t)\} \quad i = 1,\ldots,n \qquad (7.28)$$

This method has been found satisfactory in the Canada Arm Manipulator dynamics simulation.

7.5 COMPENSATION FOR ERRORS DUE TO MODEL REDUCTION BY MODAL TRUNCATION VECTORS

Use of the model reduction given by the Craig–Bampton modes, or the attachment modes, or static condensation of the component model equations, or the modal effective mass method, requires the user to decide how many vibration modes to keep. Modal truncation causes error due to incomplete representation of a deformation, and a scheme for some compensation for modes not kept was devised by Dickens [5]. To see how the method works, consider the basic problem of a small vibration under space- and time-dependent loading:

$$M\ddot{u} + Ku = F(x,y,z)\,g(t) \qquad (7.29)$$

Here, u refers to the internal dof of a component body, with its boundary dof locked; when the body is a terminal body only its attachment dof are locked. In Eq. (7.29), F(x,y,z) is a matrix representing the *spatial distribution of the load* and g(t) is a column matrix of time-functions, if various special distributions are considered. In terms of component modes ψ, and a modal coordinate ξ, the displacement is represented as

$$u = \psi\xi \qquad (7.30)$$

Recall that the associated eigenvalue problem for r number of eigenvalues and eigenvectors is given by the following, where the second equality is due to the diagonal nature of Ω^2:

$$K\psi = \Omega^2 M\psi = M\psi\Omega^2; \quad \psi = \left[\psi_1 \cdots \psi_r\right]; \quad \Omega^2 = \mathrm{diag}[\omega_1^2,\ldots,\omega_r^2] \quad (7.31)$$

Now, customarily one keeps component frequencies up to twice that of the system frequency cut-off of interest. Model reduction brings in error for not keeping all the modes. Dickens [5] devised a scheme for dynamic compensation for the modes that are not kept in the model. The scheme uses a deflection shape to represent deformation due to the spatial part of the residual load that is not represented by the modes that are kept. The idea is that the spatial distribution of the load should be representable, in the sense of a Fourier series, by a sum of modes times the modal participation factors of those modes. The spatial distribution of the force that *can* be represented by the modes retained is given by the following sequence of computational steps.

Spatial load representable by the deformation Eq. (7.30) is given from statics by

$$F = K\psi\xi \qquad (7.32)$$

From the eigenvalue solver, Eq. (7.31), Eq. (7.32) becomes

$$F = M\psi\Omega^2\xi \qquad (7.33)$$

Pre-multiplying by ψ^t:

$$\psi^T F = \psi^T M\psi\Omega^2\xi \qquad (7.34)$$

Assuming that the modes are normalized (scaled) by the mass of the structure, $\psi^t M\psi = mU$, where m is the total mass of the structure and U is a unity matrix of size given by the number of modes kept, one obtains from Eq. (7.34):

$$\psi^T F = m\Omega^2\xi \qquad (7.35)$$

Equation (7.35) leads to the modal participation factor in the spatial decomposition of load F:

$$\xi = \left[m\Omega^2\right]^{-1}\psi^T F \qquad (7.36)$$

Therefore, the spatial load represented by the modes kept follows from Eqs. (7.33) and (7.36), with subscript k indicating the contribution of the modes kept:

$$F_k = M\psi\Omega^2\left[m\Omega^2\right]^{-1}\psi^t F = M\psi m^{-1}\psi^t F \qquad (7.37)$$

Hence, the spatial distribution of the load F that *cannot* be represented by the number of modes kept, and the corresponding *static deflection* are given from Eq. (7.37), in sequence, by

$$F_{res} = [I - M\psi m^{-1}\psi^T]F$$

$$Ku_{res} = F_{res}$$

(7.38)

Dickens [6, 7] calls this displacement vector, u_{res}, the modal truncation vector associated with the spatial distribution load F. Large overall motion of flexible bodies [9], Eq. (5.16) in Chapter 5, reveals that one flexible body is subjected to a set of 12 distributed inertia loads from just the rigid body or frame rotation/translation, that is, F, u_{res} are each matrices with 12 columns. Now the spatial distribution of the inertia loads on a component undergoing large motion of the reference frame, identified in Eq. (5.16) of Chapter 5, is specified as

$$\begin{Bmatrix} f_1^* \\ f_2^* \\ f_3^* \end{Bmatrix} = -dm\begin{bmatrix} U_3 & x_1U_3 & x_2U_3 & x_3U_3 \end{bmatrix}\begin{Bmatrix} A_1 \\ \vdots \\ \vdots \\ A_{12} \end{Bmatrix}$$

(7.39)

Here, U_3 is the 3 × 3 unity matrix and $A_1,..., A_{12}$ are the time-varying components of the acceleration at a node of mass dm, located at the point with coordinates x_1, x_2, x_3 from the origin of the body coordinates. The effects of all external and working internal forces for the system are equivalently represented, to a "zeroth" order, by the rigid body inertia forces. (Note that here we have neglected rotational inertia effects which would have changed the total number of loads from 12 to 21, described in Chapter 5.) This requires the generation of 12 modal truncation vectors, and we want these vectors to be orthogonal between themselves and to the normal vibration modes. This is achieved by the following sequence of steps, including a construction of a 12 × 12 eigenvalue problem, see Eq. (7.3a), for the modal truncation vectors.

$$M^* = u_{res}^T M u_{res}$$

$$K^* = u_{res}^T K u_{res}$$

$$K^*\Phi = M^*\Phi\Omega_{MTV}^2$$

$$\Psi_{MTV} = u_{res}\Phi$$

(7.40)

Here, Ψ_{MTV}, Ω_{MTV} are called, respectively, the modal truncation vectors and the "frequencies" associated with these modal truncation vectors. Note that M and K are the original finite element model mass and stiffness matrices, and the dimensions of u_{RES}, Φ are ndof × 12 and 12 × 12, respectively.

Now the modal basis is extended by adding the modal truncation vectors to the normal vibration modes to get a more complete modal basis that compensates for the deleted modes, in two sets of column matrices:

$$[\Psi] = \begin{bmatrix} \psi & \psi_{MTV} \end{bmatrix} \tag{7.41}$$

and this modal basis $[\Psi]$ is used to apply the mode-*displacement* solution. Note that these very loads, as shown in Eq. (7.40), are also used in computing load-dependent geometric stiffness required to counteract the effect of premature linearization endemic with the use of modes, as shown in Chapter 3. Thus, a typical component model should ideally consist of a chosen number of vibration modes plus the modal truncation vectors corresponding to the spatial loads represented by Eq. (7.39). Of course, in actual practice, depending on the strength of the acceleration terms in $A_1, \ldots, A_{12}$, one can reduce the total number of modal truncation vectors.

7.6 ROLE OF MODAL TRUNCATION VECTORS IN RESPONSE ANALYSIS

Dickens et al. have shown in Refs. [6, 7] the superiority in frequency response, over a wide range of frequencies, of representing the elastic deformation by a set of vibration modes and a set of modal truncation vectors for the loads involved, as in Eq. (7.41), over the modal acceleration (MA) method. This is significant because the mode acceleration method gives a response that converges as $(1 / \omega_i^2)$ with i-th frequency modes [1, 2], with higher modes contributing less, as shown in the exposition below.

7.6.1 MODE ACCELERATION METHOD

$$M\ddot{u} + Ku = F(x,y,z)g(t)$$

$$u = K^{-1}[F(x,y,z)g(t) - M\ddot{u}]$$

$$= K^{-1}F(x,y,z)g(t) - K^{-1}\sum_{i=1}^{N} M\varphi_i\ddot{\eta}_i \tag{7.42}$$

$$= K^{-1}F(x,y,z)g(t) - \sum_{i=1}^{N} \frac{1}{\omega_i^2}\varphi_i\ddot{\eta}_i$$

where $\ddot{\eta}_i + \omega_i^2\eta_i = \varphi_i^t F(x,y,z)g(t)$

Note that the mode-displacement solution to the forced vibration problem was used in the summation, and the definition of the eigenvalues, Eq. (7.3b), was used in the last step.

Dickens et al. [7] have shown the superiority in the frequency response analysis of systems with modes augmented by modal truncation vectors, Eq. (7.41), over the mode acceleration method by the following example:

$$M = \begin{bmatrix} 1.0 & 0 & 0 & 0 \\ 0 & 1.0 & 0 & 0 \\ 0 & 0 & 1.0 & 0 \\ 0 & 0 & 0 & 0.5 \end{bmatrix}$$

$$K = 10000.0 \begin{bmatrix} 2.0 & -1.0 & 0 & 0 \\ -1.0 & 2.0 & -1.0 & 0 \\ 0 & -1.0 & 2.0 & -1.0 \\ 0 & 0 & -1.0 & 2.0 \end{bmatrix}$$

$$\psi_1 = \begin{Bmatrix} 0.39948 \\ 0.63631 \\ 0.61408 \\ 0.34183 \end{Bmatrix} \qquad \psi_{MTV} = \begin{Bmatrix} -0.53176 \\ -0.41943 \\ 0.71885 \\ 0.22164 \end{Bmatrix} \qquad R(t) = \begin{Bmatrix} 0.0 \\ 0.0 \\ 1.0 \\ 0.0 \end{Bmatrix} r(t)$$

The frequency response of the system, by keeping all modes, the modal acceleration method, and the modal truncation vector method, is shown below.

It is clear that the method of augmenting modes kept by the modal truncation vector gives a far better frequency response than the mode acceleration method. For a time-response comparison, we consider a force applied on mass #3, R(t), as shown in Figure 7.1, with the following time-dependent function that represents almost an impulse loading (Figures 7.2 and 7.3):

$$R(t) = 1.0 \quad 0 \le t \le 0.03$$

$$= 1.0 - (1/0.003)(t - 0.03) \quad 0.03 \le t \le 0.033$$

Displacement at mass #1 is shown. There is little difference between the mode acceleration and the mode plus modal truncation vector. Differences become significant with higher modal density.

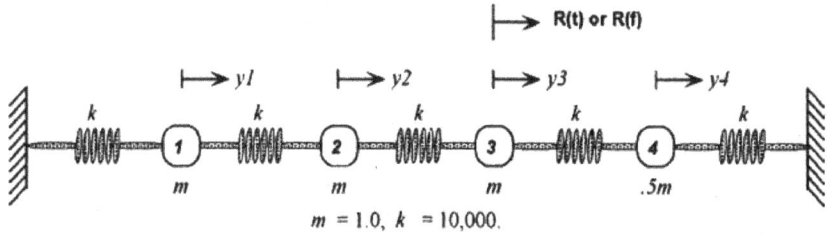

FIGURE 7.1 Simple spring-mass system with space- and time-dependent force on mass #3.

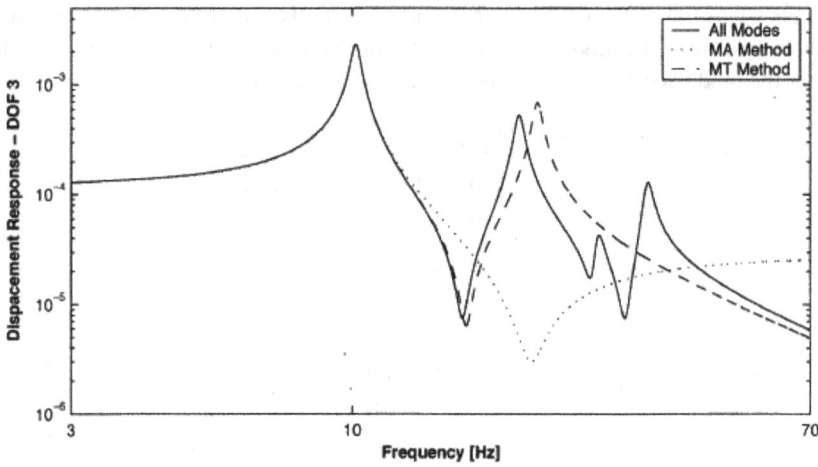

FIGURE 7.2   Frequency response of the system in Figure 7.1, with assumed 2% damping.

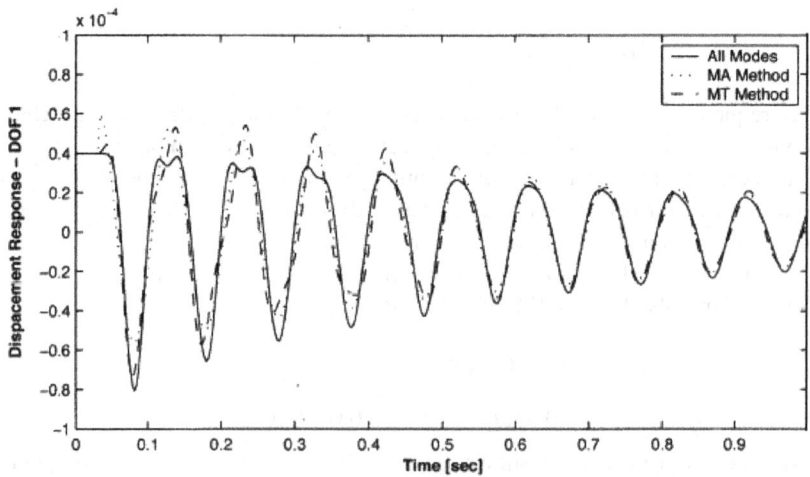

FIGURE 7.3   Time response of the system at mass #1 with loading on mass #3.

7.7   COMPONENT MODE SYNTHESIS TO FORM SYSTEM MODES

This section follows the work of Craig in Ref. [1]. Let the system be composed of two components which have a common interface, with two sets of displacement degrees of freedom, u_1, u_2. Kane's equations with undetermined multipliers, as discussed in Chapter 1, for small vibration (i.e., no large overall motion) can be shown to be given by the following dynamical equations and the common interface displacement continuity constraint:

$$\begin{bmatrix} M_1 & 0 \\ 0 & M_2 \end{bmatrix} \begin{Bmatrix} \ddot{u}_1 \\ \ddot{u}_2 \end{Bmatrix} + \begin{bmatrix} K_1 & 0 \\ 0 & K_2 \end{bmatrix} \begin{Bmatrix} u_1 \\ u_2 \end{Bmatrix} = \begin{bmatrix} C_1^t \\ C_2^T \end{bmatrix} \lambda$$

$$\begin{bmatrix} C_1 & C_2 \end{bmatrix} \begin{Bmatrix} \ddot{u}_1 \\ \ddot{u}_2 \end{Bmatrix} = \{0\}$$

(7.43)

Let the physical displacements be represented by the Craig–Bampton component modes:

$$u_1 = \psi_1 q_1$$

$$u_2 = \psi_2 q_2$$

(7.44)

Form the following triple matrix products for the two components:

$$\mu_1 = \psi_1^t M_1 \psi_1 \qquad \kappa_1 = \psi_1^t K_1 \psi_1$$

$$\mu_2 = \psi_2^t M_2 \psi_2 \qquad \kappa_2 = \psi_2^t K_2 \psi_2$$

(7.45)

Defining

$$p = \begin{Bmatrix} q_1 \\ q_2 \end{Bmatrix}; \quad \mu = \begin{bmatrix} \mu_1 & 0 \\ 0 & \mu_1 \end{bmatrix}; \quad \kappa = \begin{bmatrix} \kappa_1 & 0 \\ 0 & \kappa_1 \end{bmatrix}; \quad c = [C_1\psi_1 \quad C_2\psi_2] \quad (7.46)$$

enables one to write the dynamical and constraint equations of Eq. (7.43) as follows:

$$\mu\ddot{p} + \kappa p = c^t \sigma$$

$$c\ddot{p} = 0$$

(7.47)

Now partition the set of coordinates in p as a set of coordinates p_i that are independent, and the complement of the set p_i as dependent dof, p_d. The constraint equation in Eq. (7.47) can then be written as

$$\begin{bmatrix} c_{dd} & c_{di} \end{bmatrix} \begin{Bmatrix} p_d \\ p_i \end{Bmatrix} = 0$$

(7.48)

This is always possible because of the constraint equation in Eq. (7.48) expressing the fact that there is a common interface between the structures. Now define a reduction transformation from Eq. (7.48):

$$p = \begin{Bmatrix} p_d \\ p_i \end{Bmatrix} = \begin{bmatrix} -c_{dd}^{-1}c_{di} \\ I_{ii} \end{bmatrix} \{p_i\} \equiv S p_i \equiv S \eta$$

(7.49)

Now the dynamical equation for the *system* in Eq. (7.43) becomes

$$M\ddot{\eta} + K\eta = 0$$

(7.50)

where

$$M = S^t \mu S; \quad K = S^t \kappa S$$

(7.51)

The right-hand side zero in Eq. (7.47) is explained by the fact that

$$c\,S \equiv \begin{bmatrix} c_{dd} & c_{di} \end{bmatrix} \left\{ \begin{matrix} -c_{dd}^{-1}c_{di} \\ I_{ii} \end{matrix} \right\} = 0 \qquad (7.52)$$

When there are more than two components involved, modal synthesis for the system modes can be done by extending the same idea, by considering how the components are attached to each other. In articulated robot dynamics formulations with possibly non-linear constraints on the motion, the above approach is generalized as follows, because the formulations end in equations of the form, see Eqs. (1.92)–(1.95), as follows:

$$\left[M(q) \right] \ddot{q} = C^t(q,t)\,\lambda + G(q,\dot{q},t)$$

$$[C(q,t)]\dot{q} = B(q,t) \quad \Rightarrow \quad [C(q,t)]\ddot{q} + \dot{C}\dot{q} = \dot{B} \quad \Rightarrow \quad [C(q,t)]\ddot{q} = D(q,\dot{q},t) \qquad (7.53)$$

Here, the first equation is that of dynamics and the second string of equations are for the *constraints expressed in the velocity form*, differentiated to get the acceleration form of the constraint equations. Eliminating the constraint forces (and torques) λ, one arrives at the final form of the equations of motion (suppressing the arguments shown above):

$$M\,\ddot{q} = C^t \left[CM^{-1}C^t \right]^{-1} \left\langle D - CM^{-1}G \right\rangle + G \qquad (7.54)$$

7.8 FLEXIBLE BODY MODEL REDUCTION BY SINGULAR VALUE DECOMPOSITION OF PROJECTED SYSTEM MODES

A further reduction of the component modes is possible *after* doing component mode synthesis. One approach, based on load transformation matrices relating forces and displacements, is given in Ref. [6]. Here, we show the effectiveness of a simpler method. Consider the example of three bars, with one fixed to the ground, and the other two hinged to it, as shown in Figure 7.4. Let the elastic deformations of the three bars in Figure 7.4 be given in terms of component modes by

$$\delta_1 = \varphi_1 q_1; \quad \delta_2 = \varphi_2 q_2; \quad \delta_3 = \varphi_3 q_3 \qquad (7.55)$$

Component mode synthesis of the system would relate the component modal coordinates, q_i, to the multibody *system modal coordinates*, η, of the system, with *hinges between components locked*, in the following matrix form. The locking of hinges is a practical approach, because the frequencies and modes actually change with relative orientation change between components.

$$\left\{ \begin{matrix} q_1 \\ q_2 \\ q_3 \end{matrix} \right\} = \begin{bmatrix} A \\ B \\ C \end{bmatrix} \left\{ \eta \right\} \qquad (7.56)$$

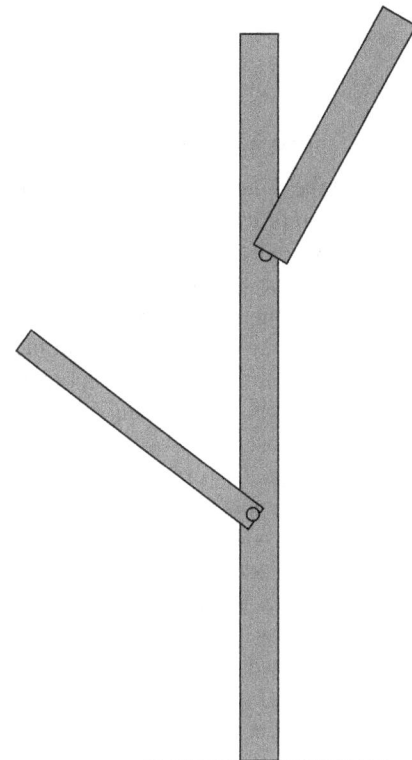

FIGURE 7.4   Three-bar system with 68 modes to demonstrate component mode selection.

Here, A,B,C are projections of system eigenvectors on the subspace of the component modal coordinates, meaning as many columns for "component mode representation" as the number of system modes. By the singular value decomposition of a matrix A, it is meant that A is broken down as shown in the first line of Eq. (7.57). (Singular value decomposition of the matrices A,B,C can be obtained by the Matlab command, SVD.)

$$\text{svd(A)} = [U_1, \Sigma_1, V_1]; \Rightarrow A = U_1 \Sigma_1 V_1^t$$

$$\text{svd(B)} = [U_2, \Sigma_2, V_2]; \qquad\qquad\qquad (7.57)$$

$$\text{svd(C)} = [U_3, \Sigma_3, V_3];$$

The singular value of a matrix means how many columns of a matrix are linearly independent. Now, keeping as many columns in U_1, U_2, U_3 as a chosen number of *largest singular values* in Σ_1, Σ_2, Σ_3, and denoting these as, $\hat{U}_1, \hat{U}_2, \hat{U}_3$, we make the deformation approximations:

$$\delta_1 = \varphi_1 q_1 \cong \varphi_1 \hat{U}_1 \eta = \hat{\varphi}_1 \eta_1$$

$$\delta_2 = \varphi_2 q_2 \cong \varphi_1 \hat{U}_2 \eta = \hat{\varphi}_2 \eta_2 \qquad (7.58)$$

$$\delta_3 = \varphi_3 q_3 \cong \varphi_3 \hat{U}_3 \eta = \hat{\varphi}_3 \eta_3$$

Here, $\hat{\varphi}_1, \hat{\varphi}_2, \hat{\varphi}_3$ may be called the reduced modal sets for components 1,2,3.

Before closing our discussion of component mode synthesis, it is noteworthy that the representation of individual components of a flexible multibody system by component modes may give rise to overall high frequencies and thus simulations may become time-consuming. Thus, for speed of simulation in certain on-line applications [9] of multibody dynamics, system modes are formed, and reduced in number, out of component modes, and kept frozen for a chosen duration, apportioning component deformations during articulation corresponding to these system modes. After the chosen interval, system modes are updated and component modes apportioned there for further articulation.

In Figure 7.5, we consider a system for applying the component model reduction by singular value decomposition of the synthesized system model, as presented by Eqs. (7.56) and (7.57). Out of 15 system modes, the SVD model reduction method kept 15 modes in matrix A of body 1, 5 modes in matrix B of body 2, and 9 modes in matrix C of body 3 (top branch) to get a (15 + 5 + 9) dof model. Figure 7.5 shows the results of comparing a 29 dof model designed to match the first 15 frequencies of a 68 dof articulated flexible three-link system. Figure 7.6 shows the result of applying the same reduction technique to an actual spacecraft application. In the figure, we overlay the

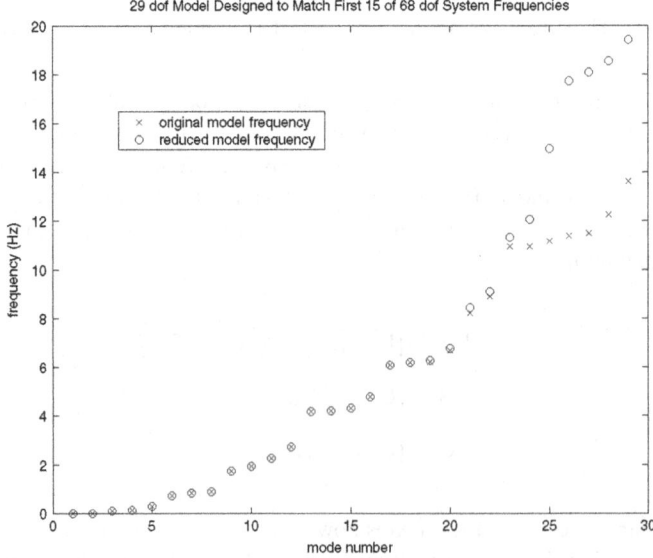

FIGURE 7.5   Twenty-nine dof model matching first 15 of 68 dof frequencies of the three-bar system.

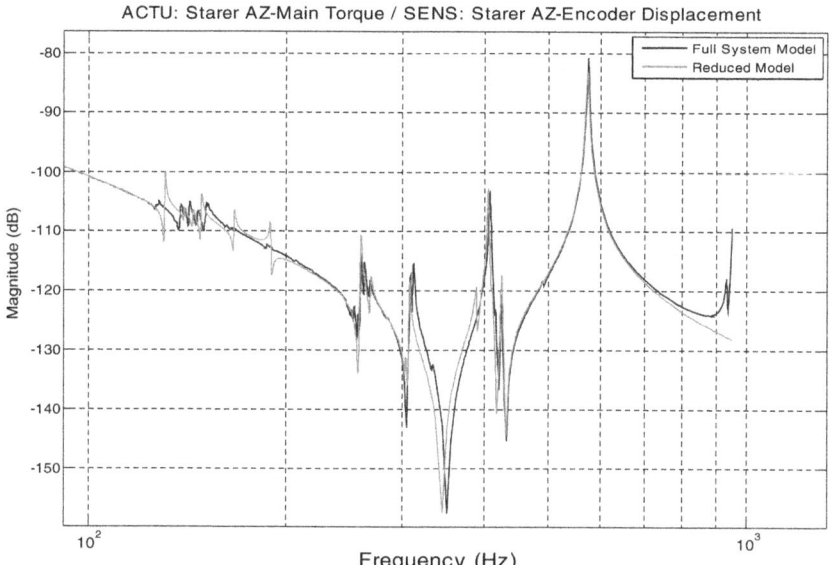

FIGURE 7.6 Bode plot of torque to angle transfer function for an actual spacecraft.

Bode plots of magnitude of a transfer function from an input torque to an output angle in an instrument mounted in the spacecraft. The curve slightly raised, in solid line, is for the full-order system, and the slightly lower curve in dashed line is for the reduced order system. This shows the effectiveness of the model reduction of the components from the full system, based on the singular values of the projection of the system modes to the component modal degrees of freedom.

7.9 DERIVING DAMPING COEFFICIENT OF COMPONENTS FROM DESIRED SYSTEM DAMPING

Follow the self-explanatory computational sequence given below, after choosing a desired system-level damping.

$$M \ddot{X} + C \dot{X} + K X = 0$$

$$X = \Phi q$$

$$\ddot{q} + \Phi^t C \Phi \dot{q} + \Omega^2 q = 0$$

$$\Phi^t C \Phi = 2 \Xi \Omega \qquad (7.59)$$

$$C = \Phi^{-1} 2 \Xi \Omega \Phi^{-1}$$

$$\Phi^{-1} = (\Phi^t \Phi)^{-1} \Phi^t$$

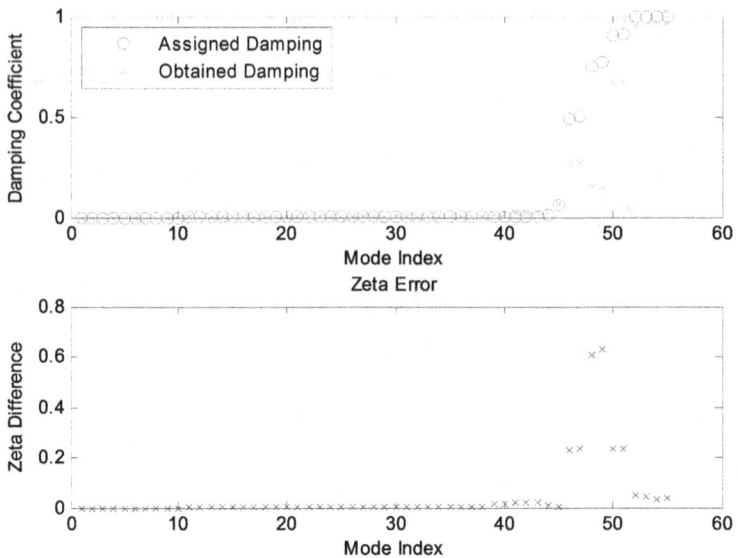

FIGURE 7.7   Results of system damping re-synthesis with derived component damping.

Note that we have used the pseudo-inverse of the system modal matrix in the last line. Component-level damping is derived for a j-th component by the selection matrix S_j that identifies the body in the system.

$$c_j = S_j C S_j^T \qquad (7.60)$$

The results of a re-synthesis of system damping with derived component damping for an actual flexible multibody spacecraft is given in Figure 7.7.

7.10   CONCLUSION

In this chapter, we have reviewed the important subjects of component mode selection, giving detailed codes for finite element formation and synthesis, and an eigenvalue problem solution. It is known that public domain software is also available, for example Ref. [10] for vibration (and buckling) analysis of assembled lattice structures with repetitive geometry and sub-structuring options. What is noteworthy here is that we have discussed both model reduction that is necessary in practice, and compensation for model reduction by modal truncation vectors. We have also discussed the role of modal truncation vectors in geometric stiffness due to rigid body inertia effects that are present with large overall motion of flexible bodies.

PROBLEM SET 7

1. Work out the steps in Ref. [1] of component mode synthesis of two components, one L/3, the other 2L/3 for a fixed-fixed beam of length L

(choose your value of L, E, I), with each component described by a few fixed-fixed modes and two constraint modes corresponding to unit displacement and unit rotation. Keep three modes for one component and six for the other.

2. Consider a simple cantilever beam with a tip load. Consider a few cantilever beam vibration modes to describe its deformation in a forced vibration analysis. Then construct a modal truncation vector corresponding to the number of vibration modes kept.

3. A massless cantilever beam of length L has two discrete masses with mass m at L/2 and L.

   Assume a transverse mode shape of the following form, and calculate the modal effective mass:

$$\varphi(x) = (x / L)^3$$

4. Repeat the example in Section 7.6, using two vibration modes and a modal truncation vector.

APPENDIX 7 MATLAB CODES FOR STRUCTURAL DYNAMICS

(Codes contributed by Dr John Dickens, Dickens Associates, Fremont, CA)

A solution to Problem 1 stated above is given below, complete with codes that would be useful for more general problems in structural dynamics. To solve Problem 1, in Matlab, "cd" to the directory where the m-files given below are typed and saved, and type problem_set_7_1 at the Matlab prompt.

In the subroutines give below, the function "problem_set_7_1.m" is the driver routine. The function "beam_2d.m" calculates the stiffness and mass for a two-dimensional beam element having transverse and rotation degrees of freedom. "beam_string.m" will create a beam model by repeating and assembling several beam elements together. "eigen_mk.m" calculates the eigenvalues/frequencies given the mass and stiffness matrices. "Blevins_beam_freqs.m" is a tabulation of the beam frequencies from Blevins [8] of interest in this analysis.

"beam_model_reduction.m" reduces a multi-element beam model using component modes, modal truncation vectors ("mtv_vectors.m"), and boundary node functions ("boundar_dof_fcn.m"). "problem_set_7_1.m" also does the synthesis both without MTVs and with MTVs. And finally, "print_checks.m" compares the frequencies of the reduced beam models, clamped-free and clamped-clamped, with analytical frequencies given by Blevins.

```
%%%%%%%%%%%%%%%%%%%%%%%%%%%%%%%%%%%
function problem_set_7_1
% Synthesis of 2 beam models. Synthesis is fixed-Fixed
% Model_1 of L/3 and 10 elements, Left Model
% Model_2 of 2L/3 and 20 elements, Right Model
% Script developed by John Dickens using Matlab 6.0.0.88 (R12)
```

```
fprintf('\nProblem Set 7.1\n');
clear
% Section = 1.1 x 5.5 rectangular, density .3
rho = .3;
E = 30E+06;
A = 1.1*5.5;
m_l = rho*A;    % mass/unit length
I = 1.1*5.5^3/12; % 1.1 x 5.5 rectangular section
EI = E*I;
L = 100; % Total Beam length
n_elt = 30;
n1_elt = n_elt/3;
n2_elt = 2*n_elt/3;
L_e = L/n_elt;
bm_elmt = beam_2d (EI,L_e,m_l, 'Consistent'); % One beam
element
% Create the full clamped-clamped beam without Substructures -------

model = beam_string(bm_elmt.stif, bm_elmt.mass, n_elt);
lst = 3:size(model.mass,1)-2; % Fix both Ends
model.stif = model.stif(lst,lst);
model.mass = model.mass(lst,lst);
sys = eigen_mk(model.mass,model.stif);
Blev_freq_sys = Blevins_beam_freqs (E,I, A,rho, L);
%--------------------------------------------------------------------
% Construct Substructure 1, Left end Fixed, n1_elt = 10
elements, Retain
% 3 modes and Include MT Vectors
chk = 1;
n1 mod = 3;
sub1 = beam_model_reduction (bm_elmt,n1_elt,n1_mod,'Left');
if chk==1%                              Check With Blevins
 lst = 1:n1_mod;
 chk_sub1_cc = sqrt(diag(sub1.rk(lst,lst)))/2/pi;
                                        % Clamped-Clamped
 chk_sub1_cf = eigen_mk(sub1.rm,sub1.rk); % Clamped-Free
 Blev_freq1 = Blevins_beam_freqs (E,I, A,rho, L/3);
 print_checks ('Sub1', n1_mod, chk_sub1_cc, chk_sub1_cf,
Blev_freq1)
end
%--------------------------------------------------------------------
% Construct Substructure 2, Right end fixed
n2_mod = 6;
sub2 = beam_model_reduction (bm_elmt,n2_elt,n2_mod,'Right');
if chk==1
 lst = 1:n2_mod;
 chk_sub2_cc = sqrt(diag(sub2.rk(lst,lst)))/2/pi;
                                        % Clamped-Clamped
 chk_sub2_cf = eigen_mk(sub2.rm,sub2.rk); % Clamped-Free
```

```
Blev_freq2 = Blevins_beam_freqs (E,I, A,rho, 2*L/3);
print_checks ('Sub2', n2_mod, chk_sub2_cc, chk_sub2_cf,
Blev_freq2)
end
% Synthesis WITHOUT MTVs ----------------------------------
% DOFs g/S1 1:n1_mod,
%     g/S2 n1_mod+(1:n2_mod)
%     2 physical
n_gen = n1_mod+n2_mod;
n_dof = n_gen+2;
lm1 = [1:n1_mod, 0 0,      n_gen+(1:2)];
lm2 = [n1_mod+(1:n2_mod), 0 0 , n_gen+(1:2)];
M1 = zeros(n_dof,n_dof); K1 = M1;
M1(lm1(lm1>0), lm1(lm1>0)) = M1(lm1(lm1>0), lm1(lm1>0)) ...
   + sub1.rm(lm1>0,lm1>0);
K1(lm1(lm1>0), lm1(lm1>0)) = K1(lm1(lm1>0), lm1(lm1>0)) ...
   + sub1.rk(lm1>0,lm1>0);
M1(lm2(lm2>0), lm2(lm2>0)) = M1(lm2(lm2>0), lm2(lm2>0)) ...
   + sub2.rm(lm2>0,lm2>0);
K1(lm2(lm2>0), lm2(lm2>0)) = K1(lm2(lm2>0), lm2(lm2>0)) ...
   + sub2.rk(lm2>0,lm2>0);
sys_1 = eigen_mk(M1,K1);
% Synthesis WITH MTVs ------------------------------------
% DOFs g/S1 1:n1_mod+2,
%     g/S2 n1_mod+2+(1:n2_mod+2)
%     2 physical
n_gen = n1_mod+2+n2_mod+2;
n_dof = n_gen+2;
lm1 = [1:n1_mod+2,      n_gen+(1:2)];
lm2 = [n1_mod+2+(1:n2_mod+2), n_gen+(1:2)];
M2 = zeros(n_dof,n_dof); K2 = M2;
M2(lm1,lm1) = M2(lm1,lm1) + sub1.rm;
M2(lm2,lm2) = M2(lm2,lm2) + sub2.rm;
K2(lm1,lm1) = K2(lm1,lm1) + sub1.rk;
K2(lm2,lm2) = K2(lm2,lm2) + sub2.rk;
sys_2 = eigen_mk(M2,K2);
% Comparison of Assemblages --------------------------------
n_fs = size(sys_2.freq,1);
n_fs = min(n_fs+2, numel(sys.freq)); % Add a couple more
System Freqs
all_fqs = zeros(n_fs,3);
nn = numel(sys_1.freq); all_fqs(1:nn,1) = sys_1.freq;
nn = numel(sys_2.freq); all_fqs(1:nn,2) = sys_2.freq;
all_fqs(:,3) = sys.freq(1:n_fs);
fprintf(['\nSummary of Synthesis Frequecies:\n', ...
   ' No. 2 Substructure 2 Substr with MTV', ...
                   '        Full Model\n']);
tbl = sprintf('%5d %19.4f %19.4f %19.4f\n',
[(1:n_fs).',all_fqs].');
tbl = strrep(tbl,'0.0000', blanks(6));
```

```
disp(tbl);
return
%%%%%%%%%%%%%%%%%%%%%%%%%%%%%%%%%%%%%%%%%eigen_mk.m follows
function [output1, output2,output3] = eigen_mk (m,k)
% [ ... ] = eigen_mk (m,k)
%
% Solve eigenvalue problem k * phi = lambda*m * phi
% and return values in ascending order.
%       lambda units are (rad/sec)**2
%Returns:
% 2 output arguments:       [ lambda, phi ] = eigen_mk(m,k)
% Single output argument:    [ out ] = eigen_mk(m,k)
% where "out" consists of:    out.lambda = lambda
%                       out.freq   = frequency
%                       out.phi    = eigenvectors
% 3 output arguments:   [ lambda, phi, freq ] = eigen_mk(m,k)

% Changed sort to algebraic rather than absolute jmd -- 04
July 2007
% Added the structured data set "out" as output option and
cleaned up
% sort of eigenvalues              jmd -- 27 July 2008

[nr,nc]=size(m);
%if nr ==1 || nc == 1; mm=diag(m);
%else           mm=m;
%end
if nr~=nc
  mm=diag(m);
else
  mm=m;
end

[vv, ww] =eig (k, mm);

% Organize in ascending order by eigenvalue
w=diag(ww);
%      t=(w.')'.*w;       % Square eigenvalues
%      [t,pointer]=sort(t);
[lambda,pointer]=sort(w);
nn=numel(w);
phi=zeros(nn,nn);
for n=1:nn ;
  if isequal(n,2); echo off; end
  temp=vv(:,pointer(n)); scale=temp.'*mm*temp;
  if (scale ~= 0.);
    scale=sqrt(1./scale);
  else msg = sprintf( ...
    'Mode %d has %6.3f generalized mass, no scaling
  possible', n, scale);
```

```
  disp(' '); disp(msg); scale=1.0;
  end
  phi(:,n)=scale*temp;
end;


freqcy = sqrt(abs(lambda))/(2*pi);
indx = find(lambda<0); % Mark negative eigenvalues as
negative Freqs

if numel(indx)>0; freqcy(indx) = -freqcy(indx); end
if (nargout == 1)
  output1.lambda = lambda;
  output1.phi = phi;
  output1.freq = freqcy;
elseif(nargout == 2)
  output1 = lambda;
  output2 = phi;
else
  output1 = lambda;
  output2 = phi;
  output3 = freqcy;
end
return
%%%%%%%%%%%%%%%%%%%%%%%%%%%%%%%%%%%%%%beam_2d.m follows
function [beam_elemt] = beam_2d (EI, L_e, m_l,Consistent)
% [beam_elemt] = beam_2d (EI, L_e, m_l, Consistent)
% Two Dimensional Beam element. Displacement and Rotation
only
% See Craig, 1981, Structural Dynamics [1], pg 387
% Input:
%  EI
%  L_e = Length of beam elemet
%  m_l = Mass/unit length = rho*A
% Output
%  beam_elemt.stif 4x4 stiffness
%     .mass 4x1 mass, no rotational Mass if only 3 input
%         arguments, otherwise consistent mass and
%         4x4 mass
%
EI_Le = EI/L_e;
ke = EI_Le * [ 12/L_e^2 6/L_e -12/L_e^2 6/L_e
          6/L_e 4 -6/L_e 2
          -12/L_e^2 -6/L_e 12/L_e^2 -6/L_e
          6/L_e 2 -6/L_e 4 ];


if nargin <4    % Lateral Lumped mass only
mL_2 = m_l*L_e/2;
me =      [ mL_2;    0;    mL_2;    0     ];
```

```
else    % Consistent mass, lateral AND rotation,
rAL_420 = m_l*L_e/420;
me = rAL_420 * [ 156 22*L_e      54      -13*L_e
         22*L_e  4*L_e^2  13*L_e    -3*L_e^2
         54 13*L_e 156 -22*L_e
         -13*L_e -3*L_e^2 -22*L_e 4*L_e^2];
end

beam_elemt.stif = ke;
beam_elemt.mass = me;
return
%%%%%%%%%%%%%%%%%%%%%%%%%%%%%%%%boundary_dof_fcn.m follows
function [bnf] = boundary_dof_fcn (stiffness, boundary_dofs)
% [bnf] = boundary_dof_fcn (stiffness, boundary_dofs)
% Input:
% stiffness    = stiffness matrix
% boundary_dofs = Degrees of freedom of the model to which
the model
%        is to be reduced using the Guyan procedure
% Output:
%    bnf.rvec = Guyan boundary function vectors

all_dofs = 1:size(stiffness,1);   % list of all the DOFs of
                                   the model
b_dofs   = boundary_dofs;         % list of the boundary
                                   DOFs
num_b    = length(b_dofs);
i_dofs   = setdiff(all_dofs, b_dofs); % list of the interior
                                       DOFs

r_vec = zeros(length(all_dofs), length(b_dofs));
                             % Initialize r_vec

%    formulate -Kii(-1)Kib and insert in r_vec
r_vec(i_dofs,:) = - stiffness(i_dofs,i_dofs) \ stiffness(i_
                 dofs, b_dofs);
r_vec(b_dofs,:) = eye(num_b,num_b); % Set the Identity
                                     matrix into
%                       boundary DOFs
bnf = r_vec;                 % Return Guyan boundary functions
                             % to calling program
return
%%%%%%%%%%%%%%%%%%%%%%%%%%%%%%%mtv_vectors.m follows
function [mtv] = mtv_vectors (sub1, sys1, bnf, b_dof)
% mtv = mtv_vectors(sub1, sys_s1, bnf, b_dof);
% Calculate MTV vectors for beam string
% Input:
%    sub1.stiff, sub1.mass,  stiffness and mass of model,
                           both full
```

```
%     sys1.phi        fixed base eigenvectors
%     bnf             boundary node functions
%     b_dof           boundary dofs
%
% MTVs are calculated from: F_residual = F - M* Phi * Phi.' F
% Note that Phi * Phi.' * M is an Identity if and only if
                           Phi is all the
% modes of the system. In that case (all the modes have been
kept,
% F_residual or u_residual are = 0), their is no need for
MTVs since all
% the Force is included in the modes (residuals are = 0).
%


lst_dof = 1:size(sub1.mass,1);     % All DOFs
i_dof = setdiff(lst_dof,b_dof);    % Interior DOFs
force = sub1.mass(i_dof,:)*bnf;    % Force Vector to Get MTVs
mass_i = sub1.mass(i_dof,i_dof);
f_res = force - mass_i*sys1.phi*sys1.phi.'*force;
u_res = sub1.stif(i_dof,i_dof)\f_res;    % Kii(-1) * F_res


% Normalization and orthgonalization to make MTVs "look like
modes"
mm    = u_res.'*mass_i*u_res;
kk    = u_res.'*f_res;
%kk   = u_res.'*sub1.stif(i_dof,i_dof)*u_res; % same values
as above
sys_mt= eigen_mk(mm,kk);
mtv = u_res * sys_mt.phi;
return
%%%%%%%%%%%%%%%%%%%%%%%%%%%%%%%%%beam_string.m follows
function [model] = beam_string (beam_stif, beam_mass, n_elt)
% [model] = beam_string (beam_stif, beam_mass)
% Input:
%   beam_stif = beam element stiffness, 4x4, u1,theta1,
                u2,theta2
%   beam_ass  = beam element mass, 4x4 or (4x1).
%   n_elt     = number of 2D beam elements to attach end-to
                end (to
%               string together)
% Output:
% model   = structured array with fields of: mass, stiffness
% model.stiff = stiffness, (n_elt+2) x (n_elt+2)
% model.mass = mass, either same size as Stiffness if mass
matrix is
%     same size as stiffness or a column vector if the beam
%     element mass, "beam_mass", is a column vector.
```

```
n_dof = 2*n_elt+2;


k = zeros(n_dof,n_dof);
lm1 = (1:4); lm = lm1;
for n = 1:n_elt
 k(lm,lm) = k(lm,lm) + beam_stif;
 lm = lm+2;
end;


lm = lm1;
if numel(beam_mass)==4;
 m = zeros(n_dof,1);
 beam_mass = reshape(beam_mass,numel(beam_mass),1);
 for n = 1:n_elt
  m(lm) = m(lm) + beam_mass;
  lm = lm+2;
 end;
else
 m = zeros(n_dof,n_dof);
 for n = 1:n_elt
  m(lm,lm) = m(lm,lm) + beam_mass;
  lm = lm+2;
 end
end


model.mass = m;
model.stif = k;
return
%%%%%%%%%%%%%%%%%%%%%%%%%%%%%print_checks.m follows
function print_checks (c_sub2, n2_mod, chk_sub2_cc, chk_
sub2_cf, Blev_freq2)


 nn = min(5,n2_mod);
 fprintf('\n%s Frequency Comparisons:\n\n', c_sub2);
 fprintf(' %s Clamped-Clamped: %f %f %f %f %f', ...
        c_sub2, chk_sub2_cc(1:nn) );
 fprintf('\n Blevins Clamped-Clamped: %f %f %f %f %f\n', ...
   Blev_freq2.clamped_clamped);
 nn = min(5,numel(chk_sub2_cf.freq));
 fprintf('\n %s Clamped-Free: %f %f %f %f %f', ...
        c_sub2, chk_sub2_cf.freq(1:nn) );
 fprintf('\n Blevins Clamped-Free: %f %f %f %f %f\n', ...
   Blev_freq2.clamped_free);
return
%%%%%%%%%%%%%%%%%%%%%%%%%%%%%Blevins_beam_freqs.m follows
function [Blev_freq] = Blevins_beam_freqs (E,I, A,rho, L)
%    [Blev_freq] = Blevins_beams_freqs (E,I, A,rho, L)
% List the Beam Frequencies from Blevins [9] for beam:
%    a. Free-Free, b. Clamped-Free, c.Clamped-Clamped
```

```
EI = E*I;
m_l = rho*A;    % mass/unit length
const = (EI/m_l)^.5/(2*pi*L^2);


% Free-Free    Blevins page 108, Table 8-1, case 1

lam_i = [4.73004 7.85320 10.99561 14.13717 17.27876];
Blev_freq.free_free    = const*lam_i.^2;

% Clamped-Free    Blevins page 108, Table 8-1, case 3

lam_i = [1.87510 4.69409 7.85476 10.99554 14.13717];
Blev_freq.clamped_free    = const*lam_i.^2;

% Clamped-Clamped Blevins [9] page 108, Table 8-1, case 7

lam_i = [ 4.73004  7.85320  10.99561  14.13717 17.27876];
Blev_freq.clamped_clamped = const*lam_i.^2;
return
%%%%%%%%%%%%%%%%%%%%%%%%%%%%%%%%%%beam_model_reduction.m follows
function sub1 = beam_model_reduction
(bm_elmt,num_elt,md_ret,clamped_end)
%
% Create a string of beam elements and reduce to HCB (Hurty-
Craig-Bampton) model.
% Used in the Example Problem in the book
% Clamped End = 'Left' or 'Right'. If Left end Clamped then
the Right end
% is the boundary for the Model reduction and Vice-versa.

% Construct Substructure Left end fixed, "num_elt" elements,
% retain "md_ret" modes with 2 MTV vectors

sub1 = beam_string(bm_elmt.stif, bm_elmt.mass, num_elt);
nr = size(sub1.mass,1);
if strcmpi(clamped_end(1),'L'); lst = (3:nr); % Fix Left end
else                lst = (1:nr-2); % Fix Right end
end
sub1.stif = sub1.stif(lst,lst);
sub1.mass = sub1.mass(lst,lst);
sub1.model = sub1;

% Modes and Truncation Vectors
n_dof1 = size(sub1.mass,1);
if strcmpi(clamped_end(1),'L');
 lst_i = 1:n_dof1-2; % Interior DOFs
 bdry_dofs= n_dof1-1:n_dof1; % Boundary DOFs, last 2 DOFs
else
 lst_i = 3:n_dof1;% Interior DOFs
```

```
bdry_dofs= 1:2;% Boundary DOFs, First 2 DOFs
end
s1_temp = eigen_mk(sub1.mass(1st_i,1st_i),sub1.stif(1st_i,1s
t_i));
s1_temp.phi = s1_temp.phi(:,1:md_ret); % Keep only "md_ret"
modes
s1_temp.lambda= s1_temp.lambda(1:md_ret);
s1_temp.freq = s1_temp.freq(1:md_ret);
bnf = boundary_dof_fcn (sub1.stif,bdry_dofs);
mtv = mtv_vectors(sub1, s1_temp, bnf, bdry_dofs);
n1 = md_ret+2+2;
t = zeros(size(sub1.stif,1),n1);
t(1st_i,1:md_ret+2) = [s1_temp.phi mtv];
t(:, md_ret+3:end)= bnf;
sub1.rt = t;
sub1.rm = t.' * sub1.mass * t;
sub1.rk = t.' * sub1.stif * t;
return
%%%%%%%%%%%%%%%%%%%%%%%%%%%%%%%%%%%
```

7A.1 Results

The results obtained are shown below for synthesis with (1) component frequencies up to twice the desired system frequencies; (2) same as (1) except that two modal truncation vectors are added; and (3) full model is kept. In substructure 1, sub1, three component modes to 275.23 Hz were retained to which two MTVs were added. Substructure 2, sub2, had six component modes to 237.26 Hz retained to which two MTVs were added.

The results show the efficiency of the modal truncation vectors approach for component modeling.

```
Sub1 Frequency Comparisons:
 Sub1 Clamped-Clamped: 50.883797 140.294855 275.233706
 Blevins Clamped-Clamped: 50.882019 140.258123 274.962433
 454.526981 678.984488
 Sub1     Clamped-Free:   7.996240 50.113198 140.349490
                          275.221136 465.580178
 Blevins  Clamped-Free:   7.996199 50.111515 140.313852
                          274.958933 454.526981

Sub2 Frequency Comparisons:
 Sub2     Clamped-Clamped: 12.720536 35.065149 68.744912
                           113.651135 169.810570
 Blevins Clamped-Clamped:  12.720505 35.064531 68.740608
                           113.631745 169.746122
 Sub2    Clamped-Free:    1.999058 12.527911 35.079014
                          68.744037 113.650974
 Blevins Clamped-Free:    1.999050 12.527879 35.078463
                          68.739733 113.631745 Table 7.1
```

TABLE 7.1

Summary of Synthesis Frequencies

No.	Substruct modes to 2*system Freq	Same substruct modes + two MTVs	Full model
1	5.6536	5.6536	5.6536
2	15.5849	15.5843	15.5843
3	30.5521	30.5518	30.5518
4	50.5180	50.5047	50.5047
5	75.5044	75.4484	75.4484
6	105.4215	105.3859	105.3859
7	140.7556	140.3228	140.3225
8	181.2012	180.2675	180.2666
9	226.2117	225.2393	225.2303
10	293.8879	276.1798	275.2310
11	573.2914	345.5828	330.2916
12		455.2700	390.4424

REFERENCES

1. Craig, R.R., Jr., (1983), *Structural Dynamics: An Introduction to Computer Methods*, Wiley.
2. Wijker, J., (2004), *Mechanical Vibrations in Spacecraft Design*, Springer.
3. Guyan, R.J., (1965), "Reduction of Stiffness and Mass Matrices", *AIAA Journal*, **3**, p. 385.
4. Ghosh, T.K., (2015), Component Model Reduction by Frequency Filtering, Unpublished Memo to NASA Johnson Space Flight Center, L3-Communications Co., Houston, TX.
5. Spanos, J.T. and Tsuha, W.S., (1991), "Selection of Component Modes for Flexible Multibody Simulation", *Journal of Guidance, Control, and Dynamics*, **14**(2), pp. 278–286.
6. Dickens, J.M. and Stroeve, A., (2000), "Modal Truncation Vectors for Reduced Dynamic Substructure Models", Paper No. AIAA-2000-1578, 41st AIAA/ASME/ AHS/ASC Structures, Structural Dynamics, and Materials Conference and Exhibit, Atlanta, GA, 3–8 April.
7. Dickens, J.M., Nakagawa, J.M., and Wittbrodt, M.J., (1997), "A Critique of Mode Acceleration and Modal Truncation Methods for Modal Response Analysis", *Computers & Structures*, **62**(6), pp. 1–14.
8. Blevins, R.D., (2001), *Formulas for Natural Frequency and Mode Shape*, Krieger Publishing.
9. Banerjee, A.K. and Dickens, J.M., (1990), "Dynamics of an Arbitrary Flexible Body in Large Rotation and Translation", *Journal of Guidance, Control, and Dynamics*, **13**(2), pp. 221–227.
10. Anderson, M.S., Williams, F.W., Banerjee, J.R., Durling, B.J., Herstorm, C.L., Kennedy, D., and Warnaar, D.B., (1986), "User Manual for BUNVIS-RG: An Exact Buckling and Vibration Program for Lattice Structures, with Repetitive Geometry and Sub-structuring Options", *NASA Technical Memorandum* 87669.

8 Block-Diagonal Mass Matrix Formulation of Equations of Motion for Flexible Multibody Systems

In Chapter 6, we presented the dynamics of an n-degree of freedom articulated flexible multibody system with equations of motion in the form $\tilde{M}\dot{U} = G$, in which $\tilde{M}$ is an $n \times n$ dense, mass matrix with time-varying elements. This means that an $n \times n$ matrix $\tilde{M}$ has to be evaluated, decomposed, and back-solved, to form $\dot{U} = \tilde{M}^{-1}G$, a task with an arithmetic operation count of n^3, at least four times per step in a fourth-order Runge–Kutta integration method. That makes the simulation expensive when n is large. In this chapter, we describe a more computationally efficient algorithm, presented in Ref. [1] and [2] for simulating motions of flexible multibody dynamic systems. As we will see, the computational savings come from an algorithm that forms a new set of matrix equations, $M\dot{U} = R$, where M is a sparse, block-diagonal matrix, meaning that one has to invert only small sub-matrices corresponding to the component bodies in the system, to compute $\dot{U}$. Here we use vibration modes to describe the flexibility of a component body, and compensate for this inherent premature linearization by including geometric stiffness due to inertia and *interbody forces*. First, we consider a simple example to see the importance of geometric stiffness due to interbody forces, as well as inertia forces, and then move on to flexible multibody systems, using a recursive formulation. Later, we consider motion constraints in a flexible multibody system with structural loops, while keeping a block-diagonal formulation. Finally, we test the predictions of the theory against results from an experimental test set-up for a deployable antenna, and considering parameter uncertainty between experiment and theory, we see very good validation of the theory.

8.1 EXAMPLE: ROLE OF GEOMETRIC STIFFNESS DUE TO INTERBODY LOAD ON A COMPONENT

Figure 8.1 shows three *massless* rigid rods, OA, AB, BC, connected at the hinges at A and B by torsional springs of stiffness k. Rod OA of length r is driven at a prescribed angular speed $\omega(t)$. Rods AB and BC are each of length L. Particles of

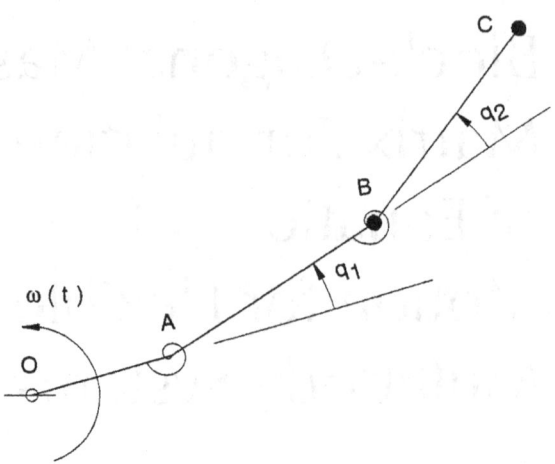

FIGURE 8.1 A system of hinged links in large rotation with small vibration.

mass m are attached at point B of AB and at C of BC. The angle between OA and AB is q_1 and the angle between AB and BC is q_2.

As has been described before, directly *linearized dynamical equations* for *small* motions in angles q_1, q_2 are obtained in Kane's method [3, 4] by keeping non-linear terms in the velocities to extract non-linear partial velocities, and *then* linearizing these partial velocities, following which velocities and accelerations can be linearized, in q_1, q_2. The linear equations for the system in Figure 8.1 can be shown, in matrix form, as follows:

$$M\ddot{Q} + KQ = F, \quad Q = [q_1, q_2]^T \tag{8.1}$$

where

$$M = mL^2 \begin{bmatrix} 5 & 2 \\ 2 & 1 \end{bmatrix}, \quad F = -m\dot{\omega}L \begin{Bmatrix} 5L + 3R \\ 2L + r \end{Bmatrix}, \quad Q = \begin{Bmatrix} q_1 \\ q_2 \end{Bmatrix}$$

$$K = \begin{bmatrix} k + 3m\omega^2 rL & m\omega^2 rL \\ m\omega^2 rL & k + m\omega^2 L(r+L) \end{bmatrix} \tag{8.2}$$

It is clear from the stiffness matrix K that effective stiffness increases with any rotation rate ω, due to centrifugal stiffening, as expected. If, on the other hand, premature linearization is done in that velocities are linearized first, and partial velocities and accelerations are determined from these linear velocities, then incorrect equations are obtained, where the error shows up in the following incorrect stiffness matrix, K_i, with M and F remaining the same as in Eq. (8.2).

$$K_i = \begin{bmatrix} k - 5m\omega^2 L^2 & -2m\omega^2 L^2 \\ -2m\omega^2 L^2 & k - m\omega^2 L^2 \end{bmatrix} \tag{8.3}$$

From Eq. (8.3), it appears that the effect of rotation is that, for *all* rotation rates ω, stiffness *decreases*. Eq. (8.3) is incorrect, because of the well-known effect of premature linearization. By the way, this error has nothing to do with Kane's *method* used for the derivation; in other words, one would get the same wrong result using the linearized velocity expression in Lagrange's equation. In Chapters 1, 5, and 6, we have seen that errors of premature linearization can be corrected, after the fact, by adding geometric stiffness due to existing inertia loads [5].

For the hinged links example problem, "existing loads" are just inertia loads due to rotation. Centrifugal force on the rod BC due to the rotating point mass m in the existing or rigid body configuration, that is, when $q_1 = q_2 = 0$, is

$$F^{BC} = m\omega^2(r + 2L) \tag{8.4a}$$

Inertia force on the rod AB consists of the inertia force due to the point mass in the mass on AB *plus the interbody force* applied by rod BC on AB (which, in the absence of any external force on BC, is the inertia force of BC), thus yielding the force magnitude on AB in the undeformed configuration as

$$F^{AB} = m\omega^2(r + L) + m\omega^2(r + 2L) \tag{8.4b}$$

Using the geometric stiffness matrix given in Ref. [3] for a bar with an *existing axial load* and small transverse displacements from the existing or reference configuration at the ends of the rods AB and BC, one can write the total potential energy associated with this *load-dependent geometric stiffness* as

$$P_g = \frac{1}{2}\left\langle \frac{F^{AB}}{L}\begin{bmatrix} 0 & Lq_1 \end{bmatrix}\begin{bmatrix} 1 & -1 \\ -1 & 1 \end{bmatrix}\begin{Bmatrix} 0 \\ Lq_1 \end{Bmatrix} + \right.$$
$$\left. \frac{F^{BC}}{L}\begin{bmatrix} Lq_1 & L(2q_1 + q_2) \end{bmatrix}\begin{bmatrix} 1 & -1 \\ -1 & 1 \end{bmatrix}\begin{Bmatrix} Lq_1 \\ L(2q_1 + q_2) \end{Bmatrix}\right\rangle \tag{8.5}$$

Here, the first term is due to force on AB, and the second term is due to force on BC. The generalized force due to this potential energy associated with geometric stiffness is

$$-\begin{Bmatrix} \dfrac{\partial P_g}{\partial q_1} \\ \dfrac{\partial P_g}{\partial q_2} \end{Bmatrix} = -\begin{bmatrix} K_g \end{bmatrix}\begin{Bmatrix} q_1 \\ q_1 \end{Bmatrix} \tag{8.6}$$

$$\begin{bmatrix} K_g \end{bmatrix} = m\omega^2 L\begin{bmatrix} 3r + 5L & r + 2L \\ r + 2L & r + 2L \end{bmatrix}$$

Now, the sum of the incorrect stiffness, due to premature linearization, given by Eq. (8.3), and the geometric stiffness due to inertia loads, including interbody force, Eq. (8.6), is

$$\left[K_i\right]+\left[K_g\right]=\begin{bmatrix} k-5m\omega^2L^2 & -2m\omega^2L^2 \\ -2m\omega^2L^2 & k-m\omega^2L^2 \end{bmatrix}+m\omega^2L\begin{bmatrix} 3r+5L & r+2L \\ r+2L & r+2L \end{bmatrix}$$

$$=\begin{bmatrix} k+3m\omega^2rL & m\omega^2rL \\ m\omega^2rL & k+m\omega^2L(r+L) \end{bmatrix} \tag{8.7}$$

This is the same stiffness matrix as in Eq. (8.2)! This clearly shows that the geometric stiffness matrix K_g in Eq. (8.6), which is based on using the inertia force on BC and the inertia force for AB in the reference (undeformed) state, when added to the incorrect stiffness matrix of Eq. (8.3) associated with premature linearization gives precisely the correct stiffness matrix of Eq. (8.2). It is obvious that the correct equations would not be obtained if the force F^{AB} on AB, see Eq. (8.4b), accounted for only the inertia of AB itself, $m\omega^2(r+L)$ and not adding the *interbody* force from BC to AB. In other words, the error of premature linearization in velocity can be compensated for interconnected bodies only by the addition of geometric stiffness due to inertia forces on the body and its outlying interconnected components. In Chapters 5 and 6, we have seen that errors due to premature linearization are analytically compensated for by load-induced stiffness. This example highlights the role of inertia-induced stiffness in determining the motion and loads on a component in a multibody system. We will pursue this idea in the context of a multibody system of arbitrarily general flexible components in large overall motion.

8.2 MULTIBODY SYSTEM WITH RIGID AND FLEXIBLE COMPONENTS

As a motivating example of a flexible multibody system, we revisit in Figure 8.2 the same example as in Figure 6.7, the space station remote manipulator system (SSRMS). It has eight links arranged in a chain with seven active joints. Typically, one end of the SSRMS latches onto the space station, while the other end grapples a payload. The manipulator can impart 6 degrees of freedom (dof) and has an *extra dof to overcome singular situations* in the control of its motion. The dynamics of the space station, the SSRMS, and payload is modeled with 13 rigid body coordinates and n flexible mode coordinates. A dense matrix formulation for (13 + n) dof, where n is the total number of component modes, will produce a (13 + n) × (13 + n) dense, time-varying mass matrix. Even for a reasonable value of the number of modes n, this makes the case for developing a sparse, block-diagonal mass matrix formulation.

Here, we review some of the basic material discussed previously for multibody systems. A system of hinge-connected rigid and flexible bodies, in a topological

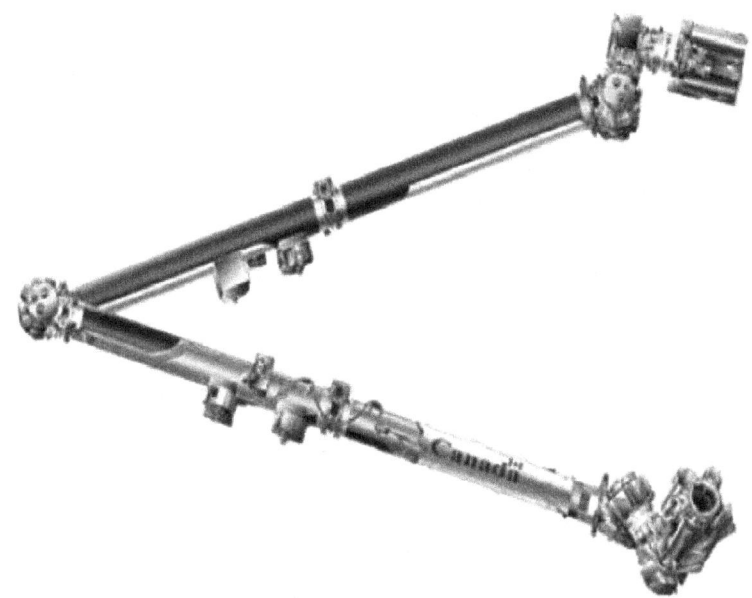

FIGURE 8.2 Example of a flexible multibody system: space station robotic manipulator system.

tree configuration, is shown in Figure 8.3a. The bodies are numbered arbitrarily from 0 to NB (NB equals 8 in Figure 8.3a) with an inertial frame labeled as body 0. Any generic body B_j has an inboard body $B_{c(j)}$ in the path going from body j to body 0. Figure 8.3b shows two adjacent bodies, B_j and $B_{c(j)}$, connected at a hinge that allows relative rotation *and* translation, between Q_j, a hinge point on B_j relative to a point P_j on body $B_{c(j)}$, and reference frames j and p(j) are fixed at points Q_j and P_j, respectively. Let τ_j denote a $(T_j \times 1)$ matrix of generalized coordinates representing T_j number of relative translational degrees of freedom of Q_j from P_j, and let θ^j denote an $(R_j \times 1)$ matrix of generalized coordinates representing R_j relative rotations of frame j with respect to frame p(j); let η^j denote M_j number of modal coordinates of body B_j. To make the formulation simple for a system of articulated flexible bodies, let us define the system generalized speeds as follows:

$$u_i^j = \dot{\tau}_i^j \quad i = 1,\ldots,T_j; \quad j = 1,\ldots,NB$$

$$u_{T_j+i}^j = \dot{\theta}_i^j \quad i = 1,\ldots,R_j; \quad j = 1,\ldots,NB$$

$$u_{T_j+R_j+i}^j = \dot{\eta}_i^j \quad i = 1,\ldots,M_j; \quad j = 1,\ldots,NB \quad (8.8)$$

$$n = \sum_{j=1}^{NB}(T_j + R_j + M_j)$$

This defines the number of degrees of freedom, n, for the system.

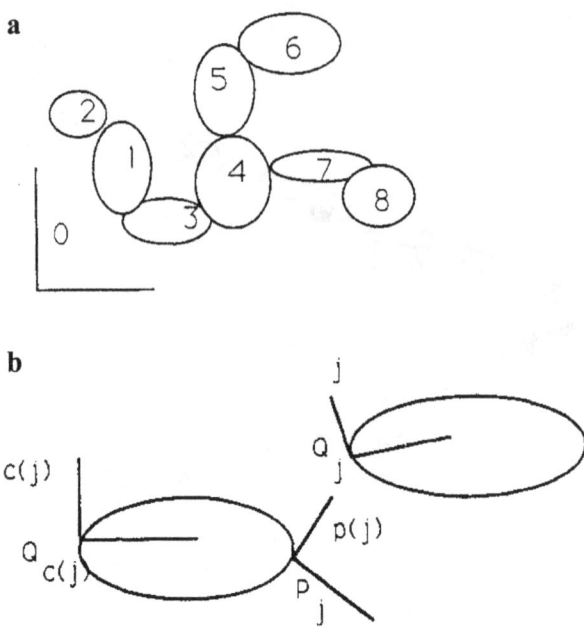

FIGURE 8.3 (a) System of rigid and flexible, hinge-connected bodies in a topological tree; and (b) two adjacent bodies B_j, $B_{c(j)}$ connected at a hinge allowing relative rotation and translation for a system of rigid and flexible bodies in a topological tree.

8.3 RECURRENCE RELATIONS FOR KINEMATICS

Making the assumption that all elastic displacements and elastic rotations are small, and referring to Figure 8.3b, the inertial angular velocity of frame j for body B_j can be expressed in the j basis by the angular velocity addition theorem, proceeding from the inertial angular velocity of frame c(j) of body $B_{c(j)}$ in basis c(j) by the following matrix expression:

$$\omega^j = C_{j,c(j)}\left[\omega^{c(j)} + \psi^{c(j)}(P_j)\dot{\eta}^{c(j)} + C_{c(j),p(j)}H^j\dot{\theta}^j\right] \tag{8.9}$$

Here, $C_{j,c(j)}$ is the direction cosine matrix in $\{\mathbf{b}^j\} = [C_{j,c(j)}]\{\mathbf{b}^{c(j)}\}$ transforming the basis vectors in frame c(j) to those in frame j; $\psi^{c(j)}(P_j)\eta^{c(j)}$ is the 3×1 vector of a small elastic modal rotation at node P_j; $C_{c(j),p(j)}$ is the basis transformation from frame p(j) to frame c(j); and $^{p(j)}\omega^j = H^j\dot{\theta}^j$, where H^j is the matrix relating the angular rotation rates to the angular velocity of frame j with respect to the frame at P_j. The velocity of the point Q_j in Figure 8.3b in the inertial frame is given by the matrix expression below, where $\phi^{c(j)}(P_j)$ is the $3 \times M_{c(j)}$ modal matrix for displacement at P_j, and $L^j\tau^j$ is the joint translation from P_j to Q_j, expressed in the p(j) basis, shown in Figure 8.3b:

$$v^{Q_j} = C_{j,c(j)} \left\langle v^{Q_{c(j)}} + \tilde{\omega}^{c(j)}[r^{Q_{c(j)}P_j} + \varphi^{c(j)}(P_j)\eta^{c(j)}] + \phi^{c(j)}(P_j)\dot{\eta}^{c(j)} \right.$$

$$\left. + C_{c(j),p(j)}L^j\dot{\tau}^j + [\tilde{\omega}^{c(j)} + \tilde{\psi}^{c(j)}(P_j)\dot{\eta}^{c(j)}]C_{c(j),p(j)}L^j\tau^j \right\rangle \tag{8.10}$$

Here, a tilde over a letter indicates a cross product in matrix notation. To illustrate, we use the notations for a vector cross product with $\omega^{c(j)}$, $\psi^{c(j)}$ (in column matrix notation) as

$$\tilde{\omega}^{c(j)} = \begin{bmatrix} 0 & -\omega_3^{c(j)} & \omega_2^{c(j)} \\ \omega_3^{c(j)} & 0 & -\omega_1^{c(j)} \\ -\omega_2^{c(j)} & \omega_1^{c(j)} & 0 \end{bmatrix}; \quad \tilde{\psi}^{c(j)}(P_j)\dot{\eta}^{c(j)} = \begin{bmatrix} 0 & -\dot{\theta}_3^e & \dot{\theta}_2^e \\ \dot{\theta}_3^e & 0 & -\dot{\theta}_1^e \\ -\dot{\theta}_2^e & \dot{\theta}_1^e & 0 \end{bmatrix}$$

$$\text{where } \begin{Bmatrix} \dot{\theta}_1^e \\ \dot{\theta}_2^e \\ \dot{\theta}_3^e \end{Bmatrix} = \left(\psi^{c(j)}(P_j) \right) \{ \dot{\eta}^{c(j)} \} \tag{8.11}$$

Now, we obtain the partial velocity of the origin of frame j with respect to the relative translation rates $\dot{\tau}^j$ at the hinge point, Q_j, and the partial angular velocity of frame j with respect to the relative rotation rates, $\dot{\theta}^j$. These can be obtained by looking at Eqs. (8.10) and (8.9) to define the partial velocity/partial angular velocity $6 \times n_j$ matrix R^j (where n_j is the total dof of the rotation and translation at hinge j):

$$R^j = \begin{bmatrix} C_{j,p(j)}L^j & 0 \\ 0 & C_{j,p(j)}H^j \end{bmatrix} \tag{8.12}$$

Here, we make the simple assumption that all generalized speeds are just the time derivatives of the relative translation of point Q_j with respect to point P_j and rotation of frame j with respect to the frame at P_j.

Note that forming partial velocities and partial angular velocities from Eqs. (8.9) and (8.10), which are linear in the modal coordinates, amounts to premature linearization [2–4], and hence, to compensate for the concomitant errors, one has to use motion-induced geometric stiffness. Following Ref. [6], we separate the inertial (N-frame) acceleration of Q_j and the angular acceleration of frame j into two groups, one involving second derivatives of the generalized coordinates, denoted subsequently by subscript 0, and the remainder acceleration and angular acceleration terms free of the second derivatives of the generalized coordinates, denoted by subscript t. This is always possible because, from Ref. [3]:

$$^N a^{Q_j} = \frac{^N d}{dt} \left(\sum_k v_k^{Q_j}\dot{q}_k + v_t^{Q_j} \right) = a_0^{Q_j} + a_t^{Q_j}$$

$$a_0^{Q_j} = \sum_k v_k^{Q_j}\ddot{q}_k; \quad a_t^{Q_j} = \sum_k \dot{q}_k \frac{^N d}{dt}\left(v_k^{Q_j} \right) + \frac{^N d}{dt}v_t^{Q_j} \tag{8.13a}$$

$$^N\mathbf{a}^j = \frac{^Nd}{dt}\left(\sum_k \omega_k^j \dot{q}_k + \omega^j_{t}\right) = \alpha_0^j + \alpha_t^j$$

(8.13b)

$$\alpha_0^j = \sum_k \omega_k^j \ddot{q}_k; \quad \alpha_t^j = \sum_k \dot{q}_k \frac{^Nd}{dt}\left(\omega_k^j\right) + \frac{^Nd}{dt}\omega_t^j$$

Here, the summations go from $k=1$ to $k=n$, n being the total number of degrees of freedom of the system. Subsequently, all terms in Eqs. (8.13a) and (8.13b) will be expressed in matrix forms, with the vector terms expressed in the body j basis.

$$\left\{\begin{matrix} a^{Q_j} \\ \alpha^j \end{matrix}\right\} = \left\{\begin{matrix} a_0^{Q_j} \\ \alpha_0^j \end{matrix}\right\} + \left\{\begin{matrix} a_t^{Q_j} \\ \alpha_t^j \end{matrix}\right\}$$

(8.13c)

Here, $a_0^{Q_j}, \alpha_0^j$ denote terms containing the derivatives of the generalized speeds in the expressions for the accelerations of the point Q_j and body j, respectively. Now, as in Ref. [6], we separate the $a_0^{Q_j}, \alpha_0^j$ terms for body j into two groups: one group indicated by a hat, representing the contribution of the derivative terms pertaining to the generalized speeds of all inboard bodies in the path from body j to body 0, and the other group for the acceleration and angular acceleration terms associated with the degrees of freedom of only hinge j.

$$\left\{\begin{matrix} a_0^{Q_j} \\ \alpha_0^j \end{matrix}\right\} = \left\{\begin{matrix} \hat{a}_0^{Q_j} \\ \hat{\alpha}_0^j \end{matrix}\right\} + \mathbf{R}^j \left\{\begin{matrix} \ddot{\tau}^j \\ \ddot{\theta}^j \end{matrix}\right\}$$

(8.14)

Differentiation of the terms in Eq. (8.10) in N, and using the notations of Eqs. (8.13a) and (8.13b) give the recursive form of the remainder acceleration term $a_t^{Q_j}$ for the point Q_j relating it to the remainder acceleration term $a_t^{Q_{c(j)}}$ for the point $Q_{c(j)}$ in Figure 8.3.

$$a_t^{Q_j} = C_{j,c(j)}\Big\langle a_t^{Q_{c(j)}} + \tilde{\alpha}_t^{c(j)}\left\{r^{Q_{c(j)}P_j} + \varphi^{c(j)}(P_j)\eta^{c(j)} + C_{c(j),p(j)}L^j\tau^j\right\}$$

$$+ \tilde{\omega}^{c(j)}\left[\tilde{\omega}^{c(j)}\left\{r^{Q_{c(j)}P_j} + \varphi^{c(j)}(P_j)\eta^{c(j)}\right\} + 2\varphi^{c(j)}(P_j)\dot{\eta}^{c(j)}\right]$$

$$+ \left[\tilde{\omega}^{c(j)} + \tilde{\psi}^{c(j)}(P_j)\dot{\eta}^{c(j)}\right]\left\{\left[\tilde{\omega}^{c(j)} + \tilde{\psi}^{c(j)}(P_j)\dot{\eta}^{c(j)}\right]C_{c(j),p(j)}L^j\tau^j\right.$$ (8.15)

$$\left. + 2C_{c(j),p(j)}L^j\dot{\tau}^j\right\} - \tilde{z}_0\,\tilde{\omega}^{c(j)}\psi^{c(j)}(P_j)\dot{\eta}^{c(j)}\Big\rangle$$

where $z_0 = C_{c(j),p(j)}L^j\tau^j$

Differentiation of Eq. (8.9) in N and use of the preceding notations in Eq. (8.13a) allow us to produce the formula for the recursive generation of the column matrix α_t^j, the remainder term of the angular acceleration of frame j in the inertial frame.

$$\alpha_t^j = C_{j,c(j)} \left\langle \alpha_t^{c(j)} + \tilde{\omega}^{c(j)} \left\{ \psi^{c(j)}(P_j)\dot{\eta}^{c(j)} + C_{c(j),p(j)}H^j\dot{\theta}^j \right\} \right.$$
$$\left. + C_{c(j),p(j)}\dot{H}^j\dot{\theta}^j + \left[\tilde{\psi}^{c(j)}(P_j)\dot{\eta}^{c(j)} \right] C_{c(j),p(j)}H^j\dot{\theta}^j \right\rangle \tag{8.16}$$

Eqs. (8.9), (8.10), (8.12), (8.15), and (8.16) are computed in what is called the first forward pass to generate the kinematical equations of the system, starting at body 1 and ending at body NB. These kinematical equations will be used for generating the dynamical equations, in a backward pass described in the next section.

8.4 CONSTRUCTION OF THE DYNAMICAL EQUATIONS IN A BLOCK-DIAGONAL FORM

It can be seen that an order-n formulation [6], producing an entirely diagonal mass matrix is not possible for flexible multibody systems because of the presence of component modal coordinates. The best that can be done is to get the dynamical equations in such a form that the mass matrix is block-diagonal. To start, we revisit Eq. (6.11) in Chapter 6, and ignoring the rotational inertia terms in it, as is usually done, write the modal equations just for body j, augmented with geometric stiffness terms, as indicated below:

$$E^j\ddot{\eta}^j + b^{jT}a^{Qj} + g^{jT}\alpha^j - \Delta_\omega^j + 2C_\omega^j + \lambda^j\eta^j + \Gamma_1^ja^{Qj} + \Gamma_2^j\alpha^j + \Gamma_3^j = \int_j \phi_e^{jT}df_e^j$$

$$\Delta_\omega^j = \left\{ \begin{array}{c} \omega^{jT}D_1\omega^j \\ \vdots \\ \vdots \\ \omega^{jT}D_{M_j}\omega^j \end{array} \right\}; \quad C_\omega^j = \left\{ \begin{array}{c} \omega^{jT}\sum_{k=1}^{M_j}d_{1k}^j\dot{\eta}_k^j \\ \vdots \\ \vdots \\ \omega^{jT}\sum_{k=1}^{M_j}d_{M_jk}^j\dot{\eta}_k^j \end{array} \right\} \tag{8.17}$$

Here, E^j is an $(M_j \times M_j)$ unity matrix for mass-normalized modes of body j; b^j, g^j are $3 \times (M_j)$ matrices; and D_i^j $(i = 1,\ldots,M_j)$ are 3×3 matrices formed out of dyadics, d_{ik}^j, $i = 3,\ldots,M_j$ for each i is a 3×1 matrix, defined in Appendix A at the end of this book. Nominal generalized structural stiffness is given by the diagonal $(M_j \times M_j)$ matrix λ_j the term in $\lambda^j\eta^j$ of Eq. (8.17). Geometric stiffness due to inertia forces [7] is represented by the terms $\Gamma_1^j, \Gamma_2^j, \Gamma_3^j$, and will be presented in detail after discussing some basic kinematics. The right hand side of Eq. (8.17) is the generalized force due to distributed external forces. Distributed inertia forces at the node at location x_i, y_i, z_i of a component are written, as before in Eq. (6.26), in terms of 12 acceleration components, as follows with U_3 being the 3×3 unity matrix, and mass m_i as the mass at node i:

$$
\begin{Bmatrix} f_1^{*i} \\ f_2^{*i} \\ f_3^{*i} \end{Bmatrix} = -m_i \begin{bmatrix} U_3 & x_i U_3 & y_i U_3 & z_i U_3 \end{bmatrix} \begin{Bmatrix} a_1^j \\ : \\ : \\ : \\ a_{12}^j \end{Bmatrix}
\tag{8.18}
$$

Here, $a_1^j, \ldots, a_{12}^j$ are defined for body j in matrix form by considering only rigid body acceleration, for an "existing state", required for computing geometric stiffness, at a generic point G_j with coordinates (x_i, y_i, z_i), based on the vector equation:

$$
{}^N\mathbf{a}^{G_j} = {}^N\mathbf{a}^{Q_j} + {}^N\boldsymbol{\alpha}^j \times \mathbf{p}^{Q_jG_j} + {}^N\boldsymbol{\omega}^j \times \left({}^N\boldsymbol{\omega}^j \times \mathbf{p}^{Q_jG_j} \right)
\tag{8.19}
$$

Here, $Q_j, \mathbf{p}^{Q_jG_j}$ are, respectively, the origin of the coordinates and the position vector from Q_j to G_j.

Elements of the matrix form of Eq. (8.19) with elements in the j basis are written after splitting the acceleration and angular acceleration terms involved, in the groups with and without second derivatives of generalized coordinates, as in Eq. (8.13c), and one gets the following:

$$
\begin{Bmatrix} a_1^j \\ a_2^j \\ a_3^j \end{Bmatrix} = \left\{ a_0^{Q_j} + a_t^{Q_j} \right\}
\tag{8.20}
$$

$$
\begin{Bmatrix} a_4^j \\ a_5^j \\ a_6^j \end{Bmatrix} = \begin{Bmatrix} z_1^j \\ z_2^j \\ z_3^j \end{Bmatrix} - \begin{bmatrix} 0 & 0 & 0 \\ 0 & 0 & -1 \\ 0 & 1 & 0 \end{bmatrix} \left\{ \alpha_0^{Q_j} + \alpha_t^{Q_j} \right\}
\tag{8.21}
$$

$$
\begin{Bmatrix} a_7^j \\ a_8^j \\ a_9^j \end{Bmatrix} = \begin{Bmatrix} z_2^j \\ z_4^j \\ z_5^j \end{Bmatrix} - \begin{bmatrix} 0 & 0 & 1 \\ 0 & 0 & 0 \\ -1 & 0 & 0 \end{bmatrix} \left\{ \alpha_0^{Q_j} + \alpha_t^{Q_j} \right\}
\tag{8.22}
$$

$$
\begin{Bmatrix} a_{10}^j \\ a_{11}^j \\ a_{12}^j \end{Bmatrix} = \begin{Bmatrix} z_3^j \\ z_5^j \\ z_6^j \end{Bmatrix} - \begin{bmatrix} 0 & -1 & 0 \\ 1 & 0 & 0 \\ 0 & 0 & 0 \end{bmatrix} \left\{ \alpha_0^{Q_j} + \alpha_t^{Q_j} \right\}
\tag{8.23}
$$

Here, we have introduced the intermediate variables:

$$\begin{bmatrix} z_1^j & z_2^j & z_3^j \\ z_2^j & z_4^j & z_5^j \\ z_3^j & z_5^j & z_6^j \end{bmatrix} = \tilde{\omega}^j \, \tilde{\omega}^j$$

$$= \begin{bmatrix} -(\omega_2^j)^2 - (\omega_3^j)^2 & (\omega_1^j)(\omega_2^j) & (\omega_1^j)(\omega_3^j) \\ (\omega_1^j)(\omega_2^j) & -(\omega_3^j)^2 - (\omega_1^j)^2 & (\omega_2^j)(\omega_3^j) \\ (\omega_1^j)(\omega_3^j) & (\omega_2^j)(\omega_3^j) & -(\omega_1^j)^2 - (\omega_2^j)^2 \end{bmatrix}$$

(8.24)

Here, the angular velocity components relate to $^N\omega^j = \omega_1^j\mathbf{b}_1^j + \omega_2^j\mathbf{b}_2^j + \omega_3^j\mathbf{b}_3^j$.

The motion-induced stiffness terms in Eq. (8.18) are now expanded upon, in terms of the generalized force due to geometric stiffness, K_{gi}^j, for unit values of the body frame acceleration components a_i^j ($i = 1,\dots,12$) in Eqs. (8.20)–(8.24), and the modal matrix ϕ^j and the modal coordinate η^j for body j are given as follows, where Γ_1^j, Γ_2^j are ($M_j \times 3$) matrices, and Γ_3^j is an ($M_j \times 1$) matrix. To simplify the algebra, we introduce the variables:

$$G_i^j = \phi^{j^T} K_{gi}^j \phi^j \quad i = 1,\dots,12$$

$$\Gamma_1^j = \begin{bmatrix} G_1^j\eta^j & G_2^j\eta^j & G_3^j\eta^j \end{bmatrix}$$

$$\Gamma_2^j = \begin{bmatrix} (G_9^j - G_{11}^j)\eta^j & (G_{10}^j - G_6^j)\eta^j & (G_5^j - G_7^j)\eta^j \end{bmatrix}$$ (8.25)

$$\Gamma_3^j = G_4^j\eta^j z_1^j + G_5^j\eta^j z_2^j + G_6^j\eta^j z_3^j + G_7^j\eta^j z_2^j + G_8^j\eta^j z_4^j + G_9^j\eta^j z_5^j$$
$$+ G_{10}^j\eta^j z_3^j + G_{11}^j\eta^j z_5^j + G_{12}^j\eta^j z_6^j$$

In terms of the acceleration grouping in Eqs. (8.13a) and (8.13b), one can rewrite the matrix form of Eq. (8.17) as below, where we have made the approximation $g^j = c^j$ in Eq. (6.11):

$$E^j\ddot{\eta}^j = A^j \begin{Bmatrix} a_0^{Q_j} \\ \alpha_0^j \end{Bmatrix} + Y_1^j$$ (8.26)

$$A^j = -\begin{bmatrix} b^{j^T} + \Gamma_1^j & c^{j^T} + \Gamma_2^j \end{bmatrix}$$ (8.27)

$$Y_1^j = A^j \begin{Bmatrix} a_t^{Q_j} \\ \alpha_t^j \end{Bmatrix} - \left(-\Delta_\omega^j + 2C_\omega^j + \Gamma_3^j + \lambda^j\eta^j - \int_j \phi_e^{j^T} df_e^j \right)$$ (8.28)

By D'Alembert's principle, the sum of the inertia and external forces and torques on body j, $f^{*j}, f_e^j, t^{*j}, t_e^j$, and the total interaction force and torque at the hinge

Q_j, from $c(j)$ to j, $f^j_{c(j)}, t^j_{c(j)}$, is zero; this is expressed by the (6×1) zero matrix equation:

$$0 = \begin{Bmatrix} f^{*j} - f^j_e - f^j_{c(j)} \\ t^{*j} - t^j_e - t^j_{c(j)} \end{Bmatrix} = M^j_1 \begin{Bmatrix} a^{Q_j}_0 \\ \alpha^j_0 \end{Bmatrix} + M^j_2 \ddot{\eta}^j + X^j - \begin{Bmatrix} f^j_{c(j)} \\ t^j_{c(j)} \end{Bmatrix} \tag{8.29}$$

As before, M^j_1, M^j_2, X^j are expressed in terms of the mass, expressed as a 3×3 diagonal matrix, and the deformation-dependent first and second moments of inertia about Q_j, or $s^{j/Q_j}, I^{j/Q_j}$, and the modal integrals, b^j, c^j, N^j (N^{j*} being the transpose of N^j).

$$M^j_1 = \begin{bmatrix} m^j & -\tilde{s}^{j/Q_j} \\ \tilde{s}^{j/Q_j} & I^{j/Q_j} \end{bmatrix} \tag{8.30}$$

$$M^j_2 = \begin{bmatrix} b^j \\ c^j \end{bmatrix} \tag{8.31}$$

$$X^j = M^j_1 \begin{Bmatrix} a^{Q_j}_t \\ \alpha^j_t \end{Bmatrix} + \begin{Bmatrix} \tilde{\omega}^j [\tilde{\omega}^j s^{j/Q_j} + 2b^j \dot{\eta}^j] - \int_j \phi^t_e df^j_e \\ \tilde{\omega}^j I^{j/Q_j} + 2N^{j*} \dot{\eta}^j \omega^j - \int_j \psi^t_e d\tau^j_e \end{Bmatrix} \tag{8.32}$$

Here ϕ_e, ψ_e are the modal elastic displacement and rotation at the node where the external force and torque, df^j_e, $d\tau^j_e$ are applied.

Now, solving for $\ddot{\eta}^j$ from Eq. (8.26) and putting the result in Eq. (8.29), one defines the following (6×6) matrix and (6×1) column matrix, recalling that for normalized modes, E^j is an $(M_j \times M_j)$ unity matrix for body j and that $A^j \neq M^j_2$:

$$M^j_3 = M^j_1 + M^j_2 [E^j]^{-1} A^j$$
$$Y^j_2 = X^j + M^j_2 [E^j]^{-1} Y^j_1 \tag{8.33}$$

One can now rewrite Eq. (8.29), using Eq. (8.33), and invoking Eq. (8.14), as follows:

$$0 = \begin{Bmatrix} f^{*j} - f^j_e - f^j_{c(j)} \\ t^{*j} - t^j_e - t^j_{c(j)} \end{Bmatrix} \equiv M^j_3 \begin{Bmatrix} a^{Q_j}_0 \\ \alpha^j_0 \end{Bmatrix} + Y^j_2 - \begin{Bmatrix} f^j_{c(j)} \\ t^j_{c(j)} \end{Bmatrix}$$

$$= M^j_3 \left\{ \begin{Bmatrix} \hat{a}^{Q_j}_0 \\ \hat{\alpha}^j_0 \end{Bmatrix} + R^j \begin{pmatrix} \ddot{\tau}^j \\ \ddot{\theta}^j \end{pmatrix} \right\} + Y^j_2 - \begin{Bmatrix} f^j_{c(j)} \\ t^j_{c(j)} \end{Bmatrix} \tag{8.34}$$

Kane's dynamical equations [3] associated with the hinge dof for a terminal body can now be written, by pre-multiplying Eq. (8.34) by the transpose of the partial velocity/partial angular velocity matrix R^j, Eq. (8.12), avoiding multiplication by the zeros in Eq. (8.12). This operation yields the equations for translational and rotational degrees of freedom at the hinge:

$$
\begin{Bmatrix} \ddot{\tau}^j \\ \ddot{\theta}^j \end{Bmatrix} = -\left[v^j\right]^{-1} R^{jT} \left[M_3^j \begin{Bmatrix} \hat{a}_0^{Q_j} \\ \hat{\alpha}_0^j \end{Bmatrix} + Y_2^j \right] + \left[v^j\right]^{-1} \begin{Bmatrix} f_h^j \\ t_h^j \end{Bmatrix}
$$

$$\text{(8.35)}$$

$$
\text{where } \begin{Bmatrix} f_h^j \\ t_h^j \end{Bmatrix} = R^{jT} \begin{Bmatrix} f_{c(j)}^j \\ t_{c(j)}^j \end{Bmatrix}
$$

Here, f_h^j, t_h^j are the $(T_j \times 1)$ hinge force and the $(R_j \times 1)$ hinge torque matrices, respectively, applied by body $c(j)$ on body j, and we have inverted the $(R_j + T_j)$ square matrix v^j:

$$
v^j = R^{jT} M_3 R^j \qquad \text{(8.36)}
$$

Using Eqs. (8.14) and (8.35) in Eq. (8.33), and introducing the notations:

$$
M = M_3^j - M_3^j R^j \left[v^j\right]^{-1} R^{jT} M_3^j
$$

$$
X = Y_2^j - M_3^j R^j \left[v^j\right]^{-1} R^{jT} Y_2^j + M_3^j R^j \left[v^j\right]^{-1} \begin{Bmatrix} f_h^j \\ t_h^j \end{Bmatrix}
$$

$$\text{(8.37)}$$

lead to a re-expression of Eq. (8.29) in the form:

$$
\begin{Bmatrix} f_{c(j)}^j \\ t_{c(j)}^j \end{Bmatrix} = \begin{Bmatrix} f^{*j} - f_e^j \\ t^{*j} - t_e^j \end{Bmatrix} = M \begin{Bmatrix} \hat{a}_0^{Q_j} \\ \hat{\alpha}_0^j \end{Bmatrix} + X \qquad \text{(8.38)}
$$

At this stage, we extend the kinematical propagation equation by using a shift operator, W^j, which was introduced in Ref. [6] for rigid bodies, to flexible bodies, relating the terms with the second derivatives in the generalized coordinates in the expressions for the acceleration of the point $Q_{c(j)}$ and P_j, and the angular acceleration of frame $c(j)$, in Figure 8.3b:

$$
\begin{Bmatrix} \hat{a}_0^{Q_j} \\ \hat{\alpha}_0^j \end{Bmatrix} = W^j \begin{Bmatrix} a_0^{Q_{c(j)}} \\ \alpha_0^{c(j)} \end{Bmatrix} + N^j \ddot{\eta}^{c(j)} \qquad \text{(8.39)}
$$

In Eq. (8.39), the shift operator W^j and the matrix N^j based on the modal matrix of body $c(j)$ for point P_j are as follows:

$$
W^j = \begin{bmatrix} C_{j,c(j)} & -C_{j,c(j)} \tilde{r}_{c(j)}^{Q_{c(j)}Q_j} \\ 0 & C_{j,c(j)} \end{bmatrix} \qquad \text{(8.40)}
$$

$$
N^j = C^j \begin{Bmatrix} \varphi^{c(j)}(P_j) - \tilde{z}_0 \, \psi^{c(j)}(P_j) \\ \psi^{c(j)}(P_j) \end{Bmatrix} \qquad \text{(8.41)}
$$

$$\mathbf{C}^j = \begin{bmatrix} \mathbf{C}_{j,c(j)} & 0 \\ 0 & \mathbf{C}_{j,c(j)} \end{bmatrix} \tag{8.42}$$

In Eq. (8.40), $\tilde{r}_{c(j)}^{Q_{c(j)}Q_j}$ is a skew-symmetric matrix formed out of the (3×1) matrix whose elements are the components of the position vector from $Q_{c(j)}$ to Q_j, resolved in the c(j) basis, and $\tilde{z}_0$ is defined in Eq. (8.15):

$$r_{c(j)}^{Q_{c(j)}Q_j} = r_{c(j)}^{Q_{c(j)}P_j} + \phi^{c(j)}(P_j)\eta^{c(j)} + \mathbf{C}_{c(j),p(j)}\mathbf{L}^j\tau^j \tag{8.43}$$

The dynamic equilibrium, Eq. (8.29), of the system of forces and moments at Q_j, expressed in the j basis, is stated by substituting Eq. (8.39) in Eq. (8.38), as the 6×1 matrix:

$$\left\{ \begin{matrix} f^{*j} - f_e^j \\ t^{*j} - t_e^j \end{matrix} \right\}_{Q_j/j} = \mathbf{M} \left\langle \mathbf{W}^j \left\{ \begin{matrix} a_0^{Q_{c(j)}} \\ \alpha_0^{c(j)} \end{matrix} \right\} + \mathbf{N}^j \ddot{\eta}^{c(j)} \right\rangle + \mathbf{X} \equiv \left\{ \begin{matrix} f_{c(j)}^j \\ t_{c(j)}^j \end{matrix} \right\} \tag{8.44}$$

The interbody force and moment applied by body j on body c(j) at P_j is obtained by replacing the force system of Eq. (8.44) from Q_j to P_j. In the c(j) basis, this is expressed as follows:

$$d_j^{c(j)} \left\{ \begin{matrix} f_{c(j)}^j \\ t_{c(j)}^j \end{matrix} \right\}_{P_j/c(j)} = d_j^{c(j)} \left[\mathbf{M} \left\langle \mathbf{W}^j \left\{ \begin{matrix} a_0^{Q_{c(j)}} \\ \alpha_0^{c(j)} \end{matrix} \right\} + \mathbf{N}^j \ddot{\eta}^{c(j)} \right\rangle + \mathbf{X} \right] \tag{8.45}$$

Here, we have used the 6×6 transformation matrix, (see Eq. [8.15] for $\tilde{z}_0$):

$$d_j^{c(j)} = \begin{bmatrix} \mathbf{C}_{c(j),j} & 0 \\ \mathbf{C}_{c(j),j}\tilde{z}_0 & \mathbf{C}_{c(j),j} \end{bmatrix} \tag{8.46}$$

Geometric stiffness due to interbody loads on body c(j) contributes to generalized active forces as follows, where only linear terms in $\eta^{c(j)}$ are kept:

$$F_{interbody}^{c(j)} = -\sum_{i=1}^{6} G_{12+i}^{c(j)} \eta^{c(j)} S_j^{c(j)} d_j^{c(j)} \left\langle \mathbf{MW}^j \left\{ \begin{matrix} a_0^{c(j)} \\ \alpha_0^{c(j)} \end{matrix} \right\} + \mathbf{X} \right\rangle \tag{8.47}$$

Here, G_{12+i}^j $(i = 1, \dots, 6)$ are six geometric stiffness matrices due to six interbody unit forces/moments, and $S_j^{c(j)}$ stands for a selection matrix generated from a 6×6 identity matrix, with a zero row number for the rotation axis number in the p(j) frame about which the axis moment, corresponding to non-working moments, cannot be transferred. Contribution to the modal equation for body c(j) comes from three sources: inertia and active forces on body c(j), structural stiffness, and geometric stiffness due to inertia loads, see Eq. (8.18), and interbody forces and the equilibrium system of forces and moments acting at Q_j. With the help of Eqs. (8.13c), (8.14), and (8.35), one gets

$$E^{c(j)}\ddot{\eta}^{c(j)} - A^{c(j)}\begin{Bmatrix} a_0^{Q_{c(j)}} \\ \alpha_0^{c(j)} \end{Bmatrix} - Y_1^{c(j)} - F_{interbody}^{c(j)} + N^{j^T}\begin{Bmatrix} f^{*j} - f_e^j \\ t^{*j} - t_e^j \end{Bmatrix}_{Q_j/j} = 0 \quad (8.48)$$

Substituting Eqs. (8.44) and (8.47) in Eq. (8.48) and collecting terms, one gets the modal equations in the same form as Eq. (8.26):

$$E^{c(j)}\ddot{\eta}^{c(j)} = A^{c(j)}\begin{Bmatrix} a_0^{Q_{c(j)}} \\ \alpha_0^{c(j)} \end{Bmatrix} + Y_1^{c(j)} \quad (8.49)$$

Here, to keep the structure in the form of Eq. (8.49), the following matrix "replacements" have been made:

$$E^{c(j)} \leftarrow E^{c(j)} + N^{j^T}MN^j$$

$$A^{c(j)} \leftarrow A^{c(j)} - N^{j^T}MW^j - \sum_{i=1}^{6} G_{12+i}^{c(j)}\eta^{c(j)}S_i^{c(j)}d_i^{c(j)}MW^j \quad (8.50)$$

$$Y_1^{c(j)} \leftarrow Y_1^{c(j)} - N^{j^T}X - \sum_{i=1}^{6} G_{12+i}^{c(j)}\eta^{c(j)}S_i^{c(j)}d_i^{c(j)}X$$

With these replacements, forces and moments acting at Q_j can be transferred to the point $Q_{c(j)}$ of body $c(j)$ and expressed in the $c(j)$ basis by using again, the shift operator of Eq. (8.40) in Eq. (8.38):

$$\begin{Bmatrix} f^{*j} - f_e^j \\ t^{*j} - t_e^j \end{Bmatrix}_{Q_{c(j)}/c(j)} = W^{j^T}\begin{Bmatrix} f^{*j} - f_e^j \\ t^{*j} - t_e^j \end{Bmatrix}_{Q_j/j} \quad (8.51)$$

To this set of forces and moments is added the resultant of all of the active and inertia forces and moments of all outboard bodies of body $c(j)$ to $Q_{c(j)}$, yielding the zero-sum (6×1) expression:

$$\sum_{j \epsilon S_0}\begin{Bmatrix} f^{*j} - f_e^j \\ t^{*j} - t_e^j \end{Bmatrix} = M_1^{c(j)}\begin{Bmatrix} a_0^{Q_{c(j)}} \\ \alpha_0^{c(j)} \end{Bmatrix} + M_2^{c(j)}\ddot{\eta}^{c(j)} + X^{c(j)} \quad (8.52)$$

Here, $M_1^{c(j)}, M_1^{c(j)}, X^{c(j)}$ come from Eqs. (8.30)–(8.32), and S_0 is the set of all bodies outward of the hinge $Q_{c(j)}$, and the following replacements state what the terms in Eq. (8.52) stand for, namely, the effect of all forces and moments due to accelerations and angular accelerations and vibrations of bodies outboard of body $c(j)$:

$$M_1^{c(j)} + W^{j^T}MW^j \rightarrow M_1^{c(j)}$$

$$M_2^{c(j)} + W^{j^T}MN^j \rightarrow M_2^{c(j)} \quad (8.53)$$

$$X^{c(j)} + W^{j^T}X \rightarrow X^{c(j)}$$

Now, the solution for $\ddot{\eta}^{c(j)}$ from Eq. (8.49) is used in Eq. (8.52), and Eq. (8.14) is invoked with j replaced by c(j), and Kane's dynamical equations associated with the translations and rotations at $Q_{c(j)}$ are written as before, $R^{c(j)}$ being the partial velocity matrix in the form of Eq. (8.14) for body c(j):

$$\left\{\begin{matrix} \ddot{\tau}^{c(j)} \\ \ddot{\theta}^{c(j)j} \end{matrix}\right\} = -\left[v^{c(j)}\right]^{-1} R^{c(j)T}\left[M_3^{c(j)}\left\{\begin{matrix} \hat{a}_0^{Qc(j)} \\ \hat{\alpha}_0^{c(j)} \end{matrix}\right\} + Y_2^{c(j)}\right] + \left[v^{c(j)}\right]^{-1}\left\{\begin{matrix} f_h^{c(j)} \\ t_h^{c(j)} \end{matrix}\right\} \quad (8.54)$$

This reproduces two cycles of the formulation of dynamical equations, covering the rotation/translation/modal degrees of freedom of two bodies, j and c(j). This pattern is repeated until the equations for body 1 are written. At this stage, the inboard body linear accelerations/angular accelerations are zero or prescribed. If the zeroth body is the inertial frame (hat quantities for inboard bodies are zero), Eq. (8.54) gives for c(j)=1 the hinge translation and rotation equations, with f_h^1, t_h^1 zero for a free system:

$$\left\{\begin{matrix} \ddot{\tau}^1 \\ \ddot{\theta}^1 \end{matrix}\right\} = -\left[v^1\right]^{-1} R^{1T} Y_2^1 + \left[v^1\right]^{-1}\left\{\begin{matrix} f_h^1 \\ t_h^1 \end{matrix}\right\} \quad (8.55)$$

Concomitantly, Eq. (8.14) reduces to

$$\left\{\begin{matrix} a_0^{Q_1} \\ \alpha_0^1 \end{matrix}\right\} = R^1\left\{\begin{matrix} \ddot{\tau}^1 \\ \ddot{\theta}^1 \end{matrix}\right\} \quad (8.56)$$

This permits the computation of $\ddot{\eta}^j$ from Eq. (8.26) for j = 1. Once the dynamical equations for the base body are formed, the rest of the dynamical equations can be obtained by a second forward pass "going up the tree", using the kinematical equations produced in the first forward pass. Thus, the overall equations are generated in three steps: a forward pass to generate the kinematical equation, a backward pass to get the dynamical equations for the first body, and finally uncovering the rest of the dynamical equations by a second forward pass.

The algorithm is block-diagonal, because it breaks the dynamical equations into sub-blocks for each body j, with the sub-blocks requiring the inversion of matrices E^j of order M_j (which is usually a unity matrix), v^j in Eq. (8.36) of order $(T_j + R_j)$. The structure of the dynamical equations solved for the second derivatives of the modal and rigid-body coordinates for the system of just one rigid body hinge connected to three terminal flexible bodies shown in Figure 8.4a, are given in Figure 8.4b.

The block-diagonal algorithm developed by the author and coded by his colleagues at Lockheed Martin Space Systems was used to simulate the Next Generation Space Telescope (NGST) and the Hubble Space Telescope shown in Figure 8.5.

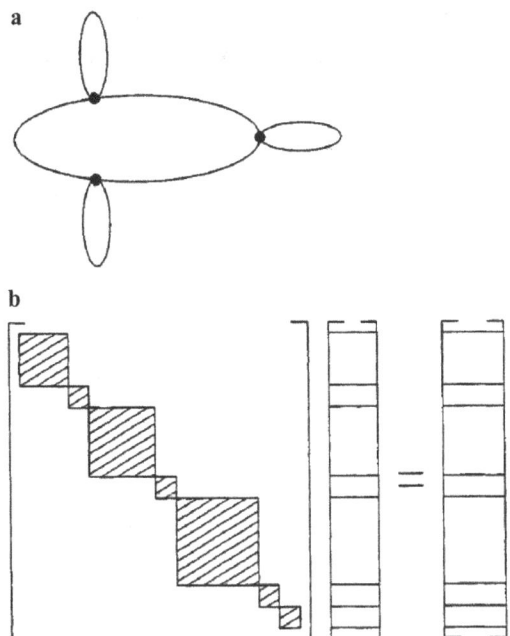

FIGURE 8.4 (a) Rigid body with three articulated flexible appendages; and (b) structure of the block dynamical equations for a rigid body connected to three hinged flexible appendages; base rigid body mass matrix at the bottom corner the system mass matrix, with large sub-matrices for modal.

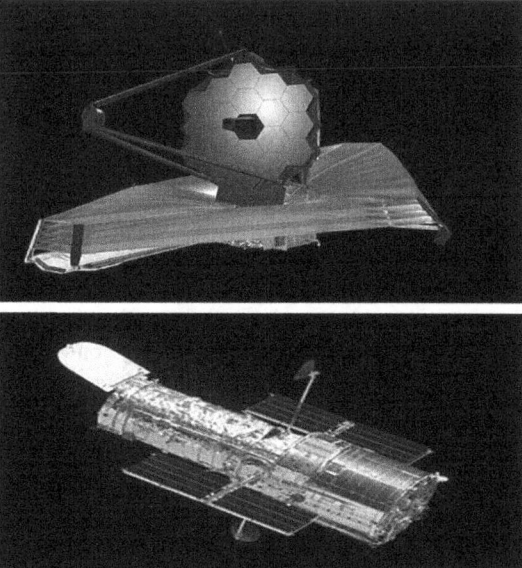

FIGURE 8.5 The Next Generation Space Telescope and the Hubble Space Telescope simulated by the block-diagonal algorithm at the Lockheed Martin Space Systems Company.

8.5   SUMMARY OF THE BLOCK-DIAGONAL
ALGORITHM FOR A TREE CONFIGURATION

First Forward Pass

Step 1. For $j = 1, ..., NB$ (number of bodies) compute ω^j from Eq. (8.9), $a_t^{Q_j}$ from Eq. (8.15), α_t^j from Eq. (8.16), E^j a unity matrix for $j = 1$, otherwise from Eq. (8.50), A^j from Eq. (8.27), Y_1^j from Eq. (8.28), M_1^j from Eq. (8.30), M_2^j from Eq. (8.31), X^j from Eq. (8.32), W^j from Eq. (8.40), and N^j from Eq. (8.41), and specify the selection matrix $S_j^{c(j)}$ for use in Eq. (8.47).

Backward Pass

Step 2. For $j = NB, ..., 1$, compute M_3^j, Y_2^j from Eq. (8.33), ν^j from Eq. (8.36), and M and X from Eq. (8.37).

Step 3. If $j = 1$, go to Step 5; otherwise compute updates defined in Eqs. (8.50) and (8.53).

Step 4. Replace j with $j - 1$ and go to Step 2.

Second Forward Pass

Step 5. For $j = 1$, evaluate in sequence Eqs. (8.55), (8.56), and (8.49). Set $j = 2$.

Step 6. Compute sequentially, Eqs. (8.39), (8.35), (8.14), and (8.49).

Step 7. If $j = NB$, stop; otherwise, replace j with $j + 1$ and go to Step 6.

8.6   NUMERICAL RESULTS DEMONSTRATING
COMPUTATIONAL EFFICIENCY

As stated before, a general-purpose flexible multibody dynamics code based on the block-diagonal algorithm given here was developed by the author and his colleagues at the Lockheed Missiles and Space Company. One application involves two wrapped-rib antennas undergoing large deformation during deployment, see Figure 8.8 An antenna was modeled with 160 spring-connected rigid segments attached to an L-shaped elastic cantilever beam, with complex interbody forces and bodies connected to one another at 2 dof hinges by non-linear springs, representing bending and torsion. Simulation of the deployment dynamics was done by the flexible multibody dynamics code based on the block-diagonal algorithm, and the algorithm involving a dense mass matrix given in Chapter 6. **Both codes produced exactly the same results, but the code using the block-diagonal algorithm given here ran 26 times faster** [12]. We present the ground test simulation results, together with the experimental results at the end of this chapter.

8.7   MODIFICATION OF THE BLOCK-DIAGONAL FORMULATION TO HANDLE MOTION CONSTRAINTS

To motivate our discussion of systems with motion constraints, we give an example of a four-bar linkage that is commonly used for solar panel deployment in spacecraft. An actual solar panel deployment simulation is reported in Ref. [9] for the Indian National Satellite (INSAT) built by the Ford Aerospace Corporation for the Indian Space Research Organization. The solar panel deployment for INSAT from the stowed to the deployed configuration, shown in Figure 8.6, shows the closed structural loop represented by the four-bar mechanism, the base, the kinematic control link, the yoke, and a little segment of the panel, making it a four-bar linkage.

When there are constraints on the motion of the system such as that represented by a closed loop of bodies, one approach that lends itself to the block-diagonal formulation is to cut the loop, impose equal and opposite unknown forces and torques that originally kept the loop closed, and solve for the unknown forces and torques by consideration of the motion constraints. Figure 8.7a and b illustrates the closed loop and its cut-loop configurations.

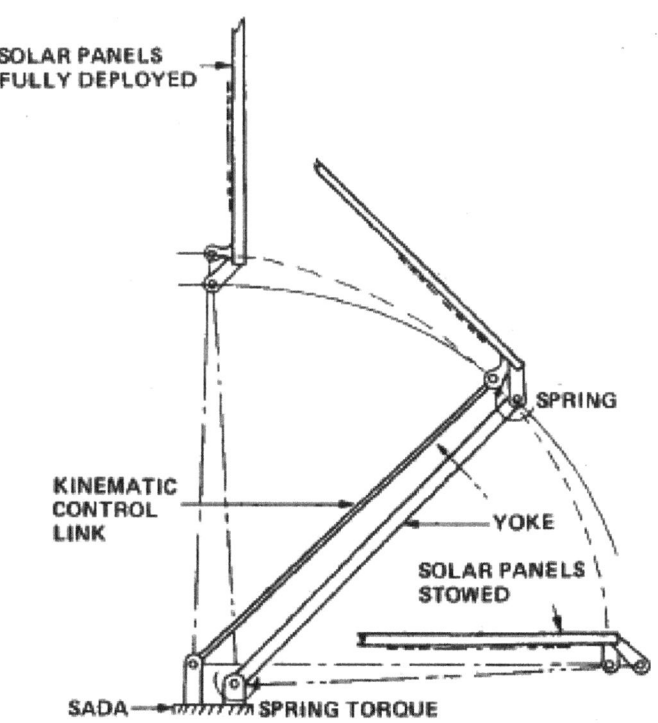

FIGURE 8.6   INSAT solar array deployment mechanism (four-bar linkage).

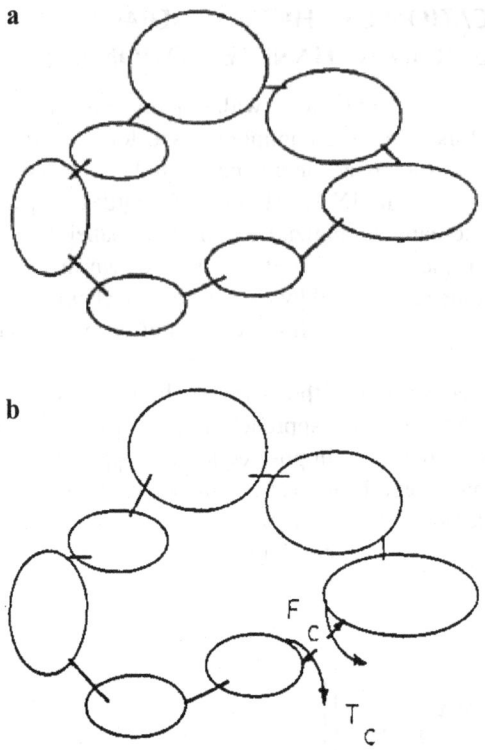

a

b

$$F$$
$$c$$

$$T$$
$$c$$

FIGURE 8.7   (a) System of bodies in a closed kinematic loop; and (b) same system with loop cut, applying equal and opposite forces and torques at hinge.

Let F_c and T_c denote the body measure numbers of the unknown constraint force and torque acting on one of the bodies at the joint that is cut, and introduce the notation:

$$\lambda = \begin{Bmatrix} F_c \\ T_c \end{Bmatrix} \tag{8.57}$$

Recognizing the body on which λ acts now as a "terminal" body j, denote

$$H_1^j = \phi_c^{j^t} - \sum_{i=1}^{6} G_{12+i}^j \eta^j S_i^j$$

$$H_2^j = \begin{bmatrix} U_3 & 0_{3\times3} \\ \tilde{r}_c^j & U_3 \end{bmatrix} \tag{8.58}$$

where G_{12+i}^j refers to the i-th generalized geometric stiffness matrix due to the interbody force and moment represented by λ; S_i^j is the 6×6 (selection) identity matrix except for a row of zeros corresponding to a rotation matrix about which moment cannot be transferred; $\phi_c^j, \tilde{r}_c^j$ refer to the modal matrix at the cut joint and the skew-symmetric matrix for the position vector from Q_j of this terminal body to the cut, respectively; and U is a 3×3 identity matrix. For the terminal body at the other side of the cut, simply reverse the sign of λ. Now with this additional set of forces and moments, Eqs. (8.26) and (8.29) become, respectively,

$$E^j \ddot{\eta}^j = A^j \left\{ \begin{matrix} a_0^{Q_j} \\ \alpha_0^j \end{matrix} \right\} + Y_1^j + H_1^j \lambda \tag{8.59}$$

$$0 = \left\{ \begin{matrix} f^{*j} - f_e^j - f_{c(j)}^j \\ t^{*j} - f_e^j - t_{c(j)}^j \end{matrix} \right\} = M_1^j \left\{ \begin{matrix} a_0^{Q_j} \\ \alpha_0^j \end{matrix} \right\} + M_2^j \ddot{\eta}^j + X^j - \left\{ \begin{matrix} f_{c(j)}^j \\ t_{c(j)}^j \end{matrix} \right\} - H_2^j \lambda \tag{8.60}$$

Here, using Eq. (8.59) in Eq. (8.60) leads to the terms defined in Eq. (8.33) as before, and on introducing the new term:

$$H_3^j = H_2^j - M_2^j [E_j]^{-1} H_1^j \tag{8.61}$$

we get the equations of motion for the hinge degrees of freedom:

$$\left\{ \begin{matrix} \ddot{\tau}^j \\ \ddot{\theta}^j \end{matrix} \right\} = -\left[v^j \right]^{-1} R^{jT} \left[M_3^j \left\{ \begin{matrix} \hat{a}_0^{Q_j} \\ \hat{\alpha}_0^j \end{matrix} \right\} + Y_2^j - H_3^j \lambda \right] + \left[v^j \right]^{-1} \left\{ \begin{matrix} f_h^j \\ t_h^j \end{matrix} \right\} \tag{8.62}$$

Note that, f_h^j, t_h^j, the working interaction forces and torques on body j are explicitly known. With this expression, the D'Alembert equation for the dynamic equilibrium of external and inertia forces and torques at the hinge Q_j is modified from Eq. (8.38) as

$$\left\{ \begin{matrix} f^{*j} - f_e^j \\ t^{*j} - t_e^j \end{matrix} \right\} = M \left\{ \begin{matrix} \hat{a}_0^{Q_j} \\ \hat{\alpha}_0^j \end{matrix} \right\} + X - H\lambda = 0 \tag{8.63}$$

Here, the definitions of M and X are the same as in Eq. (8.37), and we have further introduced:

$$H = H_3^j - M_3^j R^j [v^j]^{-1} R^{jT} H_3^j \tag{8.64}$$

Equations corresponding to the vibration modal coordinates, Eq. (8.49), for the body inboard to the cut body reduce to

$$E^{c(j)} \ddot{\eta}^{c(j)} = A^{c(j)} \left\{ \begin{matrix} a_0^{Q_{c(j)}} \\ \alpha_0^{c(j)} \end{matrix} \right\} + Y_1^{c(j)} + H_1^{c(j)} \lambda \tag{8.65}$$

Here, the notations of Eq. (8.50) are retained with no change, with geometric stiffness due to interbody forces on body c(j) accounted in the definition, using N^j from Eq. (8.41), and $d_i^{c(j)}$ from Eq. (8.46)

$$H_1^{c(j)} = \left[\sum_{i=1}^{6} G_{12+i}^{c(j)} \eta^{c(j)} S_i^{c(j)} d_i^{c(j)} - N^{j^T} \right] H \qquad (8.66)$$

Finally, using the shift operator on forces and moments from Q_j to $Q_{c(j)}$ and the updates defined in Eq. (8.53), and defining further

$$H_3^{c(j)} = H_2^{c(j)} + W^{j^T} H \qquad (8.67)$$

the procedure given earlier for deriving equations of motion for hinge dof, Eq. (8.54), is modified to yield

$$\begin{Bmatrix} \ddot{\tau}^{c(j)} \\ \ddot{\theta}^{c(j)j} \end{Bmatrix} = -\left[v^{c(j)} \right]^{-1} R^{c(j)^T} \left[M_3^{c(j)} \begin{Bmatrix} \hat{a}_0^{Qc(j)} \\ \hat{\alpha}_0^{c(j)} \end{Bmatrix} + Y_2^{c(j)} - H_3^{c(j)} \lambda \right]$$
$$+ \left[v^{c(j)} \right]^{-1} \begin{Bmatrix} f_h^{c(j)} \\ t_h^{c(j)} \end{Bmatrix} \qquad (8.68)$$

In summary, note that Eqs. (8.68) and (8.65) differ from Eqs. (8.49) and (8.54) only by the unknown constraint force and moments term, denoted by λ. These constraint force and moment components can be solved by consideration of the motion constraint equations. However, a direct substitution of the constraint equations will destroy the block-diagonal nature of the coefficient matrix of the second derivatives of the hinge degree of freedom generalized coordinates in the dynamical equations. The algorithm given in the following preserves the block-diagonal character of the equations. Define for $j = 1$:

$$f_1^1 = -\left[v^1 \right]^{-1} R^{1^T} Y_2^1 + \left[v^1 \right]^{-1} \begin{Bmatrix} f_h^1 \\ t_h^1 \end{Bmatrix}$$
$$c_1^1 = \left[v^1 \right]^{-1} R^{1^T} H_3^1 \qquad (8.69)$$

This lets us start the sequence of generating the equations of motion for body $j = 1$ with

$$\begin{Bmatrix} \ddot{\tau}^1 \\ \ddot{\theta}^1 \end{Bmatrix} = f_1^1 + c_1^1 \lambda \qquad (8.70)$$

Now, using Eq. (8.14), where the zeroth body is the inertial frame, yields

$$\begin{Bmatrix} a_0^{Q_1} \\ \alpha_0 \end{Bmatrix} = g^1 + h^1 \lambda \qquad (8.71)$$

where

$$g^1 = R^1 f_1^1$$
$$h^1 = R^1 c_1^1$$

(8.72)

Referencing Eq. (8.59) and using the definitions, again for $j = 1$:

$$f_2^1 = \left[E^1\right]^{-1}\left[A^1 g^1 + Y_1^1\right]$$
$$c_2^1 = \left[E^1\right]^{-1}\left[A^1 h^1 + H_1^1\right]$$

(8.73)

yield the equations for the modal coordinates of body 1 in the form (recall E^1 is a unity matrix for normalized modes):

$$\ddot{\eta}^1 = f_2^1 + c_2^1 \lambda$$

(8.74)

Using Eqs. (8.71) and (8.74) and the definitions:

$$e^j = W^j g^{c(j)} + N^j f_2^{c(j)}$$
$$n^j = W^j h^{c(j)} + N^j c_2^{c(j)}$$

(8.75)

in Eq. (8.39) lead to the equation:

$$\left\{ \begin{array}{c} \hat{a}_0^{Q_j} \\ \hat{\alpha}_0^{Q_j} \end{array} \right\} = e^j + n^j \lambda$$

(8.76)

When this is substituted in Eq. (8.62), and the following general notations are used:

$$f_1^j = -\left[v^j\right]^{-1} R^{j^T}\left[M_3^j e^j + Y_2^j\right] + \left[v^j\right]^{-1}\left\{ \begin{array}{c} f_h^j \\ t_h^j \end{array} \right\}$$
$$c_1^j = -\left[v^j\right]^{-1} R^{j^T}\left[M_3^j n^j - H_3^j\right]$$

(8.77)

the equation for the body j hinge dof is obtained:

$$\left\{ \begin{array}{c} \ddot{\tau}^j \\ \ddot{\theta}^j \end{array} \right\} = f_1^j + c_1^j \lambda$$

(8.78)

Using Eqs. (8.76) and (8.78) in Eq. (8.14) with the substitutions:

$$g^j = e^j + R^j f_1^j$$
$$h^j = n^j + R^j c_1^j$$

(8.79)

yields the general relation for the j-th body:

$$\left\{ \begin{matrix} a_0^{Q_j} \\ \alpha_0^j \end{matrix} \right\} = g^j + h^j \lambda \tag{8.80}$$

Finally, substituting Eq. (8.80) in Eq. (8.59) and using the notations:

$$f_2^j = \left[E^j \right]^{-1} \left[A^j g^j + Y_1^j \right]$$
$$c_2^j = \left[E^j \right]^{-1} \left[A^j h^j + H_1^j \right] \tag{8.81}$$

produce the equations for the modal coordinates for the j-th flexible body:

$$\ddot{\eta}^j = f_2^j + c_2^j \lambda \tag{8.82}$$

With generic equations, Eqs. (8.78) and (8.82), a recursive formulation of the equations of motion in the block-diagonal form is completed. Now, λ nvolved in these equations can be solved by considering the constraints. We consider *holonomic* constraints, where the constraint conditions are expressed in the form of m *algebraic* equations of loop closure as

$$f_j(q_1, \ldots, q_n) = 0, \quad j = 1, \ldots, m; \quad m < n \tag{8.83}$$

One way of using these is to differentiate Eq. (8.83) twice with respect to time, and use the resulting equations along with the dynamical equations for $(n+m)$ unknowns, $\ddot{q}_1, \ldots, \ddot{q}_n, \lambda_1, \ldots, \lambda_m$. However, this approach is known [9] to give rise to a drift with time in the constraint satisfaction at the position/orientation level. A procedure that satisfies the position, velocity, and acceleration constraints simultaneously is given below.

Two differentiations of the set of equations, Eq. (8.83), with respect to time yield:

$$J_i \dot{U}_i + J_d \dot{U}_d = -[\dot{J}_i U_i + \dot{J}_d U_d] \tag{8.84}$$

where the following notations for the Jacobian matrices J_i, J_d have been used:

$$[J_i] = \left[\frac{\partial f_j}{\partial q_i} \right]; \quad [J_d] = \left[\frac{\partial f_j}{\partial q_d} \right]; \quad U_i = \dot{q}_i; \quad U_d = \dot{q}_d \tag{8.85}$$

with the set of generalized coordinates partitioned into a subset q_i of independent generalized coordinates and a chosen subset q_d of dependent generalized coordinates, as many in number as the number of constraints, m. Now, the dynamical equations generated sequentially in Eqs. (8.78) and (8.82) for bodies $j = 1, \ldots, N$ can be stacked together and written as follows:

$$\dot{U}_i = F_i + C_i\lambda$$
$$\dot{U}_d = F_d + C_d\lambda \tag{8.86}$$

Substituting Eq. (8.86) in Eq. (8.84) provides the equations for λ the column of measure numbers of the unknown constraint forces and moments:

$$\left[J_iC_i + J_dC_d \right]\lambda = -\left\{ J_iF_i + J_dF_d + \dot{J}_iU_i + \dot{J}_dU_d \right\} \tag{8.87}$$

Using Eq. (8.87) in Eq. (8.86), the dynamical differential equations are now explicitly known. In situations where the coefficient matrix in Eq. (8.87) is singular (which should not happen for independent constraints), the constraint stabilization method of Ref. [10] may be used, or a low-order integration with recursive error correction [11] can be used. In summary, the following set of differential-algebraic equations is solved for constrained dynamical systems:

$$\dot{U}_i = F_i - C_i\left[J_iC_i + J_dC_d \right]^{-1}\left\{ J_iF_i + J_dF_d + \dot{J}_iU_i + \dot{J}_dU_d \right\} \tag{8.88}$$

$$\dot{q}_i = U_i \tag{8.89}$$

$$J_iU_i + J_dU_d = 0 \tag{8.90}$$

$$f(q_i, q_d) = 0 \tag{8.91}$$

8.8   VALIDATION OF THEORY WITH GROUND TEST RESULTS

The results reported below of a comparison of a simulation with the closed-loop block-diagonal algorithm given above and ground test measurements for a wrapped-rib antenna deployment, are taken from the conference proceedings of the American Institute of Aeronautics and Astronautics [12]. The top part of Figure 8.8 is a sketch of the flight configuration of the antenna on the spacecraft, and the bottom part of Figure 8.8 gives an overall picture of a test set-up for an antenna with the support system of an L-shaped boom and a loop-closing cable. Figure 8.9 gives the details of the bending-torsion stiffness vs. angle of twist or bending or buckling for a rib element, where an antenna rib is segmented by many ribs connected by circumferential webs. Figures 8.10–8.12 show the shoulder reaction torques in the x-, y-, and z-direction, respectively, as the antenna deploys from being initially wrapped around a spool. Figure 8.13 shows the off-load cable tension as the antenna unfurls from the spools for a closed structural loop, where the loop closes with the off-loading cable that tries to negate the gravity load on the suspended structure. While the simulation results in Figure 8.13 seem to diverge, that may be symptomatic of the fact that the cable is an on–off spring, a fact that was not modeled, and no constraint stabilization [9] was featured implemented in the formulation. All simulation results are obtained by the algorithm reported

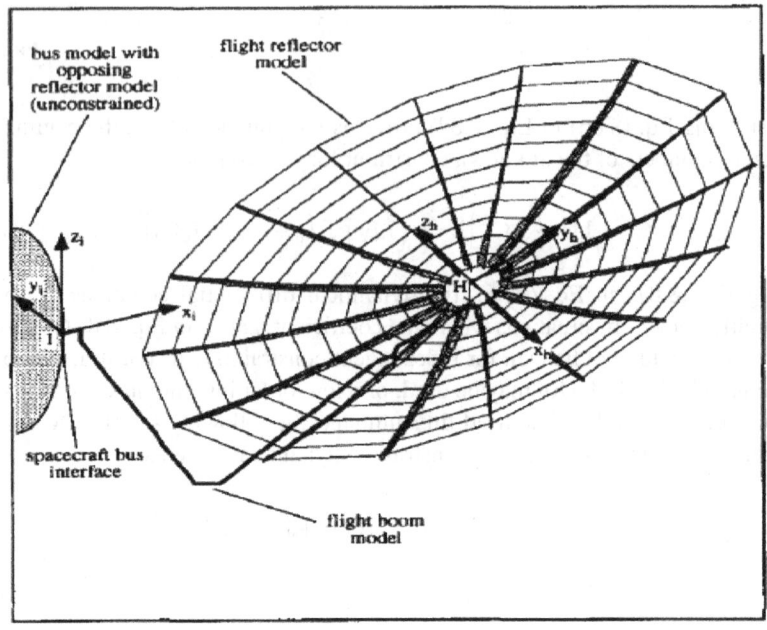

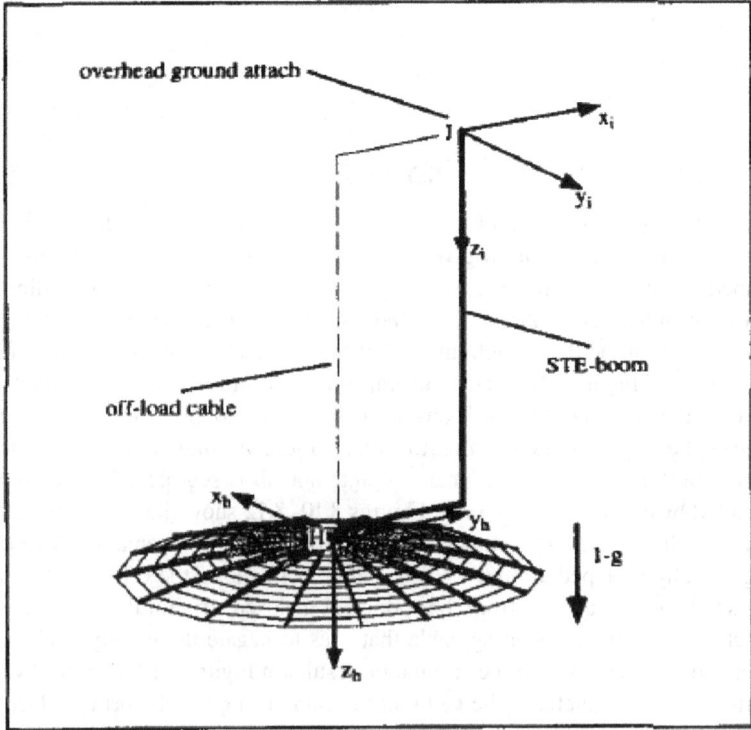

FIGURE 8.8 Flight and ground test model of antenna attached to the ceiling by an L-shaped boom and an off-load cable forming a closed-loop structure.

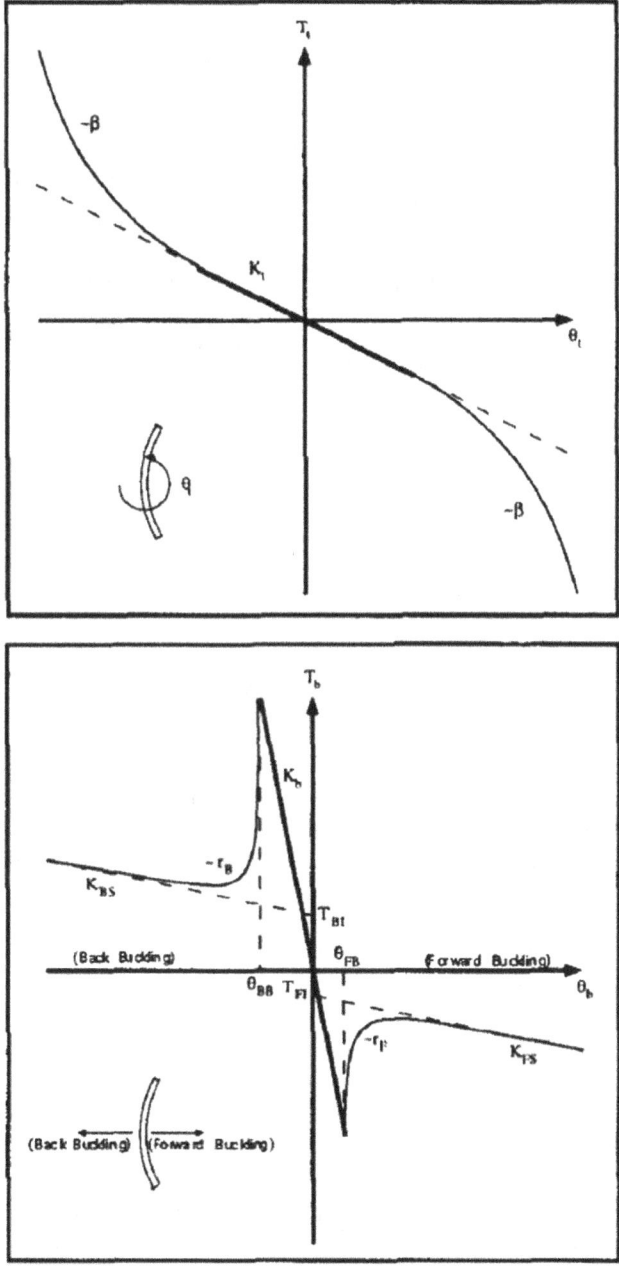

FIGURE 8.9   Rib element torsional non-linear stiffness (top figure) and bending stiffness leading to buckling (bottom figure).

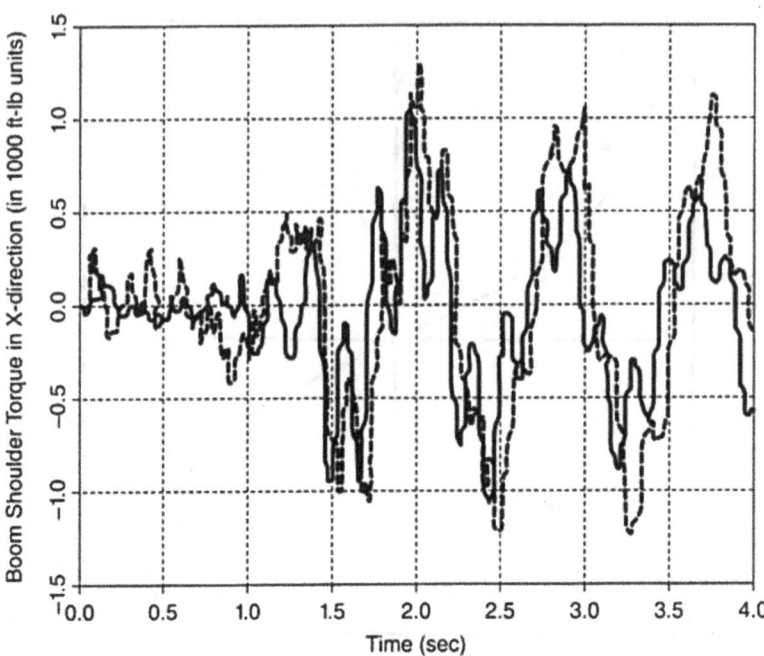

FIGURE 8.10 Boom shoulder reaction torque in x-direction for deployment of antenna as it unfurls from a spool: comparison between theory and test (test: solid line; theory: dashed line).

in this chapter. Considering the fact that the numerical values of the parameters necessary for the simulation could only be estimated approximately, the validation of the formulation with the experimental data may be deemed quite satisfactory.

8.9 CONCLUSION

A comprehensive and computationally efficient formulation has been given for the equations of large overall motion of a system of flexible bodies with motion-induced geometric stiffness and motion constraints. For an arbitrary elastic body, the method requires a one-time computation of geometric stiffness matrices due to 12 distributed inertia loadings and, at most, five point-loadings per hinge connection. By accounting for these geometric stiffness effects, the formulation retains its validity even with the use of rotation rates higher than the vibration modes frequencies retained. Use of modes by themselves would have been an act of premature linearization, showing disastrous results when the rotation rate exceeds the first bending frequency of a component. For systems with motion constraints, a block-diagonal algorithm that satisfies the position, velocity, and acceleration constraints has been given. The results of a ground test of an antenna deployment, involving a closed structural loop, show reasonable correlation between the theory and the experiment.

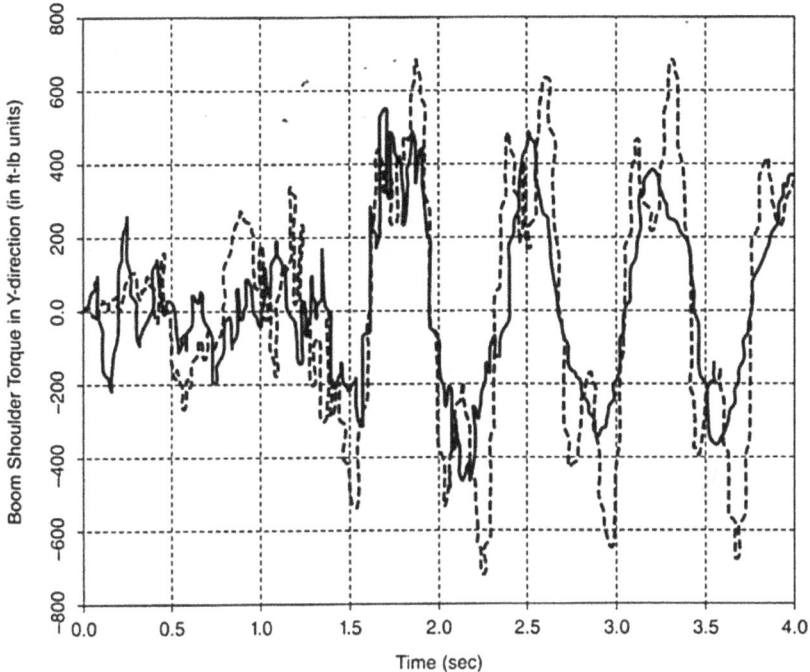

FIGURE 8.11   Boom shoulder reaction torque in y-direction for deployment of antenna as it unfurls from a spool: comparison between theory and test (test: solid line; theory: dashed line).

PROBLEM SET 8

Consider developing a Matlab code based on the block-diagonal, recursive algorithm given in this chapter.

1. (a)   Show details of the H^j matrix for a 2- or 3-dof rotational hinge between frames p(j) and j in Figure 8.3b.
    (b)   Show that the inertia force at a node located by (x_i^n, y_i^n, z_i^n) from the point Q_j in Figure 8.3b is indeed given by Eqs. (8.18) and (8.19).
    (c)   Show that the geometric stiffness terms in Eq. (8.17) are indeed given by Eq. (8.20).
    (d)   Show that the dynamic equilibrium equation, Eq. (8.24), is made up of the constituent terms in Eqs. (8.25)–(8.27).
    (e)   Show that linear and angular acceleration terms involving second derivatives of the generalized coordinates, in Eq. (8.34), are indeed given by Eqs. (8.35)–(8.37).
2. Using the algorithm given in this chapter, formulate and code the equations of motion for the two-beam exercise problem, given at the end of Chapter 6, in a block-diagonal form.

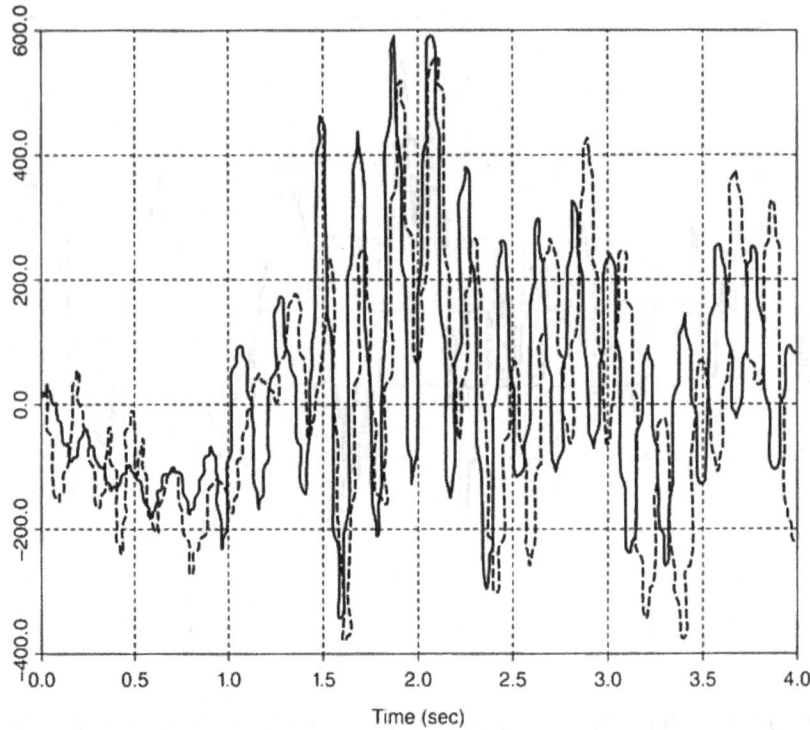

FIGURE 8.12   Boom shoulder reaction torque in z-direction for deployment of antenna as it unfurls from a spool: comparison between theory and test.

3. Consider a four-link robotic manipulator on a movable base, used in construction sites for digging soil. Once the end link digs in to scoop up the soil, it becomes a closed loop system, while it becomes a free four-link manipulator on the moving base, after the soil is scooped. Write an algorithm for the two stages of the earth-digging manipulator.

4. Formulate the block-diagonal equations for the planar swinging motion of a triple pendulum made of three hinge-connected beams of uniform mass, with each beam of length L. Code the equations and simulate the motion with your choice of parameters and initial conditions. Form the same equations with the dense matrix method of Chapter 6. Code these equations and simulate for the same parameters and initial conditions. Note the time difference in simulation.

APPENDIX 8 AN ALTERNATIVE DERIVATION OF GEOMETRIC STIFFNESS DUE TO INERTIA LOADS

(Contributed by Dr Tushar Ghosh of L3 Communications Corporation)

An interesting illustration of geometric stiffness is work done by the inertia force corresponding to the non-linear terms in the acceleration for the existing

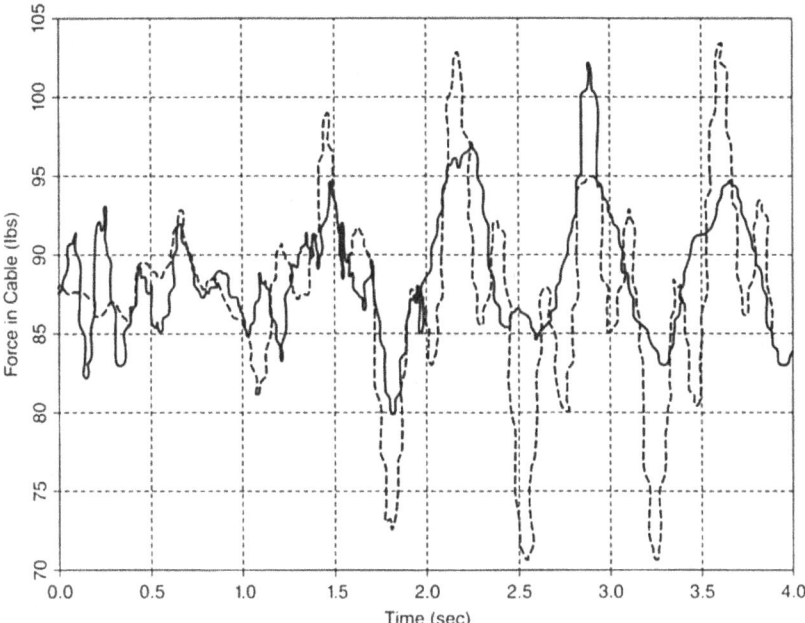

FIGURE 8.13 Off-load cable tension comparison as the antenna unfurls from a spool (test: solid line; simulation: dashed line); cable closes the structural loop.

stress state as the body goes into further deformation, is shown here for the example system of Figure 8.1.

Let $\mathbf{i}$, $\mathbf{j}$ be the unit vectors along and perpendicular to OA in Figure 8.1. The inertia force of link AB due to its lumped mass at B, and the inertia force of link BC due to its lumped mass at C are

$$\mathbf{F}^{AB} = -m(r+L)[\omega^2\,\mathbf{i} + \dot{\omega}\mathbf{j}]$$

$$\mathbf{F}^{BC} = -m(r+2L)[\omega^2\,\mathbf{i} + \dot{\omega}\mathbf{j}]$$

(A8.1)

Work done by F^{AB} due to small rotation q_1 from the reference state, and work done by F^{BC} due to small rotation (q_1+q_2) from this same reference state are, respectively:

$$\mathbf{W}^{BC} = \mathbf{F}^{BC} \cdot \left\langle \begin{array}{l} \{(L\cos q_1 - L) + [L\cos(q_1+q_2) - L]\}\,\mathbf{i} \\ + \{L\sin q_1 + L\sin(q_1+q_2)\}\,\mathbf{j} \end{array} \right\rangle$$

$$\approx -\frac{1}{2}m(r+2L)(\omega^2\,\mathbf{i} + \dot{\omega}\,\mathbf{j}) \cdot \left[\begin{array}{l} L\left\{-\dfrac{1}{2}q_1^2 - \dfrac{1}{2}(q_1+q_2)^2\right\}\mathbf{i} \\ + L\{q_1 + (q_1+q_2)\}\mathbf{j} \end{array} \right] + \cdots$$

(A8.2)

In the above expressions, only terms up to the second order in q_1, q_2 are kept because geometric stiffness is obtained from the second-order terms in the work done. The total work done by the inertia forces, from the reference configuration OAB to the deformed configuration shown in Figure 8.1 is then

$$W^{AB} + W^{BC} = \frac{1}{2} m \omega^2 L\{(r+L)q_1^2 + (r+2L)[q_1^2 + (q_1+q_2)^2]\}$$

$$= \frac{1}{2} m \omega^2 L[(3r+5L)q_1^2 + 2(r+2L)q_1q_2 + (r+2L)q_2^2] \qquad (A8.3)$$

This can be rewritten as

$$W^{AB} + W^{BC} = \frac{1}{2}\begin{bmatrix} q_1 & q_2 \end{bmatrix} m\omega^2 L \begin{bmatrix} 3r+5L & r+2L \\ r+2L & r+2L \end{bmatrix} \begin{Bmatrix} q_1 \\ q_2 \end{Bmatrix}$$

$$= \frac{1}{2}\begin{bmatrix} q_1 & q_2 \end{bmatrix} [K_g] \begin{Bmatrix} q_1 \\ q_2 \end{Bmatrix} \qquad (A8.4)$$

where we have identified the geometric stiffness matrix for the "flexible body" ABC in Figure 8.1:

$$\begin{bmatrix} K_g \end{bmatrix} = m\omega^2 L \begin{bmatrix} 3r+5L & r+2L \\ r+2L & r+2L \end{bmatrix} \qquad (A8.5)$$

This is the same result for geometric stiffness as in Eq. (8.6)!

REFERENCES

1. Banerjee, A.K., (1993), "Block-Diagonal Equations for Multibody Elastodynamics with Geometric Stiffness and Constraints", in *State of the Art Survey Lecture in Multibody Dynamics*, European Space Agency, June 1992; republished in the *Journal of Guidance, Control, and Dynamics*, 1992, **16**(6), pp. 1092–1100.
2. Banerjee, A.K., (2003), "Contributions of Multibody Dynamics to Space Flight: A Brief Review", *Journal of Guidance, Control, and Dynamics*, **26**(2), pp. 385–394.
3. Kane, T.R. and Levinson, D.A., (1985), *Dynamics: Theory and Applications*, McGraw-Hill, p. 159.
4. Kane, T.R., Ryan, R.R., Jr., and Banerjee, A.K., (1987), "Dynamics of a Cantilever Beam Attached to a Moving Base", *Journal of Guidance, Control, and Dynamics*, **10**(2), pp. 139–151.
5. Cook, R.D. (1985), Concepts and Applications of Finite Analysis, McGraw-Hill, pp. 331–341.
6. Rosenthal, D.E. (1990), "An Order-n Formulation for Robotic Systems", *Journal of Astronautical Sciences*, **38**(4), pp. 511–530.
7. Banerjee, A.K. and Lemak, M.E. (1991), "Multi-Flexible-Body Dynamics Capturing Motion Induced Stiffness", *Journal of Applied Mechanics*, **5**, pp. 766–775.

8. Banerjee, A.K. and Dickens, J.M., (1990), "Dynamics of an Arbitrary Flexible Body in Large Rotation and Translation", *Journal of Guidance, Control, and Dynamics*, **13**(2), pp. 221–227.

9. Wie, B., Furumoto, N., Banerjee, A.K., and Barba, P.M., (1986), "Modeling and Simulation of Spacecraft Solar Array Deployment", *Journal of Guidance, Control, and Dynamics*, **9**(5), pp. 593–598.

10. Baumgarte, J., (1972), "Stabilization of Constraints and Integrals of Motion in Dynamical Systems", *Computer Methods in Applied Mechanics and Engineering*, **1**, pp. 1–16.

11. Negrut, D., Jay, L.O., and Khude, N., (2009), "A Discussion of Low-Order Numerical Integration Formulas for Rigid and Flexible Multibody Dynamics", *Journal of Computational and Nonlinear Dynamics*, **4**(2), pp. 021008:1–11.

12. Lemak, M.K. and Banerjee, A.K., (1994), "Comparison of Simulation with Test of Deployment of a Wrapped-Rib Antenna", in AIAA Conference Paper AIAA-94-3577-CP., Scottsdale, Az., USA.

9 Efficient Variables, Recursive Formulation, and Multi-Loop Constraints in Flexible Multibody Dynamics

In this chapter, we undertake computationally efficient formulations of the equations of motion of complex flexible multibody systems with many motion constraints, with efficiency coming from the use of certain generalized speeds for the rotation and vibration of a flexible body, defined here. The results will show that these efficient variables reduce simulation times more than those with more commonly used motion variables. The kernel of the algorithm given here consists of deriving the equations of motion for a single flexible body described in terms of these so-called efficient variables. The single-body equations are extended to a multibody system, and the latter to closed structural loops obtained as before, by cutting the loops and exposing unknown constraint forces, in a recursive formulation; for constrained systems; the final step is to impose the constraint conditions, which together with the dynamical equations form the system of equations to be solved numerically. The development closely follows that given in Refs. [1, 2].

9.1 SINGLE FLEXIBLE BODY EQUATIONS IN EFFICIENT VARIABLES

Refer to Figure 9.1 for a schematic representation of a flexible body j with a flying reference frame, also labeled j, with its origin at the point O of the body. The body deforms with respect to this frame, and a material point P' of j moves to a point P of the body due to local elastic deformation.

Inertial N-frame velocities ${}^N\mathbf{v}^P$, ${}^N\mathbf{v}^O$ of points P and O are written in terms of the angular velocity ${}^N\boldsymbol{\omega}^j$ of the flying reference frame j in Figure 9.1, and the position vector $\mathbf{r}$ from O to P' in the undeformed configuration, and the deformation vector from P' to P given by a sum of mode shapes φ_k^j multiplied by modal coordinates, $q_k, k = 1,\ldots,n$, as follows:

$$
{}^N\mathbf{v}^P = {}^N\mathbf{v}^O + {}^N\boldsymbol{\omega}^j \times \left(\mathbf{r} + \sum_{k=1}^{n} \varphi_k^j q_k \right) + \sum_{k=1}^{n} \varphi_k^j \dot{q}_k
\tag{9.1}
$$

DOI: 10.1201/9781003231523-10

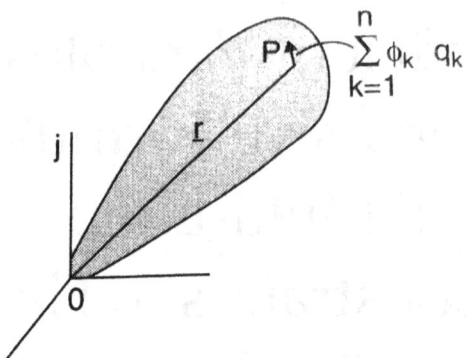

FIGURE 9.1   Schematic representation of a flexible body with a modal representation of deformation.

To make this chapter self-contained, we review here equations for a single flexible body with our choice of "efficient" generalized speeds u_i, $i = 1,\ldots,6+n$ using vectors expressed in terms of basis vectors $\mathbf{b}_i$, $i = 1,2,3$ fixed in the frame j, as in Eqs. (9.2)–(9.4):

$$u_i = {}^N\boldsymbol{\omega}^j \cdot \mathbf{b}_i, \quad i = 1,2,3 \quad \to \quad {}^N\boldsymbol{\omega}^j = u_1\mathbf{b}_1 + u_2\mathbf{b}_2 + u_3\mathbf{b}_3 \tag{9.2}$$

$$u_{3+i} = {}^N\mathbf{v}^O \cdot \mathbf{b}_i, \quad i = 1,2,3 \quad \to \quad {}^N\mathbf{v}^O = u_4\mathbf{b}_1 + u_5\mathbf{b}_2 + u_6\mathbf{b}_3 \tag{9.3}$$

We use vibration modes for body j, indicated by a superscript j, with efficient modal generalized speeds for vibration, $u_7,\ldots,u_{6+n}$, introduced in Ref. [3].

$$\sum_{k=1}^{n} \boldsymbol{\varphi}_k^j u_{6+k} = {}^N\boldsymbol{\omega}^j \times \sum_{k=1}^{n} \boldsymbol{\varphi}_k^j q_k + \sum_{k=1}^{n} \boldsymbol{\varphi}_k^j \dot{q}_k \tag{9.4}$$

Note that $\boldsymbol{\varphi}_k^j$ is a (3×1) vector with components in the x-, y-, z-axes of the j-frame.

Using Eqs. (9.2)–(9.4) in Eq. (9.1) leads to the velocity; partial velocity, and acceleration expressions:

$$^N\mathbf{v}^P = {}^N\mathbf{v}^O + {}^N\boldsymbol{\omega}^j \times \mathbf{r} + \sum_{k=1}^{n} \boldsymbol{\varphi}_k^j u_{6+k} \tag{9.5a}$$

$$^N\mathbf{v}_i^P = {}^N\mathbf{v}_i^O + {}^N\boldsymbol{\omega}_i^j \times \mathbf{r} + \delta_{i,6+k}\boldsymbol{\varphi}_k^j \quad i = 1,\ldots,6+n; \; k = 1,\ldots,n \tag{9.5b}$$

$$^N\mathbf{a}^P = {}^N\mathbf{a}^O + {}^N\boldsymbol{\alpha}^j \times \mathbf{r} + \sum_{k=1}^{n} \boldsymbol{\varphi}_k^j \dot{u}_{6+k} + {}^N\boldsymbol{\omega}^j \times \left({}^N\boldsymbol{\omega}^j \times \mathbf{r} + \sum_{k=1}^{n} \boldsymbol{\varphi}_k^j u_{6+k} \right) \tag{9.6}$$

Comparing the velocity expression in Eq. (9.5) with Eq. (9.1), we see that Eq. (9.5) is free of modal coordinates. This means that the partial velocities of P will not have any time dependency, and that has the important consequence of producing equations of motion with a mass matrix that is constant. We will see later that carrying over the idea to multibody systems also simplifies the equations, making simulations time-efficient. Dot-multiplying the terms in Eq. (9.4) by φ_i^j and integrating over the body mass, we get the kinematical equations for the derivative of the modal coordinates of body j in terms of the elastic generalized speeds, $u_7, \ldots, u_{6+n}$:

$$E_{ik}^j (u_{6+k} - \dot{q}_k) = -{}^N \omega^j \cdot \sum_{k=1}^{n} q_k \int_j \varphi_i^j \times \varphi_k^j dm; \quad E_{ik}^j = \int_{B_j} \varphi_i^j \cdot \varphi_k^j dm \quad (9.7)$$

or, writing entirely in a matrix form:

$$\dot{q} = \begin{bmatrix} -E_j^{-1} G_q^j & I \end{bmatrix} \begin{Bmatrix} \omega \\ \sigma \end{Bmatrix} \tag{9.8}$$

$$\dot{q} = [q_7, \ldots \ldots, q_{6+n}]^T; \quad \omega = [\omega_1, \omega_1, \omega_1]^T; \quad \sigma = [u_7, \ldots \ldots, u_{6+n}]^T$$

where G_q^j is an n×3 matrix written in terms of d_{ik} (involving cross product of modes) defined in the Appendix A on modal integrals at the end of the book:

$$G_q^j = \begin{bmatrix} \sum_{k=1}^{n} q_k d_{ik}^t \\ \cdot \\ \cdot \\ \cdot \\ \sum_{k=1}^{n} q_k d_{nk}^t \end{bmatrix}; \quad d_{ik} = \int_j \tilde{\phi}_i \phi_k dm \tag{9.9}$$

Following Kane's method [4], the *generalized inertia force* F_i^* corresponding to the generalized speed u_i, $i = 1, \ldots, 6 + n$ is formed by integrating the product of partial velocity with respect to the i-th generalized speed, and the acceleration of a generic particle P, expressed in Eq. (9.6):

$$-F_i^* = \int_j {}^N v_i^P \cdot {}^N a^P dm \qquad i = 6 + n \tag{9.10}$$

The integration is carried out over the mass of the entire body j. Referring to Eq. (9.5b), (9.6) for evaluating the partial and total derivatives of P, it is seen that the integrand in Eq. (9.10) is free of modal coordinates. This is, of course, true for a single flexible body, and a terminal body in a chain. The complete set of $(6+n)$ generalized inertia forces for a single-body j, can be written showing rigid–elastic coupling between the rigid (r) and elastic (e) generalized speeds for body j, as

$$-F_r^{j*} = M_1^j \begin{Bmatrix} \alpha_0^j \\ a_0^j \end{Bmatrix} + M_2^j \dot\sigma^j + X_0^j; \qquad X_0^j = M_1^j \begin{Bmatrix} \alpha_t^j \\ a_t^j \end{Bmatrix} + \left\{ \begin{array}{l} \tilde\omega^j I^{j/O} \omega^j + \sum_{i=1}^n N_i^{j*} \sigma_i^j \omega^j \\[2ex] \tilde\omega^j \tilde\omega^j S^{j/O} + \sum_{i=1}^n \tilde\omega^j b_i^j \sigma_i^j \end{array} \right\}$$

(9.11)

$$-F_e^{j*} = [M_2^j]^T \begin{Bmatrix} \alpha_0^j \\ a_0^j \end{Bmatrix} + E^j \dot\sigma^j + Y_0^j; \qquad Y_0^j = [M_2^j]^T \begin{Bmatrix} \alpha_t^j \\ a_t^j \end{Bmatrix} - N_\omega^j + \Omega_d^j$$

$$N_\omega^j = \begin{bmatrix} \omega^t N_1^j \omega \\ \vdots \\ \vdots \\ \omega^t N_n^j \omega \end{bmatrix}; \qquad \Omega_d^j = \begin{bmatrix} \omega^t \sum_{k=1}^n d_{1k}^j \sigma_k^n \\ \vdots \\ \vdots \\ \omega^t \sum_{k=1}^n d_{nk}^j \sigma_k^n \end{bmatrix}$$

(9.12)

Here, α_0^j, a_0^j in Eqs. (9.11) and (9.12) are, respectively, terms involving derivatives of the generalized speeds in expressions for the angular acceleration of frame j and the acceleration of the point O, in Figure 9.1, shown in Eq. (8.13); X_0^j, Y_0^j in Eqs. (9.11) and (9.12) involve generalized inertia force terms due to the remainder acceleration terms α_t^j, a_t^j in the rigid body angular acceleration of frame j, and the acceleration of point O, the generalized centrifugal force, and the modified Coriolis force. Note that in Eqs. (9.11) and (9.12), M_1^j, M_2^j, E^j are *constant* matrices for body j, with

$$M_1^j = \begin{bmatrix} I^{j/o} & \tilde S^{j/o} \\ -\tilde S^{j/o} & m^j U \end{bmatrix}; M_2^j = \begin{bmatrix} c^{j/O} \\ b^j \end{bmatrix}$$

(9.13)

Here, E^j is usually an $(n \times n)$ unity matrix for normalized modes; $I^{j/O}, S^{j/O}$ are, respectively, the matrices of second and first mass moments of inertia about O for body j; and modal integrals $c^{j/O}, b^j$ are given in the Appendix at the end of the book. In Eq. (9.13), U is a (3×3) unity matrix, and quantities with an overhead tilde sign indicate skew-symmetric matrices formed out of corresponding 3×1 matrices. As defined before, the generalized active force [4] due to non-conservative force vectors df^P at G for the i-th generalized speed for body j is given by

$$F_i^{nc} = \int_j \frac{{}^N d^N v^G}{du_i} \cdot df^G \quad i = 1,\ldots,6+n \tag{9.14}$$

Again, Kane has shown in Ref. [4] that if the kinematical equations relating the derivatives of the generalized coordinates are related to the generalized speeds in the form (with U a column matrix of generalized speeds):

$$\dot{q} = WU \tag{9.15}$$

then the generalized force due to a *potential function* P and a *dissipation function* D is

$$F^{cd} = -W^T \left\{ \frac{\partial P}{\partial q} + \frac{\partial D}{\partial \dot{q}} \right\} \tag{9.16}$$

If we take for P the sum of the potential energy for mass-normalized modes of frequencies in ω_n, a *diagonal matrix*, and an inertia load-dependent geometric stiffness, K_g, as in Ref. [2] and [5], and assume diagonal modal damping, then for body j:

$$P = \frac{1}{2} q^T \left[\omega_n^2 + \phi^T K_g \phi \right] q \tag{9.17}$$

$$D = \xi_n \dot{q}^T \omega_n \dot{q} \tag{9.18}$$

The generalized active force due to stiffness and damping follows from Eqs. (9.2), (9.7), and (9.16). For body j, generalized speeds for rigid body rotations and translations give rise to the (6×1) matrix:

$$F_r^{cd} = -\begin{bmatrix} -[G_q^j]^T \\ 0 \end{bmatrix} \left\{ \left[\omega_n^2 + \phi^T K_g \phi \right] q + 2\xi \omega_n \dot{q} \right\} \tag{9.19}$$

Equation (9.19) shows that our choice of generalized speeds has yielded non-zero contributions due to stiffness and damping corresponding to rigid body rotation, just as shown in Ref. [3], which is rather peculiar in coupling frame rotation/translation and elasticity. Generalized forces due to structural elasticity, geometric stiffness, and damping in the modal generalized speeds are given for body j in the customary manner using modal mass matrix, E_j:

$$F_e^{cd} = -E_j \left\{ \left[\omega_n^2 + \varphi^T K_g \varphi \right] q + 2\xi \omega_n \dot{q} \right\} \tag{9.20}$$

Generalized active forces due to forces and moments $\mathbf{F}^O$, $\mathbf{T}^O$ at a hinge at O and $\mathbf{F}^{Q_j}$, $\mathbf{T}^{Q_j}$ at a point Q_j in contact with an outboard body follow from Kane's method, as the sum of the vector dot-products:

$$F_i = \frac{^N d\mathbf{v}^O}{d u_i} \cdot \mathbf{F}^O + \frac{^N d\boldsymbol{\omega}^O}{d u_i} \cdot \mathbf{T}^Q + \frac{^N d\mathbf{v}^{Q_j}}{d u_i} \cdot \mathbf{F}^{Q_j} + \frac{^N d\boldsymbol{\omega}^{Q_j}}{d u_i} \cdot \mathbf{T}^{Q_j} \tag{9.21}$$

$$i = 1, \ldots, 6 + n$$

Using Eq. (9.7) for body j, and the modal matrix $\psi_{Q_j}^j$ for elastic rotation at the node Q_j, the point Q of body j (recall σ^j represents modal coordinate rates) leads to the *inertial* angular velocity of a nodal rigid body at Q_j, in the matrix form:

$$^N\omega^{Q_j} = {}^N\omega^j + \psi_{Q_j}^j \dot{q} = P_{Q_j}^j \omega^j + \psi_{Q_j}^j \sigma^j \tag{9.22}$$

Here, we have defined the following matrix that appears pervasively in the subsequent analysis (with U a 3×3 unity matrix and E_j usually an $n \times n$ unity matrix):

$$P_{Q_j}^j = \left[U - \psi_{Q_j}^j E_j^{-1} G_q^j \right] \tag{9.23}$$

The generalized force due to hinge torque and force at points O and Q_j of body j in Figure 9.1, for the rigid body and elastic motion, assuming that elastic displacement at O is zero, is written as

$$F_r = \begin{Bmatrix} \mathbf{T}^O \\ \mathbf{F}^O \end{Bmatrix} + Z_{Q_j} \begin{Bmatrix} \mathbf{T}^{Q_j} \\ \mathbf{F}^{Q_j} \end{Bmatrix}; \text{where } Z_{Q_j} = \begin{bmatrix} P_{Q_j}^{j^T} & \tilde{r}^{OQ_j} \\ 0 & U \end{bmatrix} \tag{9.24}$$

$$F_e = \Phi_{Q_j}^T \begin{Bmatrix} \mathbf{T}^{Q_j} \\ \mathbf{F}^{Q_j} \end{Bmatrix}; \text{where } \Phi_{Q_j}^T = \begin{bmatrix} \psi^{Q_j} & \phi^{Q_j} \end{bmatrix} \tag{9.25}$$

Kane's dynamical equations [4], $F_i + F_i^* = 0$, $i = 1, \ldots, 6 + n$, for a single free-flying flexible body j can now be formed for rigid and flexible motion variables as two sets of equations, after collecting terms from the above development:

$$M_1^j \begin{Bmatrix} \alpha_0^j \\ a_0^j \end{Bmatrix} + M_2^j \dot{\sigma}^j + X_1^j = \begin{Bmatrix} \mathbf{T}^O \\ \mathbf{F}^O \end{Bmatrix} + Z_{Q_j} \begin{Bmatrix} \mathbf{T}^{Q_j} \\ \mathbf{F}^{Q_j} \end{Bmatrix} \tag{9.26}$$

$$[M_2^j]^T \begin{Bmatrix} \alpha_0^j \\ a_0^j \end{Bmatrix} + E_j \dot{\sigma}^j + Y_1^j = \Phi_{Q_j}^T \begin{Bmatrix} \mathbf{T}^{Q_j} \\ \mathbf{F}^{Q_j} \end{Bmatrix} \tag{9.27}$$

Here, the contributions from body forces, remainder acceleration terms, and nominal and geometric stiffness and damping terms in the sequence of rotation, translation, and elastic vibration are lumped as follows, recalling that E_j is normally an $n \times n$ unity matrix:

$$X_1^j = X_0^j + \left\{ \begin{matrix} -(G_q^j)^T \\ 0 \end{matrix} \right\} \left\{ \left[\omega_n^2 + \phi^T K_g \phi \right] q + 2\xi \omega_n \dot{q} \right\} \tag{9.28}$$

$$Y_1^j = Y_0^j - E_j \left\{ \left[\omega_n^2 + \phi^T K_g \phi \right] q + 2\xi \omega_n \dot{q} \right\} \tag{9.29}$$

9.2 MULTIBODY HINGE KINEMATICS WITH EFFICIENT GENERALIZED SPEEDS

Mitiguy and Kane [6] proposed generalized speeds that are efficient in the sense that they lead to simpler equations of motion than are obtained otherwise, for a system of rigid bodies undergoing rotations about revolute joints, 2 dof Hooke's joints, and spherical joints, as shown in Figure 9.2. They show that hinge relative rotations are not the best choice, and equations of motion become simpler when the generalized speeds are chosen as the projections of the Newtonian angular velocity of a body rotating about the hinge-axis, as shown in Figure 9.2 taken from Ref. [6]. We now apply the modifications necessary for elastic joints. Consider two bodies, with reference frame j and c(j), connected by rotational *and* translational joints, with body c(j) inboard to body j along the path to body 1. Consider first that body j in Figure 9.3 is connected at P_j on c(j) by a *revolute joint* about an axis given by a vector expressed by the (3×1) column matrix h_1 in the c(j) basis. Reference [6] shows that an efficient rotational generalized speed u_{r1}^j for a *revolute joint* is such that

$$\omega^j = C_{j,c(j)} \left\{ \left[U - h_1 h_1^T \right] \omega^{Qc(j)} + u_{r1}^j h_1 \right\} \tag{9.30}$$

Here $\omega^{Qc(j)}$ is the inertial angular velocity of a nodal rigid body at $Q_{c(j)}$ in the c(j) basis, and $C_{j,c(j)}$ is the transpose of the coordinate transformation from j to the c(j) basis. Use of Eq. (9.22) in the above for an inboard body c(j) with elastic rotation at the node at Q_j yields

$$\omega^j = C_{j,c(j)} \left\{ \left[U - h_1 h_1^T \right] \left(P_{Q_j}^j \omega^j + \psi_{Q_j}^j \sigma^j \right) + u_{r1}^j h_1 \right\} \tag{9.31}$$

Recall that Eq. (9.23) defines $P_{Q_j}^j = [U - \psi_{Q_j}^j E_j^{-1} G_q^j]$. When two bodies are connected by a 2 dof rotation (Hooke's) joint, as shown in Figure 9.2, with the first rotation about the unit vector h_1 in c(j) and the second rotation about an axis vector h_2 fixed in j, the angular velocity of j by extension of Eq. (9.31) is

$$\omega^j = C_{j,c(j)} \left\{ \left[U - h_1 h_1^T - h_2 h_2^T \right] \left(P_{Q_j}^j \omega^j + \psi_{Q_j}^j \sigma^j \right) + u_{r1}^j h_1 + u_{r2}^j h_2 \right\} \tag{9.32}$$

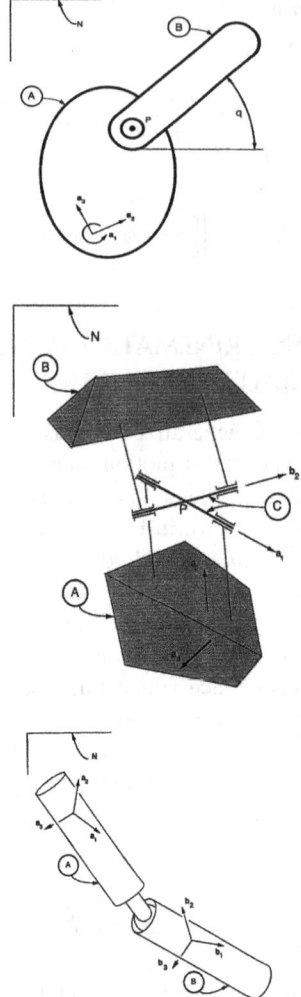

FIGURE 9.2   Efficient generalized speeds for revolute joint, Hooke's joint, and spherical joint.

The efficient generalized speeds u_{ri}, $i = 1, 2, 3$, for a spherical joint (see Figure 9.2) yield the angular velocity:

$$^{N}\omega^{j} = u_{r1}\mathbf{b}_1^{j} + u_{r1}\mathbf{b}_2^{j} + u_{r1}\mathbf{b}_3^{j} \tag{9.33}$$

If relative translation is allowed between point P_j on body $c(j)$ and point Q_j of body frame j, in a slider joint, as shown in Figure 9.3, then depending on the number of relative translational degrees of freedom, one can introduce one, two, or three generalized speeds:

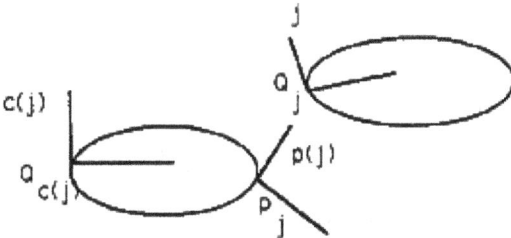

FIGURE 9.3 Two adjacent bodies connected at a hinge allowing relative rotation and translation in a system of rigid and flexible bodies, in a topological tree.

$$\delta^j = \left\{\begin{matrix} u^j_{t_1} \\ 0 \\ 0 \end{matrix}\right\} \quad \text{or} \quad \delta^j = \left\{\begin{matrix} u^j_{t_1} \\ u^j_{t_2} \\ 0 \end{matrix}\right\} \quad \text{or} \quad \delta^j = \left\{\begin{matrix} u^j_{t_1} \\ u^j_{t_2} \\ u^j_{t_3} \end{matrix}\right\} \tag{9.34a}$$

where it is defined that

$$u_{t_i} = {}^{t(j)}v^{Q_j/P_j} \cdot t^j_i \quad i = 1,2,3 \tag{9.34b}$$

Here, t^j_i are the basis vectors of a translational frame fixed on the flexible body $c(j)$ at node P_j. The velocity of point Q_j is obtained by accounting for the deformation at P_j of $c(j)$, see Eq. (5a), and the translation from P_j to Q_j (see Figure 9.3):

$$v^{Q_j} = C^t_{c(j),j}\left[v^{Q_{c(j)}} + \tilde{\omega}^{c(j)}r^{Q_{c(j)}\,P_j} + \phi^{c(j)}(P_j)\sigma^{c(j)} + \tilde{\omega}^{c(j)}C_{c(j),j}\delta^j + C_{c(j),j}\dot{\delta}^j\right] \tag{9.35}$$

The expressions for angular accelerations and linear accelerations are obtained by differentiations in the Newtonian frame of the angular velocity and the velocity. For the purpose of developing a common notation for all types of joints, we define the following relationships as in Ref. [2]. As before, acceleration terms involving derivatives of generalized speeds are indicated by subscript 0 and the remainder acceleration terms are denoted by subscript t; terms with an overhead caret refer to entities for the inboard body, and R_j denotes the $6 \times (NR+NT)$ partial angular velocity/partial velocity for the generalized speeds of the joint:

$$\left\{\begin{matrix} \alpha^j \\ a^{Q_j} \end{matrix}\right\} = \left\{\begin{matrix} \alpha^j_0 \\ a^{Q_j}_0 \end{matrix}\right\} + \left\{\begin{matrix} \alpha^j_t \\ a^{Q_j}_t \end{matrix}\right\} \tag{9.36}$$

$$\left\{\begin{matrix} \alpha^j_0 \\ a^{Q_j}_0 \end{matrix}\right\} = \left\{\begin{matrix} \hat{\alpha}^j_0 \\ \hat{a}^{Q_j}_0 \end{matrix}\right\} + R_j\left\{\begin{matrix} \dot{u}^j_r \\ \dot{u}^j_i \end{matrix}\right\} \tag{9.37}$$

$$\left\{ \begin{matrix} \hat{\alpha}_0^j \\ \hat{a}_0^{Q_j} \end{matrix} \right\} = W_j \left\{ \begin{matrix} \alpha_0^{c(j)} \\ a_0^{Q_{c(j)}} \end{matrix} \right\} + N_j \dot{\sigma}^{c(j)} \tag{9.38}$$

The matrices W_j, N_j encapsulate the details of the forward transfer of kinematical information from an inboard body $c(j)$ to its outboard body j. It can be shown that, for a revolute joint, together with a translational joint from P_j to Q_j, in Figure 9.3, Eqs. (9.37) and (9.38) end up in the following matrix forms, where $\tilde{\delta}$ is the skew-symmetric matrix corresponding to the translations δ in Eq. (9.34), and $P_{Q_j}^j$ given by Eq. (9.23):

$$W_j = \begin{bmatrix} C_{j,c(j)}\left[U - h_1 h_1^T\right]P_{Q_j}^j & 0 \\ -C_{j,c(j)}\left[\tilde{r}^{Q_{c(j)}P_j} + C_{c(j),t(j)}\tilde{\delta}^j C_{c(j),t(j)}^T P_{Q_j}^j\right] & C_{j,c(j)} \end{bmatrix} \tag{9.39}$$

$$N_j = \begin{bmatrix} C_{j,c(j)}\left[U - h_1 h_1^T\right]\psi_{P_j}^j \\ -C_{j,c(j)}\left[\phi_{P_j}^{c(j)} - C_{c(j),t(j)}\tilde{\delta}^j C_{c(j),t(j)}^T \psi_{P_j}^{c(j)}\right] \end{bmatrix} \tag{9.40}$$

Here, $\phi_{P_j}^{c(j)} \equiv \phi^{c(j)}(P_j)$ and $\psi_{P_j}^{c(j)} \equiv \psi^{c(j)}(P_j)$ with point P_j of body $c(j)$ connecting body j at Q_j with a one degree of freedom rotational joint (see Eq. (9.30)). Now define:

$$R_j = \begin{bmatrix} C_{j,c(j)}h_1 & 0 \\ 0 & C_{j,c(j)} \end{bmatrix} \tag{9.41}$$

In the context of Eq. (9.36), following Eq. (9.31) we also need

$$\alpha_t^j = C_{j,c(j)}\left\{\left[U - h_1 h_1^T\right]\left[P_{Q_j}^j \alpha_t^{c(j)} - \rho^j\right] + \dot{q}_{r1}^j \tilde{\omega}^{P_j} h_1\right\} \tag{9.42}$$

$$\rho^j = \psi_{Q_{c(j)}}^j E_{c(j)}^{-1} \dot{G}_q^j \omega^{c(j)} - \tilde{\omega}^{c(j)} \psi_{Q_{c(j)}}^j \dot{q}^{c(j)} \tag{9.43}$$

Here, we use $\dot{q}_k$ to replace q_k in Eq. (9.8) to form $\dot{G}_q^j$ in Eq. (9.43). Next, we differentiate Eq. (9.35) to get

$$a_t^{Q_j} = C_{j,c(j)}\left\{a_t^{Q_{c(j)}} - \tilde{r}^{Q_{c(j)}P_j}\alpha_t^{c(j)} + \tilde{\omega}^{c(j)}\left[\tilde{\omega}^{c(j)}r^{Q_{c(j)}P_j} + \phi^{c(j)}(P_j)\sigma^{c(j)}\right]\right.$$
$$\left. -C_{c(j),t(j)}\tilde{\delta}^j C_{c(j),t(j)}^T\left(P_Q^j\alpha_t^j - \rho^j\right) + \tilde{\omega}^j\left[\tilde{\omega}^j C_{c(j),t(j)}\delta^j + 2C_{c(j),t(j)}\dot{\delta}^j\right]\right\} \tag{9.44}$$

The expressions corresponding to Eqs. (9.39)–(9.41) for the angular acceleration for a 2 dof joint with a rotation joint, and translation, follow from Eqs. (9.32) and (9.35), yielding

$$W_j = \begin{bmatrix} C_{j,c(j)}\left[U - h_1 h_1^T - h_2 h_2^T\right]P_{Q_j}^j & 0 \\ -C_{j,c(j)}\left[\tilde{r}^{Q_{c(j)}P_j} + C_{c(j),t(j)}\tilde{\delta}^j C_{c(j),t(j)}^T P_Q^j\right] & C_{j,c(j)} \end{bmatrix} \tag{9.45}$$

$$N_j = \begin{bmatrix} C_{j,c(j)}\left[U - h_1 h_1^T - h_2 h_2^T\right]\psi_{P_j}^j \\ C_{j,c(j)}\left[\phi_{P_j}^{c(j)} - C_{c(j),t(j)}\tilde{\delta}^j C_{c(j),t(j)}^T \psi_{P_j}^{c(j)}\right] \end{bmatrix} \tag{9.46}$$

$$R_j = \begin{bmatrix} C_{j,c(j)}[h_1 \quad h_2] & 0 \\ 0 & C_{j,c(j)} \end{bmatrix} \tag{9.47}$$

The equation takes the specific form of Eq. (9.42) for two rotations q_{r1}, q_{r2}:

$$\alpha_t^j = C_{c(j),j}\left\{\left[U - h_1 h_1^T - h_2 h_2^T\right]\left(P_{Q_j}^j \alpha_t^{c(j)} - \rho^j\right) \right. \\ \left. + \left[U - h_2 h_2^T\right]\dot{q}_{r1}^j \tilde{\omega}^Q h_1 + \dot{q}_{r2}^j \tilde{\omega}^j C_{c(j),j}^T h_2\right\} \tag{9.48}$$

Similarly for body j connected to c(j) by a spherical joint and, say, a 2-dof translational joint, the angular acceleration being

$$\alpha^j = \dot{u}_{r1}j_1 + \dot{u}_{r2}j_2 + \dot{u}_{r3}j_3 \tag{9.49}$$

the kinematical transfer matrices of Eqs. (9.37) and (9.38) become

$$W_j = \begin{bmatrix} 0 & 0 \\ -C_{j,c(j)}\left[\tilde{r}^{Q_{c(j)}P_j} + C_{c(j),t(j)}\tilde{\delta}_{t(j),c(j)}^j P_{Q_j}^j\right] & C_{j,c(j)} \end{bmatrix} \tag{9.50}$$

$$N_j = \begin{bmatrix} 0 \\ C_{j,c(j)}\left[\phi_{P_j}^{c(j)} - C_{c(j),t(j)}\tilde{\delta}^j C_{c(j),t(j)}^T \psi_{P_j}^{c(j)}\right] \end{bmatrix} \tag{9.51}$$

$$R_j = \begin{bmatrix} C_{c(j),j}^t & 0 \\ 0 & C_{c(j),j}^t \end{bmatrix} \tag{9.52}$$

Multiplications with the zero blocks are to be avoided in coding these expressions.

9.3 RECURSIVE ALGORITHM FOR FLEXIBLE MULTIBODY DYNAMICS WITH MULTIPLE STRUCTURAL LOOPS

The algorithm given in Ref. [2] for a system of hinge-connected flexible bodies in a tree configuration can now be extended to systems with *many* closed structural loops by the artifice of cutting the loops, and imposing unknown constraint forces [7], while using the efficient variables and their associated kinematical transfer matrices. The general algorithm for the case of n bodies with m structural loops is best explained by following a specific example. Consider the system in Figure 9.4a, which shows five hinge-connected bodies with one closed loop with bodies 1 and 4 hinged to the ground. Figure 9.4b shows the same system after cutting

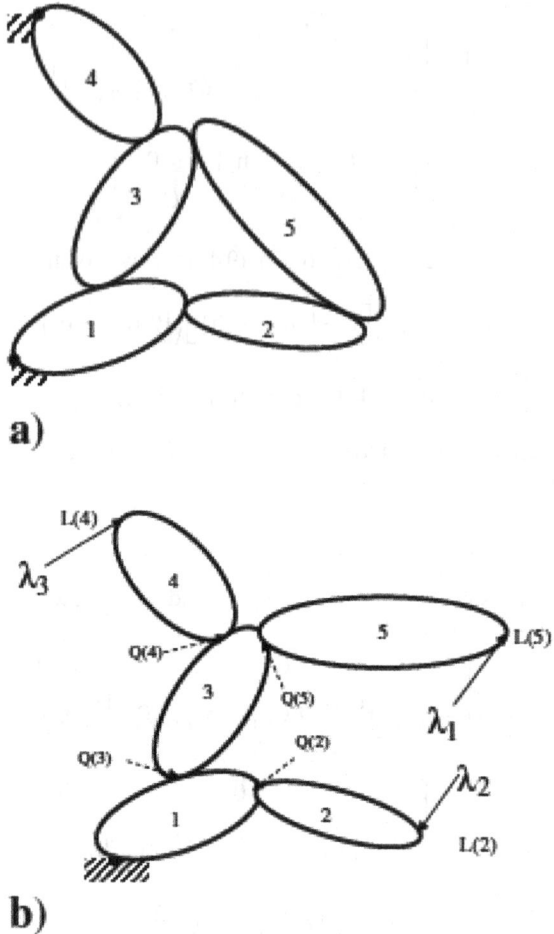

FIGURE 9.4 (a) Original constrained system and (b) system obtained by cutting structural loops and showing general (actually equal and opposite) forces and torques.

the hinges connecting body 4 to the ground, and the hinge connecting bodies 2 and 5 and replacing the action of the hinges by forces and torques; note that we have shown in Figure 9.4b, only for generality of notation, the constraint force λ_2 which actually equals $(-\lambda_1)$ for the structural loop, and constraint force λ_4 on body 4. Now we have a system with a tree structure. We will now form the dynamical equations of this system in a tree structure. The idea is to solve for the constraint forces finally by the additional equations of motion constraints.

9.3.1 FORWARD PASS

First, we do the kinematics for the system with loops cut, by recursively forming the acceleration of the hinge points and the angular accelerations of all body frames—in a forward pass. To make the presentation clear, we illustrate the process with a specific system of five bodies, where we make a joint cut at outboard link node L(5) of body 5 (note the change in notations for hinges, used here). From Eq. (9.27) (recall that superscripts refer to a body number, not a power, except in the case of inverse or transpose operations):

$$\dot{\sigma}^5 = -E_5^{-1}\left[A_5\left\{\begin{matrix}\alpha_0^5\\a_0^5\end{matrix}\right\}+Y_1^5\right]+\hat{H}_{1e}^5\lambda_1; \quad A_5 = \left[M_2^5\right]^T \tag{9.53}$$

$$\hat{H}_{1e}^5 = E_5^{-1}\Phi_{L(5)}^T \tag{9.54}$$

Note here that λ_1 stands for the column matrix of the *outboard* hinge torque and force components exposed at L(5) in Figure 9.4b. Putting Eq. (9.53) in Eq. (9.26) for $j = 5$ yields

$$M_1^5\left\{\begin{matrix}\alpha_0^5\\a_0^5\end{matrix}\right\}+X_1^5 = \left\{\begin{matrix}\tau_h^5\\f_h^5\end{matrix}\right\}+G_{51}\lambda_1 \tag{9.55}$$

Note further that τ_h^5, f_h^5 in Eq. (9.55) are components of the hinge torque and force components that do work, at the *inboard* hinge Q_5 of body 5 in Figure 9.4b.

In Eq. (9.55), the following compact replacements have been made after collecting terms:

$$M_1^5 + M_2^5 E_5^{-1}\left[M_2^5\right]^T \rightarrow M_1^5$$

$$X_1^5 + M_2^5 E_5^{-1}Y_1^5 \rightarrow X_1^5 \tag{9.56a}$$

$$Z_{L(5)} - M_2^5 E_5^{-1}\Phi_{L(5)}^T \rightarrow G_{51}$$

Note also that E_5 in Eq. (9.56) is a unity matrix if normalized vibration modes are used. Here, we have defined from Eq. (9.24):

$$Z_{L(5)} = \begin{bmatrix}P_{L(5)}^{j^T} & \tilde{r}^{Q(5)L(5)}\\0 & U\end{bmatrix} \tag{9.56b}$$

Now, using Eq. (9.37) for $j = 5$, and pre-multiplying the resulting equation by R_5^T yield the rotation and translation equations of motion from Eq. (9.55):

$$\left\{\begin{matrix}\dot{u}_r^5\\\dot{u}_t^5\end{matrix}\right\} = -v_5^{-1}R_5^T\left[M_1^5\left\{\begin{matrix}\hat{\alpha}_0^5\\\hat{a}_0^5\end{matrix}\right\}+X_1^5\right]+v_5^{-1}\left\{\begin{matrix}\tau_h^5\\f_h^5\end{matrix}\right\}+\hat{H}_{1r}^5\lambda_1 \tag{9.57}$$

where

$$v_5 = R_5^T M_1^5 R_5 \tag{9.58}$$

$$\hat{H}_{1r}^5 = v_5^{-1} R_5^T G_{51} \tag{9.59}$$

The complete set of working and non-working hinge torque and force components on body 5 (see Figure 9.4b) is obtained from Eq. (9.55), and using Eq. (9.57) in Eq. (9.37):

$$\begin{Bmatrix} T_h^5 \\ F_h^5 \end{Bmatrix} = M_5 \begin{Bmatrix} \hat{\alpha}_0^5 \\ \hat{a}_0^5 \end{Bmatrix} + X_5 - B_{31} \lambda_1 \tag{9.60}$$

after the following replacements have been made:

$$S_5 = U - M_1^5 R_5 v_5^{-1} R_5^T \tag{9.61}$$

$$M_5 = S_5 M_1^5 \tag{9.62}$$

$$X_5 = S_5 X_1^5 + M_1^5 R_5 v_5^{-1} \begin{Bmatrix} \tau_h^5 \\ f_h^5 \end{Bmatrix} \tag{9.63}$$

$$B_{31} = G_{51} - M_1^5 R_5 \hat{H}_{1r}^5 \tag{9.64}$$

The notation B_{31} in Eq. (9.60) is used to indicate the effect on inboard body 3 of λ_1 in Figure 9.4b. Bodies 2 and 4 in Figure 9.4b being terminal bodies like body 5, contribute equations similar to Eqs. (9.53)–(9.64) with two important differences in the index for the subscripts and superscripts: the body index changes from 5 to 2 or 4, and terms associated with λ_1 such as B_{31} (first subscript for inboard body number and second subscript for constraint force number) become B_{12} for λ_2 and B_{33} for λ_3 (see Figure 9.4b). The starting point for a body with outboard bodies, such as body 3 in Figure 9.4b, follows from Eq. (9.27), with Q(4), Q(5) being the points on body 3 at which bodies 4 and 5 are connected; then, imposing working and non-working torques and forces $T_h^4, F_h^4, T_h^5, F_h^5$, respectively, we have, in the format of Eq. (9.27):

$$E_3 \dot{\sigma}^3 = - \left[A_3 \begin{Bmatrix} \alpha_0^3 \\ a_0^3 \end{Bmatrix} + Y_1^3 \right] + \Phi_{Q(4)}^T \begin{Bmatrix} T_h^4 \\ F_h^4 \end{Bmatrix} + \Phi_{Q(5)}^T \begin{Bmatrix} T_h^5 \\ F_h^5 \end{Bmatrix} \tag{9.65}$$

Here, $\Phi_{Q(4)}^T, \Phi_{Q(5)}^T$ are the transpose of the modal rotation-translation matrices at Q(4), Q(5), the connection points of body 3 with its outboard bodies. Using Eq. (9.60) for body 5 and *its analogous equation for body 4* in Eq. (9.65) with subsequent use of Eq. (9.38) yields

$$\dot{\sigma}^3 = -E_3^{-1} \left[A_3 \begin{Bmatrix} \alpha_0^3 \\ a_0^3 \end{Bmatrix} + Y_1^3 \right] + \hat{H}_{1e}^3 \lambda_1 + \hat{H}_{3e}^3 \lambda_3 \tag{9.66}$$

$$\hat{H}_{1e}^3 = E_3^{-1}\Phi_{Q(5)}^T d_3^5 B_{31} \tag{9.67}$$

$$\hat{H}_{3e}^3 = E_3^{-1}\Phi_{Q(4)}^T d_3^4 B_{33} \tag{9.68}$$

Equations (9.66)–(9.68) incorporate the following replacements:

$$E_3 + \Phi_{Q(4)}^T d_3^4 M_4 N_4 + \Phi_{Q(5)}^T d_3^5 M_5 N_5 \rightarrow E_3$$

$$A_3 - \Phi_{Q(4)}^T d_3^4 M_4 W_4 - \Phi_{Q(5)}^T d_3^5 M_5 W_5 \rightarrow A_3 \tag{9.69}$$

$$Y_1^3 - \Phi_{Q(4)}^T d_3^4 X_4 - \Phi_{Q(5)}^T d_3^5 X_5 \rightarrow Y_1^3$$

Equations (9.67)–(9.69) have used the following matrix to transfer the hinge force and torque from body j to body c(j) for bodies 4 and 5 accounting for any translation δ^j at the joint, as shown in Eq. (9.70), $\tilde{\delta}^j$ being the skew-symmetric matrix corresponding to cross multiplication with vector δ^j:

$$d_{c(j)}^j = -\begin{bmatrix} C_{c(j),j} & C_{c(j),t(j)}\tilde{\delta}^j C_{t(j),j} \\ 0 & C_{c(j),j} \end{bmatrix} \tag{9.70}$$

The rotation and translation equation for body 3, modeled after Eq. (9.26), becomes

$$M_1^3 \begin{Bmatrix} \alpha_0^3 \\ a_0^3 \end{Bmatrix} + M_2^3 \dot{\sigma}^3 + X_1^3 = \begin{Bmatrix} T_h^3 \\ F_h^3 \end{Bmatrix} - Z_{Q(4)} d_3^4 \begin{Bmatrix} T_h^4 \\ F_h^4 \end{Bmatrix} - Z_{Q(5)} d_3^5 \begin{Bmatrix} T_h^5 \\ F_h^5 \end{Bmatrix} \tag{9.71}$$

Substitution from Eq. (9.60) and its counterpart for body 4, physically meaning load transfer from outward bodies, and use of the replacements:

$$M_1^3 + Z_{Q(4)} d_3^4 M_4 W_4 + Z_{Q(5)} d_3^5 M_5 W_5 \rightarrow M_1^3$$

$$M_2^3 + Z_{Q(4)} d_3^4 M_4 N_4 + Z_{Q(5)} d_3^5 M_5 N_5 \rightarrow M_2^3 \tag{9.72}$$

$$X_1^3 + Z_{Q(4)} d_3^4 X_4 + Z_{Q(5)} d_3^5 X_5 \rightarrow X_1^3$$

give rise to the equation

$$M_1^3 \begin{Bmatrix} \alpha_0^3 \\ a_0^3 \end{Bmatrix} + M_2^3 \dot{\sigma}^3 + X_1^3 = \begin{Bmatrix} T_h^3 \\ F_h^3 \end{Bmatrix} + Z_{Q(4)} d_3^4 B_{33}\lambda_3 + Z_{Q(5)} d_3^5 B_{11}\lambda_1 \tag{9.73}$$

Now putting Eq. (9.66) in Eq. (9.73) and making the following replacements:

$$M_1^3 + M_2^3 E_3^{-1} M_2^3 \rightarrow M_1^3$$

$$X_1^3 + M_2^3 E_3^{-1} Y_1^3 \rightarrow X_1^3$$

$$Z_3^4 - M_2^3 E_3^{-1}\Phi_{Q(4)}^T \rightarrow G_{33} \tag{9.74}$$

$$Z_3^5 - M_2^3 E_3^{-1}\Phi_{Q(5)}^T \rightarrow G_{31}$$

produce the following rotation and translation equations for body 3:

$$M_1^3 \begin{Bmatrix} \alpha_0^3 \\ a_0^3 \end{Bmatrix} + X_1^3 = \begin{Bmatrix} T_h^3 \\ F_h^3 \end{Bmatrix} + G_{33} d_3^4 B_{33} \lambda_3 + G_{31} d_3^5 B_{11} \lambda_1 \qquad (9.75)$$

Using Eq. (9.37) in the above equation, and subsequent pre-multiplication by R_3^t, yield the hinge rotation and translation equations for body 3:

$$\begin{Bmatrix} \dot{u}_r^3 \\ \dot{u}_t^3 \end{Bmatrix} = -v_3^{-1} R_3^T \left[M_1^3 \begin{Bmatrix} \hat{\alpha}_0^3 \\ \hat{a}_0^3 \end{Bmatrix} + X_1^3 \right] + v_3^{-1} \begin{Bmatrix} \tau_h^3 \\ f_h^3 \end{Bmatrix} + \hat{H}_{3r}^3 \lambda_3 + \hat{H}_{1r}^3 \lambda_1 \qquad (9.76)$$

Here, we have used the notations:

$$\hat{H}_{3r}^3 = v_3^{-1} R_3^T G_{33} d_3^4 B_{33} \qquad (9.77)$$

$$\hat{H}_{1r}^3 = v_3^{-1} R_3^T G_{31} d_3^5 B_{31} \qquad (9.78)$$

The derivatives of the generalized speeds for body 3 having been expressed, as in Eqs. (9.66) and (9.76), the working and non-working hinge torque and force on body 3 are obtained from Eq. (9.71) as

$$\begin{Bmatrix} T_h^3 \\ F_h^3 \end{Bmatrix} = M_3 \begin{Bmatrix} \hat{\alpha}_0^3 \\ \hat{a}_0^3 \end{Bmatrix} + X_1^3 - B_{11} \lambda_1 - B_{13} \lambda_3 \qquad (9.79)$$

Here, again, the following substitutions have been incorporated in sequence:

$$S_3 = I - M_1^3 R_3 v_3^{-1} R_3^T$$

$$M_3 = S_3 M_1^3$$

$$X_1^3 = S_3 X_1^3 + M_1^3 R_3 v_3^{-1} \begin{Bmatrix} \tau_h^3 \\ f_h^3 \end{Bmatrix} \qquad (9.80)$$

$$B_{13} = S_3 d_3^4 B_{33}$$

$$B_{11} = S_3 d_3^5 B_{31}$$

The dynamics of body 1 in Figure 9.4b is influenced by hinge torques and forces from both body 3 and body 2. With body 2 now being a terminal body, derivatives of its generalized speeds expressed just for body 5, its hinge loads can be written as in the development of Eq. (9.60) as

$$\begin{Bmatrix} T_h^2 \\ F_h^2 \end{Bmatrix} = M_2 \begin{Bmatrix} \hat{\alpha}_0^2 \\ \hat{a}_0^2 \end{Bmatrix} + X_2 - B_{12} \lambda_2 \qquad (9.81)$$

With body 1 being an interior body, the development of its dynamical equations is similarly identical to that of body 3. Thus, we have the sequence of replacements (see Eq. [9.69]), implying load from outboard bodies to an inner body:

$$E_1 + \Phi_{Q(2)}^T d_1^2 M_2 N_2 + \Phi_{Q(3)}^T d_1^3 M_3 N_3 \rightarrow E_1$$

$$A_1 - \Phi_{Q(2)}^T d_1^2 M_2 W_2 - \Phi_{Q(3)}^T d_1^3 M_3 W_3 \rightarrow A_1 \qquad (9.82)$$

$$Y_1^1 - \Phi_{Q(2)}^T d_1^2 X_2 - \Phi_{Q(3)}^T d_1^3 X_3 \rightarrow Y_1^1$$

Here, Q(2) and Q(3) are points on body 1 to which body 2 and body 3 are connected. The dynamical equations for the modal coordinates of body 1 follow from the pattern as before:

$$\dot{\sigma}^1 = -E_1^{-1} \left[A_1 \left\{ \begin{array}{c} \alpha_0^1 \\ a_0^1 \end{array} \right\} + Y_1^1 \right] + \hat{H}_{1e}^1 \lambda_1 + \hat{H}_{2e}^1 \lambda_2 + \hat{H}_{3e}^1 \lambda_3 \qquad (9.83)$$

$$\hat{H}_{1e}^1 = E_1^{-1} \Phi_{Q(3)}^T d_1^3 B_{13} \qquad (9.84)$$

$$\hat{H}_{2e}^1 = E_1^{-1} \Phi_{Q(2)}^T d_1^2 B_{12} \qquad (9.85)$$

$$\hat{H}_{3e}^1 = E_1^{-1} \Phi_{Q(3)}^T d_1^3 B_{13} \qquad (9.86)$$

Again, making the load-transfer indicators similar to those in Eq. (9.72), that is

$$M_1^1 + Z_{Q(2)} d_1^2 M_2 W_2 + Z_{Q(3)} d_1^3 M_3 W_2 \rightarrow M_1^1$$

$$M_2^1 + Z_{Q(2)} d_1^2 M_2 N_2 + Z_{Q(3)} d_1^3 M_3 N_2 \rightarrow M_2^1 \qquad (9.87)$$

$$X^1 + Z_{Q(2)} d_1^2 X_2 + Z_{Q(3)} d_1^3 X_3 \rightarrow X^1$$

and replacements required in this case similar to Eq. (9.74), namely,

$$M_1^1 + M_2^1 E_1^{-1} A_1 \rightarrow M_1^1$$

$$X^1 + M_2^1 E_1^{-1} Y_1^1 \rightarrow X^1$$

$$Z_1^3 - M_2^1 E_1^{-1} \Phi_{Q(3)}^T \rightarrow G_{11} \qquad (9.88)$$

$$Z_1^2 - M_2^1 E_1^{-1} \Phi_{Q(2)}^T \rightarrow G_{12}$$

$$G_{11} \rightarrow G_{13}$$

one obtains the equations for the rigid body rotation and translation of body 1:

$$\left\{ \begin{array}{c} \dot{u}_r^1 \\ \dot{u}_t^1 \end{array} \right\} = -v_1^{-1} R_1^1 X^1 + v_1^{-1} \left\{ \begin{array}{c} \tau_h^1 \\ f_h^1 \end{array} \right\} + \hat{H}_{1r}^1 \lambda_1 + \hat{H}_{2r}^1 \lambda_2 + \hat{H}_{3r}^1 \lambda_3 \qquad (9.89)$$

Here, we have used the notations:

$$v_1 = R_1^T M_1^1 R_1 \tag{9.90}$$

$$\hat{H}_{1r}^1 = v_1^{-1} R_1^T G_{11} d_1^3 B_{11} \tag{9.91}$$

$$\hat{H}_{2r}^1 = v_1^{-1} R_1^T G_{12} d_1^2 B_{12} \tag{9.92}$$

$$\hat{H}_{3r}^1 = v_1^{-1} R_1^T G_{13} d_1^3 B_{13} \tag{9.93}$$

9.3.2 FORWARD PASS

To unscramble the dynamical equations for the rest of the bodies, we rewrite Eq. (9.89) as

$$\begin{Bmatrix} \dot{u}_r^1 \\ \dot{u}_t^1 \end{Bmatrix} = F_{rf}^1 + H_{1r}^1 \lambda_1 + H_{2r}^1 \lambda_2 + H_{3r}^1 \lambda_3 \tag{9.94}$$

Here, F_{rf}^1, H_{1r}^1 stand for

$$F_{rf}^1 = -v_1^{-1} R_1^T Y_2^1 + v_1^{-1} \begin{Bmatrix} \tau_h^1 \\ f_h^1 \end{Bmatrix} \tag{9.95}$$

$$H_{ir}^1 = \hat{H}_{ir}^1 \quad i = 1, 2, 3 \tag{9.96}$$

If body 1 is connected to the ground, then kinematical equations in Eq. (9.37) reduce to

$$\begin{Bmatrix} \alpha_0^1 \\ a_0^1 \end{Bmatrix} = R_1^T \begin{Bmatrix} \dot{u}_r^1 \\ \dot{u}_t^1 \end{Bmatrix} \tag{9.97}$$

Now with the use of the following defining notations:

$$S_f^1 = R_1 F_{rf}^1 \tag{9.98}$$

$$S_{ci}^1 = R_1 H_{ir}^1 \quad i = 1, 2, 3 \tag{9.99}$$

$$F_{ef}^1 = E_1^{-1} \left\{ Y_1^1 + A_1 S_f^1 \right\} \tag{9.100}$$

Equation (9.83) is written in terms of the elastic generalized speeds for body 1 as

$$\dot{u}_e^1 = F_{ef}^1 + H_{1e}^1 \lambda_1 + H_{2e}^1 \lambda_2 + H_{3e}^1 \lambda_3 \tag{9.101}$$

Here, the following update of the constraint force coefficients has been made:

$$H_{ie}^1 = \hat{H}_{ie}^1 + E_1^{-1}A_1S_{ci}^1 \quad i = 1,2,3 \tag{9.102}$$

The dynamical equations for body 1, as in Eqs. (9.94) and (9.101), can be used to form the equations for the rest of the bodies with the help of the kinematical transfer matrices in Section 9.2. Thus, appealing to Eq. (9.38) for body 2, and defining

$$\gamma_f^2 = W_2S_f^1 + N_2F_{ef}^1 \tag{9.103}$$

$$\gamma_{ci}^2 = W_2S_{ci}^1 + N_2H_{ie}^1, \quad 1,2,3 \tag{9.104}$$

one gets

$$\begin{Bmatrix} \hat{\alpha}_0^2 \\ \hat{a}_0^2 \end{Bmatrix} = \gamma_f^2 + \gamma_{c1}^2\lambda_1 + \gamma_{c2}^2\lambda_2 + \gamma_{c3}^2\lambda_3 \tag{9.105}$$

Specifying $\hat{H}_{ir}^2 = 0$, $i = 1,2,3$, unless a non-zero value of the rigid body base motion is prescribed, as can be the case here for the example system in Figure 9.2 for $\hat{H}_{ir}^2$, we have

$$H_{ir}^2 = \hat{H}_{ir}^2 - v_2^{-1}R_2^TM_1^2\gamma_{ci}^2 \quad i = 1,2,3 \tag{9.106}$$

Derivatives of the rigid body generalized speeds for body 2 then follow with

$$F_{rf}^2 = -v_2^{-1}R_2^T\left(X^2 + M_2^1\gamma_f^2\right) + v_2^{-1}\begin{Bmatrix} \tau_2 \\ f_2 \end{Bmatrix} \tag{9.107}$$

$$\begin{Bmatrix} \dot{u}_r^2 \\ \dot{u}_t^2 \end{Bmatrix} = F_{rf}^2 + H_{1r}^2\lambda_1 + H_{2r}^2\lambda_2 + H_{3r}^2\lambda_3 \tag{9.108}$$

Now, appealing to Eq. (9.37) for body 2 in the cut-out system of Figure 9.4b forms

$$S_f^2 = \gamma_f^2 + R_2F_{rf}^2 \tag{9.109}$$

$$S_{ci}^2 = \gamma_{ci}^2 + R_2H_{ir}^2 \quad i = 1,2,3 \tag{9.110}$$

$$\begin{Bmatrix} \alpha_0^2 \\ a_0^2 \end{Bmatrix} = S_f^2 + S_{c1}^2\lambda_1 + S_{c2}^2\lambda_2 + S_{c3}^2\lambda_3 \tag{9.111}$$

This allows one to revisit the dynamical equations for the modal generalized speeds for body 2, specifying as before:

$$\hat{H}_{ie}^2 = 0 \quad i = 1,2,3 \tag{9.112}$$

unless a non-zero value of the relevant prescribed motion is prescribed. Now update

$$H_{ie}^2 = \hat{H}_{ie}^2 + E_2^{-1} A_2 S_{ci}^2 \quad i = 1, 2, 3 \tag{9.113}$$

to finally present, for this example system:

$$F_{ef}^2 = E_2^{-1} \left(Y_1^2 + A_2 S_f^2 \right) \tag{9.114}$$

$$\dot{u}_e^2 = F_{ef}^2 + H_{1e}^2 \lambda_1 + H_{2e}^2 \lambda_2 + H_{3e}^2 \lambda_3 \tag{9.115}$$

Equations for rigid and elastic generalized speeds for bodies 3, 4, and 5 in Figure 9.4b follow by continuing the recursion in the forward pass. An algorithm for the general case, of nb bodies with m loops, now emerges by induction from this representative example. A pseudo-code for the general case is given in the Appendix at the end of this chapter.

9.4   EXPLICIT SOLUTION OF DYNAMICAL EQUATIONS USING MOTION CONSTRAINTS

So far, the dynamical equations have been formed by retaining non-working constraint forces that are as yet unknown. Motion constraint conditions are now stated, in their acceleration forms, as many in number as there are components of unknown constraint forces, and these two sets of equations, that is, the dynamical equations and the constraint equations, are solved together. Of course, the constraint conditions are specific to a problem. For a loop formed by a chain of articulated flexible bodies with end-points of body 1 and body n pinned, cutting the loop at L(n) of body n and using the dynamical equations to construct the angular acceleration and the acceleration at the cut of L(n) yield the strikingly simple form of the acceleration constraint:

$$\left[Z_{L(n)}^T S_c^n + \Phi_{L(n)} H_e^n \right] \lambda = \begin{Bmatrix} \alpha^{L(n)} \\ a^{L(n)} \end{Bmatrix} - \begin{Bmatrix} \alpha_t^{L(n)} \\ a_t^{L(n)} \end{Bmatrix} - Z_{L(n)}^T S_f^n + \Phi_{L(n)} F_{ef}^n \tag{9.116}$$

The bottom three of six rows of this matrix equation corresponding to the acceleration constraint can be used to solve for λ. This is because the accelerations, and not the angular accelerations at the joint which may be a revolute or other type of rotational joint, must be the same. For the purposes of assigning initial conditions with the constraints, one must also write the position and velocity constraints.

Position constraints are generally of the non-linear functional form, $f(q)=0$, where q are the generalized coordinates, as shown previously, and must be solved by Newton's method, for dependent coordinates in terms of the independent coordinates [8]. Velocity constraints for a loop are linear, and can be written as the matrix equation, $[A_c(q)]\{U\} = 0$, where U is the column matrix of system generalized speeds. The constraint matrix $A_c(q)$ can be filled by recursively forming its entries corresponding to the generalized speeds U_j that represent the rotational, translational, and modal motion of body j, as follows, when one writes the angular velocity and velocity of the link node L(n) of body n that is constrained in rotation and translation. Note the use of kinematical quantities defined in generating the dynamical equations.

$$\begin{Bmatrix} \omega_{L(n)} \\ v_{L(n)} \end{Bmatrix} = \left[Z_{L(n)} R_n \Phi_{L(n)} \right] \{U_n\} + Z_{L(n)} \left[W_n R_{c(n)} N_n \right] \{U_{c(n)}\}$$

$$+ Z_{L(n)} W_n \left[W_{c(n)} R_{c^2(n)} N_{c(n)} \right] \{U_{c^2(n)}\} + \dots\dots; \quad c^2(n) \equiv c(c(n))$$

(9.117)

Recall that in contrast to the block-diagonal nature [2] of the mass matrices involved in the recursive algorithm given above, a non-recursive formulation with augmented generalized speeds available in Refs. [7, 8] produces the coupled dynamical and constraint equations:

$$[M(q)]\{\ddot{q}\} = \{C(q,\dot{q})\} + [A_c(q)]^T \{\Lambda\}$$

$$[A_c(q)]\{\ddot{q}\} = \{D(q,\dot{q})\}$$

(9.118)

In this customary formulation, M(q) in Eq. (9.118) is a dense, time-varying, square matrix of size equal to the augmented degrees of freedom of the system, and hence getting $\ddot{q}$ is computationally more expensive compared to the algorithm given here. Of course, constraints in both these approaches can be "stabilized", with no drift, by the procedures given in Chapter 8, with as accurate constraint satisfaction as desired.

9.5 COMPUTATIONAL RESULTS AND SIMULATION EFFICIENCY FOR MOVING MULTI-LOOP STRUCTURES

The numerical solutions of the differential equations of dynamics with non-linear algebraic constraint equations are most accurately done by using differential algebraic solvers, such as DASSL [9]. In the results presented next, we use an alternative procedure [13] for solving the dynamical equations together with the acceleration form of the constraint differential equations, by the well-known Kutta–Merson integrator, and then over-riding the solution of dependent generalized coordinates, by their solution obtained by Newton's method, in terms of the independent coordinates. A word of caution is that this approach is not immune to singularity, which happens when two constraint equations are identical, and this needs to be checked [11–13].

9.5.1 SIMULATION RESULTS

Large angle slewing of a *single* flexible body, a solar sail spacecraft with 6 rigid body dof and 44 vibration modes, is simulated first for a bang-bang torque

TABLE 9.1

Comparison of Simulation Times by Three Different Formulations

Method I	Block-diagonal mass with efficient generalized speeds	140 CPU sec
Method II	Block-diagonal mass formulation and customary speeds	667 CPU sec
Method III	Dense mass matrix formulation and customary speeds	699 CPU sec

application, using *three formulations*: one following the efficient generalized speeds for a single flexible body, called Method I here, given in Section 9.1, and the other two using Kane's method, one using the block-diagonal formulation with customary generalized speeds [6], called Method II, and the other using the dense-matrix formulation of chapter 6, called Method III, the latter two using the generalized speeds, $u_i = \dot{q}_i, i = 1,\ldots,6+n$, where q_i includes x-, y-, z-coordinates and the three Euler angles. Table 9.1 shows the comparison of simulation times.

Figures 9.5 and 9.6 actually show *three* plots from the three formulations. The results were identical for a 300 sec simulation.

We now consider the same multibody system considered before and reproduced here in Figure 9.7 for a system of flexible bodies connected by a revolute joint, a Hooke's joint with 2 rotational degrees of freedom, and a spherical ball joint. Each body is modeled by 10 vibration modes. Figure 9.8 shows the *internal torque* required to prescribe zero rotation for locking 1 rotational degree of freedom of the ball joint, *by the same three methods*, the analysis requiring a simple prescribed motion modification of the recursive algorithm of Ref. [2].

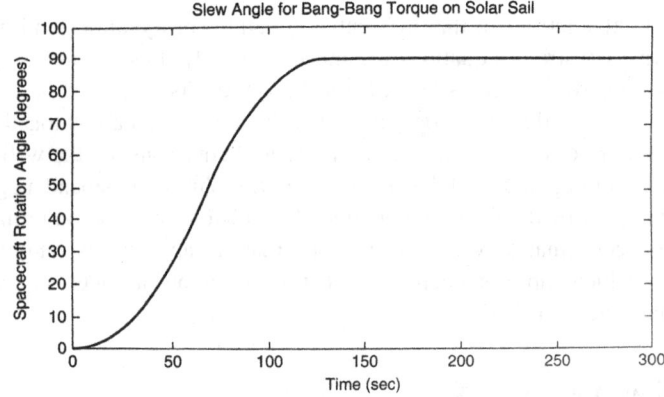

FIGURE 9.5 Solar sail slew angle vs. time for a 90° slewing maneuver due to bang-bang torque, showing indistinguishable results obtained by three simulation methods.

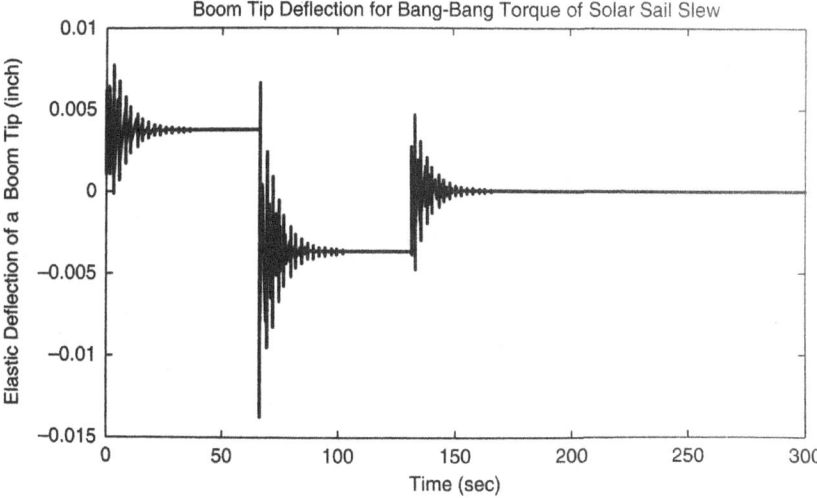

FIGURE 9.6 Boom tip deflection vs. time for bang-bang torque for solar sail slewing by three methods.

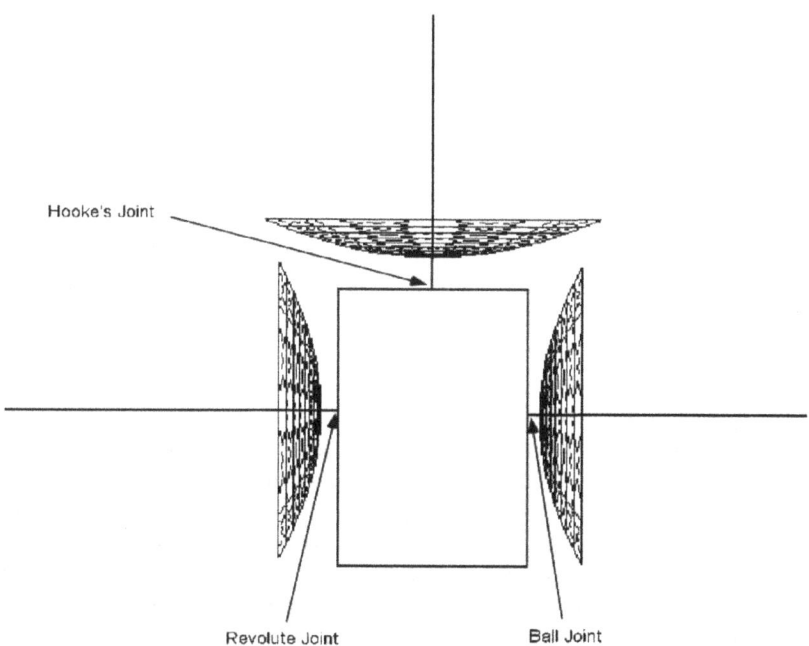

FIGURE 9.7 Flexible spacecraft connected to three flexible antennas via Hooke's joint, revolute joint, and ball joint.

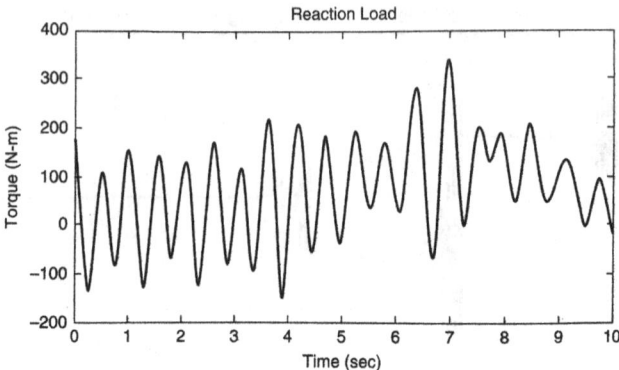

FIGURE 9.8 Internal reaction torque corresponding to prescribed motion (locking 1 rotational dof of one antenna) by three methods.

TABLE 9.2
Simulation Time Comparison for 10 sec with Three Formulations

Method I	Recursive formulation with efficient generalized speeds	5.2 CPU sec
Method II	Recursive formulation and customary generalized speeds	7.6 CPU sec
Method III	Non-recursive formulation, customary generalized speeds	10.2 CPU sec

The CPU (computer processing unit) times required for the three different formulations for a 10 sec simulation of this $(6+3+2+1+4\times10)$ or a 52 degrees of freedom system, with 10 modes for each flexible body, are as shown in Table 9.2.

Thus, in this example a recursive method using an efficient generalized speed is about two times faster than real time. The next two examples we discuss are for systems with motion constraints. Figure 9.9 gives a stroboscopic plot of a whirling chain of five flexible trusses connected by spherical joints in a three-dimensional, constrained motion with both ends in grounded sockets. Each truss has 3 relative rotational dof, and two of them have 4 modes and the remaining three have 12 modes.

Figure 9.10 shows the plots of a large angle rotation between bodies 4 and 5 from the left end. We have labeled the plots with indicators such as "EKane" for the extended-Kane algorithm of Ref. [7] and "ONeff" for the present (block-diagonal efficient formulation) method. The plots are identical to the naked eye. Figure 9.11 shows the time histories of the first five modal coordinates of body 5. The results look identical, with negligible numerical discrepancies.

Table 9.3 shows the computer times taken by the two methods. The table shows that the computational advantage of the present method increases with the increased number of modes per truss.

FIGURE 9.9 Stroboscopic plot from a simulation of a whirling chain of five flexible trusses each connected by a ball joint, with end points in grounded sockets.

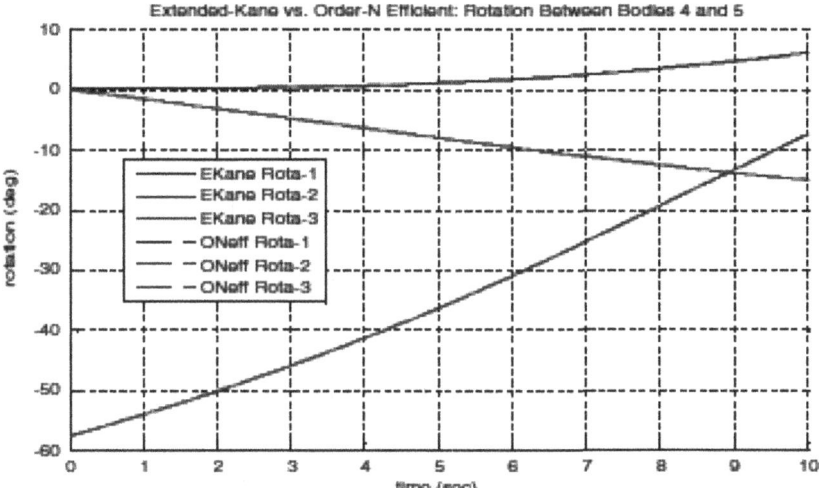

FIGURE 9.10 Plots of three rotations vs. time, between bodies 4 and 5, given by the extended Kane method (with an undetermined multiplier) of Ref. [7] and the present efficient order-n (actually block-diagonal) method.

Figure 9.12 shows an example of a three-dimensional, multi-loop, flexible structure mechanism, considered to illustrate an application of the main algorithm presented in this chapter. The two parts in Figure 9.13 show the rotations and errors between the two solutions, based on the present theory and the extended Kane method of Ref. [7]. A comparison of the modal coordinate solutions for a sample body, body 4, is shown in Figure 9.14.

Finally, the two parts of Figure 9.15 show the computed constraint force and loop closure error for loop 3.

Table 9.4 shows the computational performance of the recursive algorithm with efficient variables for the flexible multi-loop system of Figure 9.11, having three loops and nine constraint equations. In comparison with the algorithm of Ref. [7], it is clear that the overall combination of the block-diagonal algorithm

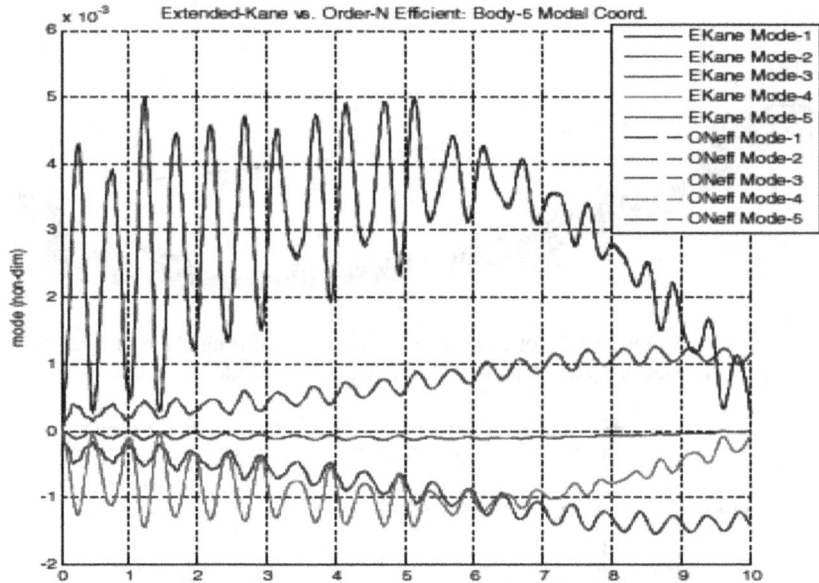

FIGURE 9.11 Comparison of solutions for the first five modal coordinates of body 5 by the extended Kane method [7] and the recursive method with efficient generalized speeds.

TABLE 9.3
CPU Comparison of Block-Diagonal vs. Extended Kane Formulation

Number of modes	Number of generalized speeds	CPU sec for extended Kane	Ratio of block-diagonal vs. E-Kane
4	35	40.75	1.01
12	75	128.61	2.49

and the choice of efficient motion variables produces a significantly better performance, as the number of modes per body is increased.

It should be noted finally that more recent research [12, 13] on the numerical solution of the dynamics of multi-loop structures has been done, by solving the differential-algebraic equations [10] of motion and the constraint equations in finite difference form recursively. The method may give slightly better results in constraint maintenance than that shown here in the example.

PROBLEM SET 9

The algorithm given in this chapter on multiloop constraints is quite complex. The basic idea for analyzing closed structural loops is to cut the loop and treat the system as an open-loop system with equal and opposite unknown forces and torques that are determined by adding constraint equations. Try coding the equations in Matlab for the system shown in Figure 9.4a.

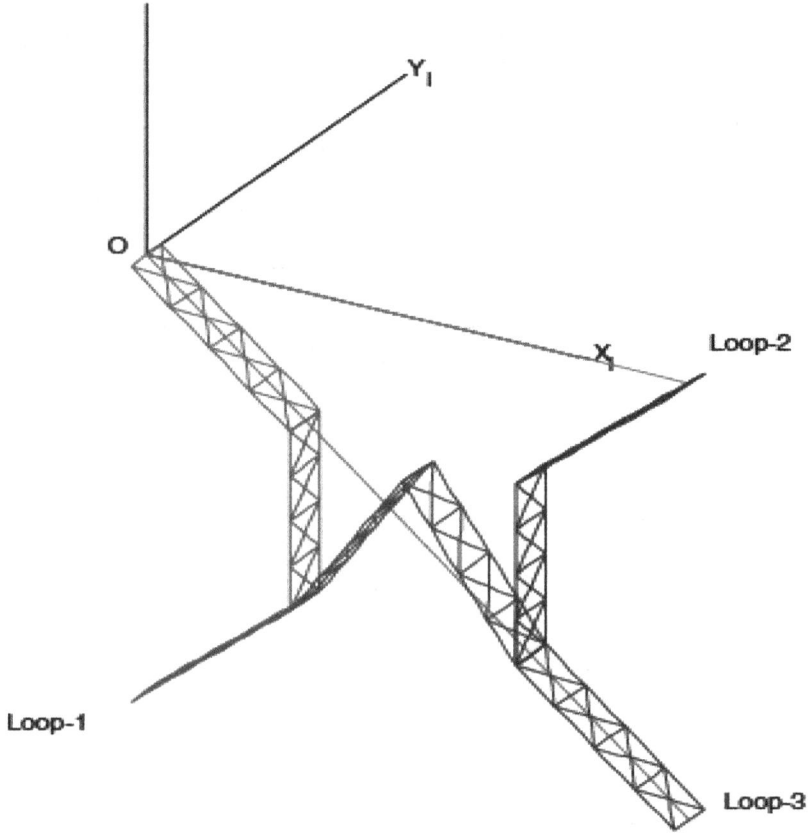

FIGURE 9.12   Flexible three loops structure with eight ball-jointed trusses in large motion.

1. Consider the planar motion of *three rigid rods* connected by revolute joints and acted on only by gravity. Derive the recursive equations of motion in two ways: (1) using $u_i = \dot{\theta}_i, i = 1, 2, 3$, where θ_1 is the angle from the vertical for the first link, and θ_2, θ_3 are the relative angles; and (2) using $u_1 = \dot{\theta}_1, u_2 = \dot{\theta}_1 + \dot{\theta}_2, u_3 = \dot{\theta}_1 + \dot{\theta}_2 + \dot{\theta}_3$. Show that the latter set of definitions is more efficient in the sense that it produces simpler equations.

2. Repeat the above exercise with the third link considered flexible, with two flexible cantilever modes, one using $u_1 = \dot{\theta}_1, u_2 = \dot{\theta}_1 + \dot{\theta}_2, u_3 = \dot{\theta}_1 + \dot{\theta}_2 + \dot{\theta}_3 + \sum_{i=1}^{2} \varphi_i \dot{q}_i$, ϕ_i being the i-th mode, and the other using Eq. (9.4) for an efficient choice of modal generalized speed. Uses for ϕ_i in the cantilever modes are given in Chapter 5.

3. Consider the analysis of the slider crank mechanism in Chapter 1, by starting with a model of a double pendulum. Use efficient generalized speeds for the two links, and then impose the constraint condition that the end of

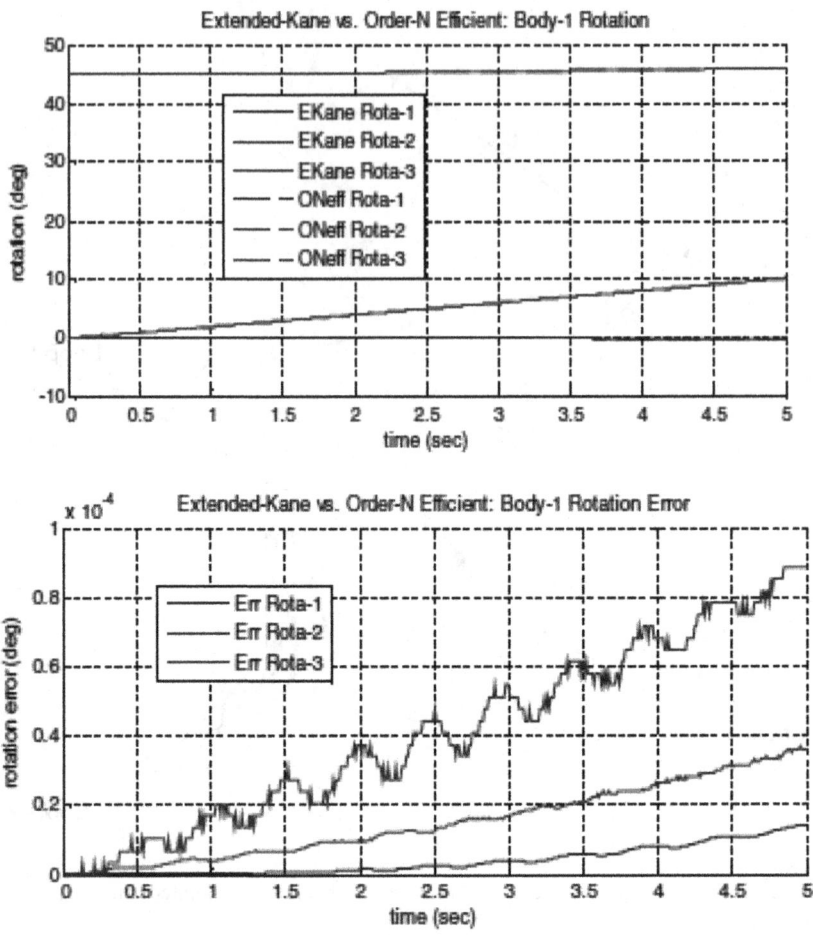

FIGURE 9.13 Comparison of rotation angles and errors for the ball-and-socket joint for body 1 by the efficient recursive block-diagonal formulation and the extended Kane method [7].

the second link must move on a horizontal plane. Derive the equations of motion with the constraint, introducing a Lagrange multiplier to represent the force normal to the horizontal plane. Now consider the crank rod and the connecting rod as bending beams and perform an analysis of a system with a closed structural loop. Choose your own parameters to represent the length, mass of the links elastic modulus, and the area moment of inertia, and simulate the motion given an initial velocity of the links.

4. Consider a "four bar" linkage, made up of three beams, with the first beam connected to the ground by a pin joint, the second beam connected to the first beam at its end by a pin, and the third beam connected to the second beam by a pin and closing a structural loop by being connected

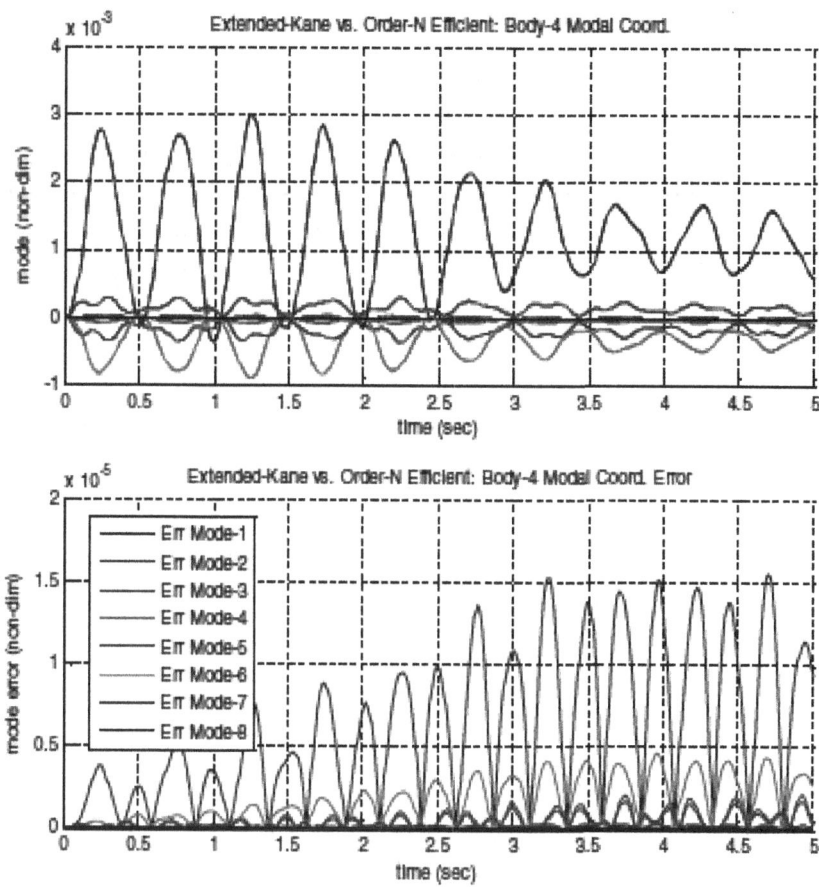

FIGURE 9.14 Comparison of modal coordinates for body 4 and error time history by two methods.

to the ground by another pin. Consider each beam of a different length L; assign 1 rotational degree of freedom and one elastic mode sin (πx/L) for each beam. Write two constraint equations for the horizontal and vertical directions, describing the loop closure. Derive the equations of motion, and numerically solve the differential algebraic equations involved.

APPENDIX 9 PSEUDO-CODE FOR CONSTRAINED NB-BODY M-LOOP RECURSIVE ALGORITHM IN EFFICIENT VARIABLES

9A.1 BACKWARD PASS

Here, we provide the extensions necessary beyond the backward pass for unconstrained systems that are given in Ref. [2]. The following segment is invoked during the backward pass for each body after all its mass updates are completed. All

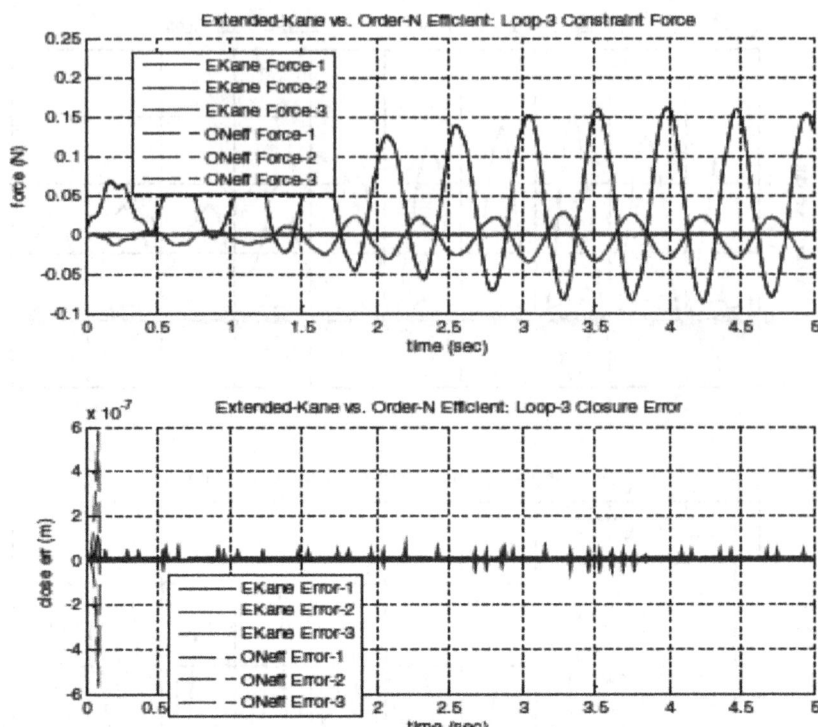

FIGURE 9.15 Constraint force (N) and loop closure error (m) vs. time, for loop 3.

bodies with cut joints become terminal bodies for which the modal mass matrix does not require an update and hence its inverse is computed only once. In the following, we use the notion of a direct path as going from a particular cut joint to body 1. Also Q(jp) is a point on a flexible body j, to which outboard body jp is hinged, and L(j) is the point on body j where the cut occurs. The pseudo-code algorithm description follows.

TABLE 9.4

Simulation Time Comparison of Order-n Algorithm with Efficient Generalized Speeds vs. Extended Kane Theory for a Multibody System with Three Structural Loops

Number of modes	Number of generalized speeds	CPU sec for 10 sec extended Kane	Ratio of block-diagonal over extended Kane
0	24	6.28	0.969
4	56	57.57	1.516
8	88	909.29	3.053
12	120	2484.3	4.455

For j going from nb to 1 **Do** the loop:

If j lies in the direct path for loop k

$$\hat{H}_{ek}^{j} = E_j^{-1}\Phi_{L(j)}^{T} \quad k = 1,\ldots,m \tag{A9.1}$$

$$G_j = Z_{L(j)} - M_2^j \hat{H}_{ek}^{n} \quad k = 1,\ldots,m \tag{A9.2}$$

$$\hat{H}_{rk}^{n} = v_j^{-1}R_j^{T}G_j \tag{A9.3}$$

$$B_{c(j)k} = G_j - M_1^j R_j \hat{H}_{rk}^{j} \quad k = 1,\ldots,m \tag{A9.4}$$

Else
jp = o(j), outboard body j, and assuming a translation joint at jp of δ^{jp}:

$$d_j^{jp} = \begin{bmatrix} C_{j,jp} & C_{j,t(jp)}\tilde{\delta}^{jp}C_{j,t(jp)}^{T} \\ 0 & C_{j,jp} \end{bmatrix} \tag{A9.5}$$

$$\hat{H}_{ek}^{j} = E_j^{-1}\Phi_{Q(jp)}^{j}d_j^{jP}B_{jk} \quad k = 1,\ldots,m \tag{A9.6}$$

$$G_j = Z_{jp} - M_2^j E_j^{-1}\Phi_{Q(jP)}^{j} \tag{A9.7}$$

$$\hat{H}_{rk}^{j} = v_j^{-1}R_j^{T}G_j d_j^{jP}B_{jk} \quad k = 1,\ldots,m \tag{A9.8}$$

$$B_{c(j)k} = G_j d_j^{jP}B_{jk} - M_1^j R_j \hat{H}_{rk}^{j} \quad k = 1,\ldots,m \tag{A9.9}$$

Continue

9A.2 Forward Pass

Initialize by setting $H_{rk}^{j} = H_{ek}^{j} = 0$, $j = 1,\ldots,nb$; $k = 1,\ldots,m$.
Refer to Eqs. (9.95) and (9.100) for F_{rf}^{1}, F_{ef}^{1}.

$$H_{rk}^{1} = \hat{H}_{rk}^{1} \quad k = 1,\ldots,m \tag{A9.10}$$

$$S_f^{1} = R_1 F_{rf}^{1} \tag{A9.11}$$

$$S_{ck}^{1} = R_1 H_{rk}^{1} \quad k = 1,\ldots,m \tag{A9.12}$$

$$H_{ek}^{1} = \hat{H}_{ek}^{1} + E_j^{-1}A_1 S_{ck}^{1} \quad k = 1,\ldots,m \tag{A9.13}$$

The dynamical equations for the rigid and elastic degrees of freedom for body 1 then become

$$\dot{u}_r^1 = F_{rf}^1 + \sum_{k=1}^m H_{rk}^1 \lambda_k \tag{A9.14}$$

$$\dot{u}_e^1 = F_{ef}^1 + \sum_{k=1}^m H_{ek}^1 \lambda_k \tag{A9.15}$$

For $j = 2 \longrightarrow nb$

$$\gamma_f^j = W_j S_f^{c(j)} + N_j F_{ef}^{c(j)} \tag{A9.16}$$

$$\gamma_{ck}^j = W_j S_{ck}^{c(j)} + N_j H_{ek}^{c(j)} \quad k = 1,\dots,m \tag{A9.17}$$

$$F_{rf}^j = -v_j^{-1} R_j^T \{ Y_1^j + M_j^1 \gamma_f^j \} + v_j^{-1} \begin{Bmatrix} \tau_j \\ f_j \end{Bmatrix} \tag{A9.18}$$

$$H_{rk}^j = \hat{H}_{rk}^j - v_j^{-1} R_j^T M_1^j \gamma_{ck}^j \quad k = 1,\dots,m \tag{A9.19}$$

$$S_f^j = \gamma_f^j + R_j F_{rf}^j \tag{A9.20}$$

$$S_{ck}^j = \gamma_{ck}^j + R_j H_{rk}^j \quad k = 1,\dots,m \tag{A9.21}$$

$$F_{ef}^j = E_j^{-1} \{ Y_1^j + M_2^j S_f^j \} \tag{A9.22}$$

$$H_{ek}^j = \hat{H}_{ek}^j + E_j^{-1} M_2^j S_{ck}^j \quad k = 1,\dots,m \tag{A9.23}$$

The dynamical equations for body j in rigid and elastic degrees of freedom are now given as

$$\dot{u}_r^j = F_{rf}^j + \sum_{k=1}^m H_{rk}^j \lambda_k \tag{A9.24}$$

$$\dot{u}_e^j = F_{ef}^j + \sum_{k=1}^m H_{ek}^j \lambda_k \tag{A9.25}$$

Continue forward pass until j = nb.

ACKNOWLEDGEMENT

The author is grateful to the Management at the Advanced Technology Center, Lockheed Missiles and Space System Company, where this work [1] was done, for permission to include this material in this book.

REFERENCES

1. Banerjee, A.K. and Lemak, M.E., (2007), "Recursive Algorithm with Efficient Variables for Flexible Multibody Dynamics with Multiloop Constraints", *Journal of Guidance, Control, and Dynamics*, **30**(3), pp. 780–790.
2. Banerjee, A.K., (1992), "Block-Diagonal Equations for Multibody Elastodynamics with Geometric Stiffness and Constraints", in *State of the Art Survey Lecture in Multibody Dynamics*, European Space Agency, June; republished in *Journal of Guidance, Control, and Dynamics*, **16**(6), 1993, pp. 1092–1100.
3. D'Eleuterio, G.M.T. and Barfoot, T.D., (1999), "Just a Second, We'd Like to Go First: A First Order Discretized Formulation for Elastic Multibody Dynamics", in *Proceedings of the 4th International Conference on Dynamics and Control of Structures in Space*, edited by C.L. Kirk and R. Vignjevic, Cranfield University, Bedfordshire, UK, May 24–28.
4. Kane, T.R. and Levinson, D.A., (1985), *Dynamics, Theory and Applications*, McGraw-Hill.
5. Banerjee, A.K. and Dickens, J.M., (1990), "Dynamics of an Arbitrary Flexible Body in Large Rotation and Translation", *Journal of Guidance, Control, and Dynamics*, **13**(2), pp. 221–227.
6. Mitiguy, P.C. and Kane, T.R., (1996), "Motion Variables Leading to Efficient Equations of Motion", *International Journal of Robotics Research*, **15**(5), pp. 522–532.
7. Wang, J.T. and Huston, R.L., (1987), "Kane's Equations with Undetermined Multipliers: Application of Constrained Multibody Systems", *Journal of Applied Mechanics*, **54**(2), pp. 424–429.
8. Amirouche, F.M.L., (1992), *Computational Methods in Multibody Dynamics*, Prentice-Hall, pp. 434–460.
9. Brennan, K.E., Campbell, S.L., and Petzold, L., (1989) *Numerical Solutions of Initial Value Problems in Differential-Algebraic Equations*, Elsevier Publications.
10. Kane, T.R., Likins, P.W., and Levinson, T.R., (1983), *Spacecraft Dynamics*, McGraw-Hill.
11. Negrut, D., Jay, L.O., and Khude, N., (2009), "A Discussion of Low-Order Numerical Integration Formulas for Rigid and Flexible Multibody Dynamics", *Journal of Computational and Nonlinear Dynamics*, **4**(4).
12. Simeon, B., (2013), *Computational Flexible Multibody Dynamics: A Differential Algebraic Approach*, Springer.
13. Lemak, M.E., (1992), "Enforcement of Configuration Satisfaction Concurrent with the Numerical Solution of Multi-Flexible-Body Equations of Motion", Engineer's Thesis, Stanford University.

10 An Order-n Formulation for Beams with Undergoing Large Deflection and Large Base Motion

The large deflection of beams with respect to reference frames that are undergoing large motion is normally analyzed by non-linear finite element methods. In this chapter, we describe the large deflection of beams undergoing large overall motion, in a plane, by modeling the beam as a free-flying system of many rigid rods connected by springs at revolute joints. At first, this may seem to be a very crude way to model the large deflection of beams. As the results in this chapter will show, the rigid element model can actually be as accurate as a non-linear finite element model, and with a good algorithm can be more efficient in reducing simulation time. Note that a representation of the deflection of a beam by its vibration modes is not justified for large deflection. The algorithm used here is a specific simplification of the block-diagonal formulation given in Chapter 8. We demonstrate here only the case of planar bending, but three-dimensional bending and twisting can be handled by changing the orientation of the rotation axes at joints. Simulation based on the rigid segment model is made time-efficient by the so-called order-n formulation, where the computational effort increases linearly with n, the degree of freedom, and not as n^3 as in a non-recursive multibody formulation , given in Chapter 6. A side benefit of the order-n formulation is a way of computing *internal loads* in a system of bodies connected in a topological tree. Two numerical integration schemes are considered, Kutta–Merson and Newmark. The Newmark integration scheme is linked here with the order-n equations. We show that the order-n formulation with discrete rigid segments is more efficient, time-wise, than a non-linear finite element formulation, even with a fourth-order integrator, to get the same level of accuracy as a finite element formulation which needs to use a second-order integrator to keep out instantaneous high frequency modes. The material reported here is based on Refs. [1, 2].

10.1 DISCRETE MODELING FOR LARGE DEFLECTION OF BEAMS

The example system under consideration is the shuttle-antenna system shown in Figure 10.1, where, in an experimental study of waves in space plasma

DOI: 10.1201/9781003231523-11

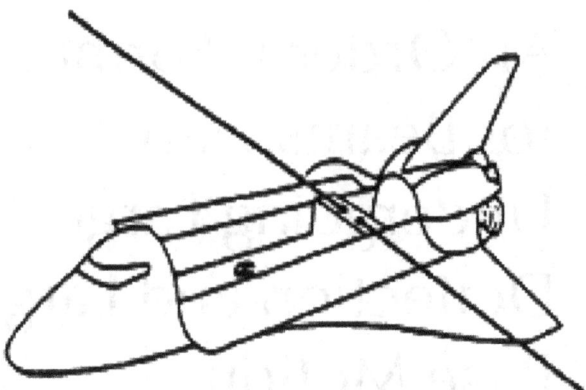

FIGURE 10.1   Shuttle WISP antenna system.

(WISP), the shuttle rotates slowly, carrying two 150 m long antennas. We treat the shuttle as a rigid body, while the long antennas are treated as beams of high flexibility.

Figure 10.2 shows a particular way of discretizing beams, with rigid rods connected by torsional springs. This rigid finite segment method has been shown in Ref. [3] to adequately represent the bending of beams. We use the same model here for large deflections. Kruszewski et al. [3] use EI/L as the stiffness coefficient for all the springs, where E is the elastic modulus, I the cross-sectional area moment of inertia, and L the length of a segment. Here, spring coefficients consistent with the Bernoulli–Euler beam theory are derived. The required spring coefficient σ at any joint can be computed by equating the bending moment at that location, due to a tip load P, to the restoring moment of the spring at that joint. Thus, in Figure 10.2, at joint 1 at a distance of L/2 from the fixed end, the spring constant is found as follows (the segment of length L/2 is used to retain the cantilever boundary condition of the beam):

$$\sigma_1\theta_1 = P\left(nL - \frac{L}{2}\right) \tag{10.1}$$

One computes θ_1 from the deflection δ_2 at the second joint from the fixed end in Figure 10.2:

$$\theta_1 = \frac{\delta_2}{L} \tag{10.2}$$

In general, deflection δ due to a tip load, at a distance x from the fixed end, is given by linear beam theory as

$$\delta = \left(\frac{P}{6EI}\right)x^2(3nL - x) \tag{10.3}$$

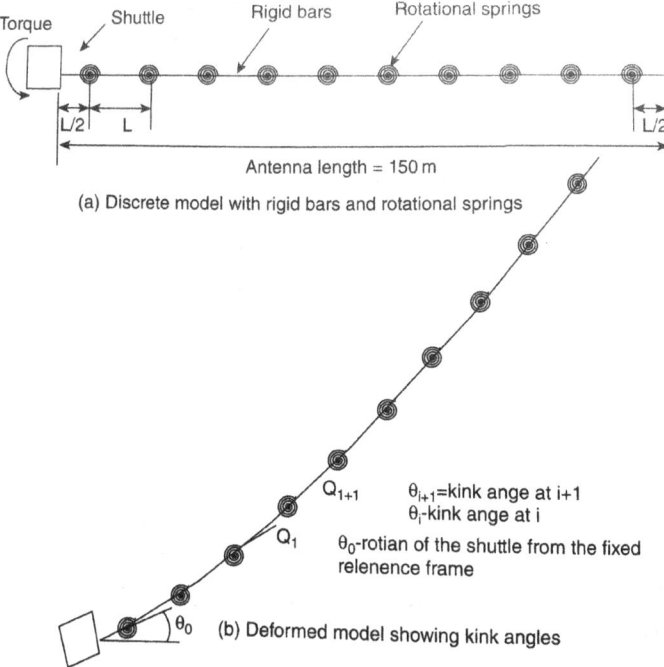

(a) Discrete model with rigid bars and rotational springs

(b) Deformed model showing kink angles

FIGURE 10.2 Beam modeled with rigid rods connected by rotational springs.

Evaluating the deflection at a location $x = 3L/2$, and substituting in Eqs. (10.1) and (10.2) yield

$$\sigma_1 = \frac{8}{9}\left(\frac{EI}{L}\right) \qquad (10.4)$$

For joint 2, the relation between angle θ_2 and the deflections at joints 2 and 3 is

$$\theta_2 = \frac{(\delta_3 - 2\delta_2)}{L} \qquad (10.5)$$

and the bending moment equation at joint 2 is

$$\sigma_2\theta_2 = P\left(nL - \frac{3L}{2}\right) \qquad (10.6)$$

Using Eq. (10.3) for the distances to joints 2 and 3 to evaluate δ_2, δ_3, and substitution in Eqs. (10.5) and (10.6) lead to

$$\sigma_2 = \frac{48n - 72}{42n - 71}\frac{EI}{L} \qquad (10.7)$$

For a general joint i, the relative angle is given by the *central finite difference* formula:

$$\theta_i = \frac{(\delta_{i+1} - 2\delta_i + \delta_{i-1})}{L} \tag{10.8}$$

and the moment equilibrium requires

$$\sigma_i \theta_i = P\left[nL - \left(i - \frac{1}{2}\right)L\right] \tag{10.9}$$

Evaluation of δ_{i-1}, δ_i, δ_{i+1} from Eq. (10.3) and substitution in Eqs. (10.8) and (10.9) give

$$\sigma_i = \frac{EI}{L} \quad i = 3, \ldots, n \tag{10.10}$$

Equations (10.4), (10.7), and (10.10) thus list all the necessary spring constants.

10.2 MOTION AND LOADS ANALYSIS BY THE ORDER-N FORMULATION

Motion analysis by an order-n or O(n) formulation, where the system "mass matrix" requires operation counts of only order n, is well known [4–6], and is reviewed here not only to formulate the equations, but also to bring out the features needed for the determination of loads. A system of bodies connected in a topological tree configuration by single dof rotational joints is considered, with the bodies numbered as $j = 1, \ldots, n$. With a slight change of notion here from before, we call a generic body j as connected to its inboard body i at the hinge Q_j. The formulation given here follows the work of Rosenthal, Ref. [5], for which the basis is Kane's equations. To demonstrate the process, and restricting ourselves to planar motion, we choose the generalized speed u_j:

$$u_j = {}^i\omega^j \cdot \lambda_j \quad j = 1, \ldots, n \tag{10.11}$$

Here, ${}^i\omega^j$ stands for the angular velocity vector of the outward body j with respect to its inboard body i, and λ_j is the unit vector along the axis of rotation at the revolute joint. The angular velocity of j in the inertial frame N is related to the angular velocity of i by the angular velocity addition theorem:

$$^N\omega^j = {}^N\omega^i + u_j\lambda_j \tag{10.12}$$

The kinematical equation, defining generalized speeds u_j, is chosen simply as

$$u_j = \dot{q}_j \quad j = 1, \ldots, n \tag{10.13}$$

where q_j is the angle of rotation of a line fixed in body j with respect to a line fixed in body i. The velocity of the hinge point Q_j is related to the velocity Q_i inboard body anchor point i as

$$^N\mathbf{v}^{Q_j} = {}^N\mathbf{v}^{Q_i} + {}^N\boldsymbol{\omega}^i \times \mathbf{r}^j \tag{10.14}$$

where $\mathbf{r}^j$ is the position vector from Q_i to Q_j. Equations (10.12) and (10.14) give rise to the following expressions for the partial angular velocity [7] of body j and the partial velocity of Q_i, with respect to the j-th generalized speed u_j as

$$^N\boldsymbol{\omega}_j^j = \lambda_j \tag{10.15}$$

$$^N\mathbf{v}_j^{Q_j} = 0 \tag{10.16}$$

As in previous chapters, the angular acceleration for body j is split into two groups, one group involving derivatives of the generalized speeds, and the other free of such derivatives:

$$^N\boldsymbol{\alpha}^j = {}^N\boldsymbol{\alpha}_0^j + {}^N\boldsymbol{\alpha}_t^j \tag{10.17}$$

As before, the first group, involving derivatives of generalized speeds, is further split as

$$^N\boldsymbol{\alpha}_0^j = {}^N\hat{\boldsymbol{\alpha}}_0^j + {}^N\boldsymbol{\omega}_j^j \dot{u}_j \tag{10.18}$$

with the hat quantities referring to angular acceleration terms containing derivatives of the generalized speeds for the inboard body i:

$$^N\hat{\boldsymbol{\alpha}}_0^j = {}^N\boldsymbol{\alpha}_0^i \tag{10.19}$$

The second group of terms in Eq. (10.17), the remainder acceleration terms, is zero and true only for planar motion, that is,

$$^N\boldsymbol{\alpha}_t^j = 0 \tag{10.20}$$

Similarly, all acceleration terms are split into terms involving derivatives of generalized speeds and the remainder terms in the acceleration:

$$^N\mathbf{a}^{Q_j} = {}^N\mathbf{a}_0^{Q_j} + {}^N\mathbf{a}_t^{Q_j} \tag{10.21}$$

$$^N\mathbf{a}_0^{Q_j} = {}^N\hat{\mathbf{a}}_0^{Q_j} \tag{10.22}$$

$$^N\hat{\mathbf{a}}_0^{Q_j} = {}^N\mathbf{a}_0^{Q_i} + \boldsymbol{\alpha}_0^i \times \mathbf{r}^j \tag{10.23}$$

$$^N\mathbf{a}_t^{Q_j} = {}^N\mathbf{a}_t^{Q_i} + {}^N\boldsymbol{\alpha}_t^i \times \mathbf{r}^j + {}^N\boldsymbol{\omega}^i \times ({}^N\boldsymbol{\omega}^i \times \mathbf{r}^j) \tag{10.24}$$

Now all vectors pertaining to body j are expressed in the vector basis of body j, and their measure numbers are shown in matrix forms. Thus, a matrix R_j of partial velocity and partial angular velocity in the j basis for the joint frame at Q_j is obtained by inspection of Eqs. (10.15) and (10.16), as the 6×1 matrix of scalar elements:

$$R_j = \left\{ \begin{array}{c} 0 \\ \lambda_j \end{array} \right\} \tag{10.25}$$

so that Eqs. (10.22) and (10.18) are written in matrix form as

$$\left\{ \begin{array}{c} ^N\mathbf{a}_0^{Q_j} \\ ^N\boldsymbol{\alpha}_0^j \end{array} \right\} = \left\{ \begin{array}{c} ^N\hat{\mathbf{a}}_0^{Q_j} \\ ^N\hat{\boldsymbol{\alpha}}_0^j \end{array} \right\} + R_j \dot{u}_j \tag{10.26}$$

As in previous chapters, Eqs. (10.19) and (10.23) are rewritten in terms of a matrix shift operator, which relates kinematical variables from the outboard body to an inboard body:

$$\left\{ \begin{array}{c} ^N\hat{\mathbf{a}}_0^{Q_j} \\ ^N\hat{\boldsymbol{\alpha}}_0^j \end{array} \right\} = \mathbf{W}^j \left\{ \begin{array}{c} ^N\mathbf{a}_0^{Q_i} \\ ^N\boldsymbol{\alpha}_0^i \end{array} \right\} \tag{10.27}$$

Here, using the skew-symmetric matrix $\tilde{r}^j$ corresponding to vector cross product with r^j a (3×1) matrix, and C_{ij} as the direction cosine matrix transforming the basis vectors of body j to the basis vectors in body i, the shift operator becomes

$$\mathbf{W}^j = \begin{bmatrix} C_{ij}^t & -C_{ij}^t \tilde{r}^j \\ 0 & C_{ij}^t \end{bmatrix} \tag{10.28}$$

At this stage, we can form all the kinematical variables we need for all the bodies, going along a "forward pass" from body 1 to body n.

Dynamical equations are generated in two steps: a "backward pass" starting from body n and ending in body 1, uncovering the dynamical equations for body 1, after which, a second forward pass is undertaken, where the rest of the dynamical equations are unraveled. For a generic body j, we write the D'Alembert equations of dynamic equilibrium in terms of the inertia forces and torques and the resultant of the external forces and moments about the hinge Q_j as follows:

$$\left\{ \begin{array}{c} f^{*j} - f_e^j \\ t^{*j} - t_e^j \end{array} \right\} \equiv M^j \left\{ \begin{array}{c} ^N\mathbf{a}_0^{Q_j} \\ ^N\boldsymbol{\alpha}_0^j \end{array} \right\} + X^j \tag{10.29}$$

$$\mathbf{M}^j = \begin{bmatrix} \mathbf{m}^j & -\tilde{\mathbf{s}}^{j/Q_j} \\ \tilde{\mathbf{s}}^{j/Q_j} & \mathbf{I}^{j/Q_j} \end{bmatrix} \tag{10.30}$$

$$\mathbf{X}^j = \mathbf{M}^j \left\{ \begin{matrix} {}^N\mathbf{a}_t^{Q_j} \\ {}^N\boldsymbol{\alpha}_t^j \end{matrix} \right\} + \left\{ \begin{matrix} \tilde{\boldsymbol{\omega}}^j\tilde{\boldsymbol{\omega}}^j\mathbf{s}^{j/Q_j} - \mathbf{f}_e^j \\ \tilde{\boldsymbol{\omega}}^j\mathbf{I}^{j/Q_j}\boldsymbol{\omega}^j - \mathbf{t}_e^j \end{matrix} \right\} \tag{10.31}$$

Here, the matrices $\mathbf{M}^j$, $\mathbf{X}^j$ are defined in terms of the mass $\mathbf{m}^j$, the first and second moments of inertia about the hinge Q_j, that is, $\mathbf{I}^{j/Q_j}$, $\mathbf{s}^{j/Q_j}$, with $\tilde{\mathbf{s}}^{j/Q_j}$ the skew-symmetric form of cross product with $\mathbf{s}^{j/Q_j}$ as used before. Note that $\tilde{\boldsymbol{\omega}}^j$ is the skew-symmetric matrix formed for cross product out of the three elements in the vector $\boldsymbol{\omega}^j$. Kane's equations [7] associated with the hinge degrees of freedom for a *terminal* body j are obtained by pre-multiplying the force and moment terms in Eq. (10.29), augmented by any hinge spring torque $\mathbf{t}_h^j$, by transposing the partial velocity/partial angular velocity matrix in Eq. (10.25).

$$\ddot{\boldsymbol{\theta}}^j = -\left[\mathbf{v}^j\right]^{-1}\left[\mathbf{R}^j\right]^T\left[\mathbf{M}^j\left\{\begin{matrix}{}^N\hat{\mathbf{a}}_0^{Q_j} \\ {}^N\hat{\boldsymbol{\alpha}}_0^j\end{matrix}\right\} + \mathbf{X}^j\right] + \left[\mathbf{v}^j\right]^{-1}\left\{\begin{matrix}0 \\ \mathbf{t}_h^j\end{matrix}\right\} \tag{10.32}$$

Here we have introduced the matrix:

$$\mathbf{v}^j = [\mathbf{R}^j]^T\mathbf{M}^j\mathbf{R}^j \tag{10.33}$$

Using Eqs. (10.26) and (10.32) in Eq. (10.29), and introducing the notations:

$$\mathbf{M} = \mathbf{M}^j - \mathbf{M}^j\mathbf{R}^j[\mathbf{v}^j]^{-1}[\mathbf{R}^j]^T\mathbf{M}^j$$

$$\mathbf{X} = \mathbf{X}^j - \mathbf{M}^j\mathbf{R}^j[\mathbf{v}^j]^{-1}[\mathbf{R}^j]^T\mathbf{X}^j + \mathbf{M}^j\mathbf{R}^j[\mathbf{v}^j]^{-1}\left\{\begin{matrix}0 \\ \mathbf{t}_h^j\end{matrix}\right\} \tag{10.34}$$

lead to a re-expression of Eq. (10.29) as

$$\left\{\begin{matrix}\mathbf{f}^{*j} - \mathbf{f}_e^j \\ \mathbf{t}^{*j} - \mathbf{t}_e^j\end{matrix}\right\} = \mathbf{M}\left\{\begin{matrix}\hat{\mathbf{a}}^{Q_j} \\ \hat{\boldsymbol{\alpha}}_0^j\end{matrix}\right\} + \mathbf{X} \tag{10.35}$$

The system of inertia and external forces and their moments at and about Q_j, expressed in the j basis, is obtained by substituting Eq. (10.27) in Eq. (10.35), to get the dynamic equilibrium terms represented by the left-hand side of the following equation:

$$\left\{\begin{matrix}\mathbf{f}^{*j} - \mathbf{f}_e^j \\ \mathbf{t}^{*j} - \mathbf{t}_e^j\end{matrix}\right\}_{Q_j/j} \equiv \mathbf{MW}^j\left\{\begin{matrix}\mathbf{a}_0^{Q_{c(j)}} \\ \boldsymbol{\alpha}_0^{c(j)}\end{matrix}\right\} + \mathbf{X} \tag{10.36}$$

Now, the forces and moments acting at Q_j can be equivalently transferred at and about Q_i, and expressed in the i basis by using the shift transform of Eq. (10.27):

$$\begin{Bmatrix} f^{*j} - f_e^j \\ t^{*j} - t_e^j \end{Bmatrix}_{Q_i/i} = [W^j]^T \begin{Bmatrix} f^{*j} - f_e^j \\ t^{*j} - t_e^j \end{Bmatrix}_{Q_j/j} \tag{10.37}$$

To this set of forces and moments is added the resultant of all active and inertia forces and moments of body i about Q_i, yielding the dynamic equilibrium (zero-sum):

$$\begin{Bmatrix} f^{*i} - f_e^i \\ t^{*i} - t_e^i \end{Bmatrix} \equiv \sum_{k \in S_i} \begin{Bmatrix} f^{*k} - f_e^k \\ t^{*k} - t_e^k \end{Bmatrix} = M^i \begin{Bmatrix} {}^N a_0^{Q_i} \\ {}^N \alpha_0^i \end{Bmatrix} + X^i \tag{10.38}$$

Here, S_i is the set of all bodies *outboard* of hinge Q_i that pass on their cumulative loads on body i, and we have used the following replacements:

$$M^i + [W^j]^T M W^j \rightarrow M^i \tag{10.39}$$

$$X^i + [W^j]^T X \rightarrow X^i \tag{10.40}$$

Equation (10.38) is of the same form as Eq. (10.29). This pattern is repeated until the equations for the base body, body 1, are explicitly written. At that stage, the inboard body acceleration and angular acceleration are either zero or prescribed. Assigning the zeroth body as the inertial frame, Eq. (10.32) reduces for j = 1 to

$$\ddot{\theta}^1 = -[v^1]^{-1}[R^1]^T X^1 + [v^1]^{-1} \begin{Bmatrix} 0 \\ t_h^1 \end{Bmatrix} \tag{10.41}$$

and concomitantly, Eq. (10.26) yields

$$\begin{Bmatrix} a_0^{Q_1} \\ \alpha_0^1 \end{Bmatrix} = R^1 \begin{Bmatrix} 0 \\ \ddot{\theta}^1 \end{Bmatrix} \quad \left(\ddot{\theta}^1 = -\left[v^1\right]^{-1}\left[R^1\right]^T X^1 + \left[v^1\right]^{-1} \begin{Bmatrix} 0 \\ t_h^1 \end{Bmatrix} \right) \tag{10.42}$$

Once the dynamical equations for body 1 are formed, the rest of the dynamical equations are uncovered in a second forward pass using the kinematical propagation equations previously established.

Computation of internal loads is a bonus feature attendant with the order-n algorithm given above. This is obtained from the key equation, Eq. (10.38), which shows the resultant of all external and inertia forces and torques outboard of any joint. To compute, for example, the loads at joint 2 from the fixed end of the system (see Figure 10.2) one proceeds in the second forward pass to obtain $\ddot{\theta}_2$ and uses Eq. (10.29), specialized as

$$\sum_{j \in S_2} \begin{Bmatrix} f^{*j} - f_e^j \\ t^{*j} - t_e^j \end{Bmatrix} = M_1^2 \begin{Bmatrix} {}^N a_0^{Q_2} \\ {}^N \alpha_0^2 \end{Bmatrix} + X^2 \tag{10.43}$$

Again, recall that superscript 2, including that in $\tilde{r}^2$ below, does not mean squaring, but just refers to quantities associated with body 2. Equation (10.43) is precisely what we need for the loads except that the force system is shifted to the root of the cantilever beam in our example by

$$\begin{Bmatrix} f_s \\ b_m \end{Bmatrix} = \begin{bmatrix} C_{12} & 0 \\ \tilde{r}^2 C_{12} & C_{12} \end{bmatrix} \left\langle \sum_{j \in S_2} \begin{Bmatrix} f^{*j} - f_e^j \\ t^{*j} - t_e^j \end{Bmatrix} \right\rangle \tag{10.44}$$

Here, S_2 is the set of all bodies outboard of joint 2; the shear force and bending moment, f_s, b_m, respectively, are due to forces to the right of joint 2. Finally, the additional force and moment, due to inertia of the part of the beam that represents body 1 (see Figure 10.2), is added to the shear force and moment from Eq. (10.44) to get the total shear and moment at the root of the cantilever.

Coding: A Fortran code for the order-n formulation is given in the Appendix at the end of the book.

10.3   NUMERICAL INTEGRATION BY THE NEWMARK METHOD

The order-n equations generated in the preceding section for large bending are *non-linear ordinary differential equations* of the form:

$$\ddot{q} = f(\dot{q}, q) \tag{10.45}$$

In rigid body dynamics, integration of these equations is customarily done by using the Runge-Kutta–Merson method. In structural dynamics involving mass, stiffness, and damping matrices, the differential equations tend to be stiff due to a mixture of high and low frequencies, and traditionally *implicit* integration methods such as the Newmark-δ method are used. Here, the Newmark method [8] is modified for the order-n formulation as follows. The equations for velocity and displacement are written in Newmark integration for a *time-step* h with a chosen value of δ as

$$\dot{q}_{t+h} = \dot{q}_t + [(1-\delta) \ddot{q}_t + \delta \ddot{q}_{t+h}]h \tag{10.46}$$

$$q_{t+h} = q_t + \dot{q}_t h + [(0.5 - \alpha)\ddot{q}_t + \alpha \ddot{q}_{t+h}]h^2 \tag{10.47}$$

The order-n dynamical equations of Eq. (10.45) are discretized, for example, as

$$\ddot{q}_{t+h} = f(\dot{q}_{t+h}, q_{t+h}) \tag{10.48}$$

Upon substitution of the following notations:

$$a_1 = \delta / (\alpha h); \quad a_2 = (1 - \delta / \alpha)$$

$$a_3 = [(1-\delta) - (0.5 - \alpha)\delta / \alpha]h$$

$$a_4 = (0.5 - \alpha)h^2; \quad a_5 = \alpha h^2; \quad a_6 = \delta h \tag{10.49}$$

$$G_1 = -a_1 q_t + a_2 \dot{q}_t + a_3 \ddot{q}_t \qquad (10.50)$$

$$G_2 = q_t + h\dot{q}_t + a_4 \ddot{q}_t \qquad (10.51)$$

Angular rates and displacements, with a choice of $\delta = 2\alpha$, Eqs. (10.46) and (10.47), are expressed by the equations:

$$\dot{q}_{t+h} = a_1 q_{t+h} + G_1 \qquad (10.52)$$

$$q_{t+h} - G_2 - a_5 f\{(a_1 q_{t+h} + G_1), q_{t+h}\} = 0 \qquad (10.53)$$

Equation (10.53) is a set of *implicit, non-linear* algebraic equations in the angular displacement, q_{t+h}, and can only be solved by the *Newton iteration method* using the Jacobian matrix, where I is an (n × n) identity matrix for n q's in Eq. (10.45):

$$J = \left[I - a_6 \frac{\partial f}{\partial \dot{q}_{t+h}} - a_5 \frac{\partial f}{\partial q_{t+h}} \right] \qquad (10.54)$$

Convergence of the Newton iterations is accepted when the norm of the correction vectors between successive iterations for the root of the vector function $f(\dot{q}, q)$ in Eq. (10.45) is less than some small number, like 10^{-6}.

10.4 LARGE DEFORMATION DYNAMICS USING THE NON-LINEAR FINITE ELEMENT METHOD

In this section, we give a brief outline of the finite element method used as an independent, standard approach to solve large deformation problems in non-linear elastodynamics, to check the results of the order-n formulation presented in this chapter. The finite element code used for this purpose is NEPSAP [10] (non-linear elastic plastic structural analysis program), a comprehensive system of linear/non-linear, static/dynamic analysis code developed at Lockheed, Sunnyvale, California. A number of static and dynamic problems have been solved with NEPSAP and compared with known exact solutions for its validations. For the WISP antenna problem, it is the geometric non-linearity that is of interest because of the large deformation. Two types of non-linearity are significant here: (a) the use of a complete kinematic expression for finite strains, i.e., the inclusion of non-linear terms in the strain–displacement relations; and (b) the use of the deformed configuration of the body to obtain the equations of motion which, as a consequence, become dependent on the total deformation. The non-linear strain–tensor relation is shown below, with $u_{i,j}$ the partial derivative of deformation u_i with respect to the spatial variable, x_j, and so on:

$$\varepsilon_{ij} = \frac{1}{2}[(u_{i,j} + u_{j,i}) + (u_{m,i} u_{m,j}^1 + u_{m,i}^1 u_{m,j}) + (u_{m,i} u_{m,j})] \qquad (10.55)$$

Here, the terms without superscript denote the incremental values involved in the motion of a body from a configuration at state 1 to an infinitesimally close state 2. In this definition, terms with superscript 1 denote the known values at state 1, and repetition in m indicates a tensor sum over m in the range 1,2,3. There are three groups of terms in Eq. (10.55); the first one is the conventional small displacement strain, while the second one represents the effect of large displacements manifested in terms of gradients of known displacements at state 1, and the final term represents the non-linear part of the strain–displacement relation. The first two terms are linearly dependent on the incremental deformations u, whereas the last term varies quadratically as the incremental deformations. For many structural analysis problems, it is essential to consider both of these sources of non-linearities. The effects of both geometric non-linearities on the behavior of systems are quite easily incorporated in the general formulation, and the equations to be solved in non-linear elasto-statics are the linearized incremental equations of equilibrium in the form:

$$[K_t]\{\Delta u\} = \{\Delta P\} \qquad (10.56)$$

where K_t is the instantaneous stiffness, or "tangent stiffness" including geometric stiffness, and Δu, ΔP are the incremental displacement and incremental load, respectively. The total displacements are computed by adding the incremental solution to the previously computed total values. Particularly relevant to the WISP problem considered here is the large displacement analysis presented by Bisshopp and Drucker [9], exact matching with their results having been obtained using the Lockheed code NEPSAP. The dynamics problem, of course, has inertia (and damping, neglected here) forces in addition to the elastic forces represented by Eq. (10.56), and so, the dynamic incremental equation becomes

$$M\Delta\ddot{u} + K_t\Delta u = \Delta P \qquad (10.57)$$

where the incremental displacements, velocities, and accelerations occur during a time-interval Δt, from time t to t + Δt. The Newmark integration scheme is invoked, with Eq. (10.57) transformed into a simple set of linear algebraic equations of the form given below, representing Newton iterations required in the integration, where equilibrium is considered at the time step ahead:

$$\tilde{K}_t\Delta u^{(k)} = \Delta P^{(k)} \qquad (10.58)$$

$$\tilde{K}_t = K_t + \frac{4}{\Delta t^2}M \qquad (10.59)$$

$$\Delta P^{(k)} = R^2 - F^{2,(k-1)} - M\left(\frac{4}{\Delta t^2}\Delta u^{(k-1)} - \frac{4}{\Delta t}\dot{u}^1 - \ddot{u}^1\right) \qquad (10.60)$$

Here, k is the *iteration counter*; superscripts 1 and 2 denote variables at t and t + Δt; and R, F are the total load and the equilibrium load, respectively. For a good discussion of the details of the integration scheme for non-linear finite element analysis, see Bathe [8].

10.5   COMPARISON OF THE NUMERICAL PERFORMANCES OF THE ORDER-N FORMULATION AND THE FINITE ELEMENT FORMULATION

As we have done several times before, we consider a spin-up problem for the shuttle-antenna system for a crucial test. The shuttle has a centroidal moment of inertia of 10^7 kg-m^2 and two 150 m long antennas are attached at points (0.5 m, −10 m; −0.5 m, −10 m) from the mass center of the shuttle. The antennas are beams of flexural rigidity 1676 N-m^2 and of mass density 0.335 kg/m. In terms of differential equations, this is not a "stiff" problem, with the ratio of maximum frequency to the minimum frequency of 20. Order-n equations for the proposed spring-connected rigid finite element model are integrated by using both the Newmark integrator [8] and a variable step fourth-order Kutta–Merson integrator. Simulation results are given for a torque pulse of 2687 N-m acting for 20 sec on the shuttle. Figure 10.3 shows the rotation angle of the shuttle vs. time during spin-up. It is clear that the reference frame for the non-linearly elastic antenna beams goes through a large angle of rotation. In the sequel, we use as the truth model the NEPSAP [10] finite element solution obtained by the Newmark integration, to assess the accuracy of the rigid segment order-n solution. Figure 10.4 establishes the accuracy of the NEPSAP solution itself for a cantilever beam, against the classical solution by Bisshopp and Drucker [9] for the normalized displacement of the beam with a tipload.

Now we present the results for spin-up for both the rigid element order-n and NEPSAP models [10]. Figures 10.5 and 10.6 present the results for discretizing each beam into 10 elements, each 15 m long. Order-n solutions by the Newmark method for step sizes 1.0 sec and 0.5 sec are overlaid on the NEPSAP solution for a step size of 1.0 sec. It is seen that while the general behavior of the beam tip deflection and bending moment as given by the order-n spring-connected rigid element model agrees quite well with the non-linear finite element solution, the rigid element model predicts a 4% higher peak deflection and an 8% lower peak moment. Slightly better results are expected by tightening the error tolerance in Newmark integration or reducing the length of the rigid segments.

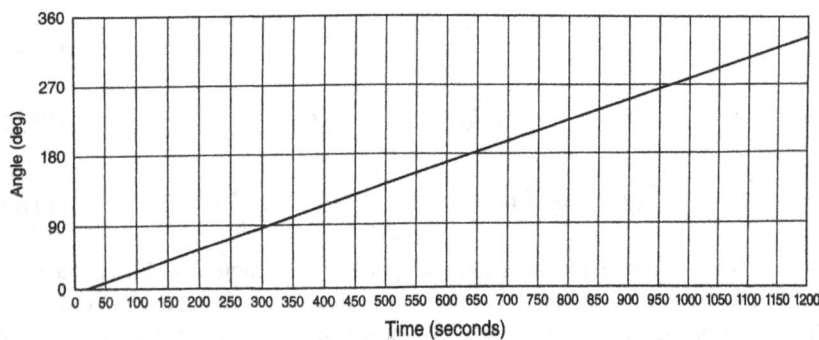

FIGURE 10.3   Shuttle roll rotation in degrees vs. time during spin-up.

The computer time for a Newmark integration for a 1200 sec rotational motion is given in Table 10.1. It is seen that the order-n solution is almost 25 times faster than the solution given by the finite element method, while the two results in Figures 10.5 and 10.6 are quite close to each other.

A Newmark integration of the order-n dynamical equations is done with an update of the associated Jacobian matrix (that does not change much) every 100 sec. It is clear that while the rigid element formulation with the Newmark integration is computationally very efficient, it is not as accurate, as Figure 10.6 shows.

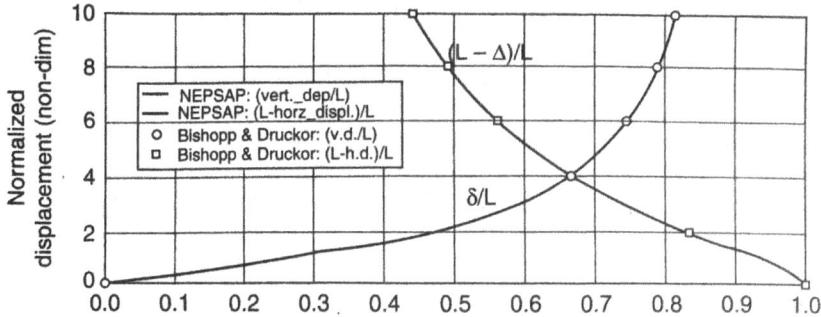

FIGURE 10.4 Normalized displacement comparison for the NEPSAP code, with classical Bisshopp and Drucker [9] solution for tip displacement.

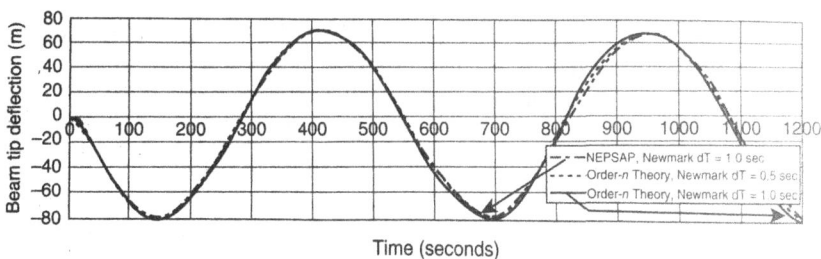

FIGURE 10.5 Tip deflection (m) vs. time, 10 element non-linear finite element method and rigid segment O(n) solutions.

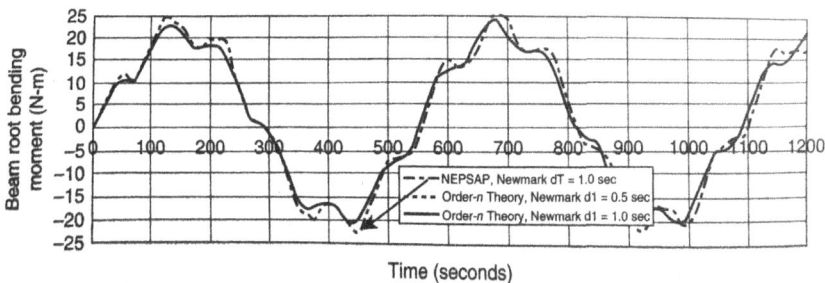

FIGURE 10.6 Beam root bending moment (N-m) vs. time, 10 element non-linear finite element method and rigid segment order-n solutions.

TABLE 10.1

CPU Times for Order-n and Finite Element Large Deflection Solutions

Formulation	Number of elements	Step size (sec)	CPU time (sec)
Order-n	10	1.0	24
Order-n	10	0.5	40.9
Finite element	10	1.0	890

To address the accuracy question even further, the order-n equations were integrated with a fourth-order variable step Kutta–Merson ordinary differential equation solver, with a maximum step size of 1.0 sec, and the results compared better with the finite element solution using the Newmark integrator for 1.0 step size, for both models using 20 elements. Figure 10.7 shows the almost perfect agreement between the two tip deflection solutions.

Figure 10.8 shows that the difference in moment loads computation with the two models is smaller than with less number of elements. Finally, a comparison of the shear force at the fixed end of the cantilever, obtained by the rigid element and non-linear finite element methods, for the 20 rigid element and non-linear finite element models, is given in Figure 10.9, showing some difference as the time increases.

The computer solution time for Figures 10.7 and 10.8, for the order-n solution with the rigid element model and the NEPSAP finite element model using the Newmark integrator, is compared in Table 10.2, showing, as in Table 10.1, the computational efficiency of the rigid segment order-n method over the finite element method. The order-n solution is almost six times faster!

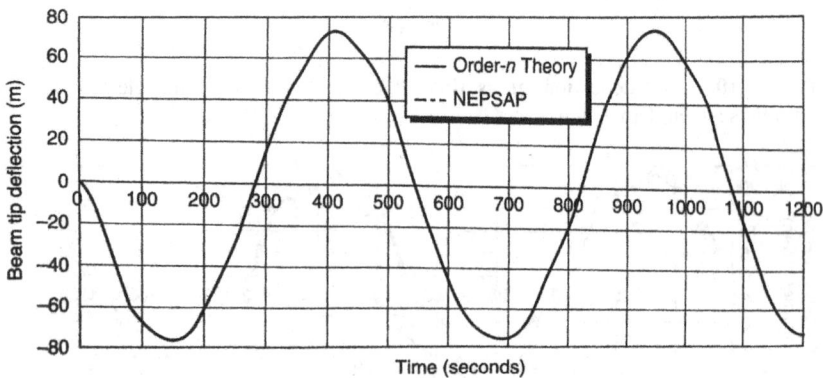

FIGURE 10.7 Beam tip deflection (m) vs. time by rigid segment order-n and Runge–Kutta integrator vs. non-linear finite element solution with Newmark integrator, both using 20 elements.

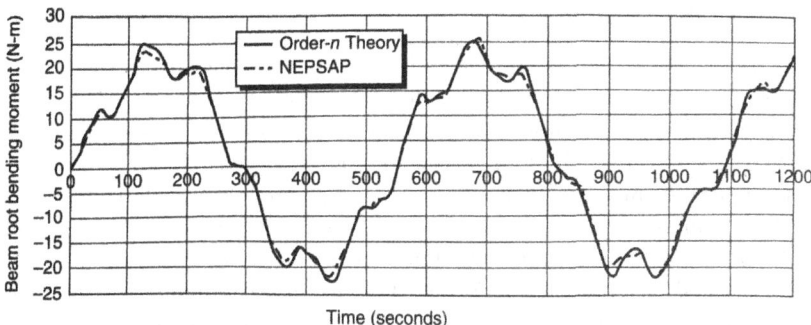

FIGURE 10.8 Beam root bending moment (N-m) vs. time, by rigid segment order and Runge–Kutta integrator and non-linear finite element solutions with Newmark integrator, both using 20 elements.

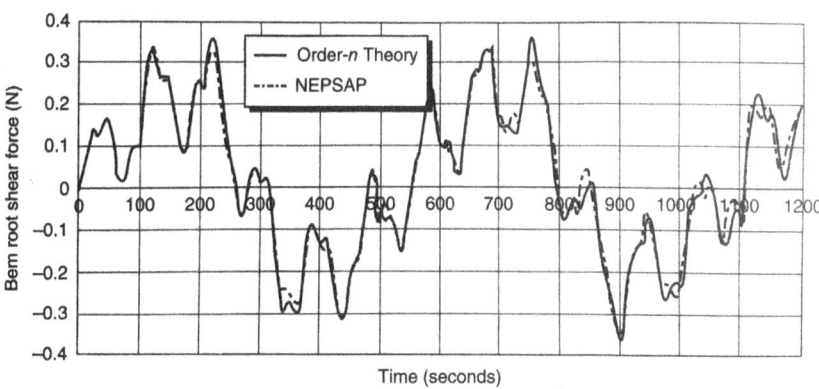

FIGURE 10.9 Beam root shear force (N) vs. time, 20 element, rigid element order-n and non-linear finite element methods.

TABLE 10.2

CPU Times for Order-n and Finite Element Large Deflection Solutions, Same Step Size

Formulation	Number of elements	Step size (sec)	CPU (sec)
Order-n	20	1.0	182
Finite element	20	1.0	1160

10.6 CONCLUSION

The results show that use of spring coefficients obtained from linear beam theory for a spring-connected rigid element model gives results for very large deflection, as good as those produced by the non-linear finite element solution. The order-n formulation given for the rigid segment model provides computation of internal loads like shear force and bending moment as natural by-products of the algorithm. The solution procedure indicates that while the rigid segment order-n formulation is very much superior to the finite element method in terms of computing time, it requires a higher-order integrator to give comparable levels of accuracy with a finite element solution. A Fortran code given as Appendix D at the end of the book can be used for this simulation.

ACKNOWLEDGMENT

Dr S. Nagarajan performed the numerical simulations based on the non-linear finite element code, NEPSAP, which he and Dr P. Sharifi developed for Lockheed Missiles and Space Company, and provided the basis of the comparison with results for the order-n algorithm given in this chapter. The author is grateful for his help.

PROBLEM SET 10

1. Code the recursive order-n algorithm given in Section 10.3, and use it to represent the motion of a system of massive links connected by revolute joints, and hanging in a uniform gravity field. Consider an initial value of the angles to represent a curved swing, and consider rate damping due to air on each link. Choose your own parameter values and simulate the ensuing motion.

2. Derive the vibration equations of the rotary system model of the WISP shuttle-antenna system shown below, and calculate the time period T of the first vibration mode for your value of J, m, L, k. For a minimum time slew of the undeformed system through an angle of 90 degrees, the minimum time T_{min} due to a bang-bang torque τ_{max} can be computed as $T_{min} = \sqrt{\pi J / (2\tau_{max})}$. Simulate the response of the system for this bang-bang torque. There is a way [11] to suppress the resulting vibrations of period T by a convolution of the bang-bang torque with a three-impulse sequence of (0.25, 0.5, 0.25) coming at 0, T/2, T. This produces the *shaped torque* scaled to *unit* torque magnitude, shown in Figure 10.10; actual torque has to be multiplied by this scaling. Obtain simulation results with this torque time history.

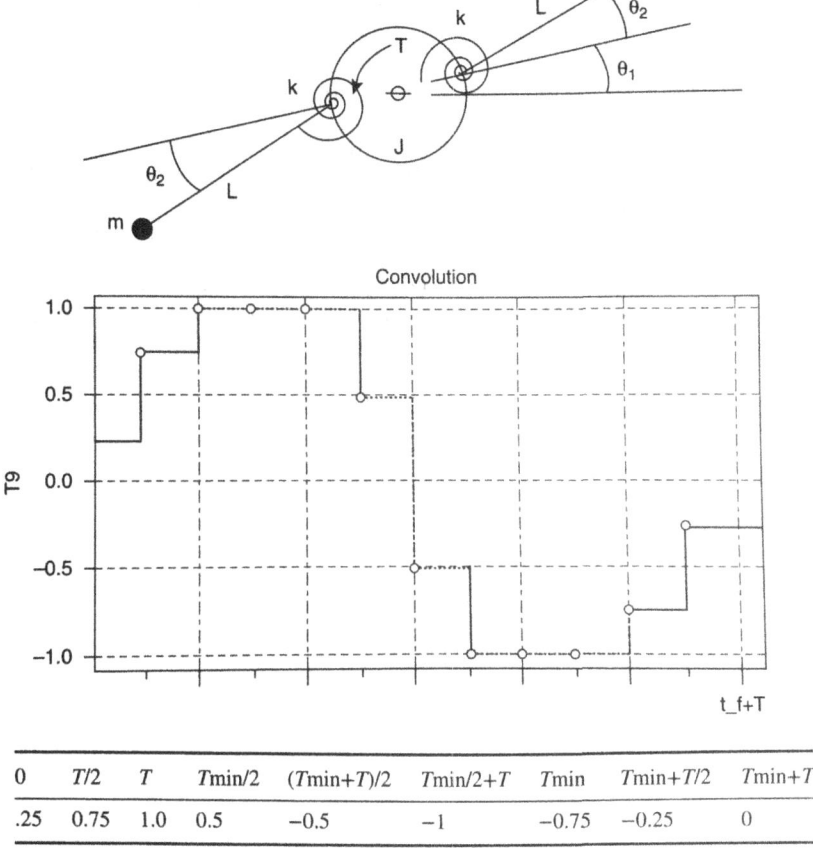

0	$T/2$	T	$T\text{min}/2$	$(T\text{min}+T)/2$	$T\text{min}/2+T$	$T\text{min}$	$T\text{min}+T/2$	$T\text{min}+T$
.25	0.75	1.0	0.5	−0.5	−1	−0.75	−0.25	0

FIGURE 10.10 Convolution of a bang-bang command with a three-impulse sequence for suppression of vibration of period T extends the duration of the command by T.

REFERENCES

1. Banerjee, A.K., (1993), "Dynamics and Control of the WISP shuttle-antenna system", *Journal of the Astronautical Sciences*, **41**, pp. 73–90.
2. Banerjee, A.K. and Nagarajan, S., (1997), "Efficient Simulation of Large Overall Motion of Beams Undergoing Large Deflection", *Multibody System Dynamics*, **1**, pp. 113–126.
3. Kruszewski, J., Gawronski, W., Wittbrodt, E., Najbar, F., and Grabowski, S., (1975), *Metoda Sztywnych Elementow Skonczonvch (Rigid Finite Element Method)*, Arkady Warszawa (in Polish).
4. Hollerbach, J.M., (1980), "A Recursive Lagrangian Formulation of Manipulator Dynamics and a Comparative Study of Dynamics Formulation complexity", *IEEE Transactions on Systems, Man, and Cybernetics*, **SMC-10**(11), pp. 730–736.

5. Rosenthal, D.E., (1990), "An Order-n Formulation for robotic Systems", *Journal of Astronautical Sciences*, **38**(4), pp. 511–530.

6. Keat, J.E., (1990), "Multibody System Order-n Formulation Based on Velocity Transform Method", *Journal of Guidance, Control, and Dynamics*, **13**(2), pp. 207–212.

7. Kane, T.R. and Levinson, D.A., (1985), *Dynamics: Theory and Applications*, McGraw-Hill.

8. Bathe, K.J (1982), *Finite Element Procedure in Engineering Analysis*, Prentice-Hall.

9. Bisshopp, K.E. and Drucker, D.C., (1945), "Large Displacement of Cantilever Beams", *Quarterly of Applied Mathematics*, **3**, pp. 272–275.

10. Nagarajan, S. and Sharifi, P., (1980), *NEPSAP Theory Manual*, Lockheed Missiles & Space Co.

11. Singer, N.C., and Seering, W.P., (1990), "Preshaping Command Inputs to Reduce System Vibration", *Journal of Dynamic Systems, Measurement and Control*, **112**(March), pp. 76–82.

11 Deployment/Retraction of Beams and Cables from Moving Vehicles

Small Deflection Analysis, and Variable-N Order-N Formulations for Large Deflection

This chapter describes the development and application of computationally efficient formulations to the extrusion and retraction of elastic beams from a moving base such as a spacecraft, and of a cable being reeled in or out from a winch on a ship, to control an underwater maneuvering vehicle. Here, deployment and retraction are handled for an n degree of freedom (dof) model by simply varying the number n of discretized bodies out, with careful state updates to maintain continuity of motion as the beam or cable goes from one value of n to another. The material is based on the author's papers [1–3], while Ref. [4] is a related material.

Extrusion of beams from a stowed configuration is necessary for spacecraft deploying a solar panel or an antenna. First, we consider the case when small deflection models, that accompany slow deployment rates, are sufficient. The deployment of a cable from a ship to an underwater vehicle, with its own control system to maneuver a device to do sea-floor mine searches, requires a constrained variable-n order-n algorithm, which is described next. Finally, we consider large deflections for a beam or cable during deployment/retraction.

11.1 SMALL DEFLECTION ANALYSIS OF BEAM EXTRUSION/ RETRACTION FROM A ROTATING BASE

A method [1] of analysis for boom deployment or retrieval from a rotating base is presented here, for the case when small deflections are expected due to the slow motion of the base. The analysis is carried out on a substitute system for a deploying/retracting beam coming out of a rotating base B. Two linear torsion

DOI: 10.1201/9781003231523-12

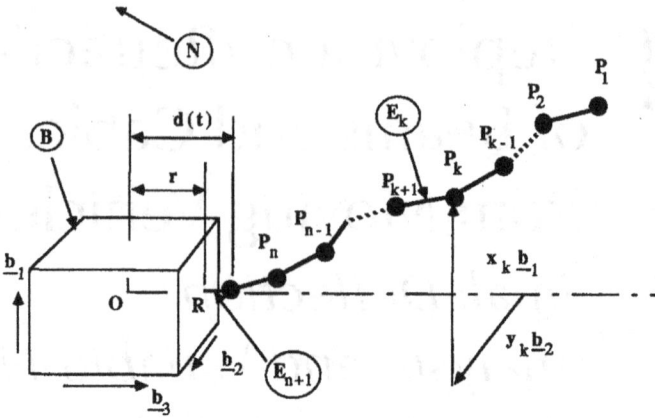

FIGURE 11.1 Discrete particle system in convection, to model deployment/retrieval.

springs of modulus KX_i, KY_i $(i = 1,...,n)$ connecting two links of L of mass m_i and moments of inertia EI, EJ are used for the discretization.

The main idea is to work with n sets of modal coordinates of vibration modes for the current number n of a mass-spring set, out of the base, as shown in Figure 11.1, while a convection of the system is prescribed by a function of time d(t), to describe deployment/retrieval.

Here, point O is fixed in a Newtonian frame N. Rigid massless links $E_i, i = 1,...,n+1$, each of length L, connected to each other by ball-and-socket joints, have particles P_i. The rest of the parameters can be calculated in the process shown in Chapter 10, as follows:

$$m_1 = m_{n+2} = \frac{\rho L}{2(n+1)L}; \quad m_i = \frac{\rho L}{(n+1)L} \quad (i = 2,...,n+1)$$

$$KX_i = EI/L; \quad KY_i = EJ/L \quad (i = 1,...,n-2)$$

$$KX_{n-1} = \frac{6(n-1)EI}{(3n-8)L}; \quad KY_{n-1} = \frac{6(n-1)EJ}{(3n-8)L} \tag{11.1}$$

$$KX_n = \frac{3nEI}{2(3n+1)L}; \quad KY_n = \frac{3nEJ}{2(3n+1)L}$$

The location of a particle P_k is

$$x_k = \mathbf{r}^{OP_k} \cdot \mathbf{b}_1; \quad y_k = \mathbf{r}^{OP_k} \cdot \mathbf{b}_2 \quad (k = 1,...,n) \tag{11.2}$$

The angular velocity of body B is prescribed with functions, $\omega_1(t), \omega_2(t), \omega_3(t)$:

$$^N\omega^B = \omega_1\mathbf{b}_1 + \omega_2\mathbf{b}_2 + \omega_3\mathbf{b}_3 \tag{11.3}$$

The principal purpose of the algorithm is to determine x_k, y_k by modal expansion as

$$x_k = \sum_{j=1}^{v} A_{kj} q_j \quad k = 1,\ldots,n$$

$$y_k = \sum_{j=1}^{v} B_{kj} q_j \quad k = 1,\ldots,n$$

(11.4)

To this end, we follow the steps in sequence:

Step 1. We find the vibration modes, with $\omega_1(t)$, $\omega_2(t)$, $\omega_3(t)$ set equal to zero, holding $d(t)$ as equal to a constant such that P_{n+1} is outside the physical confines of B. Letting n^* be the largest value of n of interest, we determine for $n = 1,\ldots, n^*$, the natural frequencies $\Omega_i, i = 1,\ldots, v$ where v is the number of vibration modes kept, with the modal matrix normalized to the mass of the n particles. Let A_{kj}, B_{kj} $(k = 1,\ldots,n; \; j = 1,\ldots,v)$ be those elements of the modal matrix associated with x_k, y_k, respectively, where the mode shape matrix, initial deformations, and rates of change of x_k, y_k are as follows:

$$\phi_n = \begin{bmatrix} A_{11} & \cdot & \cdot & A_{1v} \\ \cdot & \cdot & \cdot & \cdot \\ \cdot & \cdot & \cdot & \cdot \\ A_{n1} & \cdot & \cdot & A_{nv} \\ B_{11} & \cdot & \cdot & B_{1v} \\ \cdot & \cdot & \cdot & \cdot \\ \cdot & \cdot & \cdot & \cdot \\ B_{n1} & \cdot & \cdot & B_{nv} \end{bmatrix}; \quad q_n = \begin{Bmatrix} q_1 \\ \\ \\ q_v \end{Bmatrix};$$

(11.5)

$$\delta_n(0) = \begin{Bmatrix} x_1(0) \\ \cdot \\ \cdot \\ x_n(0) \\ y_1(0) \\ \cdot \\ \cdot \\ y_n(0) \end{Bmatrix}; \quad v_n(0) = \begin{Bmatrix} \dot{x}_1(0) \\ \cdot \\ \cdot \\ \dot{x}_n(0) \\ \dot{y}_1(0) \\ \cdot \\ \dot{y}_n(0) \end{Bmatrix}$$

Step 2. Then we compute the following eight "modal sums" for $n = 1,\ldots, n^*$:

$$A_i = \sum_{k=1}^{n} m_k A_{ki} \qquad (i = 1, \ldots, v)$$

$$B_i = \sum_{k=1}^{n} m_k B_{ki} \qquad (i = 1, \ldots, v)$$

$$R_{ij} = \sum_{k=1}^{n} m_k A_{ki} A_{kj} \qquad (i, j = 1, \ldots, v)$$

$$S_{ij} = \sum_{k=1}^{n} m_k B_{ki} B_{kj} \qquad (i, j = 1, \ldots, v)$$

$$T_{ij} = \sum_{k=1}^{n} m_k (B_{ki} A_{kj} - A_{ki} B_{kj}) \qquad (i, j = 1, \ldots, v) \qquad (11.6)$$

$$U_{ij} = \sum_{k=1}^{n} m_k (A_{ki} B_{kj} + B_{ki} B_{kj}) \qquad (i, j = 1, \ldots, v)$$

$$C_{kij} = \sum_{r=k}^{n} [(A_{ri} - A_{r+1,i})(A_{rj} - A_{r+1,j}) + (B_{ri} - B_{r+1,i})(B_{rj} - B_{r+1,j})]/L$$

$$(k = 1, \ldots, n; i, j = 1, \ldots, v)$$

$$E_{ij} = \sum_{k=1}^{n} m_k C_{kij} \qquad (i, j = 1, \ldots, v)$$

Step 3. Now we evaluate for $n = 1, \ldots, n^*$, the following:

$$G_{ij} = 2\omega_3 T_{ij} \qquad (i,j = 1, \ldots, v)$$

$$K_{ij} = \dot{\omega}_3 T_{ij} + \omega_1 \omega_2 U_{ij} - \left(\omega_2^2 + \omega_3^2\right) R_{ij} - \left(\omega_3^2 + \omega_1^2\right) S_{ij} +$$

$$\left(\omega_1^2 + \omega_2^2\right) \sum_{k=1}^{n} m_k C_{kij} \left[d + (n-k+1)L\right] - \ddot{d} E_{ij} \qquad (i,j = 1, \ldots, v)$$

$$\qquad (11.7)$$

$$f_i = -\left\{ \begin{array}{l} (\dot{\omega}_2 + \omega_3 \omega_1) \sum_{k=1}^{n} m_k A_{ki} \left[d + (n-k+1)L\right] - \\[2mm] (\dot{\omega}_1 - \omega_2 \omega_3) \sum_{k=1}^{n} m_k B_{ki} \left[d + (n-k+1)L\right] \\[2mm] +2\dot{d}\left(\omega_2 A_i - \omega_1 B_i\right) \end{array} \right\} \qquad (i = 1, \ldots, v)$$

Step 4. For n=1,..., n*, form $\phi_n, \delta_n, v_n, q(0), \dot{q}(0)$ as follows: Use Eq. (11.5) to indicate the initial values of $\delta_n(0), v_n(0)$. Then, by using the pseudo-inverse of ϕ_n, one gets

$$q(0) = [\phi_n^T \phi_n]^{-1} \phi_n^T \delta_n(0)$$
$$\dot{q}(0) = [\phi_n^T \phi_n]^{-1} \phi_n^T v_n(0)$$

(11.8)

Step 5. Find specifically,

$$[q_1(0),...,q_v(0)]^T = q(0)$$
$$[\dot{q}_1(0),...,\dot{q}_v(0)]^T = \dot{q}(0)$$

(11.9)

Step 6. After assigning values to the modal damping factor, $\varsigma_i, (i = 1,...,v)$, one sets up the ordinary differential equations of motion:

$$\ddot{q}_i + \sum_{j=1}^{v} \left[G_{ij}\dot{q}_j + K_{ij}q_j \right] + \Omega_i^2 q_i + 2\varsigma_i \Omega_i \dot{q}_i = f_i \quad (i = 1,...,v)$$

(11.10)

Now solve them in the time interval, $t_0 \leq t \leq t_0 + T$, with T such that for a link of length L to be out:

$$L = \int_{t_0}^{t_0+T} \dot{d}(t)dt$$

(11.11)

where $t_0 = 0$ at the beginning of the extrusion of a link. If $n < n*$, then n links have been extruded, and we replace n with n + 1 in Eqs. (11.5) and (11.8) the terminal values from the preceding stage of extrusion while setting $x_{n+1}, y_{n+1}, \dot{x}_{n+1}, \dot{y}_{n+1}$ equal to zero. For retraction, one simply reverses the process.

Step 7. To find the tip deflection, evaluate x_1, y_1 from Eq. (11.4) as

$$x_1 = \sum_{j=1}^{v} A_{1j}q_j$$
$$y_1 = \sum_{j=1}^{v} B_{1j}q_j$$

(11.12)

Step 8. The dynamic bending moment at the root of R of the beam is given by Eq. (11.13), where r denotes the distance OR in Figure 11.1.

$$M_1 = -\sum_{k=1}^{n} m_k[d-r+(n-k+1)L]\{(\dot{\omega}_3+\omega_1\omega_2)\sum_{j=1}^{v}A_{kj}q_j$$

$$-(\omega_1^2+\omega_3^2)\sum_{j=1}^{v}B_{kj}q_j + 2\omega_3\sum_{j=1}^{v}A_{kj}\dot{q}_j$$

$$+\sum_{j=1}^{v}B_{kj}\ddot{q}_j - 2\omega_1\dot{d}-(\dot{\omega}_1-\omega_2\omega_3)[d+(n-k+1)L]\}$$

$$-\sum_{k=1}^{n}m_k\{(\omega_1^2+\omega_2^2)[d+(n-k+1)L]-\ddot{d}\}\sum_{j=1}^{v}B_{kj}q_j \qquad (11.13)$$

$$M_2 = \sum_{k=1}^{n} m_k[d-r+(n-k+1)L]\{-(\dot{\omega}_3+\omega_1\omega_2)\sum_{j=1}^{v}B_{kj}q_j$$

$$-(\omega_2^2+\omega_3^2)\sum_{j=1}^{v}A_{kj}q_j - 2\omega_3\sum_{j=1}^{v}B_{kj}\dot{q}_j$$

$$+\sum_{j=1}^{v}A_{kj}\ddot{q}_j - 2\omega_1\dot{d}+(\dot{\omega}_2+\omega_3\omega_1)[d+(n-k+1)L]\}$$

$$+\sum_{k=1}^{n}m_k\{(\omega_1^2+\omega_2^2)[d+(n-k+1)L]-\ddot{d}\}\sum_{j=1}^{v}A_{kj}q_j$$

11.1.1 RATIONALE

The non-linear expression for the velocity of a generic particle P_k is

$$^{N}\mathbf{v}^{P_k} = {}^{N}\boldsymbol{\omega}^{B}\times\left\{\sum_{j=1}^{v}\left(A_{kj}\mathbf{b}_1+B_{kj}\mathbf{b}_2\right)q_j + [d+L\sum_{i=k}^{n}J_i^{0.5}]\mathbf{b}_3\right\}$$

$$+\sum_{j=1}^{v}\left(A_{kj}\mathbf{b}_1+B_{kj}\mathbf{b}_2\right)\dot{q}_j$$

$$+\left(\dot{d}-L\sum_{l=k}^{n}J_i^{-0.5}\sum_{i=1}^{v}\sum_{j=1}^{v}N_{mij}\dot{q}_i q_j\right)\mathbf{b}_3 \quad (k=1,\dots,n) \qquad (11.14)$$

where

$$J_i = 1-\left(\frac{x_i-x_{i+1}}{L}\right)^2-\left(\frac{y_i-y_{i+1}}{L}\right)^2 \qquad (i=1,\dots,n)$$

$$N_{mij} = \left(\frac{A_{mi}-A_{m+1,i}}{L}\right)\left(\frac{A_{mj}-A_{m+1,j}}{L}\right)+$$

$$\left(\frac{B_{mi}-B_{m+1,i}}{L}\right)\left(\frac{B_{mj}-B_{m+1,j}}{L}\right) \qquad (m,i,j=1,\dots,n)$$

The non-linear partial velocity is obtained from the above as

$$^N\mathbf{v}_i^{P_k} = A_{ki}\mathbf{b}_1 + B_{ki}\mathbf{b}_2 - L\sum_{m=k}^{n}\sum_{j=1}^{v}J_i^{-0.5}N_{mij}q_j\mathbf{b}_3 \quad (k = 1,\ldots,n; \, i = 1,\ldots,v) \qquad (11.15)$$

As per Kane's rules of direct linearization [5], without deriving nonlinear equations, we linearize the non-linear partial velocity expression, and then linearize the expression of velocity and differentiate that to get linear acceleration.

$$\hat{J}_i = 1 \quad (i = 1,\ldots,n)$$

$$^N\tilde{\mathbf{v}}_i^{P_k} = A_{ki}\mathbf{b}_1 + B_{ki}\mathbf{b}_2 - \sum_{j=1}^{v}C_{mij}q_j\mathbf{b}_3 \quad (k = 1,\ldots,n; i = 1,\ldots,v)$$

$$C_{mij} = \sum_{l=k}^{n} L \left[\left(\frac{A_{mi} - A_{m+1,i}}{L}\right)\left(\frac{A_{mj} - A_{m+1,j}}{L}\right) + \left(\frac{B_{mi} - B_{m+1,i}}{L}\right)\left(\frac{B_{mj} - B_{m+1,j}}{L}\right) \right] \quad (m = 1,\ldots,n; i,j = 1,\ldots,v) \quad (11.16)$$

$$^N\tilde{\mathbf{v}}^{P_k} = {^N}\boldsymbol{\omega}^B \times \left\{ \sum_{j=1}^{v}(A_{kj}\mathbf{b}_1 + B_{kj}\mathbf{b}_2)q_j + [d + L(n-k+1)]\mathbf{b}_3 \right\}$$

$$+ \sum_{j=1}^{v}(A_{kj}\mathbf{b}_1 + B_{kj}\mathbf{b}_2)\dot{q}_j + \dot{d}\mathbf{b}_3 \quad (k = 1,\ldots,n) \qquad (11.17)$$

Differentiating the linearized velocity to get the linearized acceleration and form a linearized generalized inertia force yields:

$$^N\tilde{\mathbf{a}}^{P_k} = \frac{^B d \, ^N\tilde{\mathbf{v}}^{P_k}}{dt} + {^N}\boldsymbol{\omega}^B \times {^N}\tilde{\mathbf{v}}^{P_k} \quad (k = 1,\ldots,n) \qquad (11.18)$$

$$F_i^* = -\sum_{k=1}^{n} m_k \, ^N\tilde{\mathbf{a}}^{P_k} \cdot {^N}\tilde{\mathbf{v}}_i^{P_k} \quad (i = 1,\ldots,v) \qquad (11.19)$$

The linearized active force due to structural stiffness and damping is

$$\tilde{F}_i = -\Omega_i^2 q_i - 2\zeta_i\Omega_i \qquad (11.20)$$

Then, we complete forming Kane's linear equations of motion:

$$\tilde{F}_i + \tilde{F}_i^* = 0 \qquad (11.21)$$

Equation (11.21) leads to Eq. (11.10), the equations of motion for the small deflection analysis for beam extrusion/retraction.

11.2 SIMULATION RESULTS

We coded the steps in the theory to do a simulation, first for the spin-up problem of the 150 m long WISP antenna beam, described in Chapter 10. Let

$$^N\boldsymbol{\omega}^B = \omega_1\mathbf{b}_1 + \omega_2\mathbf{b}_2 + \omega_3\mathbf{b}_3,$$

$$\omega_1 = \frac{0.125}{57.3 \times 300}\left[t - \frac{150}{\pi}\sin\frac{\pi t}{150}\right] \text{rad/sec} \quad t \le 300\,\text{sec}; \quad \omega_2 = \omega_3 = 0. \quad (11.22a)$$

This "spin-up" simulation result is shown in Figure 11.2. It compares a *simpler version of the continuum beam in spin-up*, Ref. [5], with the present theory, with n* = 29, segment length L = 5 m. The result applies to the case when the shuttle rotates at 1 deg/sec about an axis normal to the direction of extrusion. As can be seen, the two curves agree very well.

Next, we simulate the main cases of extrusion and retraction, with a time history from 15 m to 150 m and vice versa, completed in 1387 sec, when the rate for extrusion/retraction is prescribed as follows:

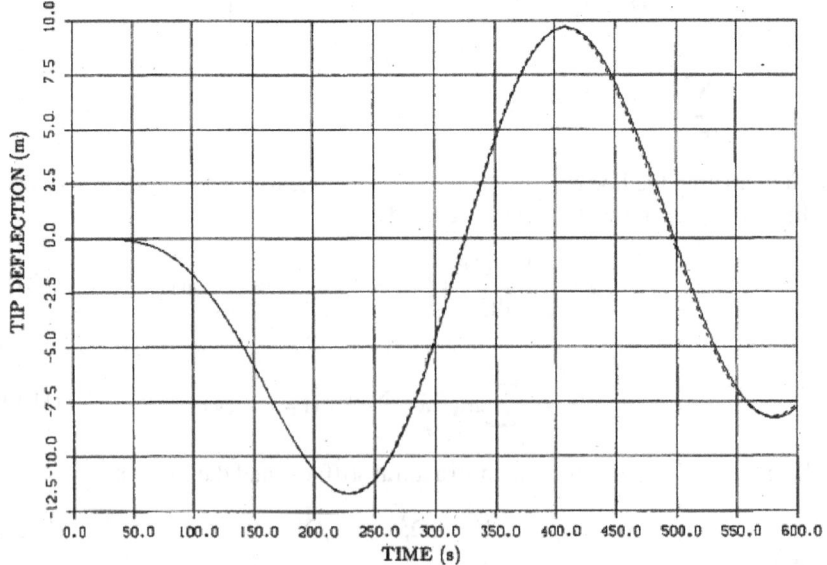

FIGURE 11.2 Spin-up result comparison between present discretized beam theory and a continuum beam theory (Reproduced with permission from AIAA).

$$\dot{d} = 0.133 \left[1 - \left(\frac{t}{1500} \right)^3 \right] \quad \text{for extrusion;}$$

$$\dot{d} = -0.133 \left[1 - \left(\frac{1500 - t}{1500} \right)^3 \right] \quad \text{for retraction} \tag{11.22b}$$

Two vibration modes are used in each plane, i.e., $\nu = 4$, $\zeta_1 = \zeta_2 = 0.05$. The allowable bending moment for this beam is 135 N-m. The results are given in Figures 11.3 and 11.4. It is seen that, as long as the angular velocity of the base B is perpendicular to $\mathbf{b}_3$, the direction of extrusion/retrieval, beam deflections occur only in the plane perpendicular to the angular velocity vector and can be attributed to the Coriolis effect. Thus, their magnitude may be expected to be proportional both to the base spin rate and the extrusion/retraction speed. From simulations, it is seen that tip deflections can become quite large, particularly during retraction, approaching the limit of applicability of linear elasticity associated with the use of vibration modes. The algorithm given above also accommodates base motions more general than the one just considered. In particular, it applies when the angular velocity of B in N has a large component perpendicular to the direction of extrusion/retraction, and a small component along this direction (see Figure 11.1), perhaps due to an attitude control system malfunction.

To get an idea of deflections under this condition, we ran a simulation with body angular velocity components, $\omega_1 = \sqrt{1/1.01} \deg/\sec; \omega_2 = 0; \omega_3 = 0.1\omega_1$. The results on deflections and the bending moment are shown in Figures 11.5 and 11.6. Note that Figure 11.5 suggests that there can be a potentially objectionable whirl subsequent to the extrusion when a small angular velocity component of the base exists in the direction of extrusion.

11.3   DEPLOYMENT OF A CABLE FROM A SHIP TO A MANEUVERING UNDERWATER SEARCH VEHICLE: USE OF A CONSTRAINED ORDER-N FORMULATION

This section is based on Ref. [2], and deals with the deployment of a cable reeled out of a winch on a ship to a maneuvering vehicle searching for mines on the sea floor. Figure 11.7 is a sketch of the system, where an unmanned underwater vehicle (UUV) has its own control system, and the length of the cable connecting it to the ship above has to be extended or shortened to conform to a control command from the UUV; this is what constitutes the constraint. Efficient modeling of the dynamics of a cable being reeled out of a moving ship to an underwater maneuvering vehicle requires a *constrained order-n formulation* that can handle both large three-dimensional curvature of the cable and end motion constraints. We review the details in the following sub-sections.

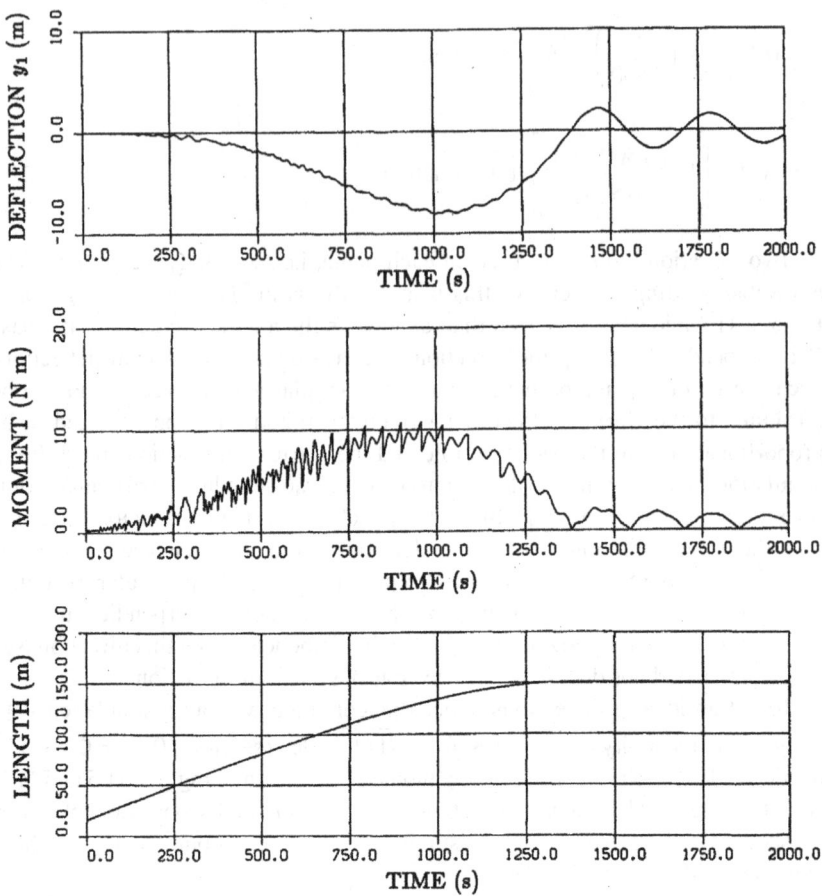

FIGURE 11.3 Extrusion of a beam from 15 m to 150 m, with angular velocity of the base perpendicular to the direction of extrusion (case of slow deployment rate).

11.3.1 CABLE DISCRETIZATION AND VARIABLE-N ORDER-N ALGORITHM FOR CONSTRAINED SYSTEMS WITH CONTROLLED END BODY

The algorithm used is that given in Chapter 8, with a necessary modification to enforce the end control constraint. To make this chapter self-contained in the order-n aspect, only the relevant details of the constrained order-n algorithm are reviewed here. The cable is modeled as a chain of rods with 2 degree of free-dom rotational joints having soft rotational stiffness (an assumption for a cable in water) to be able to produce a three-dimensional shape with the cable. Figure 11.8 is a sketch of the discretization of the cable, with link 1 closest to the ship.

Let λ be the unknown tension on the cable applied by the unmanned underwa-ter vehicle. Then, considering the free-body diagram of the n-th rod connected to the UUV, and summing inertia and external forces on it, excluding the force

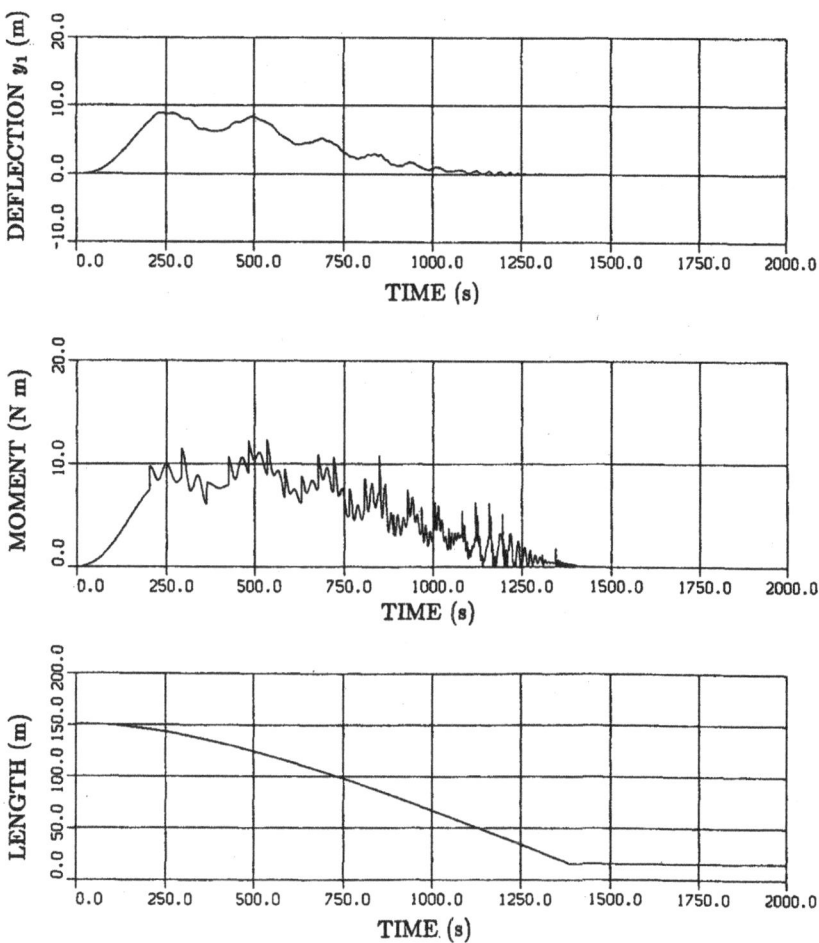

FIGURE 11.4   Retraction of beam from 150 m to 15 m, with angular velocity of the base perpendicular to the direction of retraction (Reproduced with permission from AIAA).

and torque from the $(n - 1)$-th rod, and taking the moment of these forces about the hinge point Q_n, in Figure 11.8, one has the following equilibrium equations:

$$\left\{ \begin{matrix} f^{*n} - f_e^n - \lambda \\ t^{*n} - t_e^n - \tilde{L}\lambda \end{matrix} \right\} \equiv M^n \left\{ \begin{matrix} a_0^{Q_n} \\ \alpha_0^n \end{matrix} \right\} + X^n - H^n\lambda \qquad (11.23)$$

Note that end body force λ may not be aligned with the position vector from the end of link n of length L to Q_n, and all vectors are resolved in body n basis for their matrix forms. In Eq. (11.23), f^{*n}, f_e^n are the inertia force and other external forces as noted above, and t^{*n}, t_e^n are the inertia and external torques (including the hinge spring torque) on the n-th body, and we have defined

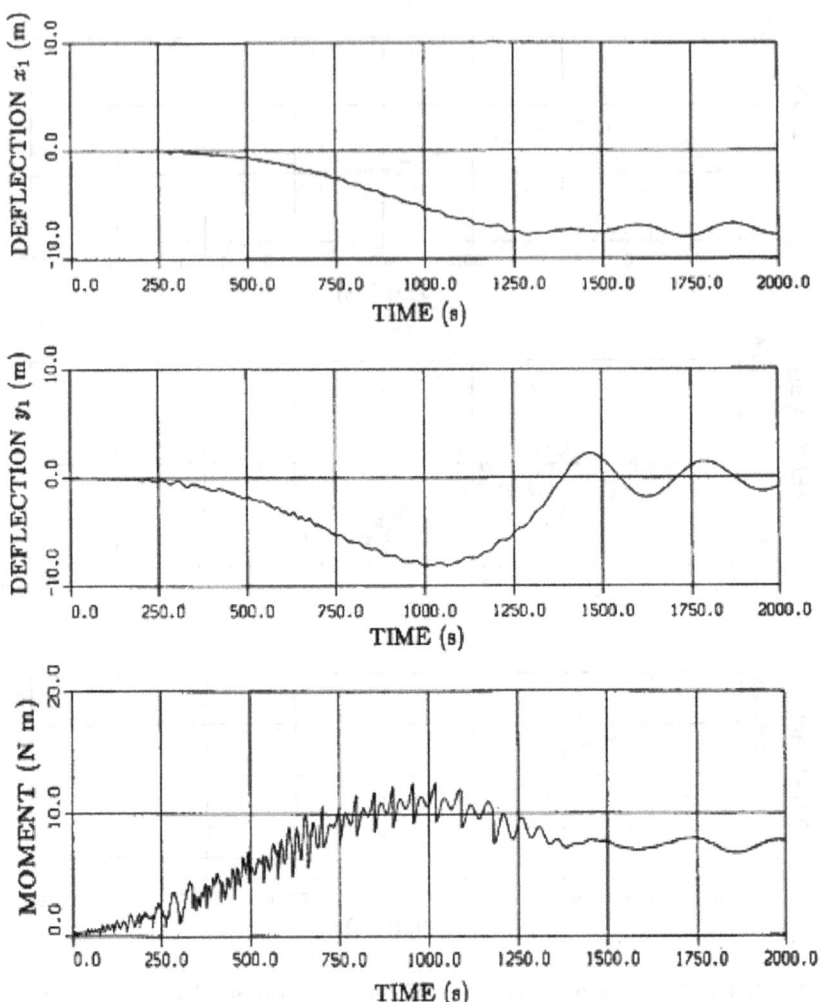

FIGURE 11.5 Beam deflection at tip in two axes and bending moment time history with prescribed extrusion; angular velocity of base *not* perpendicular to extrusion line (Reproduced with permission from AIAA).

$$M^n = \begin{bmatrix} m & -\tilde{S}^n \\ \tilde{S}^n & I^n \end{bmatrix}, \qquad H^n = \begin{Bmatrix} U \\ \tilde{L} \end{Bmatrix}$$

$$X^n = M^n \begin{Bmatrix} a_t^{Q_n} \\ \alpha_t^n \end{Bmatrix} + \begin{Bmatrix} \tilde{\omega}^n \, \tilde{\omega}^n \, S^n - f_e^n \\ \tilde{\omega}^n \, I^n \, \omega^n - t_e^n \end{Bmatrix}$$

(11.24)

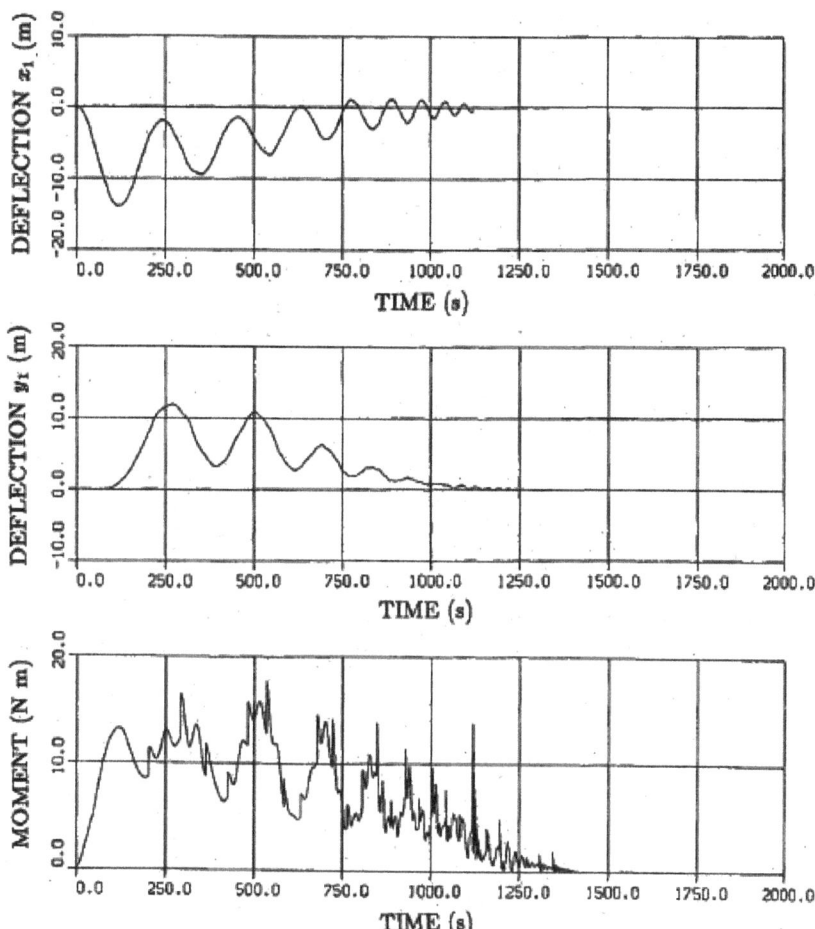

FIGURE 11.6   Beam deflection and moment time history at tip in two axes and with prescribed retraction; angular velocity of base *not* perpendicular to retraction line (Reproduced with permission from AIAA).

As before, S^n and I^n are the first and second moments of inertia about Q_n; U is 3 × 3 unity matrix; and ω^n is the inertial angular velocity of body n in its own basis. Again, a tilde denotes a skew-symmetric matrix formed out of (3 × 1) matrices, for matrix representation of vector cross products. Here, $\tilde{L}$ is formed out of a column matrix whose first element is L, the length of the rod, the other two element being zero. Symbols in Eq. (11.23) with zero subscript represent all terms that involve second derivatives of the general coordinates, and those in Eq. (11.24) with subscript t denote remainder acceleration terms in the expression for the acceleration of the hinge point Q_n and the angular acceleration of body n, as has been used in earlier chapters:

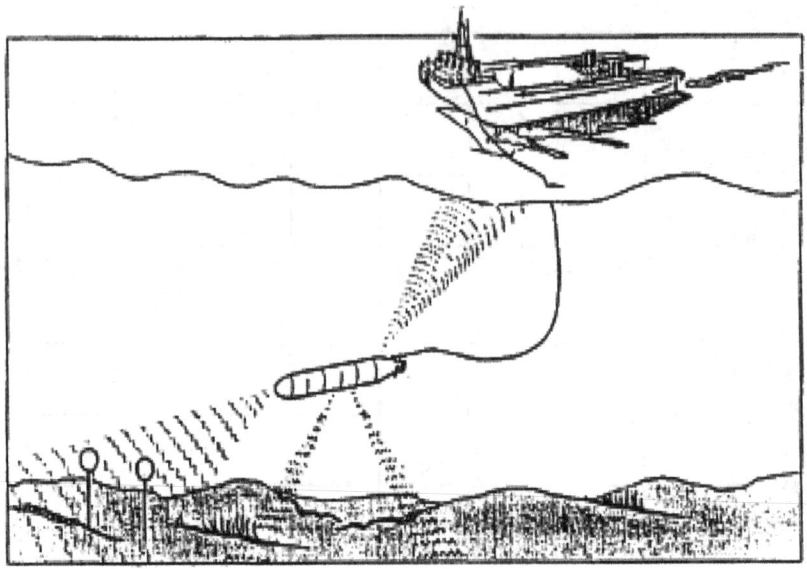

FIGURE 11.7 Sketch of a ship towing an underwater maneuvering vehicle carrying out a sea-floor mine search (Reproduced with permission from AIAA).

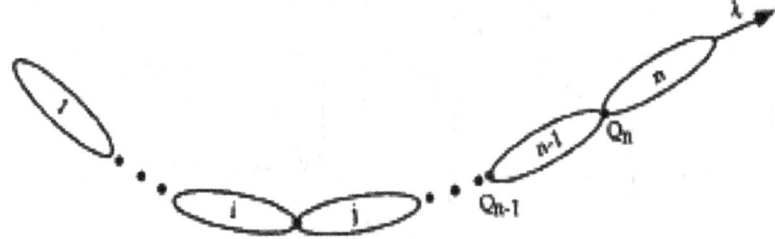

FIGURE 11.8 Discretization of the cable with 2 dof rotational joints with soft springs.

$$\left\{ \begin{matrix} a^{Q_n} \\ \alpha^n \end{matrix} \right\} = \left\{ \begin{matrix} a_0^{Q_n} \\ \alpha_0^n \end{matrix} \right\} + \left\{ \begin{matrix} a_t^{Q_n} \\ \alpha_t^n \end{matrix} \right\} \qquad (11.25)$$

Again, exposing the contributions of the 2 degree of freedom relative rotation rates at the body n hinge, one can further write

$$\left\{ \begin{matrix} a_0^{Q_n} \\ \alpha_0^n \end{matrix} \right\} = \left\{ \begin{matrix} \hat{a}_0^{Q_n} \\ \hat{\alpha}_0^n \end{matrix} \right\} + R^n \ddot{\theta}^n \qquad (11.26)$$

where the hat notation encompasses the group of acceleration and angular acceleration terms involving $\ddot{\theta}^1, \ldots, \ddot{\theta}^{n-1}$, where $\ddot{\theta}^i$ may itself have two elements, and

R^n is the 6 ×2 matrix of the partial velocity of Q_n and the partial angular velocity of body n. Substituting Eq. (11.26) into Eq. (11.23) and pre-multiplying by $(R^n)^T$, one forms Kane's dynamical equations with undetermined multipliers [9, 10], associated with θ^n, in which non-working constraint forces are automatically eliminated (a property of Kane's equations).

$$\ddot{\theta}^n = -v_n^{-1}R^{n^T}\left\{M^n\begin{pmatrix}\hat{a}_0^{Q_n}\\ \hat{\alpha}_0^n\end{pmatrix} + X^n - H^n\lambda\right\} + v_n^{-1}\tau^n \tag{11.27}$$

where

$$v_n = R^{n^T}M^nR^n \tag{11.28}$$

and τ^n is the two-component hinge torque applied by body (n − 1) on body n. Using Eq. (11.27) in Eq. (11.23) gives the D'Alembert equation, of forces and moments.

$$\begin{Bmatrix}f^{*n} - f_e^n - \lambda\\ t^{*n} - t_e^n - \tilde{L}\lambda\end{Bmatrix} \equiv M\begin{Bmatrix}\hat{a}_0^{Q_n}\\ \hat{\alpha}_0^n\end{Bmatrix} + X^n - H\lambda \tag{11.29}$$

Here, the following notations have been used with U now representing a 6 × 6 identity matrix:

$$M = PM^n, \qquad P = U - M^nR^nv_n^{-1}R^{n^T} \tag{11.30}$$

$$X = PX^n, \qquad H = PH^n$$

Now, as has been shown in previous chapters, kinematics relating the motion of inboard (n − 1) to that of body n means a transfer of motion represented by a shift transform W^n:

$$\begin{Bmatrix}\hat{a}_0^{Q_n}\\ \hat{\alpha}_0^n\end{Bmatrix} = W^n\begin{Bmatrix}a_0^{Q_{n-1}}\\ \alpha_0^{n-1}\end{Bmatrix}$$

$$W^n = \begin{bmatrix}C_{n,n-1} & -C_{n,n-1}\tilde{L}\\ 0 & C_{n,n-1}\end{bmatrix} \tag{11.31}$$

Here, $\{b^n\} = [C_{n,n-1}]\{b^{n-1}\}$ is the coordinate transformation matrix relating n-body basis vectors to (n − 1) body basis vectors. The same shift transform serves to shift the forces and moments from the hinge Q_n to a statically equivalent system of force and moment at the hinge Q_{n-1}:

$$\begin{Bmatrix}f^{*n} - f_e^n - \lambda\\ t^{*n} - t_e^n - \tilde{L}\lambda\end{Bmatrix} \equiv W^{n^T}\left\{MW^n\begin{pmatrix}a_0^{Q_{n-1}}\\ \alpha_0^{n-1}\end{pmatrix} + X - H\lambda\right\} \tag{11.32}$$

Now considering the free-body diagram of the $(n-1)$th body and adding the inertia and all external forces, including hinge forces from body $n-2$ to body $n-1$ with the force system represented by Eq. (11.32) and, recalling that non-working forces are eliminated, one gets

$$\begin{Bmatrix} f^{*n-1} - f_e^{n-1} \\ t^{*n-1} - t_e^{n-1} \end{Bmatrix} \equiv M^{n-1} \begin{pmatrix} a_0^{Q_{n-1}} \\ \alpha_0^{n-1} \end{pmatrix} + X^{n-1} - H^{n-1}\lambda \tag{11.33}$$

To keep Eq. (11.33) in its form, one has made the new updates, which physically means a transfer of loads from all outboard bodies:

$$M^{n-1} + W^{n^T} M W^n \rightarrow M^{n-1}$$

$$X^{n-1} + W^{n^T} X \rightarrow X^{n-1} \tag{11.34}$$

$$H^{n-1} = W^{n^T} H$$

One can see that Eq. (11.33) is just like Eq. (11.29). This completes a cycle that is repeated by going backward from the body n, the maneuvering end bond body, to bodies $n-2$, $n-3$, etc., in succession, until one encounters body 1, for which the inboard motion, the ship's motion in this case, is assumed to be known. The equations for the hinge degrees of freedom for the first link connected to the ship, in this case, can be written as follows:

$$\ddot{\theta}^1 = f_1^1 + c_1^1 \lambda \tag{11.35}$$

where

$$f_1^1 = -v_1^{-1} R^{1^T} X^1 + v_1^{-1} \tau^1 \tag{11.36}$$

$$c_1^1 = v_1^{-1} R^{1^T} H^1$$

In terms of Eqs. (11.35) and (11.36), Eq. (11.26) yields for body 1:

$$\begin{Bmatrix} a_0^{Q_1} \\ \alpha_0^1 \end{Bmatrix} = g^1 + h^1 \lambda \tag{11.37}$$

where

$$g^1 = R^1 f_1^1 \tag{11.38}$$

$$h^1 = R^1 c_1^1$$

Equation (11.31) now yields, for a generic j-th body:

$$\begin{Bmatrix} \hat{a}_0^{Q_j} \\ \hat{\alpha}_0^j \end{Bmatrix} = e^j + n^j \lambda \qquad (11.39)$$

where the kinematical shift operation from body i, the inboard body of j, to body j produced

$$e^j = W^j g^i$$
$$n^j = W^j h^i \qquad (11.40)$$

This defines the generic hinge rotation equation:

$$\ddot{\theta}^j = f_1^j + c_1^j \lambda \qquad (11.41)$$

where

$$f_1^j = -v_j^{-1} R^{jT} \{ M^j e^j + X^j \}$$
$$c_1^j = -v_j^{-1} R^{jT} \{ M^j n^j - H^j \} \qquad (11.42)$$

Use of Eqs. (11.41) and (11.42) in Eq. (11.26), along with new denotations:

$$g^j = e^j + R^j f_1^j$$
$$h^j = n^j + R^j c_1^j \qquad (11.43)$$

provides the general relation:

$$\begin{Bmatrix} a_0^{Q_j} \\ \alpha_0^j \end{Bmatrix} = g^j + h^j \lambda \qquad (11.44)$$

The stage is now set for uncovering all the dynamical equations in a so-called *second forward pass*, going from body 2 to body n. This completes the review of the order-n algorithm for constrained systems, as applied to this problem. These equations have to be solved together with a consideration of the constraint, which is due to the underwater maneuvering vehicle having its own independent control system. Before doing that however, we will take up the issue of specifying external forces on the underwater cable.

11.3.2 HYDRODYNAMIC FORCES ON THE UNDERWATER CABLE

For steady flow, the force exerted by a fluid on a submerged accelerating body is composed of two components, one depending on fluid friction and the other on inertia of the displaced fluid [11]. The transverse oscillatory force on the cable due to vortex shedding is neglected here for simplicity. For cylindrical bodies, the force due to viscous drag is given by

$$\mathbf{f}_d = -0.5\rho_w DL \left[\pi C_f |v_1| v_1 \mathbf{b}_1 + C_{dn}(v_2 \mathbf{b}_2 + v_3 \mathbf{b}_3)\sqrt{v_2^2 + v_3^2} \right] \qquad (11.45)$$

Here, ρ_w is the density of sea water; D is the cable diameter; L is the cable segment length; C_f is the axial friction coefficient; C_{dn} is the nominal drag coefficient; and v_1, v_2, v_3, are the three components of the body relative to the water, taking into account water current effects. The force required to move the displaced fluid is treated as an apparent added mass at the center of the link in the inertia computation [10]:

$$m' = L(m + C_m \rho_w \pi D^2 / 4) \qquad (11.46)$$

where m is the mass per unit length of the cable; L is the discretized segment length; and C_m is the fluid mass coefficient. Note that Eq. (11.46), while simplifying matters, does not rigorously treat the physics of fluid inertia in the sense that only acceleration components perpendicular to the cable will affect the apparent fluid mass. In addition to these effects, the force of buoyancy reduces the force of gravity of the rod element, giving rise to a force of "wet weight" approximation given by

$$\mathbf{f}_w = -\pi DL (\rho_c - \rho_w) g \mathbf{n}_2 \qquad (11.47)$$

where ρ_c, ρ_w g, $\mathbf{n}_2$ are, respectively, the cable density, water density, acceleration due to gravity, and the inertial gravity direction unit vector. Finally, we consider the force at the end of the cable due to mass gain or loss associated with deployment or retraction, respectively. This force is in the nature of a thrust, and is represented as a force acting on the link closest to the ship, as follows:

$$\mathbf{f}_t^1 = -(\pi / 4)D^2 \rho_c \dot{L}[(v_1^1 - \dot{L})\mathbf{b}_1^1 + v_2^1 \mathbf{b}_2^1 + v_3^1 \mathbf{b}_3^1] \qquad (11.48)$$

where $\dot{L}, v_1^1$, and $\mathbf{b}_1^1$ are, respectively, the deployment rate, the axial component of the velocity vector, and the unit vector along the axis for body 1, and so on.

11.3.3 NON-LINEAR HOLONOMIC CONSTRAINT, CONTROL-CONSTRAINT COUPLING, CONSTRAINT STABILIZATION, AND CABLE TENSION

With one end of the cable attached to the ship, the other end is constrained to move with the UUV, which has its own model and embedded propulsion control system as it goes in search of underwater mines. This implies the existence of a holonomic constraint, which means some of the generalized coordinates are dependent (indicated by subscript d in the following) on the independent ones (denoted by subscript i). This, in turn, means that Eq. (11.41) can be written in a column matrix form as

$$\ddot{\theta}_i = F_i + C_i \lambda \tag{11.49}$$

$$\ddot{\theta}_d = F_d + C_d \lambda \tag{11.50}$$

The constraint force components represented in λ can be determined by solving Eqs. (11.49) and (11.50) simultaneously with the constraint equations. Now the constraint equation can be written in terms of an error in the position vectors from the inertial origin to the UUV measured in two ways: (1) once to the ship and then along the cable, and (2) to the UUV directly, as in the error vector equation:

$$e = \left(p^s + \sum_{i=1}^{n} L_i b_1^i \right) - p^{UUV} = 0 \tag{11.51}$$

Here p^s and p^{UUV} stand for the position vectors from an inertial origin to the ship and to the UUV, respectively. Equation (11.51) is non-linear in the sets of angles between the links, θ_i, θ_d. Although a differential-algebraic equation solver may be used to solve and satisfy Eqs. (11.49)–(11.51), this takes away from the efficiency of the order-n formulation that can be used in an on-line control system. One alternative is to differentiate Eq. (11.51) twice and solve the resulting equation together with Eqs. (11.49) and (11.51). This, of course, gives rise to the well-known problem of constraint violation with time, which necessitates constraint stabilization [12]. Here, we use a *proportional-derivative-integral* correction of error, which is known to mitigate steady-state error in control. Using the *scalar* form of the equation in Eq. (11.51) in the inertial basis, and using their derivative and integral, one obtains

$$\ddot{e} + k_p e + k_d \dot{e} + k_i \int_0^t e(\tau) d\tau = 0 \tag{11.52}$$

Equation (11.52) requires computation of the velocity and acceleration of the point P of the cable that is attached to the UUV and comparing these with, respectively, the velocity and acceleration of the UUV. Writing the acceleration of P in matrix form as

$$a^P = J_i \ddot{\theta}_i + J_d \ddot{\theta}_d + a_t^P \tag{11.53}$$

where J_i and J_d are Jacobian terms that are really partial velocities, with all vectors resolved in the inertial basis, and substituting Eqs. (11.49) and (11.50) in Eq. (11.53), one obtains from Eq. (11.52), the following matrix equation to determine the cable tension λ:

$$[J_i C_i + J_d C_d]\lambda = a^{UUV} - \left\{ \begin{array}{l} J_i F_i + J_d F_d + a_t^P + k_p e + k_d (v^P - v^{UUV}) \\ + k_i \displaystyle\int_0^t e(\tau)d\tau \end{array} \right\} \quad (11.54)$$

where v^P and v^{UUV} are the velocities of the cable end points P and the UUV, respectively. The acceleration of the UUV, a^{UUV} in Eq. (11.54), is given as an output of the control system for the depth and speed control of the UUV. This control system treats the force applied by the cable, λ, as a disturbance to the system, and thus the output acceleration becomes a function of λ. Because of the non-linearities of the signal limiters used in the implementation of the control laws, see Figure 11.9, the functional dependence $a^{UUV}(\lambda)$ of the output acceleration on λ is non-linear. Thus, Eq. (11.54) is a non-linear equation in λ, and must be solved iteratively, as follows:

$$[J_i C_i + J_d C_d]\lambda_{n+1} = a^{UUV}(\lambda_n) - [J_i F_i + J_d F_d + a_t^P$$
$$+ k_p e + k_d (v^P - v^{UUV}) + k_i \int_0^t e(\tau)d\tau] \quad (11.55)$$

Details of the control winch mechanization of a non-linear control with a limiter itself are not described here, but are given in Ref. [2]. Figure 11.9 gives an overview of the control system.

For our purpose of describing the cable dynamics during deployment and retrieval, it suffices to say that it is a hardware realization of the cable rate control law given by

$$\dot{L} \approx K_v (L_c - L) \quad (11.56)$$

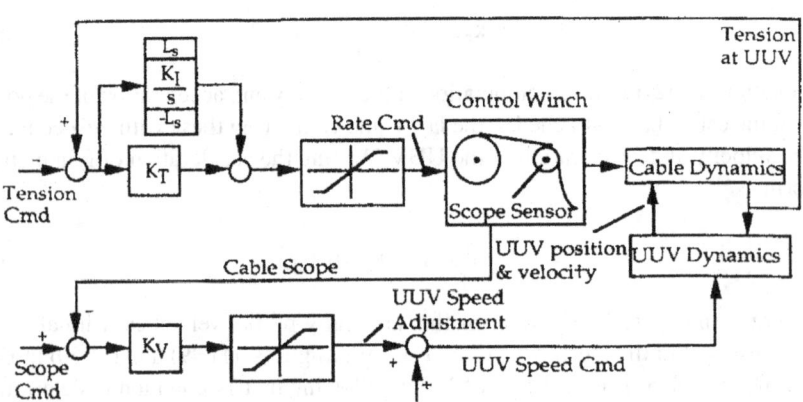

FIGURE 11.9 Cable length management non-linear control law block diagram.

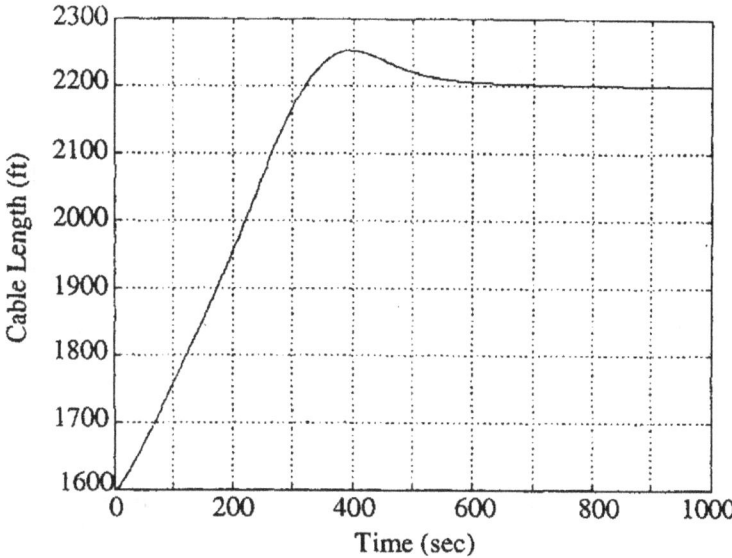

FIGURE 11.10 Cable length vs. time during deployment of underwater vehicle.

Here, L_c and L are, respectively, the commanded and actual "scope" (submerged length of cable). Equation (11.56) describes a stable system for all values of K_y, with a time constant of $(1/K_y)$; K_y is set to 0.01, so that for small errors in length (<150 ft), correction would be effective in 300 sec. The effect of the rate limiter is felt beyond this 150 ft error. The simulation results of cable deployment and retrieval are described next.

11.4 SIMULATION RESULTS

As has been said, the actual application of the cable dynamics algorithm given above is to an underwater vehicle carrying out sea-floor mine detection, with the UUV having its own controller, while being attached to the ship by means of a cable. Simulations of the time histories of cable length, cable tension, and constraint length violation error are shown in Figures 11.10–11.12, respectively. Figure 11.13 shows the shape of the cable with deployment. Some typical results for the retrieval of the cable, namely, of the cable length and cable tension are given in Figures 11.14 and 11.15.

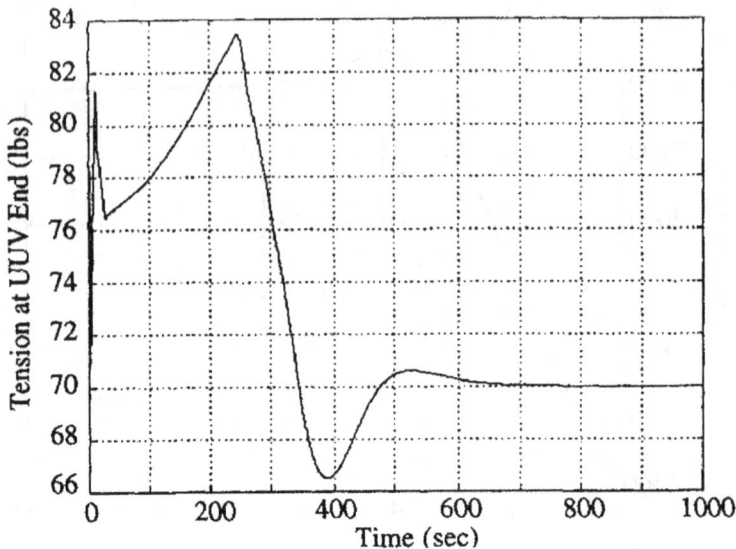

FIGURE 11.11   Cable tension vs. time for deployment of underwater vehicle.

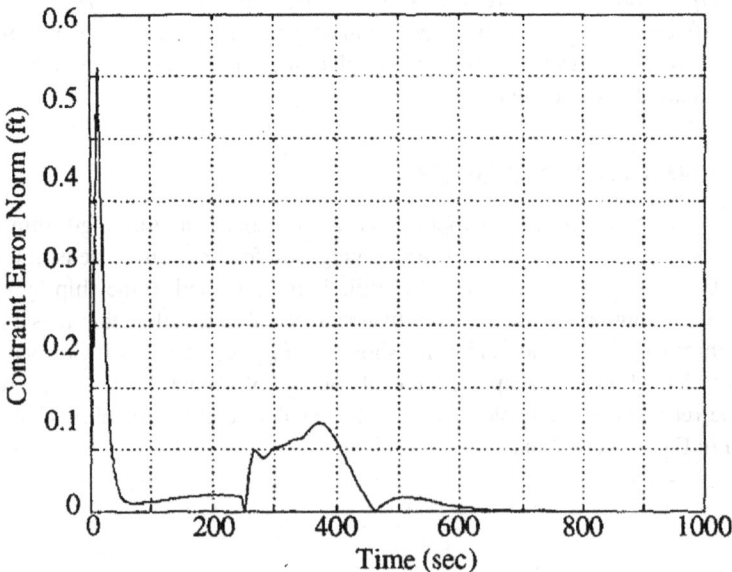

FIGURE 11.12   Constraint length error norm vs. time for deployment of underwater vehicle.

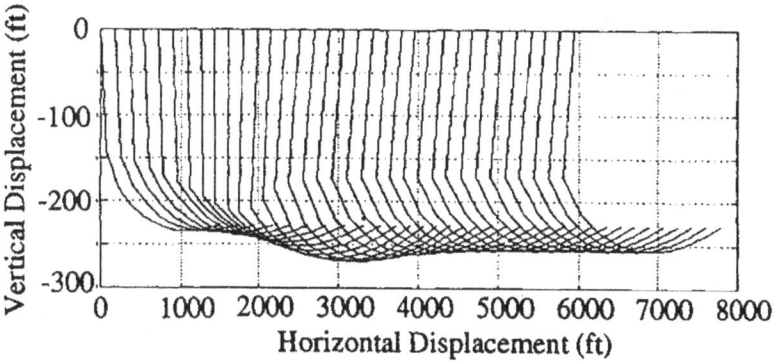

FIGURE 11.13 Cable shape during the deployment of the underwater vehicle.

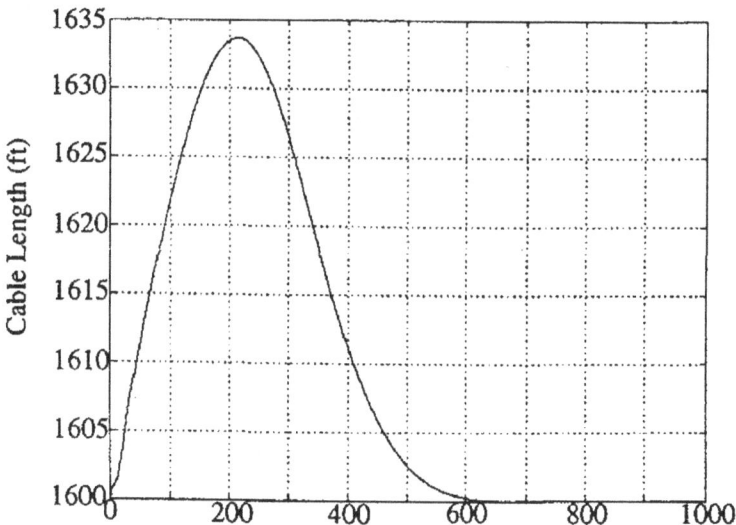

FIGURE 11.14 Cable length during retrieval of the underwater vehicle.

11.5 CASE OF LARGE BEAM DEFLECTION DURING DEPLOYMENT/RETRACTION

The process of extrusion or retraction of a beam possibly undergoing large deflection from a moving base is idealized by the deployment of a discretized chain of rigid elements represented in Figure 11.16, where the body B rotates in a prescribed manner in a Newtonian reference frame N, with the beam anchored at a moving point A at distance d, from a point O fixed in B. Specifying the deployment or retraction rate of the beam amounts to specifying d as a function of time, with the number of beam segments updated when a whole segment is deployed or retracted. In Figure 11.16, the discrete links are connected by rotational springs, as modeled in Ref. [3].

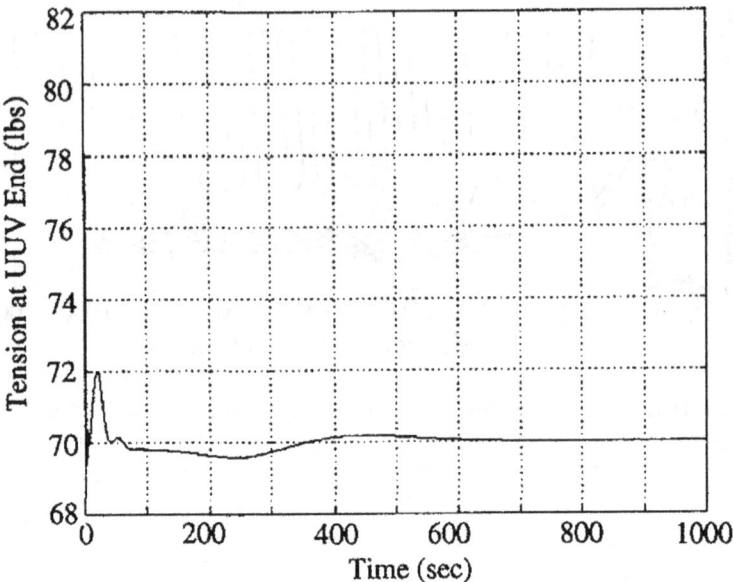

FIGURE 11.15   Cable tension during retrieval of the underwater vehicle.

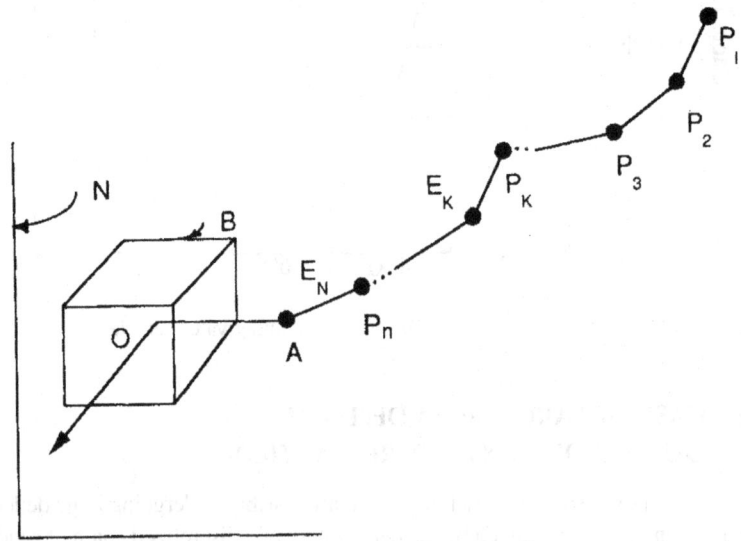

FIGURE 11.16   Discretized representation of a beam being extruded from a rotating base.

For convenience, the beam is discretized by massless links with lumped mass at the ends of the links. Torsion springs resist relative rotation between links, and the stiffness coefficients are obtained by equating the deflections due to a tip load with that of a Bernoulli–Euler beam, as in Chapter 10, where the following relations are given, with ρ and EI/L as the linear mass density and the flexural rigidity of the beam, respectively, with i referring to a generic link, each link of length L:

$$m_i = \rho L \quad (i = 1,\ldots,n) \tag{11.57}$$

$$k_1 = EI/L$$

$$k_2 = 6(n-1)EI/[(3n-1)L]$$

$$k_i = k_1 \quad i = 1,\ldots,n-1 \tag{11.58}$$

$$k_n = k_2 + (k_1 - k_2)d/L$$

Here d is the length of the part of link just being extruded. Note that Eq. (11.58) represents a linear interpolation in the stiffness value at the point A in Figure 11.16 for the n-the spring between its values at the beginning and end of extrusion of link n + 1. Reference [1] used a beam model given by three modes, and based on the results given in this chapter, the present discretized rigid element model seems to be a good approximation.

11.5.1 DEPLOYMENT/RETRACTION FROM A ROTATING BASE

11.5.1.1 Initialization Step

Consider only revolute joints for simplicity, and specify a starting value of n for the number of bodies, numbering them from the free end as 1, 2, 3,..., as in Figure 11.6. Assign inertia properties and stiffness as per Eqs. (11.57) and (11.58). Initialize the generalized coordinates q_k, and generalized speeds defined as $u_k = \dot{q}_k, (k = 1,\ldots n)$ for all the hinge relative angles. Prescribe the base motion and the extrusion rate, i.e., the rate of the change of distance d from O to A in Figure 11.1, as a function of time.

11.5.1.2 Forward Pass in an Order-n Formulation

Order-n formulation [12] for an n degree of freedom system is used for rapid execution time. Here, we review the steps in the algorithm particularized for the chain system. Consider the two bodies (chain links) in Figure 11.17.

Step 1. Define the basis transformation matrix C_{ik} between body k and its inboard body i along the path to B, for all element k. The elements of C_{ik} are

$$C_{ik}(l,m) = b_i(l) \cdot b_k(m) \quad (k = n,\ldots,1; \; l,m = 1,2,3) \tag{11.59}$$

where $b_i(m)$ is the m-th component of the basis vector of the i-th body frame.

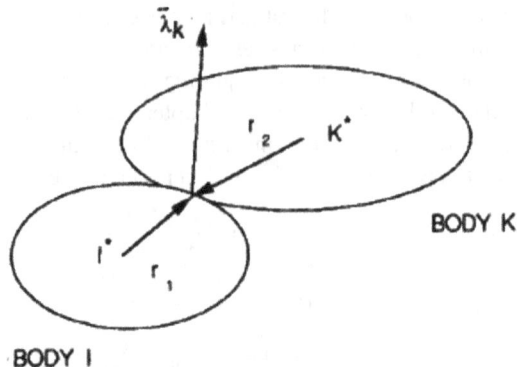

FIGURE 11.17 Two hinged bodies i and k in a chain, with body k outboard of body i.

Step 2. Form the angular velocity vector ${}^N\boldsymbol{\omega}^k$ of the k-th body in the k basis and define the corresponding 3×1 matrices ${}^N\omega^k$ (k=1,…, n) using the k-th body vector basis components (note *body n is nearest to B*):

$$
{}^N\omega^n = {}^N\omega^B + u_n\lambda_n
$$

$$
{}^N\omega^k = {}^N\omega^i + u_k\lambda_k \quad (k = n-1,\ldots,1)
$$

(11.60)

Here, λ_k is the unit vector along the rotation axis for the k-th hinge and B is the reference body, to which body n is hinged and whose motion is prescribed. Note that the hinge-axis vectors may not be restricted to be parallel; this feature allows three-dimensional motion.

Step 3. Form the partial angular velocity [5] of the k-th body with respect to the i-th generalized speeds in the k basis, from Eq. (11.60):

$$
{}^N\omega_k^k = \lambda_k \quad (k = n,\ldots,1)
$$

(11.61)

and form the 3×1 matrix with measure numbers for the k-th body,

Step 4. From the kinematical relation relating the inertial velocity of k* in Figure 11.2 to the joint between body k and body i, express the partial velocity of the mass center of the k-th body with respect to the generalized speed u_k in the k basis, in the following sequence.

$$
{}^N v^{k*} = {}^N v^{i*} + {}^N\omega^i \times \mathbf{r}_1 - {}^N\omega^k \times \mathbf{r}_2 \quad k = (n-1,\ldots,1)
$$

$$
{}^N v^{n*} = {}^N v^A + {}^N\omega^n \times \mathbf{r}^{An*}
$$

(11.62a)

The minus sign arises because $\mathbf{r}_2^k$ is directed to the i–k contact point opposite to the outward direction, and goes from the mass center of body k *inward* to the k-th joint, see Figure 11.16. Note $\mathbf{r}^{An*}$ is the position vector from the contact point A at the end of the deploying segment to the mass center n* of segment of n (segment AP_n in Figure 11.16). Form the 3×1 partial velocity [6] matrices v_k^{k*} (k = 1,…,n) out of the following vector equation:

$$^N\mathbf{v}_k^{k*} = -\,^N\boldsymbol{\omega}_k^k \times \mathbf{r}_2^k \quad (k = n,\ldots,1) \tag{11.62b}$$

Step 5. Express the remainder angular acceleration vector α_t^k from the angular acceleration of the k-th body, defined by the following equations:

$$^N\boldsymbol{\alpha}^k = \sum_{j=1}^{n} {}^N\boldsymbol{\omega}_j^k \dot{u}_j + \boldsymbol{\alpha}_t^k \quad (k = n,\ldots,1)$$

$$^N\boldsymbol{\alpha}_t^n = {}^N\boldsymbol{\alpha}^B \tag{11.63a}$$

$$^N\boldsymbol{\alpha}_t^k = {}^N\boldsymbol{\alpha}_t^i + {}^N\boldsymbol{\omega}^k \times (u_k\boldsymbol{\lambda}_k) \quad (k = n-1,\ldots,1)$$

Here, Eq. (11.63a) defines α_t^k in the k basis, and forms the corresponding (3×1) matrix $^N\alpha_t^k$. Note that $^N\alpha^B$ is the prescribed angular acceleration of the body B, out of which the beam is being deployed. Now, in the k basis, form the 3×1 matrices α_t^k $(k = n,\ldots,1)$ from the vector equations, Eq. (11.63a):

$$^N\alpha^k = \sum_{j=1}^{n} {}^N\omega_j^k \dot{u}_j + \alpha_t^k \quad (k = n,\ldots,1)$$

$$^N\alpha_t^n = \alpha^B \tag{11.63b}$$

$$^N\alpha_t^k = \alpha_t^i + {}^N\tilde{\omega}^k \lambda_k u_k \quad (k = n-1,\ldots,1)$$

Step 6. Express the remainder acceleration term, $^N a_t^{k*}$, of the acceleration of the mass center k* of k, and those of the inboard body i as follows, for k* directed as in Figure 11.16:

$$^N\mathbf{a}^{k*} = \sum_{j=1}^{n} {}^N\mathbf{v}_j^{k*} \dot{u}_j + {}^N\mathbf{a}_t^{k*} \quad (k = n,\ldots,1)$$

$$^N\mathbf{a}_t^{n*} = {}^N\mathbf{a}^B - {}^N\boldsymbol{\omega}^n \times ({}^N\boldsymbol{\omega}^n \times \mathbf{r}_2^n) \tag{11.64a}$$

$$^N\mathbf{a}_t^{k*} = {}^N\mathbf{a}_t^{i*} + {}^N\boldsymbol{\alpha}_t^i \times \mathbf{r}_1^k + {}^N\boldsymbol{\omega}^i \times ({}^N\boldsymbol{\omega}^i \times \mathbf{r}_1^k)$$
$$- {}^N\boldsymbol{\alpha}_t^k \times \mathbf{r}_2^k - {}^N\boldsymbol{\omega}^k \times ({}^N\boldsymbol{\omega}^k \times \mathbf{r}_2^k) \quad (k = n-1,\ldots,1)$$

The matrix forms expressed in the k basis of these vector equations are

$$^N\mathbf{a}^{k*} = \sum_{j=1}^{n} {}^N\mathbf{v}_j^{k*} \dot{u}_j + {}^N\mathbf{a}_t^{k*} \quad (k = n,\ldots,1)$$

$$^N\mathbf{a}_t^{n*} = {}^N\mathbf{a}^B - {}^N\tilde{\omega}^{n\,N}\tilde{\omega}^n \mathbf{r}_2^n \tag{11.64b}$$

$$^N\mathbf{a}_t^{k*} = {}^N\mathbf{a}_t^{i*} - \tilde{\mathbf{r}}_1^{k\,N}\boldsymbol{\alpha}_t^i + {}^N\tilde{\omega}^{i\,N}\tilde{\omega}^i \mathbf{r}_1^k + \tilde{\mathbf{r}}_2^{k\,N}\boldsymbol{\alpha}_t^k - {}^N\tilde{\omega}^{k\,N}\tilde{\omega}^k \mathbf{r}_2^k$$
$$(k = n-1,\ldots,1)$$

Step 7. Represent the external force F_k and the external torque T_k as 3×1 matrices corresponding to vectors in the k basis, and with I^k the moment of inertia matrix of body k, form 3×1 matrices, F_t^{*k}, T_t^{*k}, corresponding to the vector equations:

$$\mathbf{F}_t^{*k} = m_k\,{}^N\mathbf{a}_t^{k*} - \mathbf{F}^k \quad (k = n,\dots,1)$$

$$\mathbf{T}_t^{*k} = \mathbf{I}^k \cdot {}^N\mathbf{\alpha}_t^k + {}^N\mathbf{\omega}^k \times (\mathbf{I}^k \cdot {}^N\mathbf{\omega}^k) - \mathbf{T}^k \quad (k = n,\dots,1) \tag{11.65a}$$

$$F_t^{*k} = m_k\,{}^N a_t^{k*} - F^k \quad (k = n,\dots,1)$$

$$T_t^{*k} = I^k\,{}^N\alpha_t^k + {}^N\tilde{\omega}^k I^k \omega^k - T^k \quad (k = n,\dots,1) \tag{11.65b}$$

Step 8. Now form the 6×6 matrix M_k and the 6×1 matrices X_k, Y_k (k = n,...,1):

$$M_k = \begin{bmatrix} m_k U & 0 \\ 0 & I^k \end{bmatrix}$$

$$X_k = \left\{ \begin{matrix} F_t^{*k} \\ T_t^{*k} \end{matrix} \right\} \tag{11.66}$$

$$Y_k = \left\{ \begin{matrix} {}^N v_k^{k*} \\ {}^N \omega_k^k \end{matrix} \right\}$$

Here, U is a 3×3 unity matrix, and 0 represents a 3×3 null matrix.

11.5.1.3 Backward Pass

Step 1. Set $k = 1$, $M = M_k$, and $X = X_k$.

Step 2. Form and store the 6×1 matrix Z_k and the scalars, m_{kk} and f_k, as follows:

$$Z_k = MY_k$$

$$m_{kk} = Y_k^T Z_k \tag{11.67}$$

$$f_k = Y_k^T X + \tau_k$$

Here, τ_k is the hinge torque applied by body i on body k at the k-th hinge, and superscript T indicates the transpose of the matrix. Form the 6×6 augmented mass matrix and the 6×1 augmented remainder inertia force column matrix X:

$$M = M - [Z_k Z_k^T]/m_{kk}$$

$$X = X - Z_k (f_k / m_{kk}) \tag{11.68}$$

Step 3. Define the 6×6 shift transform matrix W between bodies i and k:

$$W = \begin{bmatrix} C_{ik}^T & -\tilde{r}^{i*k*}C_{ik}^T \\ 0 & C_{ik}^T \end{bmatrix} \tag{11.69}$$

Here, $\tilde{r}^{i*k*}$ is the skew-symmetric matrix for a cross product with the components of the position vector from the mass center $i*$ of body i inboard to the mass center $k*$ of body k in the k basis. Next, transfer the inertia force and the inertia torque of the k-th body, represented by the augmented mass and remainder inertia force matrices, M and X of Eq. (11.68) to the i-th body and add to the inertia force matrices for body i, forming new matrices M, K:

$$M = M_i + W^T M W$$

$$X = X_i + W^T X \tag{11.70}$$

Step 4. Increase k by 1 and repeat steps 2 and 3 until m_{nn} and f_n are formed.

11.5.1.4 Forward Pass

Step 5. Set $k = n - 1$, and $i = n$. Form the following:

$$\dot{u}_n = -f_n/m_{nn}$$

$$\left\{ \begin{matrix} {}^N a_0^{i*} \\ {}^N \alpha_0^i \end{matrix} \right\} = Y_n \dot{u}_n \tag{11.71}$$

Step 6. Form two 6×1 matrices and the scalar derivative $\dot{u}_k$ in the sequence shown below, with Z_k, f_k defined in Eq. (11.67), recalling that ${}^N \hat{a}_0^k, {}^N \hat{\alpha}_0^k$ mean the acceleration/angular acceleration of the inboard body i for the outboard body k.

$$\left\{ \begin{matrix} {}^N \hat{a}_0^k \\ {}^N \hat{\alpha}_0^k \end{matrix} \right\} = W \left\{ \begin{matrix} {}^N a_0^{i*} \\ {}^N \alpha_0^i \end{matrix} \right\}$$

$$\dot{u}_k = -\left([Z_k]^T \left\{ \begin{matrix} {}^N \hat{a}_0^k \\ {}^N \hat{\alpha}_0^k \end{matrix} \right\} + f_k \right) / m_{kk} \tag{11.72}$$

$$\left\{ \begin{matrix} {}^N a_0^{k*} \\ {}^N \alpha_0^k \end{matrix} \right\} = \left\{ \begin{matrix} {}^N \hat{a}_0^{k*} \\ {}^N \hat{\alpha}_0^k \end{matrix} \right\} + Y_k \dot{u}_k$$

Step 7. Decrease the indices k and i by 1, and repeat step 6 until $\dot{u}_1$ is formed.

11.5.1.5 Extrusion/Retraction Step

If one link has been extruded, replace n by $n + 1$ and assume that $q_{n+1} = u_{n+1} = 0$. For retraction, replace n by $n - 1$ when one link is pulled in, and update the state as described below to maintain continuity in displacements and velocities. Reset the values of the beam stiffness based on Eq. (11.68) for the current value of n. Go to Step 1 of "Forward Pass". For retraction, an update of the dynamical states is needed as one link goes in. The situation is shown in

Figure 11.18. The initial values of the generalized coordinates and generalized speeds for a new set of links have to be computed on the basis of the continuity of the physical displacements and velocities of the new system and the old system. This requires the following two sets of equations for displacement and velocity to be satisfied, going from n to n − 1 links in retraction, with superscript (+) indicating an unknown initial value in the n − 1 system, to the n-system with superscript (−) for a known end value.

$$\sum_{j=i}^{n-1} L \sin\left(\sum_{k=j}^{n-1} q_k^+\right) = \sum_{j=i}^{n} L\left(\sin \sum_{k=j}^{n} q_k^-\right) \qquad (i = 1,\ldots,n-1) \quad (11.73a)$$

$$\sum_{j=i}^{n-1} L \cos\left(\sum_{k=j}^{n-1} q_k^+\right)\left(\sum_{m=j}^{n-1} \dot{q}_m^+\right) = \sum_{j=i}^{n} L \cos\left(\sum_{k=j}^{n} q_k^-\right)\left(\sum_{m=j}^{n} \dot{q}_m^-\right)$$

$$(i = 1,\ldots,n-1)$$

$$(11.73b)$$

The velocity equations in Eq. (11.73b) are linear. The displacement equations, given first in Eq. (11.73a), are non-linear in the new generalized coordinates, of the form $f_i(q_1^+,\ldots,q_{n-1}^+) = 0$, $i = 1,\ldots,n-1$, and they can be solved for the angles, by Newton's iterative method for solving non-linear equations, where the associated Jacobian matrix turns out to be triangular because the deflection at any revolute joint does not depend on the motion of the outboard links:

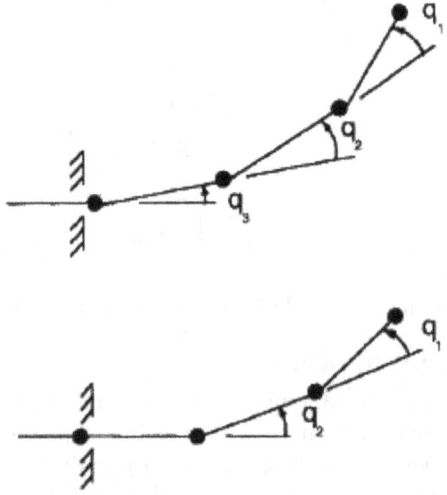

FIGURE 11.18   Schematic representation of jump discontinuity as one link goes in during retraction.

$$\left[\frac{\partial f_i}{\partial q_j^+}\right]\left(\{q_j^+\}_{i+1} - \{q_j^+\}_i\right) = -\{f_i\}, \quad i = 1,\dots,n-1 \tag{11.74a}$$

$$\frac{\partial f_i}{\partial q_j^+} = 0 \quad j > i \tag{11.74b}$$

$$= L\cos\sum_{k=1}^{n-1} q_k^+ \quad (i = 1,\dots,n-1) \tag{11.74c}$$

$$= \frac{\partial f_i}{\partial q_{j-1}^+} + L\cos\sum_{k=j}^{n-1} q_k^+ \quad (i = 1,\dots,n-1;\ j = i+1,\dots,n-1) \tag{11.74d}$$

The equation for the new derivatives of the generalized coordinates, Eq. (11.74b), is actually linear, with the coefficient matrix precisely the Jacobian matrix of Eqs. (11.73a) and (11.73b) and is easily solved.

The Newton–Raphson method for iteratively solving nonlinear equations in the matrix form is used to calculate the updated states:

$$\left[\frac{\partial f}{\partial q}\right]_i\left(\{q\}_{i+1} - \{q\}\right) = -\{f\}_i \tag{11.75}$$

Iteration is stopped when the right hand side of Eq. (11.75) is nearly zero.

This basic update of generalized coordinates and their derivatives is valid for both extrusion and retraction. Note that no initial value update is needed for extrusion since Eqs. (11.74a) and (11.74b), with superscripts (+) and (–) interchanged, are identically satisfied as the system grows from (n – 1) to n elements, with the new element having zero transverse displacement and velocity.

11.6 NUMERICAL SIMULATION OF EXTRUSION AND RETRACTION

The variable-n order-n algorithm is now used to simulate extrusion/retraction of the WISP antenna from a slowly rolling shuttle. The antenna is a tube of 0.0636 m diameter, 0.00254 m wall thickness, and modulus of elasticity 1.96491×10^{10} N/m^2. Fully deployed, the antenna is 150 m long. The shuttle is assumed to rotate such that $\omega^B = \Omega(t)\mathbf{b}_3$; here, however, it is assumed that $\Omega = 1$ deg/sec about an axis normal to the direction of extrusion/retraction. Extrusion from 5 m to 150 m in 750m seconds is prescribed at the rate:

$$\dot{d} = 0.26666\left[1 - \left(\frac{t}{750}\right)^3\right] \quad t \le 750 \text{ m/sec} \tag{11.76a}$$

and retraction from 150 m to 5 m in 750m seconds is prescribed at the rate:

$$\dot{d} = 0.26666\left[1-\left(1-\frac{t}{750}\right)^3\right] \quad t \le 750 \text{ m/sec} \qquad (11.76b)$$

Figure 11.19a and b shows the transverse beam deflections for extrusion at the beam tip and at a point 50 m from the tip, respectively; the associated length time history is in Figure 11.19c. Note that the beam tip deflection around 130 m is about 16 m. This is already a large deflection, beyond the range of the linear beam theory, so that any modal superposition would have been inapplicable, whereas our formulation

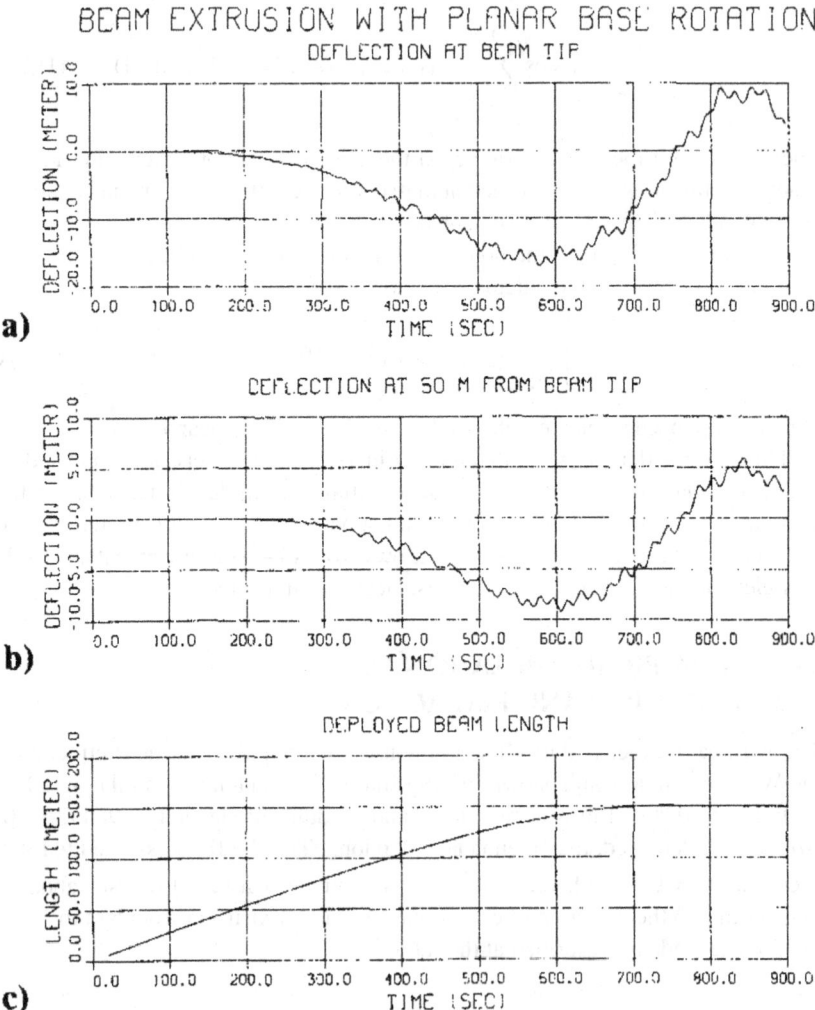

FIGURE 11.19 Deflection (m) at beam tip, 50 m from tip, and length (m) deployed for extrusion of a beam from a rotating base; beam deflection at 130 m is 16 m which is already large; Coriolis loading causes beam to lag.

does not have any such limitation. Checking up to 750 sec, we see that the extrusion from a rotating base makes the beam *lag* its original undeformed configuration; this is because of *Coriolis inertia* loading; also note that the lumped mass at 50 m from the beam tip starts late in its deflection because it does not come out until the time that depends on the extrusion rate. Figure 11.20a and b shows the deflection results for retraction, and Figure 11.20c shows the length vs. time. From Figure 11.20a and b, note that the tip deflection is 23 m when the beam instantaneous length is 133 m; this is a larger deflection than that obtained with extrusion. Here, the beam *leads* the undeformed deformation, again due to the direction of the Coriolis inertia load. The

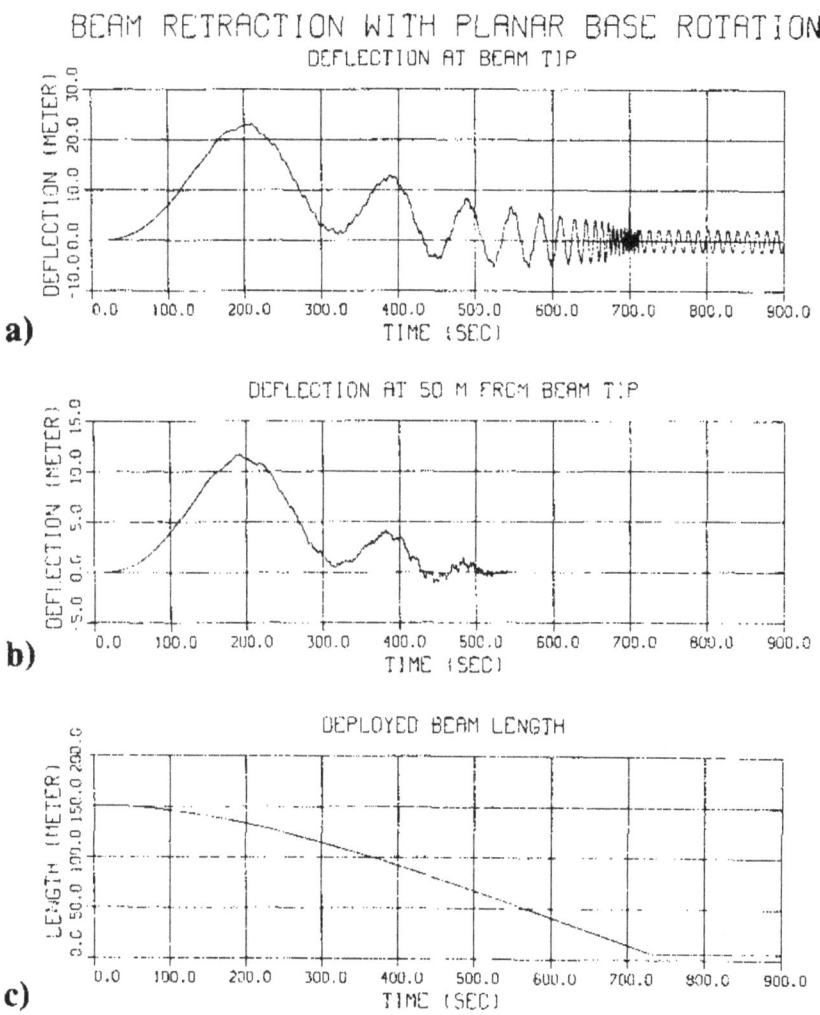

FIGURE 11.20 Deflection (m) at beam tip, at 50 m from tip, and length (m) out for retraction of a beam into a rotating base large beam tip deflection 23 m when it is retracted to 133 m (frequency increase toward end of retraction mimics spaghetti eating!).

frequency increase as the length gets shorter reminds one of "spaghetti eating"! The oscillation of the point 50 m from the tip stops as the point goes into the containing body, whereas for the tip mass of the terminal oscillations, Figure 11.20 shows the transverse vibrations of a stubby beam. The results of Figure 11.20 show that the deflections of a retracting beam do not necessarily grow unbounded; this is in contrast to the case of a tether which has a tendency to wrap up around the exiting body. Because a tether can be thought of as a beam of zero bending stiffness, the method given in this chapter can be used to analyze the retraction of a tether into a rotating body. Figure 11.21 shows the deflections and instantaneous length of a cable at half

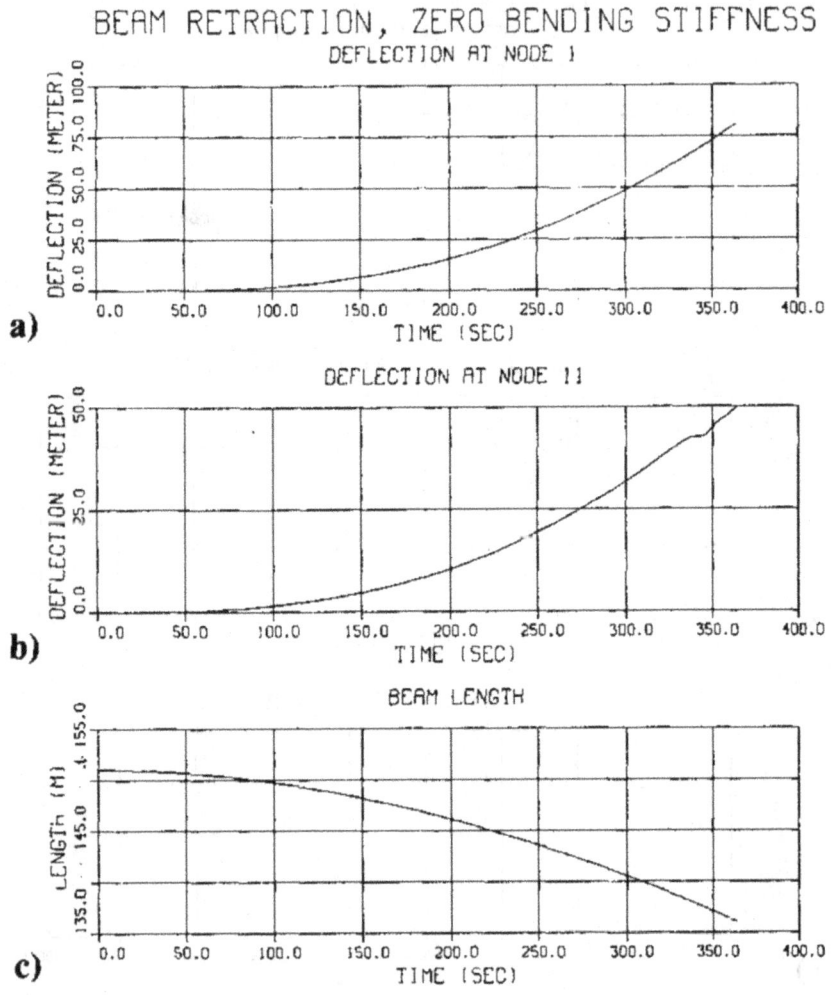

FIGURE 11.21 Deflection (m) of zero bending stiffness beam at tip, at 50 m from the tip, and length (m) of a cable being retracted (at half the rate used for Figure 11.19) with planar rotation.

the retraction rate used for Figure 11.20. It is seen that the cable has very large deflections with less than 15 m of retraction completed, and the cable tends to fly off. This indicates that it is the bending stiffness of a beam that prevents its deflection from going unbounded. Finally, we show some results for spin-up of the shuttle body in three dimensions as in

$$^{N}\omega^{B} = 0.001\mathbf{b}_1 + \frac{0.0174533}{600}\left[t - \frac{300}{\pi}\sin\left(\frac{\pi t}{300}\right)\right]\mathbf{b}_3 \quad t \le 600 \tag{11.77}$$

$$= 0.001\mathbf{b}_1 + 0.0174533\mathbf{b}_3 \quad t > 600$$

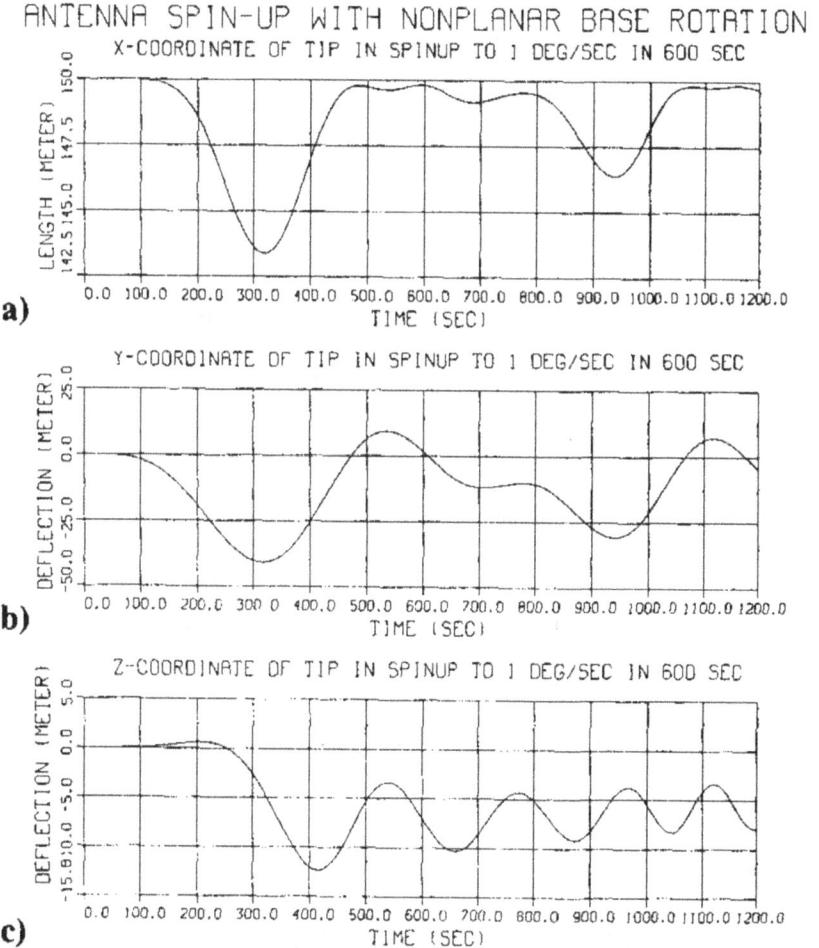

FIGURE 11.22 Three-dimensional X-, Y-, Z-deflection of antenna due to non-planar spin-up of base to 1 deg/sec in 600 sec.

The three-dimensional response of the beam is obtained by modeling two degrees of rotation at the joints. Figure 11.22a shows the axial foreshortening time history, and Figure 11.22b and c shows the transverse tip deflections.

11.7 CONCLUSION

It has been shown that the problem of the extrusion and retraction of a beam from a rotating base can be done by discretizing the beam and using vibration modes for the varying number n of beam segments deployed or retracted, when the deployment rates are small, where the number n is increased or decreased by 1. For large deployments, an order-n solution is proposed, in which initial values of the generalized coordinates and generalized speeds for a new set of links have to be computed from the old set on the basis of the continuity of the physical displacements and velocities of the new system and the old system. For the deployment/retraction of a cable, with one end of the cable attached to the ship, the other end is constrained to move with the UUV, which has its own model and embedded propulsion control system as it goes in search of underwater mines, the constrained order-n formulation is shown to be quite effective.

REFERENCES

1. Banerjee, A.K. and Kane, T.R., (1989), "Extrusion of a Beam from Rotating Base", *Journal of Guidance, Control, and Dynamics*, **12**(2), pp. 140–146.
2. Banerjee, A.K. and Do, V.N., (1994), "Deployment Control of a Cable Connecting a Ship to an Underwater Vehicle", *Journal of Guidance, Control, and Dynamics*, **17**(6), pp. 1327–1332.
3. Banerjee, A.K., (1992), "Order-n Formulation of Extrusion of a Beam with Large Bending and Rotation", *Journal of Guidance, Control, and Dynamics*, **15**(1), pp. 121–127.
4. Banerjee, A.K. and Nagarajan, S., (1997), "Efficient Simulation of Large Overall Motion of Beams Undergoing Large Deflection", *Multibody System Dynamics*, **1**, pp. 113–126.
5. Kane, T.R., Ryan, R.R., and Banerjee, A.K., (1987), "Dynamics of a Cantilever Beam Attached to a Moving Base", *Journal of Guidance, Control, and Dynamics*, **10**(2), pp. 139–151.
6. Kane, T.R. and Levinson, P.W., (1985), *Dynamics, Theory and Application*, McGraw-Hill.
7. Banerjee, A.K., (1993), "Block-Diagonal Equations of Multibody Elasto-dynamics with Geometric Stiffness and Constraints", *Journal of Guidance, Control, and Dynamics*, **16**(6), pp. 221–227.
8. Banerjee, A.K. and Dickens, J.M., (1990), "Dynamics of an Arbitrary Flexible Body in Large Rotation and Translation", *Journal of Guidance, Control, and Dynamics*, **13**(2), pp. 140–146.
9. Wang, J.T. and Huston, E.L., (1987), "Kane's Equations with Undetermined Multipliers: Application to Constrained Multibody Systems", *Journal of Applied Mechanics*, **54**, pp. 424–429.
10. Huston, R.L. and Kamman, J.W., (1981), "A Representation of Fluid Forces in Segment Cable Models", *Computers and Structures*, **14**, pp. 281–287.

11. Myers, J.J., Holm, C.H., and McAllister, R.F. (eds), (1969), *Handbook of Ocean and underwater Engineering*, McGraw-Hill.
12. Rosenthal, D., (Oct.–Dec., 1990), "An Order-N Formulation for Robotic Systems", *Journal of Astronautical Sciences*, **38**(4), pp. 511–530.
13. Baumgarte, J., (1972), "Stabilization of Constraint and Integrals of Motion in Dynamical Systems", *Computer Methods in Applied Mechanics and Engineering*, **1**, pp. 1–16.

12 Flexible Rocket Dynamics, Using Geometric Softness and a Block-Diagonal Mass Matrix

12.1 INTRODUCTION

In this chapter, we present a computationally efficient formulation that can be used for a flexible rocket with a swiveling thruster nozzle used for directional control. The dynamics of systems with variable mass has been treated in the literature with a simplified particle approach giving rise to the classical rocket equation [1], and a more complex control volume approach [2] with the flow of liquid propellant out of a solid base. It has been shown in Ref. [3] that the equations derived with such flow considerations reduce to the equations given by the particle approach under the assumption of a constant mass depletion rate. Reference [4] analyzes the vibration problem of a rocket. In this chapter, we use the methods in Refs. [5, 6] dynamical equations with a block-diagonal mass matrix for large motion featuring geometric softness of a flexible rocket. The analysis given here is principally based on Refs. [7, 8]. Error in premature linearization, inherent with the use of modes [9], is compensated for by augmenting the structural stiffness with geometric stiffness, softness in this case, due to the compressive system of axial inertia and thrust forces; this feature enables the analyst to predict buckling of a slender rocket (see Chapter 5). Mode selection for rockets is an art, as has been noted for rockets and the space shuttle solid rocket booster in Refs. [10, 11]. Here, for simplicity, we use free-free modes of a slender beam to characterize the vibration of a rocket. Mode selection strategies based on solving the modal eigenvalue problem for various "stages" of rocket mass, as discussed in Ref. [11], are more appropriate. Discussion of multistage rocket equation is given in Ref. [13]. Equations developed here, however, should be useful as an adequate approximation of the dynamics of flexible multistage rockets, particularly for computationally efficient, model-based on-line non-linear control design.

DOI: 10.1201/9781003231523-13

12.2   KANE'S EQUATION FOR A VARIABLE MASS FLEXIBLE BODY

Consider a system of nodal particles $P_k, k = 1, \ldots, n$, each particle losing mass at the rate of $\dot{m}^k$, and let the motion of this system be described by v generalized speeds. Nodal particle P_k of mass m^k at time t is subjected to an external force F^k, and at time $t + \Delta t$ it acquires a velocity $v^k + \Delta v^k$ by ejecting a particle of mass $(-\dot{m}^k \Delta t)$, where $\dot{m}^k$ is the uniform mass loss rate, with an ejection velocity v_e^k relative to P_k. Acceleration of the ejected mass due to the changed velocity being $(v_e^k / \Delta t)$ implies a force imparted to the ejected particle as

$$F_i^k = \frac{(-\dot{m}^k \Delta t) v_e^k}{\Delta t} = -\dot{m}^k v_e^k \tag{12.1}$$

By reaction, $-F_i^k$ acts on the particle that remains, P_k, in addition to any external force F^k. Now appealing to Newton's law in the inertial frame N, one gets the vector form of the classical equation for a particle losing mass:

$$m^k \frac{{}^N dv^k}{dt} = F^k + \dot{m}^k v_e^k \tag{12.2}$$

Here, the second term on the right expresses thrust. Note that this derivation, by Kane [5], differs from those based on the conservation of momentum given in many texts. Ge and Cheng [6] dot-multiplied the vector equations, Eq. (12.2), by Kane's partial velocity [12] vector v_r^k of P_k with respect to the r-th generalized speed of the *system*, to form what they called *extended Kane's equations* for the system of variable mass with n degrees of freedom:

$$F_r + F_r^* + F_r^{**} = 0 \quad (r = 1, \ldots, n) \tag{12.3}$$

The first term here is the generalized active force due to gravity and aerodynamics, the second term is the generalized inertia force, and the generalized active force due to thrust:

$$F_r = \sum_{k=1}^{v} {}^N v_r^k \cdot F^k \quad (r = 1, \ldots, n) \tag{12.4}$$

$$F_r^* = \sum_{k=1}^{v} {}^N v_r^k \cdot (-m^k \, {}^N a^k) \quad (r = 1, \ldots, n) \tag{12.5}$$

$$F_r^{**} = \sum_{k=1}^{v} {}^N v_r^k \cdot (\dot{m}^k \, {}^N v_e^k) \quad (r = 1, \ldots, n) \tag{12.6}$$

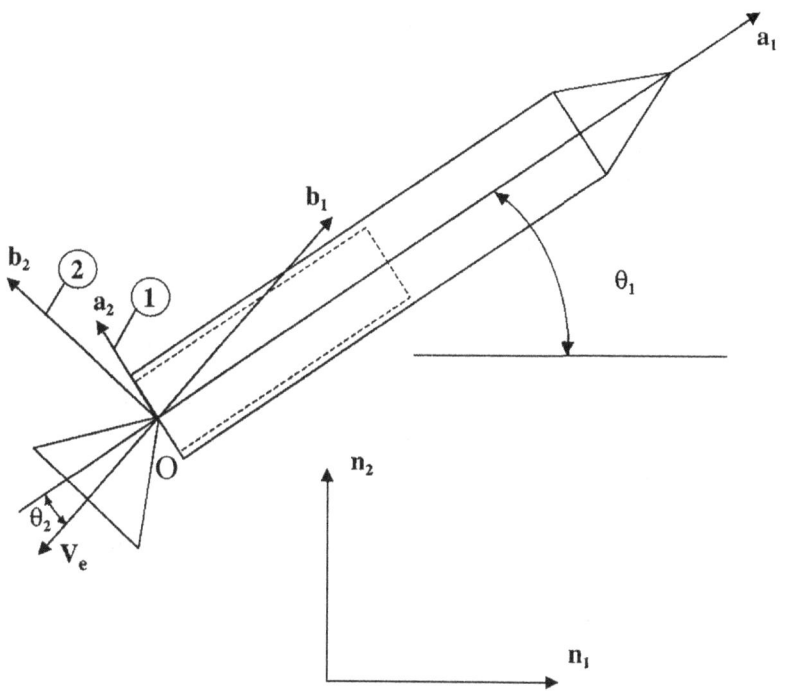

FIGURE 12.1 Planar view of a rocket with a gimbaled nozzle engine.

At this stage, we consider a special case of a flexible rocket body with a rigid engine that swivels in a prescribed gimbal motion for directional control. The system with basis vectors fixed in the rocket undeformed form and the engine are shown in Figure 12.1.

Let O be the point where the engine joins the rocket, and the velocity of O in N, the angular velocity $\boldsymbol{\omega}^1$ of the rocket frame-1 in inertial frame N, and the velocity of a generic particle k in the flexible rocket body be given as follows, in terms of $(6 + \mu)$ generalized speeds, where μ is the number of vibration modes of the rocket body B:

$$^N\mathbf{v}^O = u_1\mathbf{b}_1 + u_2\mathbf{b}_2 + u_3\mathbf{b}_3$$

$$^N\boldsymbol{\omega}^1 = u_4\mathbf{b}_1 + u_5\mathbf{b}_2 + u_6\mathbf{b}_3$$

$$^N\mathbf{v}^k = {}^N\mathbf{v}^O + {}^N\boldsymbol{\omega}^1 \times (\mathbf{p}^k + \mathbf{d}^k) + \sum_{i=1}^{\mu} \boldsymbol{\varphi}_i^k u_{6+i} \qquad (12.7)$$

Here, we have expressed the deformation $\mathbf{d}^k$ at the point with a position vector $\mathbf{p}^k$ from the fixed point O in frame-1 to the particle P_k, in terms of a number μ of modes $\boldsymbol{\varphi}_i^k$, modal coordinates, and associated generalized speeds defined as

$$\mathbf{d}^k = \sum_{i=1}^{\mu} \varphi_i^k q_i \quad k = 1, \ldots, n \tag{12.8}$$

$$u_{6+i} = \dot{q}_i$$

The partial velocity of P_k for the r-th generalized speed follows from Eq. (12.7) as

$$^N\mathbf{v}_r^k = {^N}\mathbf{v}_r^O + {^N}\boldsymbol{\omega}_r^1 \times (\mathbf{p}^k + \mathbf{d}^k) + \varphi_i^k \delta_{r-6,i} \quad (r = 1, \ldots, 6 + \mu; i = 1, \ldots, \mu) \tag{12.9}$$

Here, the Kronecker delta is associated with the generalized speeds $u_r, r = 1, \ldots, 6 + \mu$. Note that Eq. (12.7) uses linear vibration modes and this linearization is inherently premature, leading to error in results. As has been shown before [9], premature linearization error can be compensated a posteriori by including geometric stiffness due to associated inertia loads. Before we do that, we evaluate the acceleration of a generic particle P_k.

$$^N\mathbf{a}^k = {^N}\mathbf{a}^O + {^N}\boldsymbol{\alpha}^1 \times (\mathbf{p}^k + \mathbf{d}^k) + \sum_{i=1}^{\mu} \varphi_i^k \dot{u}_{6+i} + {^N}\boldsymbol{\omega}^1 \times \begin{bmatrix} ^N\boldsymbol{\omega}^1 \times (\mathbf{p}^k + \mathbf{d}^k) \\ +2\sum_{i=1}^{\mu} \varphi_i^k u_{6+i} \end{bmatrix} \tag{12.10}$$

Kane's equations for a flexible body losing mass can now be derived by substituting Eqs. (12.9) and (12.10) into Eqs. (12.3)–(12.6). Equation (12.3) can be separated into translation, rotation, and vibration equations depending on which generalized speed, r, in the partial velocity equation, Eq. (12.9), is non-zero in the dot-multiplications of Eqs. (12.3)–(12.6). The *translation equations* of the flexible rocket are obtained for $r = 1, 2, 3$ by dot-multiplying the first term in the right-hand side of Eq. (12.9) with Eq. (12.10) for n particles, and using modal sums $\mathbf{s}^k, \mathbf{b}_i^k$ given in Section 12.2. Note that the modal sums representing modal integrals given in Appendix A at the end of the book, should be strictly computed for the instantaneous mass and modes, but are rarely done so.

$$^N\mathbf{v}_r^O \cdot \left\langle \sum_{k=1}^{n} \left\{ \begin{matrix} m^k {\,}^N\mathbf{a}^O + {^N}\boldsymbol{\alpha}^1 \times \mathbf{s}^k + \sum_{i=1}^{\mu} \mathbf{b}_i^k \dot{u}_{6+i} \\ + {^N}\boldsymbol{\omega}^1 \times \left[{^N}\boldsymbol{\omega}^1 \times \mathbf{s}^k + 2\sum_{i=1}^{\mu} \mathbf{b}_i^k u_{6+i} \right] \\ -\dot{m}^k {\,}^N\mathbf{v}_e^k \end{matrix} \right\} \right\rangle = {^N}\mathbf{v}_r^O \cdot (\mathbf{F}^{\text{ext}}) \tag{12.11}$$

$$r = 1, 2, 3$$

Here, $\mathbf{F}^{ext}$ is the total external force, combining gravity and aerodynamics, replaced at the point O. These equations agree with those for a system with relative particle motion derived in Ref. [1]. The rotation equations follow by dot-multiplying the second term on the right side of Eq. (12.9) with Eq. (12.10), and are

$$
{}^N\boldsymbol{\omega}_r^1 \cdot \left\langle
\begin{array}{l}
\displaystyle\sum_{k=1}^{n} m^k(\mathbf{p}^k+\mathbf{d}^k)\times[{}^N\mathbf{a}^O+{}^N\boldsymbol{\alpha}^1\times(\mathbf{p}^k+\mathbf{d}^k) \\[2mm]
\displaystyle +\sum_{i=1}^{\mu}\varphi_i^1\ddot{u}_{6+i}]-\dot{m}^k\,{}^N\mathbf{v}_e^k \\[2mm]
\displaystyle +{}^N\boldsymbol{\omega}^1\times\left\{{}^N\boldsymbol{\omega}^1\times(\mathbf{p}^k+\mathbf{d}^k)+2\sum_{i=1}^{\mu}\varphi_i^1\dot{u}_{6+i}\right\}
\end{array}
\right\rangle
$$

$$
= {}^N\boldsymbol{\omega}_r^1\cdot[\sum_{k=1}^{n}(\mathbf{p}^k+\mathbf{d}^k)\times\mathbf{F}_k^{ext}] \quad r=4,5,6
\tag{12.12}
$$

Note that the right-hand side term is the generalized force due to the moment of $\mathbf{F}^{ext}$ about O, the swivel point of the rocket engine, Figure 12.1. Now, of all the terms in Eq. (12.12), we give special attention to the following three terms, neglecting squares of modal coordinates and products of modal coordinates and their derivatives, associated with modes:

1. $\displaystyle {}^N\boldsymbol{\omega}_r^1\cdot\sum_{k-1}^{n}m^k(\mathbf{p}^k+\mathbf{d}^k)\times[{}^N\boldsymbol{\alpha}^1\times(\mathbf{p}^k+\mathbf{d}^k)]$

$$
= {}^N\boldsymbol{\omega}_r^1\cdot\sum_{k=1}^{n}m^k[(\mathbf{p}^k+\mathbf{d}^k)\cdot(\mathbf{p}^k+\mathbf{d}^k)\mathbf{U}-(\mathbf{p}^k+\mathbf{d}^k)(\mathbf{p}^k+\mathbf{d}^k)]\cdot{}^N\boldsymbol{\alpha}^1
\tag{12.13}
$$

$$
\approx {}^N\boldsymbol{\omega}_r^1\cdot<\mathbf{I}_r^{B/O}+\sum_{k=1}^{\mu}[\mathbf{N}_k^1+\mathbf{N}_k^{1\,T}]q_k^1>\cdot{}^N\boldsymbol{\alpha}^1={}^N\boldsymbol{\omega}_r^1\cdot[\mathbf{I}^{B/O}\cdot{}^N\boldsymbol{\alpha}^1]
$$

2. $\displaystyle \boldsymbol{\omega}_r^1\cdot\sum_{k=1}^{n}m^k(\mathbf{p}^k+\mathbf{d}^k)\times\{\boldsymbol{\omega}^1\times[\boldsymbol{\omega}^1\times(\mathbf{p}^k+\mathbf{d}^k)]\}$

$$
= \boldsymbol{\omega}_r^1\cdot\left\langle\boldsymbol{\omega}^1\times\left[\sum_{k=1}^{n}m^k\begin{Bmatrix}(\mathbf{p}^k+\mathbf{d}^k)\cdot(\mathbf{p}^k+\mathbf{d}^k)\bar{\mathbf{U}} \\ -(\mathbf{p}^k+\mathbf{d}^k)(\mathbf{p}^k+\mathbf{d}^k)\end{Bmatrix}\right]\cdot\boldsymbol{\omega}^1\right\rangle
\tag{12.14}
$$

$$
= \boldsymbol{\omega}_r^1\cdot\left\langle\boldsymbol{\omega}^1\times\bar{\mathbf{I}}^{B/O}\cdot\boldsymbol{\omega}^1\right\rangle
$$

3. $^{N}\omega_{r}^{1} \cdot < \sum_{k=1}^{n} m^{k}(\mathbf{p}^{k} + \mathbf{d}^{k}) \times [^{N}\omega^{1} \times \{2 \sum_{i=1}^{\mu} \varphi_{i}^{k} u_{6+i}\}]$

$\approx {}^{N}\omega_{r}^{1} \cdot < 2 \sum_{k=1}^{n} m^{k} \sum_{i=1}^{\mu} [(\varphi_{i}^{k} \cdot \mathbf{p}^{k})\mathbf{U} - \varphi_{i}^{k}\mathbf{p}^{k})]u_{6+i} > \cdot {}^{N}\omega^{1}$ (12.15)

$= 2 {}^{N}\omega_{r}^{1} \cdot [\sum_{k=1}^{\mu} \mathbf{N}_{k}^{1\,T} u_{6+i}] \cdot {}^{N}\omega^{1}]$

Crucial to the derivation of Eqs. (12.13)–(12.15) are the vector-dyadic identities:

$$\mathbf{A} \times (\mathbf{B} \times \mathbf{C}) = (\mathbf{A} \cdot \mathbf{C})\mathbf{B} - \mathbf{C}(\mathbf{A} \cdot \mathbf{B}) = [(\mathbf{C} \cdot \mathbf{A})\mathbf{U} - \mathbf{CA}] \cdot \mathbf{B}$$
(12.16)
$$\mathbf{A} \cdot (\mathbf{B} \times \mathbf{C}) = \mathbf{B} \cdot (\mathbf{C} \times \mathbf{A}) = \mathbf{C} \cdot (\mathbf{A} \times \mathbf{B})$$

Using Eqs. (12.13)–(12.15) in Eq. (12.12) gives the rotational equations of the rocket, body 1:

$$^{N}\omega_{r}^{1} \cdot \left\langle \begin{array}{l} \mathbf{s}_{0}^{1} \times {}^{N}\mathbf{a}^{O} + \mathbf{I}^{B/O} \cdot {}^{N}\alpha^{1} + \sum_{i=1}^{\mu} \mathbf{c}_{k}^{1}\ddot{u}_{6+i} + {}^{N}\omega^{1} \times \left(\mathbf{I}^{B/O} \cdot {}^{N}\omega^{1}\right) \\ + 2 \sum_{i=1}^{\mu} \mathbf{N}_{i}^{1T} u_{6+i} \cdot {}^{N}\omega^{1} - \dot{m}^{k}\mathbf{s}_{0}^{1} \times {}^{N}\mathbf{v}_{e}^{1} \end{array} \right\rangle$$
(12.17)

$$= {}^{N}\omega_{r}^{1} \cdot \left[\sum_{k=1}^{n} m^{k}(\mathbf{p}^{k} + \mathbf{d}^{k}) \times (\mathbf{F}_{k}^{ext}) \right] \quad r = 4,5,6$$

Note that, after all, Eq. (12.17) is the same equation as Eq. (6.11), with two crucial differences: the modal integrals should be evaluated with instantaneous mass and the presence of term with the mass loss rate, $\dot{m}^{k}$. Here, we have defined the deformation-dependent moment of inertia and modal dyadics by modal sums representing modal integrals given in Appendix A at the end of the book, for the rocket about O, to be used later in their matrix forms.

$$\mathbf{I}^{B/O} = \mathbf{I}_{r}^{B/O} + \sum_{k='1}^{\mu} [\mathbf{N}_{k}^{1} + \mathbf{N}_{k}^{1\,T}]q_{k}^{1}$$
(12.18)

$$\mathbf{N}_{i}^{1} = \sum_{k=1}^{n} \sum_{i=1}^{\mu} \{\mathbf{p}^{k} \cdot \varphi_{i}^{k}\mathbf{U} - \mathbf{p}^{k}\varphi_{i}^{k}\}u_{6+i}$$
(12.19)

In Eq. (12.17), the term $\left[-\sum_{k=1}^{n} \dot{m}^{k}\mathbf{s}^{1} \times {}^{N}\mathbf{v}_{e}^{k} \right]$, $\mathbf{s}^{1}$ being the first mass moment of the rocket about the point O where the engine is attached, represents the *jet damping torque for a flexible rocket.*

The vibration equation for the flexible rocket given below accounts for thrust effects, and it follows naturally from Kane's equations by dot-multiplying the third term in Eq. (12.9) with Eq. (12.10), and using the second equation in Eq. (12.16), and summing over all particles, k:

$$\left|\begin{array}{l} \mathbf{b}_{r-6}^1 \cdot {}^N\mathbf{a}^O + \mathbf{c}_{r-6}^1 \cdot {}^N\boldsymbol{\alpha}^1 + \displaystyle\sum_{i=1}^{\mu} \mathbf{e}_{r-6,i}^1 \, \dot{u}_{6+i} \\[2ex] -{}^N\boldsymbol{\omega}^1 . \mathbf{D}_{r-6}^1 \cdot {}^N\boldsymbol{\omega}^1 + 2\displaystyle\sum_{j=1}^{\mu} \mathbf{d}_{r-6,j}^k u_{6+j} \cdot {}^N\boldsymbol{\omega}^1 \\[2ex] -\boldsymbol{\varphi}_{r-6}^k \cdot \displaystyle\sum_{k=1}^{n} \dot{m}^k \, {}^N\mathbf{v}_e^k \end{array}\right\rangle = \sum_{k=1}^{n} \boldsymbol{\varphi}_{r-6}^k \cdot (\mathbf{F}_k^{ext}) \qquad (12.20)$$

$$r = 7,\dots 6+\mu$$

In Eq. (12.20) we have used the modal integral,

$$\mathbf{D}_{r-6} = \mathbf{N}_{r-6}^1 + \sum_{k=1}^{n} m^k \sum_{s=1}^{\mu} q_s [(\boldsymbol{\varphi}_s^k \cdot \boldsymbol{\varphi}_{r-6}^k)\mathbf{U} - \boldsymbol{\varphi}_s^k \boldsymbol{\varphi}_{r-6}^k] \dots 6+\mu \qquad (12.21)$$

To compensate for premature linearization due to the use of vibration modes, $\boldsymbol{\varphi}_i^k$, one must consider geometric stiffness, softness in this case because of a compressive system of loads, thrust f and axial inertia force (ignoring the drag for now) in dynamic equilibrium, we compute the generalized force associated with the *load-dependent geometric stiffness* [14] matrix:

$$S_{rj}^g = \Phi_r{}^1 K_g \Phi_j f \qquad (12.22)$$

Here, K_g is the geometric stiffness matrix due to the thrust along the axis $\mathbf{b}_1$ given by

$$f = \sum_{k=1}^{n} \dot{m}^k \, {}^N\mathbf{v}_e^k \cdot \mathbf{b}_1 \qquad (12.23)$$

and the right-hand side of Eq. (12.20) changes, including the geometric stiffness, the structural stiffness and damping, and the generalized force due to external forces lumped at O:

$$\sum_{k=1}^{n} \boldsymbol{\varphi}_{r-6}^k \cdot (\mathbf{F}_k^{ext}) \Rightarrow \sum_{j=1}^{\mu} S_{r-6,j}^g q_j \, f - m\omega_{n,r-6}^2 \, q_{r-6} - 2\xi_{n,r-6} \, m\omega_{n,r-6} \, \dot{q}_{r-6}$$

$$+ \sum_{k=1}^{n} \boldsymbol{\varphi}_{r-6}^k \cdot (\mathbf{F}_k^{ext}); \quad r = 7,\dots,6+\nu \qquad (12.24)$$

Mass loss and thrust cause changes in the assumed modes and frequencies. While the natural frequencies increase with mass loss, the effect of axial thrust is to lower the frequencies [6], as can be observed in Eq. (12.24). Here, we make an approximation that the structural stiffness stays the same while the mass varies according to some interpolation function as assumed later. Of course, one can work with interpolated values of mass and modal integrals that are stored for modes for various fill levels of liquid fuel, as is reported in Ref. [11]. Before closing this section, we note some of the approximations of the particle-based formulation given above. The most obvious omission is the consideration of the flow of the combustion gases inside the rocket, which gives rise to reaction forces and moments due to the Coriolis inertia effect and unsteady flow. However, for solid propellant launch vehicles, these effects are negligible, and even for liquid-engine rockets the major flow effect can be approximated as that due to the burned gases expelled through the nozzle.

12.3   MATRIX FORM OF THE EQUATIONS FOR VARIABLE MASS FLEXIBLE BODY DYNAMICS

The equations written above are in vector notation. To implement these in an algorithm, we need the scalar version; for this, we *express all vectors in the frame-1 basis*. This process results in the representation of the translation and rotation equations, Eqs. (12.11) and (12.17), as

$$M_1 \left\{ \begin{matrix} {}^N a^0 \\ {}^N \alpha^1 \end{matrix} \right\} + A_1^t \ddot{q} + X_1 = \left\{ \begin{matrix} 0 \\ 0 \end{matrix} \right\} \tag{12.25}$$

$$M_1 = \begin{bmatrix} mU & -\tilde{s} \\ \tilde{s} & I^1 \end{bmatrix} \tag{12.26}$$

$$A_1^t = \begin{bmatrix} b \\ c \end{bmatrix} \tag{12.27}$$

$$X_1 = \left\{ \begin{matrix} {}^N \tilde{\omega}_1 ({}^N \tilde{\omega}_1 s + 2b\dot{q}) - \dot{m}^k v_e - F^{ext} \\ {}^N \tilde{\omega}^1 I^1 {}^N \omega^1 + 2 \sum_{i=1}^{\mu} N_i^{1T} u_{6+i} {}^N \omega^1 - \dot{m}^k \tilde{r}_e v_e - T^{ext} \end{matrix} \right\}$$

$$\dot{q} = [u_7, \ldots, u_{6+\mu}]^T \tag{12.28}$$

Here, U stands for a 3×3 unity matrix, m, s, and I are, respectively, the instantaneous mass and the matrices of instantaneous mass moment and moment of inertia about O; F^{ext} is the external force and T^{ext} is the external torque resolved in the body-1 basis; and ω_1 is a (3×1) matrix of angular velocity of the rocket frame; a tilde sign over a symbol denotes a skew-symmetric matrix representation of the corresponding vector in a cross product. Thus $\tilde{r}^e$ is made out of the component of the vector from rocket mass center to the exit point of the jet. Furthermore, we have used the following summations, where ϕ^k, p^k, d^k are 3×1 matrices of the mode shape, position vector, and deformation vector of Eq. (12.8), respectively, and N_i, I_i, q_i are the i-th modal dyadic, inertia dyadic, and modal coordinate, respectively:

$$b = \sum_{k=1}^{n} m^k \varphi^k \qquad (12.29)$$

$$c = \sum_{k=1}^{n} m^k \tilde{p}^k \varphi^k \qquad (12.30)$$

$$s = \sum_{k=1}^{n} m^k p^k + \sum_{i}^{\mu} b_i q_i \qquad (12.31)$$

$$N_i^1 = \sum_{k=1}^{n} m^k \left[(p^k \cdot \varphi_i^k)U - p^k \varphi_i^k \right] \quad i = 1,\ldots,\mu \qquad (12.32)$$

$$I^{B/O} = \sum_{k=1}^{n} m^k [(p^{k^T} p^k)U - p^{k^T} p^k] + \sum_{k=1}^{\mu} [N_k + N_k^T]q_k \qquad (12.33)$$

It is assumed here that all nodal particles, P_k, have the same uniform mass reduction rate and ejection velocity, as in a process of sublimation. The vibration equations, Eq. (12.20), have the matrix representation, with associated matrices:

$$\begin{bmatrix} b_1^T & c_1^T \\ : & : \\ : & : \\ b_\mu^T & c_\mu^T \end{bmatrix} \begin{Bmatrix} {}^N a^O \\ {}^N \alpha^1 \end{Bmatrix} + [E_1]\{\ddot{q}\} + Z_1 = 0 \qquad (12.34)$$

$$E_1 = \sum_{k=1}^{v} \varphi^{k^T} m^k \varphi^k \qquad \begin{bmatrix} \text{unity diagonal matrix for normalized} \\ \text{orthogonal modes} \end{bmatrix} \qquad (12.35)$$

With f as the thrust force, and φ_i^k denoting the 3×1 matrix of the i-th mode at the k-th particle, we have

$$Z_1 = Y_1 + (m\Omega^2 - K^g f)q + 2m\xi\Omega\dot{q} - \sum_{k=1}^{v} \varphi^{k^T} F^{ext} \qquad (12.36)$$

$$Y_1 = \left\{ \begin{array}{c} -[^N\omega^1]^T D_1{}^N\omega^1 + 2\sum_{j=1}^{\mu} d_{1j}^t \dot{q}_j{}^N\omega^1 \\ \vdots \\ \vdots \\ \vdots \\ -[^N\omega^1]^T D_\mu{}^N\omega^1 + 2\sum_{j=1}^{\mu} d_{\mu j}^t \dot{q}_j{}^N\omega^1 \end{array} \right\} - \left[\begin{array}{c} \sum_{k=1}^{\mu} \phi_1^{k^T} \\ \vdots \\ \vdots \\ \sum_{k=1}^{\mu} \phi_\mu^{k^T} \end{array} \right] \dot{m}^k \{v_e\} \quad (12.37)$$

Here, we have allowed for a three-component exit velocity $\{v_e\}$ in the rocket body B basis, and used the notations:

$$D_r = N_r + \sum_{k=1}^{n} \sum_{i=1}^{\mu} q_i [\phi_i^{k^T} \phi_r^k U - \phi_i^k \phi_r^k] m^k \quad (r = 1, \ldots, \mu) \quad (12.38)$$

$$d_{rj} = \sum_{k=1}^{n} m^k \sum_{j=1}^{\mu} \tilde{\phi}_r^k \phi_j^k \quad (r, j = 1, \ldots, \mu) \quad (12.39)$$

12.4 BLOCK-DIAGONAL ALGORITHM FOR A FLEXIBLE ROCKET WITH A SWIVELING NOZZLE

A flexible rocket with a swiveling engine represents two articulated bodies, with the rocket being a flexible body and the engine usually modeled as a rigid body. The overall model is $(8 + \mu)$ degrees of freedom (dof), with the rocket having $(6 + \mu)$ dof and the engine assigned 2 dof of relative rotation, a dense matrix formulation may be expensive at least for an on-line model-based control design using a number (μ) of modes. A block-diagonal mass matrix formulation can be revisited for this purpose. When the nozzle or engine motion is prescribed (as is realizable by a high-bandwidth controller), this algorithm has to be modified. Thus, if the attachment point O of the rocket and nozzle in Figure 12.1 has a velocity vector $^N v^O$ with generalized speeds u_1, u_2, u_3, the acceleration of O can be expressed in the rocket frame, written in matrix notation as

$$^N a^O = ^N\dot{v}^O + ^N\tilde{\omega}^1{}^N v^O \quad (12.40)$$

with u_4, u_5, u_6 being the generalized speeds for the angular velocity $^N\omega^1$ of the rocket body frame-1, the angular acceleration of frame-1 has the matrix components:

$$^N\alpha^1 = \begin{bmatrix} \dot{u}_4 & \dot{u}_5 & \dot{u}_6 \end{bmatrix}^T \quad (12.41)$$

The nozzle has its reference frame, frame-2, with a fixed point that coincides with O of the flexible body. Frame-2 for the nozzle is oriented with respect to frame-1 by a body 2-3 rotation through commanded time histories of the nozzle orientation angles, θ_1, θ_2. The angular velocity of frame-2 is given in body-2 basis by the matrix:

$$^N\omega^2 = C_{12}^T \, ^N\omega^1 + R \begin{Bmatrix} \dot{\theta}_1 \\ \dot{\theta}_2 \end{Bmatrix} \tag{12.42}$$

$$R = \begin{bmatrix} \sin\theta_2 & 0 \\ \cos\theta_2 & 0 \\ 0 & 1 \end{bmatrix} \tag{12.43}$$

For prescribed relative rotations θ_1, θ_2, the angular acceleration of frame-2, the engine in Figure 12.1, is split into two terms, as done before, one involving derivatives of the generalized speeds of body-1 basis rotation, and the other a remainder term involving prescribed rotation rate derivatives:

$$^N\alpha^2 = \, ^N\alpha_0^2 + \, ^N\alpha_t^2 \tag{12.44}$$

$$^N\alpha_0^2 = C_{12}^T \, ^N\alpha^1 \tag{12.45}$$

$$^N\alpha_t^2 = [\dot{R} + C_{12}^T \, ^N\tilde{\omega}^1 C_{12} R] \begin{Bmatrix} \dot{\theta}_1 \\ \dot{\theta}_2 \end{Bmatrix} + R \begin{Bmatrix} \ddot{\theta}_1 \\ \ddot{\theta}_2 \end{Bmatrix} \tag{12.46}$$

The block-diagonal algorithm for prescribed nozzle motion is started by first applying Kane's equations to the dynamics of the nozzle. To this end, we first write the equilibrium system of inertia and external forces and torques about O on body 2 in its own basis, including unknown forces and torques of interaction from body 1, as follows, with s^2, I^2 meaning first and second moments of inertia of the nozzle (body 2) about its hinge:

$$\begin{Bmatrix} f^{*2} - f^{ext2} \\ t^{*2} - t^{ext2} \end{Bmatrix} \equiv M_2 \begin{Bmatrix} ^N a^O \\ ^N\alpha^2 \end{Bmatrix} + \begin{Bmatrix} ^N\tilde{\omega}^2 \, ^N\tilde{\omega}^2 s^2 - f^{ext2} \\ ^N\tilde{\omega}^2 I^2 \, ^N\tilde{\omega}^2 - t^{ext2} \end{Bmatrix} \qquad M_2 = \begin{bmatrix} m_2 & 0 \\ 0 & I_2 \end{bmatrix} \tag{12.47}$$

In this case, body 2 is the nozzle, and f^{ext2}, t^{ext2} are the force and torque on the nozzle (body 2). In view of Eqs. (12.44) and (12.45), one can write

$$\begin{Bmatrix} ^N a^O \\ ^N\alpha^2 \end{Bmatrix} = W \begin{Bmatrix} ^N a^O \\ ^N\alpha^2 \end{Bmatrix} + \begin{Bmatrix} 0 \\ ^N\alpha_t^2 \end{Bmatrix} \tag{12.48}$$

$$W = \begin{bmatrix} C_{12}^T & 0 \\ 0 & C_{12}^T \end{bmatrix} \tag{12.49}$$

so that Eq. (12.46) is rewritten for body 2 (nozzle) in basis 2 with superscript 2 as follows:

$$\begin{Bmatrix} f^{*2} - f^{ext2} \\ t^{*2} - t^{ext2} \end{Bmatrix} \equiv M_2 W \begin{Bmatrix} {}^N a^0 \\ {}^N \alpha^2 \end{Bmatrix} + X_2 \tag{12.50a}$$

Here

$$X_2 \equiv M_2 \begin{Bmatrix} 0 \\ {}^N \alpha_t^2 \end{Bmatrix} + \begin{Bmatrix} {}^N \tilde{\omega}^2 \tilde{\omega}^2 s^2 - f^{ext2} \\ {}^N \tilde{\omega}^2 I^2 \omega^2 - t^{ext2} \end{Bmatrix} \tag{12.50b}$$

Equation (12.50a) is expressed in the body-1 basis by using the W-transformation in Eq. (12.48):

$$\begin{Bmatrix} f^{*2} - f^{ext2} \\ t^{*2} - t^{ext2} \end{Bmatrix} = W^T M_2 W \begin{Bmatrix} a^0 \\ \alpha^2 \end{Bmatrix} + W^T X_2 \tag{12.51}$$

In preparation for considering inertia forces and torques on the flexible body, body 1, we first write the solution of Eq. (12.25), using Eqs. (12.27) and (12.36), formally as

$$\ddot{q} = -E_1^{-1} \left\{ A_1 \begin{pmatrix} {}^N a^0 \\ {}^N \alpha^2 \end{pmatrix} + Z_1 \right\} \tag{12.52}$$

Here, E_1 is an $n \times n$ unity diagonal matrix with instantaneous mass m on the diagonals, for the usual orthogonal modes, and Z_1 is given by Eq. (12.36); interaction forces and moments do not appear on the right-hand side due to the assumption that the elastic deformation at O is zero. Now, the D'Alembert equilibrium system of inertia and external forces on body 1, including interaction forces and torques from body 1, can be expressed in the body-1 basis as per Eq. (12.25):

$$\begin{Bmatrix} f^{*1} - f^{ext1} \\ t^{*1} - t^{ext1} \end{Bmatrix} \equiv M_1 \begin{Bmatrix} {}^N a^0 \\ {}^N \alpha^1 \end{Bmatrix} + A_1^T \ddot{q} + X_1 \tag{12.53}$$

Using Eq. (12.52) in Eq. (12.53) and using the notations, with E_i^{-1} usually an unity diagonal matrix, yield

$$\hat{M}_1 = M_1 - A_1^T E_1^{-1} A_1 \tag{12.54}$$

$$\hat{X}_1 = X_1 - A_1^T E_1^{-1} Z_1 \tag{12.55}$$

This gives the resultant of all inertia and external forces/torques due to gravity and aerodynamic drag on body 1 in basis 1:

$$\begin{Bmatrix} f^{*1} - f^{ext1} \\ t^{*1} - t^{ext1} \end{Bmatrix} \equiv \hat{M}_1 \begin{Bmatrix} {}^N a^0 \\ {}^N \alpha^1 \end{Bmatrix} + \hat{X}_1 \tag{12.56}$$

Finally, the resultant of the inertia and external forces for the two-body system in the dynamic system, with interaction forces/torques cancelled, is obtained by adding, for two bodies, Eqs. (12.53) and (12.56):

$$\left[W^T M_2 W + \hat{M}_1\right] \begin{Bmatrix} _N a^O \\ _N \alpha^1 \end{Bmatrix} + \left\{W^T X_2 + \hat{X}_1\right\} = 0 \qquad (12.57)$$

The solution of Eq. (12.57) for a^O, α^1 leads to the explicit statement of the kinematical differential equations, Eqs. (12.40) and (12.41) of the frame, and the vibration equation, Eq. (12.34) for the flexible rocket. This completes the block-diagonal algorithm for a flexible rocket with prescribed nozzle rotation.

12.5 NUMERICAL SIMULATION OF PLANAR MOTION OF A FLEXIBLE ROCKET

The block-diagonal algorithm is numerically implemented here just for a planar motion of a flexible rocket. For simplicity, the rocket body is represented by the continuum model, as against a discrete set of particles, of a *free-free beam* [15] (even though cantilevered to the aft motor dome where the nozzle connects to the main body). This makes the modal sums as integrals of the mode shape functions for each λ_i, or i-th root of the characteristic equation of a free-free beam of length L. *Only two modes* are kept in the simulation.

$$\phi_i(x) = \cosh(\lambda_i x / L) + \cos(\lambda_i x / L) - \sigma_i[\sinh(\lambda_i x / L) + \sin(\lambda_i x / L)]$$

$$i = 1,2$$

$$(12.58)$$

$$\sigma_i = \frac{\cosh\lambda_i - \cos\lambda_i}{\sinh\lambda_i - \sin\lambda_i} \quad i = 1,2 \qquad (12.59)$$

One can evaluate the following time-varying modal integrals:

$$m_i = \int_0^{m(t)} \phi_i^2 dm \quad i = 1,2 \ \left[\text{modal effective mass, used to select modes}\right] \quad (12.60)$$

$$b_{2i} = \int_0^{m(t)} \phi_i dm = 2m_i\sigma_i / \lambda_i \quad i = 1,2 \qquad (12.61)$$

$$c_i = \int_0^{m(t)} x\phi_i dm = 2m_i L / \lambda_i^2 \quad i = 1,2 \qquad (12.62)$$

Here, $m(t) = m_0 - \dot{m}t$ can be taken for simplicity, or by Hermite interpolation from the final mass m_f, at time t_f, where the instantaneous mass is interpolated as $m(t) = m_f\, f(t)$, as in Ref. [11]:

$$
f(t) = \left\{
\begin{aligned}
&\frac{m_0}{m_f} - \frac{\dot{m}_0\, t}{m_f} + \left(\frac{t}{t_f}\right)^2 \left[3 + 2\frac{\dot{m}_0\, t_f}{m_f} - 3\frac{m_0}{m_f}\right] \\
&- \left(\frac{t}{t_f}\right)^3 \left[2 + \frac{\dot{m}_0\, t_f}{m_f} - 2\frac{m_0}{m_f}\right]
\end{aligned}
\right\}
\tag{12.63}
$$

Before passing, it should be noted that one can also work with a mean value of the mass, and use modes for the rocket structure with this mean mass value [15]. Matrices A_1, E_1, M_1, X_1, given by Eqs. (12.27), (12.35), (12.26), and (12.28), are as follows:

$$
A_1 = \begin{bmatrix} 0 & b_{21} & g_1 \\ 0 & b_{22} & g_2 \end{bmatrix}
\tag{12.64}
$$

$$
E_1 = \begin{bmatrix} m_1 & 0 \\ 0 & m_1 \end{bmatrix}
\tag{12.65}
$$

$$
M_1 = \begin{bmatrix} m_1 & 0 & -s_2 \\ 0 & m_1 & s_1 \\ -s_2 & s_1 & J_1 \end{bmatrix}
\tag{12.66}
$$

(J_1 is the moment of inertia of the rocket about an axis perpendicular to the plane in planar rotation.)

$$
X_1 = \left\{
\begin{aligned}
&-\dot{\theta}_1^2 s_1 - 2\dot{\theta}_1 \sum_{i=1}^{2} b_i \dot{q}_i - \dot{m}v_e \cos\theta_2 - f^{ext1} \\
&-\dot{\theta}_1^2 s_2 - \dot{m}v_e \sin\theta_2 - f^{ext2} \\
&\dot{m}v_e \left[\sum_{i=1}^{2} \phi_i\left(\frac{L}{2}\right)q_i \cos\theta_2 - \frac{L}{2}\sin\theta_2\right] - \tilde{r}_c f^{ext1}
\end{aligned}
\right\}
\tag{12.67}
$$

Here, L is the length of the rocket body, and gimbal angles, θ_1, θ_2, Eq. (12.25) defines

$$
\left\{\begin{matrix} s_1 \\ s_2 \end{matrix}\right\} = \left\{\begin{matrix} m_1 L/2 \\ b_{21}q_1 + b_{22}q_2 \end{matrix}\right\}
\tag{12.68}
$$

The two components of the external force and torque due to aerodynamics and gravity are expressed in the rocket body frame as

$$
\begin{Bmatrix} f^{ext2} \\ f^{ext2} \\ t^{ext} \end{Bmatrix} = 0.5\rho(\dot{x}^2 + \dot{y}^2)A_s \begin{Bmatrix} C_x \\ C_y \\ L_m C_m + 0.0212 L\alpha \end{Bmatrix} - m_1 g \begin{Bmatrix} \sin\theta_1 \\ \cos\theta_1 \\ 0.5L\cos\theta_1 \end{Bmatrix} \tag{12.69}
$$

Here, A_s, L_m, C_x, C_y, C_m, are the aerodynamic surface area, the characteristic length for moment, and three aerodynamic coefficients, respectively, given as functions of the angle of attack α. Matrix Z_l is given by Eq. (12.36), incorporating Eq. (12.19):

$$
Z_l = \begin{Bmatrix} m_1(\omega_1^2 - \dot{\theta}_1^2)q_1 \\ m_2(\omega_2^2 - \dot{\theta}_2^2)q_2 \end{Bmatrix} + \frac{6\dot{m}V_e}{5L} \begin{Bmatrix} [\phi_1(L)]^2 q_1 + \phi_1(L)\phi_2(L)q_2 \\ [\phi_2(L)]^2 q_2 + \phi_1(L)\phi_2(L)q_1 \end{Bmatrix}
$$
$$
- f_{ext}^{(2)} \begin{Bmatrix} \phi_1(L/2) \\ \phi_2(L/2) \end{Bmatrix} \tag{12.70}
$$

Here, the second column matrix, involving the multiplier $\dot{m}$ represents the geometric stiffness effect due to axial thrust on a cantilever beam [13]; with $\dot{m} < 0$, "stiffness" actually becomes softness in this context. For the nozzle oriented as in Figure 12.1, matrices M_2, X_2 follow from Eqs. (12.21) and (12.23) for rigid bodies:

$$
M_2 = \begin{bmatrix} m_2 & 0 & 0 \\ 0 & m_2 & -m_2 r_2 \\ 0 & -m_2 r_2 & J_2 \end{bmatrix} \tag{12.71}
$$

$$
X_2 = \begin{Bmatrix} 0 \\ -m_2 r_2 \\ J_2 \end{Bmatrix} \ddot{\theta}_2 + \begin{Bmatrix} m_2(\dot{\theta}_1 + \dot{\theta}_2)^2 r_2 + m_2 g \sin(\theta_1 + \theta_2) \\ m_2 g \cos(\theta_1 + \theta_2) \\ -m_2 g r_2 \cos(\theta_1 + \theta_2) \end{Bmatrix} \tag{12.72}
$$

(J_2 is the moment of inertia of an engine about the axis perpendicular to the plane of rotation.) The direction cosine matrix of Eq. (12.48), referring to Figure 12.1, becomes

$$
W = \begin{bmatrix} \cos\theta_2 & \sin\theta_2 & 0 \\ -\sin\theta_2 & \cos\theta_2 & 0 \\ 0 & 0 & 1 \end{bmatrix} \tag{12.73}
$$

Open-loop control in the form of prescribed gimbal motion for θ_2 of the nozzle is taken as follows, with the engine shutting off at $t = 60$:

$$
\theta_2 = 2.618 \times 10^{-3} t \quad t < 10
$$
$$
\theta_2 = 2.618 \times 10^{-3} t \left[1 - \frac{t-10}{60} \right] + 5.236 \times 10^{-3} \sin\frac{\pi(t-10)}{30} \quad 10 < t < 70 \tag{12.74}
$$

Now, the matrices in Eqs. (12.54) and (12.55) can be formed. Equation (12.53) is set up with two acceleration components, where

$$
\left\{ \begin{matrix} a^o \\ \alpha^1 \end{matrix} \right\} = \left\{ \begin{matrix} \ddot{x}_1 - \dot{y}_1\dot{\theta}_1 \\ \ddot{y}_1 + \dot{x}_1\dot{\theta}_1 \\ \ddot{\theta}_1 \end{matrix} \right\} \tag{12.75}
$$

Finally, the gimbal torque required to realize the prescribed gimbaled-engine motion is obtained by reducing Eq. (12A.6) given in the Appendix:

$$
\tau_c = -J_2\ddot{\theta}_2 - \begin{bmatrix} 0 & -m_2r_2 & J_2 \end{bmatrix} W \left\{ \begin{matrix} a^o \\ \alpha^1 \end{matrix} \right\} + m_2gr_2\cos(\theta_1 + \theta_2) \tag{12.76}
$$

The numerical simulation was carried out with the following data for an actual system:

$m_1 = 4400\ \text{slug}, m_2 = 90\ \text{slug}, L = 60\ \text{ft}, \dot{m} = -39\ \text{slug/sec},$

$v_e = -I_{sp}g; I_{sp} = 267; J_2 = 200\ \text{slug}-\text{ft}^2, r_2 = 0.5\ \text{ft}, Q = 2425\ \text{lb/ft}^2,$

$A_s = 46\ \text{ft}^2, L_m 21.4\ \text{ft}; \text{beam EI} = 1.14\text{x}10^9\ \text{lb}-\text{ft}^2,$

$\Omega_1 = 1.05\ \text{Hz}, \Omega_2 = 6.6\ \text{Hz};$ \hfill (12.77)

aerodynamic coefficients :

$C_x = 0.0115\alpha - 0.420, C_y = -0.0425\alpha, C_m = 0.149\alpha,$

$\rho = .002377\exp(-4.15\text{x}10^{-5}\ y_i)$

Three separate simulations were done: for a rigid rocket, for a rocket with low flexibility (36 times the value of the EI factor quoted above), and for actual flexibility. The results are shown in Figures 12.2–12.9, following the gimbal angle prescribed motion of Figure 12.2. Figures 12.3 and 12.4 show the differences in heights reached and the horizontal distance, with implications on possible landing differences. The pitch angle history in Figure 12.5 shows that a gyro placed on a flexible rocket would read a different value of the pitch from that of a rigid rocket, and thus requires control differently. How bending deformations vary along the length of the rocket are shown in Figure 12.6; note that deflection is negative corresponding to a positive nozzle angle, which is an inertia effect. Figures 12.7 and 12.8 show the effect on the natural frequency of the thrust, highlighting the *softening effect of thrust*. Naturally, frequencies increase with mass reduction, with our assumption that structural stiffness remains unchanged. Other results are also relevant and interesting. Figure 12.9 shows the torque required to produce the prescribed gimbal rotation in Eq. (12.74), with a high frequency jitter reflected by the flexibility of the rocket body.

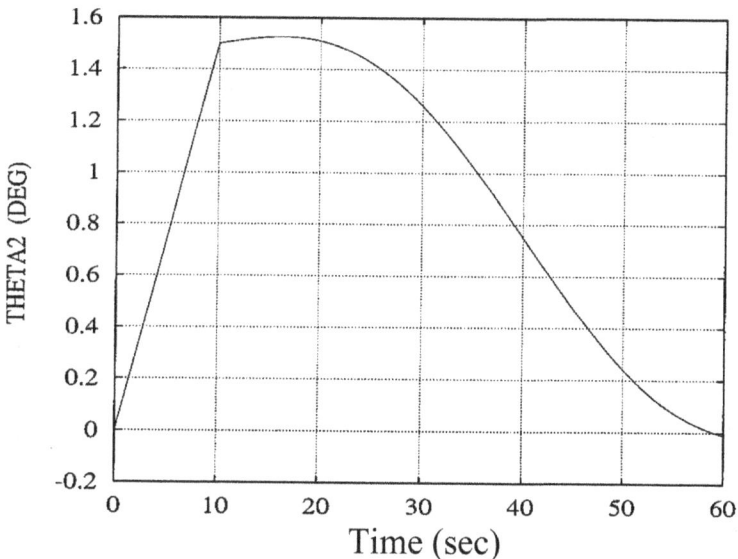

FIGURE 12.2   Prescribed motion of rocket nozzle angle.

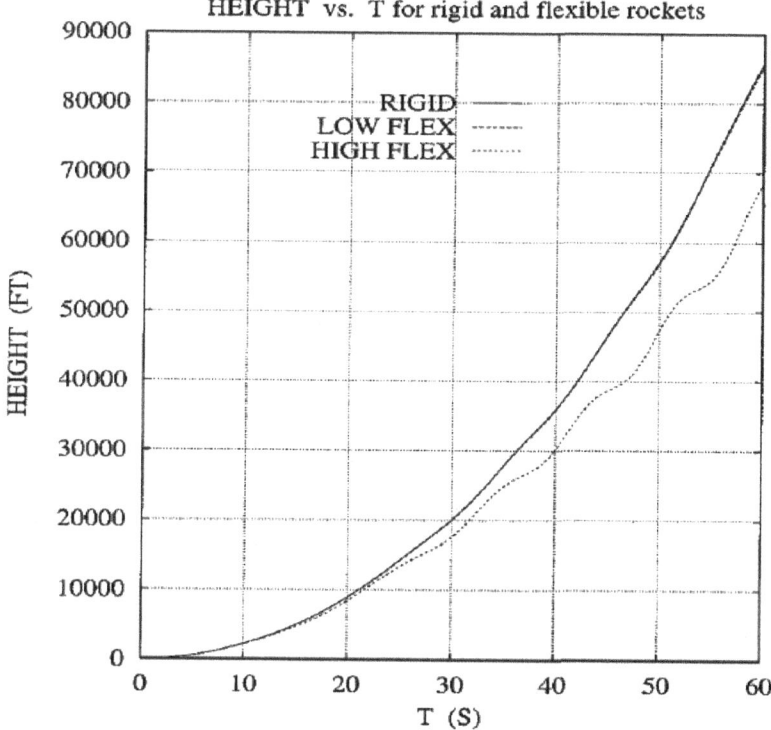

FIGURE 12.3   Rocket altitude vs. time for three models.

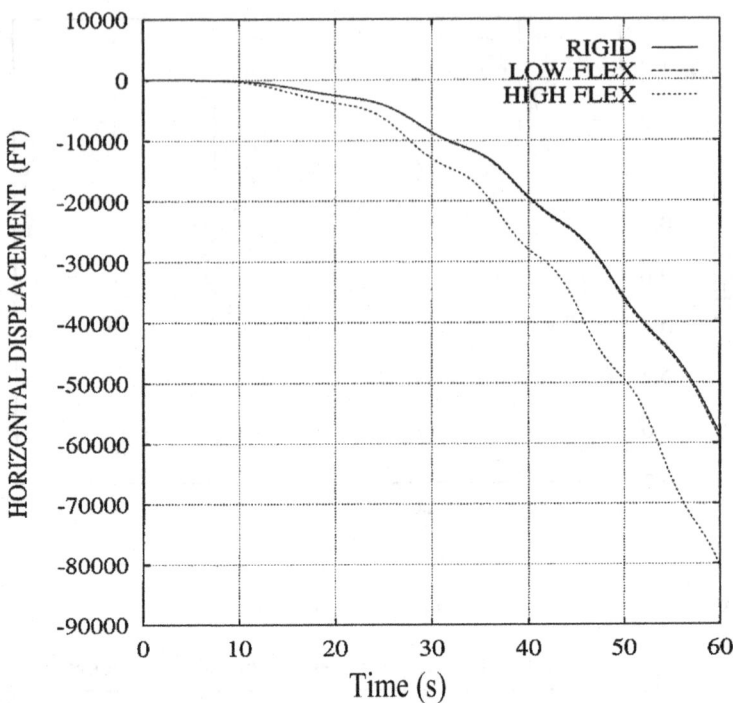

FIGURE 12.4 Rocket horizontal displacement vs. time for three models.

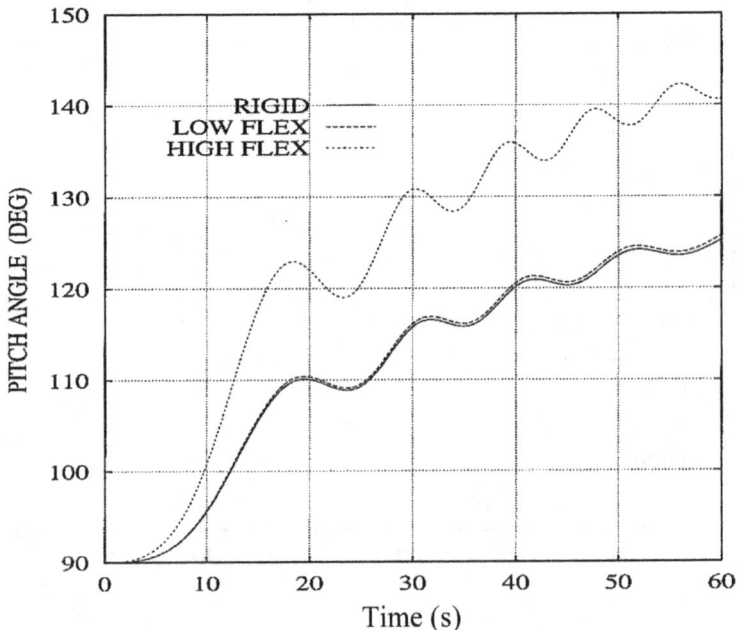

FIGURE 12.5 Rocket pitch angle vs. time for three models.

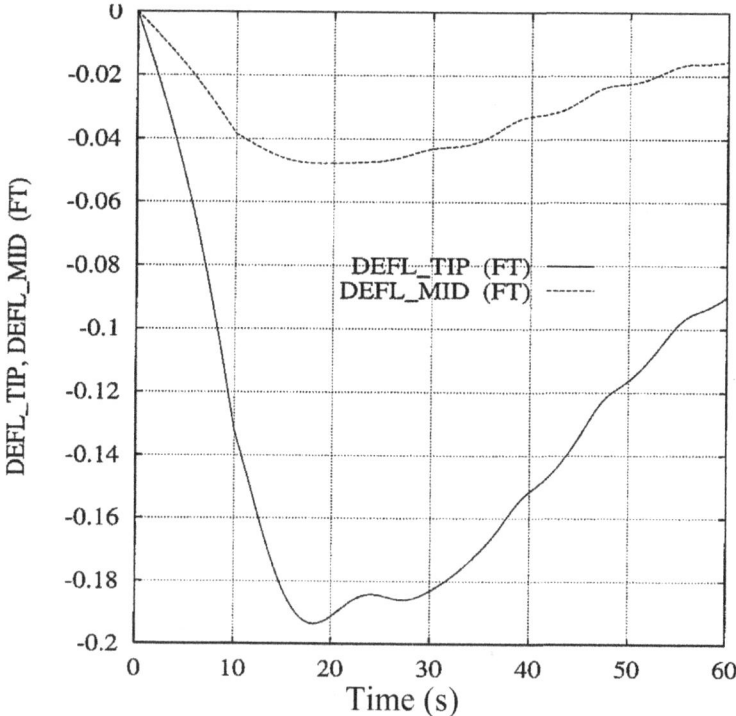

FIGURE 12.6 Tip and mid-point deflection of the high-flexibility rocket model.

12.6 CONCLUSION

The results show the substantial effect of flexibility on flight behavior for control of a rocket, with the nozzle angle time history prescribed as realizable by a high gain control system. In particular, Figures 12.3–12.5 show that the horizontal and vertical motion and the pitch angle are quite different for a flexible body model from those for a rigid body model. The fidelity of the rocket trajectory and the desired impact point thus demand the need in rocket control simulations for a model of flexible body with mass loss. The order n algorithm lends itself to a computationally efficient on-line non-linear dynamics model for control design of the rocket.

ACKNOWLEDGMENT

The author is grateful to Dr C.J. Chang and Andrew Imbrie of Lockheed Missiles and Space Company, Sunnyvale, California, for providing the data for an actual rocket test flight. The results shown here were reported to have compared quite well with the output of a low flexibility test rocket flight measurements. I am also grateful to my colleague, David Levinson, who showed me how to use his code, Autolev, to obtain results by zeroing out the flexibility terms. Finally thanks are due to my long-time friend, Dr Tushar Ghosh of L3 Communications for information in Ref. [11] about the practice in updating modes.

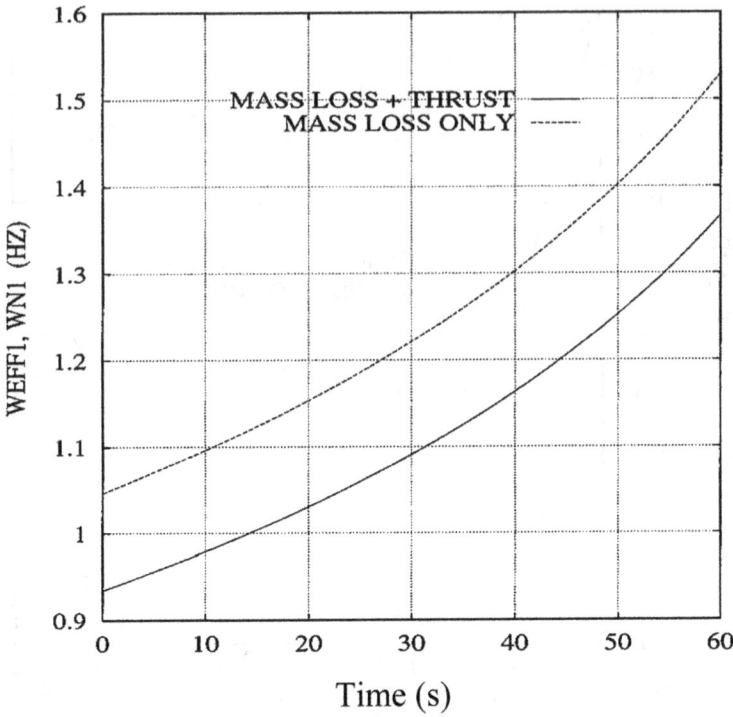

FIGURE 12.7 First mode frequency vs. time for high flexibility rocket with/without thrust.

PROBLEM SET 12

When a non-spinning rocket rotates about a transverse axis, the ejected gas gives it a reaction torque that is opposite to the direction of rotation. If the moment of inertia of the rocket about its mass center is mk^2, and if the external torque is zero, show by equating torque to the rate of change of angular momentum that the angular velocity of the rocket is given by $\omega = \omega_0 [m/m_0]^{(L^2/k^2-1)}$, where ω_0, m_0, L, and k are, respectively, the initial angular velocity, initial mass, length from mass center to nozzle exit, and radius of gyration. This is a simple analysis of jet damping for a rigid rocket.

1. Simulate the motion of a variable mass *single* flexible body, for data in Eq. (12.72).
2. Assume that the rocket is a variable mass single rigid body, simulate motion for the data in Eq. (7.2), and overlay the results against those in the Example Problem.

Recall that Newton's law is strictly, $\mathbf{F} = \mathbf{ma}$, for *constant mass*. In Ref. [5], using Newton's law, Kane shows that the rocket equation considering the mass inside

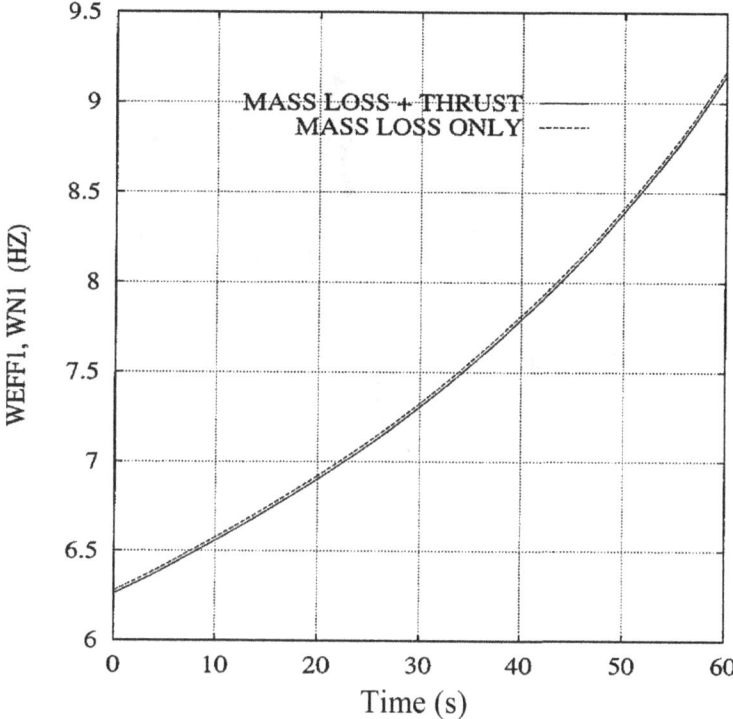

FIGURE 12.8 First mode frequency vs. time for low flexibility rocket with/without thrust.

and ejected, for a body filled with a liquid propellant, modeled as a continuum, is given by $\mathbf{F} = (M + m_0 - \mu t)^R \mathbf{a}^{P*} + \mu \mathbf{v}_e$, where $\mathbf{F}$ is the external force on the rocket body, M is the mass of the rocket body, m_0 is the mass of the propellant at t = 0, μ is the mass flow rate, and $^R\mathbf{a}^{P*}, \mathbf{v}_e$ are the acceleration of the rocket body and the exit velocity of the ejected mass. Derive this equation of motion.

APPENDIX 12 ALGORITHM FOR DETERMINING TWO GIMBAL ANGLE TORQUES FOR THE NOZZLE FOR THE EXAMPLE PROBLEM

A summary of the block-diagonal algorithm for a rocket driven by two gimbal angle torques τ_1, τ_2 driving two gimbal degrees of freedom in pitch-yaw nozzle angles θ_1, θ_2 is given below.

1. Compute M_1, X_1 from Eqs. (12.66) and (12.67); M_2, X_2 from Eqs. (12.71) and (12.72); and Z_1 from Eq. (12.70).
2. Introduce in terms of Eq. (12.43) the 6 × 2 partial velocity matrix:

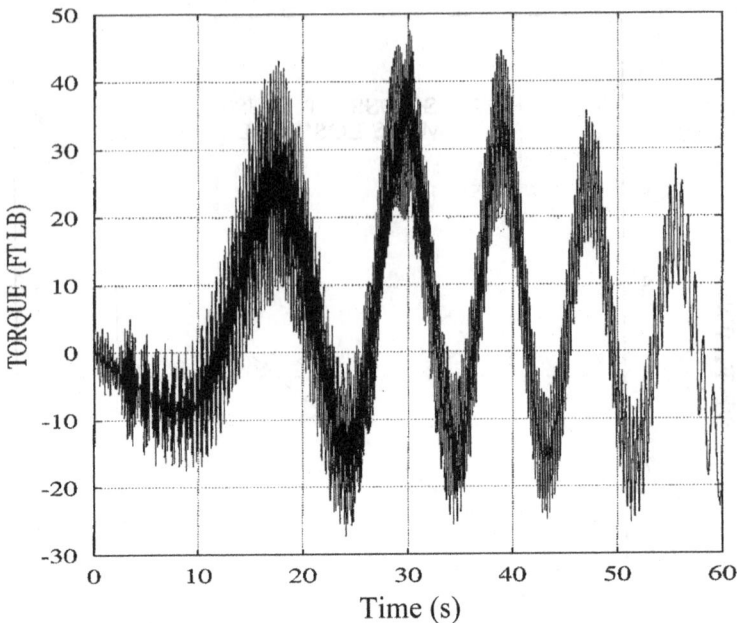

FIGURE 12.9 Gimbal torque for prescribed nozzle motion for a high flexibility rocket.

$$\Pi = \begin{bmatrix} 0 \\ R \end{bmatrix} \qquad (12A.1)$$

and define for the inertia matrix I_2 of the gimbal:

$$\mu = R^t I_2 R \qquad (12A.2)$$

3. Compute for a 6 × 6 unity matrix U:

$$M = [U - M_2\Pi\mu^{-1}\Pi^T]M_2 \qquad (12A.3)$$

$$X = [U - M_2\Pi\mu^{-1}\Pi^T]X_2 - M_2\Pi\mu^{-1}\begin{Bmatrix} \tau_1 \\ \tau_2 \end{Bmatrix} \qquad (12A.4)$$

4. Solve the linear algebraic equation, Eq. (12.57):

$$\left[W^T M W + \hat{M}_1\right]\begin{Bmatrix} a^o \\ \alpha^1 \end{Bmatrix} + \left\{W^t X + \hat{X}_1\right\} = 0 \qquad (12A.5)$$

5. The gimbal dynamics equations are obtained, following the trail from Eqs. (12.45), (12.43), (12.44), (12.42), and (12.46):

$$\begin{Bmatrix} \ddot{\theta}_1 \\ \ddot{\theta}_2 \end{Bmatrix} = -\mu^{-1} \left\{ \Pi^T \left[M_2 W \begin{pmatrix} a^0 \\ \alpha^1 \end{pmatrix} + X_2 \right] - \begin{Bmatrix} \tau_1 \\ \tau_2 \end{Bmatrix} \right\} \tag{12A.6}$$

6. The rocket vibration equations are given by Eq. (12.52).

REFERENCES

1. Thomson, W.T., (1963), *Introduction to Space Flight*, Wiley, pp. 230–236.
2. Meirovitch, L., (1970), "General Motion of a Variable Mass Flexible Rocket with Internal Flow", *Journal of Spacecraft and Rockets*, **7**(2), pp. 186–195.
3. Eke, F.O. and Wang, S.-M., (Dec., 1994), "Equations of Motion of Two-Phase Variable Mass Systems with Solid Base", *Journal of Applied Mechanics*, **61**(4), 855–860.
4. Joshi, A., (1995), "Free Vibration Characteristics of Variable Mass Rockets Having Large Axial Thrust/Acceleration", *Journal of Sound and Vibration*, **187**(4), pp. 727–736.
5. Kane, T.R., (1961), "Variable Mass Dynamics?", *Bulletin of Mechanical Engineering Education*, **2**(20), pp. 62–65.
6. Ge, Z.M. and Cheng, Y.H., (June, 1982), "Extended Kane's Equations for Nonholonomic Variable Mass System", *Journal of Applied Mechanics*, **49**(2), pp. 429–431.
7. Banerjee, A.K., (2000) "Dynamics of a Variable-Mass, Flexible-Body System", *Journal of Guidance, Control, and Dynamics*, **23**(3), pp. 501–508.
8. Banerjee, A.K. and Lemak, M.E., (2008), "Dynamics of a Flexible Body with Rapid Mass Loss", in Proceedings of the International Congress on Theoretical and Applied Mechanics, Australia.
9. Banerjee, A.K. and Dickens, J.M., (1990), "Dynamics of an Arbitrary Flexible Body in Large Rotation and Translation", *Journal of Guidance, Control, and Dynamics*, **13**(6), pp. 221–227.
10. Craig, R.R., Jr., (1981), *Structural Dynamics: An Introduction to Computer Methods*, Wiley.
11. Quiocho, L.J., Ghosh, T.K., Frenkel, D., and Huynh, A., (2010), "Mode Selection Techniques in Variable Mass Flexible Body Modeling", in AIAA Modeling and Simulation Technologies Conference, 2–5 August, Toronto, Canada. Paper No. AIAA 2010-7607.
12. Kane, T.R., and Levinson, D.A., (1985), Dynamics: Theory and Applications, McGraw-Hill.
13. Peraire, J. and Widnall, S., (Fall, 2008), "Variable Mass Systems: The rocket Equation", in *MIT Lecture Notes: L14*, Version 2.0, Unpublished Class-Notes.
14. Cook, R.D., (1985), *Concepts and Applications of Finite Element Analysis*, McGraw-Hill, pp. 331–341.
15. Blevins, R.D., (1979), *Formulas for Natural Frequency and Mode Shape*, Krieger Publishing Co.

13 Large Amplitude Fuel Slosh in Spacecraft in Large Overall Motion

This chapter is based on Ref. [1], analyzing large amplitude fuel slosh in a flexible spacecraft undergoing large overall motion. Fuel slosh is a disturbance to the attitude control or orbit transfer of a spacecraft, and the problem becomes severe when multiple tanks are used. The control design itself must take into account the rigid–elastic coupling of the overall spacecraft, with the frequencies and "modes" of small amplitude fuel slosh. Much work has been done on these topics. While Refs. [2] and [3] are general overviews, Refs. [4–7] are samples of work dealing particularly with the problem of fuel slosh. Of course, the classic reference for treating small amplitude fuel slosh as an offset pendulum is Ref. [7]. Here, we not only consider the large overall motion of a spacecraft with large amplitude fuel slosh, but also determine the force imparted by the fuel on the wall of the tank. Kane's method [8] with undetermined multipliers [9] is used to develop the equations of motion and find out the constraining force. To put the entire problem in context, we consider a spacecraft equipped with three reaction wheels, two flexible solar panels, and four fuel tanks. Simulation results are given for various levels of fill-fraction of the fuel in the tanks. An important part of the work, not reported in Ref. [1], is the derivation of the linearized dynamical equations for small amplitude fuel slosh in a tank; this is useful for control design, and is presented in the Appendix of this chapter.

13.1 MODELING LARGE AMPLITUDE SLOSHING FUEL AS PARTICLES CRAWLING ON WALLS OF FOUR TANKS IN A SPACECRAFT WITH FLEXIBLE SOLAR PANELS

Consider the physical model sketched in Figure 13.1. The spacecraft system consists of a rigid hub, three-axis reaction control wheels, four fuel tanks, and two rigidly attached rectangular flexible solar panels. We define six generalized speeds for translation of the mass center B* of the spacecraft B, and the rotation of B, in the body basis $\mathbf{b}_1$, $\mathbf{b}_2$, $\mathbf{b}_3$ in the Newtonian reference frame, N:

$$u_i = {}^N\mathbf{v}^{B*} \cdot \mathbf{b}_i \quad i = 1,2,3 \tag{13.1}$$

$$u_{3+i} = {}^N\boldsymbol{\omega}^B \cdot \mathbf{b}_i \quad i = 1,2,3 \tag{13.2}$$

DOI: 10.1201/9781003231523-14

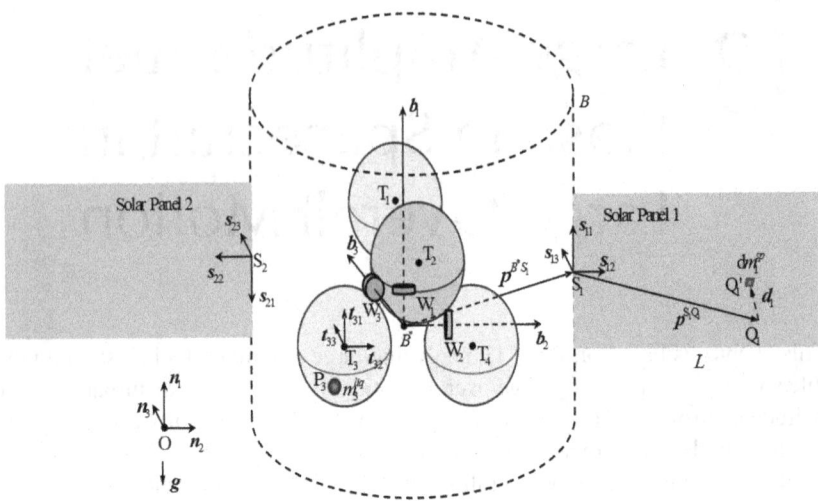

FIGURE 13.1 Sketch of a flexible spacecraft with four fuel tanks, three reaction wheels, and two solar panels.

These imply the following:

$$^N\mathbf{v}^{B*} = u_1\mathbf{b}_1 + u_2\mathbf{b}_2 + u_3\mathbf{b}_3 \tag{13.3}$$

$$^N\boldsymbol{\omega}^B = u_4\mathbf{b}_1 + u_5\mathbf{b}_2 + u_6\mathbf{b}_3 \tag{13.4}$$

The generalized speeds for the three-axis reaction wheels and their angular velocities are

$$u_{6+i} = {}^B\boldsymbol{\omega}^{W_i} \cdot \mathbf{b}_i \quad i = 1, 2, 3 \tag{13.5}$$

$$^N\boldsymbol{\omega}^{W_1} = {}^N\boldsymbol{\omega}^B + {}^B\boldsymbol{\omega}^{W_1} = (u_4 + u_7)\mathbf{b}_1 + u_5\mathbf{b}_2 + u_6\mathbf{b}_3 \tag{13.6}$$

$$^N\boldsymbol{\omega}^{W_2} = {}^N\boldsymbol{\omega}^B + {}^B\boldsymbol{\omega}^{W_2} = u_4\mathbf{b}_1 + (u_5 + u_8)\mathbf{b}_2 + u_6\mathbf{b}_3 \tag{13.7}$$

$$^N\boldsymbol{\omega}^{W_3} = {}^N\boldsymbol{\omega}^B + {}^B\boldsymbol{\omega}^{W_3} = u_4\mathbf{b}_1 + u_5\mathbf{b}_2 + (u_6 + u_9)\mathbf{b}_3 \tag{13.8}$$

The sloshing fuel in one tank is modeled as a particle of lumped mass crawling on the inside of a tank wall. Refer to Figure 13.1. Let the position vector of the mass center of the k-th tank T_k^* from the spacecraft body mass center B*, and that of the k-th fuel particle P_k from T_k^* be expressed for (k = 1,2,3,4) as follows, where O is the origin of the inertial frame N:

$$\mathbf{p}^{OP_k} = \mathbf{p}^{OB*} + \mathbf{p}^{B*T_k^*} + \mathbf{p}^{T_k^*P_k}$$

$$\mathbf{p}^{B*T_k^*} = X_k\mathbf{b}_1 + Y_k\mathbf{b}_2 + Z_k\mathbf{b}_3 \quad k = 1, 2, 3, 4 \tag{13.9}$$

$$\mathbf{p}^{T_k^*P_k} = x_k\mathbf{b}_1 + y_k\mathbf{b}_2 + z_k\mathbf{b}_3$$

To compute the force imparted by the fuel on the tank wall, we introduce additional degrees of freedom for the fuel "particle", P_k. Thus, we introduce three generalized speeds for the four x-, y-, z-motion of each particle P_k, $k = 1,2,3,4$; the idea is to use these extra generalized speeds to determine the force on the tank when these degrees of freedom are constrained out to be zero. Derivatives of the position vectors from the tank center T_k^* to the particles P_k in the spacecraft frame B define the generalized speeds:

$$u_{6+3k+i} = \frac{^B d\mathbf{p}^{T_k^* P_k}}{dt} \cdot \mathbf{b}_i = {}^B \mathbf{v}^{P_k} \cdot \mathbf{b}_i \qquad k = 1,2,3,4; \ i = 1,2,3 \quad (13.10a)$$

This means

$$^B \mathbf{v}^{P_k} = u_{7+3k}\mathbf{b}_1 + u_{8+3k}\mathbf{b}_2 + u_{9+3k}\mathbf{b}_3 \qquad k = 1,2,3,4 \quad (13.10b)$$

Thus, 12 generalized speeds, $u_{10}, \ldots, u_{21}$ (four of which will be eliminated by constraints), are assigned to the three-dimensional motions of the mass particles representing fuel slosh in the four tanks. The velocity of P_k in N is expressed by the theorem of a particle moving on a moving body [8], where $^N \mathbf{v}^{\bar{P}_k}$ is the velocity of the "image" point $\bar{P}_k$ on the body B which is instantaneously coincident with the moving particle, P_k.

$$^N \mathbf{v}^{P_k} = {}^N \mathbf{v}^{\bar{P}_k} + {}^B \mathbf{v}^{P_k}$$

$$= {}^N \mathbf{v}^{B^*} + {}^N \boldsymbol{\omega}^B \times (\mathbf{p}^{B^* T_k} + \mathbf{p}^{T_k P_k}) + {}^B \mathbf{v}^{P_k}$$

$$= \mathbf{b}_1[u_2 + u_5(Z_k + z_k) - u_6(Y_k + y_k) + u_{7+3k}] \qquad (13.11)$$

$$= \mathbf{b}_2[u_2 + u_6(X_k + x_k) - u_4(Z_k + z_k) + u_{8+3k}]$$

$$= \mathbf{b}_3[u_3 + u_4(Y_k + y_k) - u_5(X_k + x_k) + u_{9+3k}] \qquad k = 1,2,3,4$$

Let the transverse deformation at a generic point G on a solar panel be given by ν number of vibration modes:

$$\delta^G = \sum_{i=1}^{\nu} \varphi_i^G \eta_i \quad (13.12)$$

Suppose we take $\nu = 3$, i.e., three modes per solar panel (allowing each panel to respond differently), and define the modal generalized speeds, for the panels P1, P2,...:

$$u_{21+i} = \dot{\eta}_i^{P1} \quad i = 1,2,3$$

$$u_{24+i} = \dot{\eta}_i^{P2} \quad i = 1,2,3 \qquad (13.13)$$

This defines a 27 degrees of freedom system. The velocities of the attachment points of the solar panels to the bus, and generic points G of P1, P2, are

$$^N\mathbf{v}^{S1} = {^N}\mathbf{v}^{B*} + {^N}\boldsymbol{\omega}^B \times \mathbf{p}^{B*S_1}$$

$$^N\mathbf{v}^{G/P1} = {^N}\mathbf{v}^{S1} + {^N}\boldsymbol{\omega}^B \times \left[\mathbf{p}^{S_1\bar{G}/P1} + \sum_{i=1}^{\nu} \varphi_i^{P1}\eta_i^{P1} \right] + \sum_{i=1}^{\nu} \varphi_i^{P1} u_{21+i}$$

$$^N\mathbf{v}^{S2} = {^N}\mathbf{v}^{B*} + {^N}\boldsymbol{\omega}^B \times \mathbf{p}^{B*S_2}$$ (13.14)

$$^N\mathbf{v}^{G/P2} = {^N}\mathbf{v}^{S2} + {^N}\boldsymbol{\omega}^B \times \left[\mathbf{p}^{S_2\bar{G}/P2} + \sum_{i=1}^{\nu} \varphi_i^{P2}\eta_i^{P2}) \right] + \sum_{i=1}^{\nu} \varphi_i^{P2} u_{24+i}$$

Kane's method of formulating the equations of motion requires forming the partial velocities in N of the spacecraft mass center B*, generic points G/P1, G/P2 of the two solar panels, and the four lumped-mass particles, P_1, P_2, P_3, P_4 representing the fuels in the tanks, and the partial angular velocities in N of B, W1, W2, W3. Then, we find the accelerations of B*, G/P1, G/P2, P_1, P_2, P_3, P_4 in N and the angular accelerations in N of the bodies B, W1, W2, W3, and the reference frames of the solar panels.

Generalized inertia forces, defined by Kane [8], are formed for this system with $\nu = 3$:

$$
F_r^* = -\left\langle
\begin{array}{l}
^N\mathbf{v}_r^{B*} \cdot (m^B\,{}^N\mathbf{a}^{B*}) + {}^N\boldsymbol{\omega}_r^B \cdot [\mathbf{I}^{B/B*} \cdot {}^N\boldsymbol{\alpha}^B + {}^N\boldsymbol{\omega}^B \times (\mathbf{I}^{B/B*} \cdot {}^N\boldsymbol{\omega}^B)] \\[2ex]
+ \displaystyle\sum_{i=1}^{3} {}^N\boldsymbol{\omega}_r^{W_i} \cdot [\mathbf{I}^{W_i/W_i*} \cdot {}^N\boldsymbol{\alpha}^{W_i} + {}^N\boldsymbol{\omega}^{W_i} \times (\mathbf{I}^{W_i/W_i*} \cdot {}^N\boldsymbol{\omega}^{W_i})] \\[2ex]
+ \displaystyle\sum_{k=1}^{4} {}^N\mathbf{v}_r^{P_k} \cdot (m\,{}^N\mathbf{a}^{P_k}) \\[2ex]
+ \displaystyle\int_{P1} \left[\begin{array}{l} {}^N\mathbf{v}_r^{S1} + {}^N\boldsymbol{\omega}_r^B \times \left(\mathbf{p}^{S_1\bar{G}/P1} + \displaystyle\sum_{i=1}^{\nu} \varphi_i^{\bar{G}/P1}\eta_i^{P1} \right) \\[2ex] + \displaystyle\sum_{i=1}^{\nu} \varphi_i^{\bar{G}/P1}\delta_{r,21+i} \end{array} \right] \cdot {}^N\mathbf{a}^{G/P1} dm \\[4ex]
+ \displaystyle\int_{P2} \left[\begin{array}{l} {}^N\mathbf{v}_r^{S2} + {}^N\boldsymbol{\omega}_r^B \times \left(\mathbf{p}^{S_2\bar{G}/P2} + \displaystyle\sum_{i=1}^{\nu} \varphi_i^{\bar{G}/P2}\eta_i^{P2} \right) \\[2ex] + \displaystyle\sum_{i=1}^{\nu} \varphi_i^{\bar{G}/P2}\delta_{r,24+i} \end{array} \right] \cdot {}^N\mathbf{a}^{G/P2} dm
\end{array}
\right\rangle
$$

(13.15)

Note that the two integrals in Eq. (13.15) give rise to the so-called modal integrals, needed for any flexible body, defined in the Appendix at the end of this book. This work involves expressing the accelerations of G/P1, G/P2, and taking their dot-products with the partial velocities of G/P1, G/P2. Inescapably, the work involved in carrying out the operations required in Eq. (13.15) is massive, as has been shown several times in the book. Evaluation of the terms in Eq. (13.15) is skipped here for the sake of brevity in this chapter, but, of course, this had to be done and coded in a computer program to get the simulation results reported here.

13.1.1 GENERALIZED ACTIVE FORCES

The generalized force contributions for the r-th generalized speed are attributable to gravity on body B at B* (with lumped mass for the wheels and the four slosh particles) and the two solar panels, the reaction wheel torques on B from the wheels due to the relative angular velocity of the three wheels with respect to the body B, four sets of constraint forces (normal force and friction forces) applied by body B on the particles P_1, P_2, P_3, P_4 representing the lumped fuel in the four tanks, and the structural stiffness of the two solar panels. The spacecraft in orbit does not involve rapid motions in translation and rotation, and thus a customary analysis [11] can be used, neglecting any motion-induced stiffness [12]. Generalized active forces due to gravity, contact force $\mathbf{F}_c^{B/P_k}$ between the tank and the fuel particle P_i, and elasticity of the solar panels for the system are evaluated as

$$F_r = {}^N\mathbf{v}_r^{B^*} \cdot \mathbf{F}_g^{B^*} + \sum_{i=1}^{3} {}^B\boldsymbol{\omega}_r^{W_i} \cdot \mathbf{T}^{B/W_i} + \sum_{k=1}^{4} [{}^N\mathbf{v}_r^{P_k} \cdot \mathbf{F}_g^{P_k} + {}^B\mathbf{v}_r^{P_k} \cdot \mathbf{F}_c^{B/P_k}]$$

$$+ \int_{P1} {}^N\mathbf{v}_r^{G/P1} \cdot d\mathbf{F}_g^{G/P1} + + \int_{P2} {}^N\mathbf{v}_r^{G/P2} \cdot d\mathbf{F}_g^{G/P2} - \sum_{i=1}^{3} [\Lambda_{i/P1}^2 \eta_i^{P1} \delta_{r,21+i} - \Lambda_{i/P2}^2 \eta_i^{P2} \delta_{r,24+i}]$$

$$r = 1,\ldots,27$$

$$(13.16)$$

13.2 GENERALIZED FORCE DUE TO NORMAL FORCES ENFORCING CONSTRAINED MOTION OF LUMPED-MASS PARTICLES CRAWLING ON AN ELLIPTICAL SURFACE

The set of four constraint forces $\mathbf{F}_c^{B/P_k}$ applied by the tank body B on the four lumped-mass particles representing fuel slosh are determined by the method of Kane's equations with undetermined multipliers [9]. The fuel modeled as a particle crawls on the wall of the fuel tank. It is constrained to move on the path

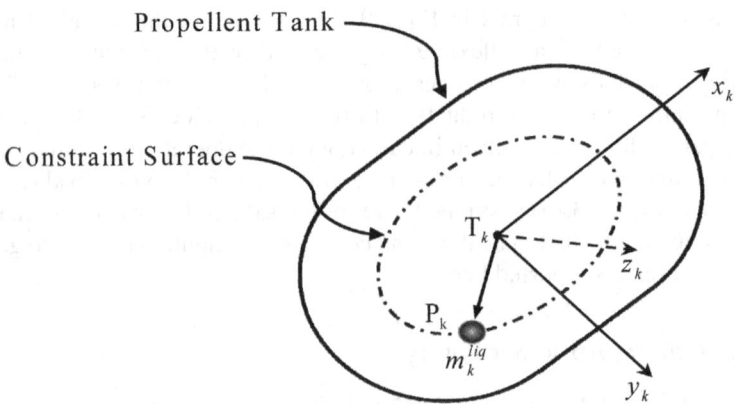

FIGURE 13.2 Sketch of the constraint surface parallel to the fuel tank wall, showing one particle representing the liquid fuel as a lumped mass, crawling on the inside surface of the tank.

described by the center-line of the mass center of the fluid slug. A conceptual sketch of the situation is shown in Figure 13.2.

For spherical and spherical-crown tanks, both of which are used in spacecraft, the tank surface, S_k, can be approximated as an ellipsoid with the geometry represented for the k-th tank, T_k, with the coordinates x_k, y_k, z_k, locating the k-th particle at time t, as follows:

$$S_k(x,y,z) = \frac{x_k^2}{a_k^2} + \frac{y_k^2 + z_k^2}{b_k^2} - 1 = 0 \tag{13.17}$$

For a spherical tank, $a_k = b_k$. As the k-th particle P_k is constrained to move on the constraint surface, that is, it cannot dig into the wall, for its x-, y-, z-motion, four sets of velocity constraints arise for k = 1,2,3,4 from the four tanks. From the extra generalized speeds introduced in Eq. (13.10), in order to free up the motion of the particles, the constraint equations are

$$\sum_{i=1}^{3} u_{6+3k+i} \mathbf{b}_i \cdot \mathbf{n}_k = 0 \quad i = 1,2,3; \quad k = 1,2,3,4 \tag{13.18}$$

Here, $\mathbf{n}_k$ is the unit vector along the outward normal to the surface, S_k, and its relation to the gradient vector ∇S_k of the surface at the points, P_k, is expressed as

$$\mathbf{n}_k = \frac{1}{2}\nabla S_k = \frac{x_k}{a_k^2}\mathbf{b}_1 + \frac{y_k}{b_k^2}\mathbf{b}_2 + \frac{z_k}{b_k^2}\mathbf{b}_3 \quad k = 1,2,3,4 \tag{13.19}$$

Define

$$\alpha_k = \frac{x_k}{a_k^2}; \quad \beta_k = \frac{y_k}{b_k^2}; \quad \gamma_k = \frac{z_k}{b_k^2} \quad k = 1,2,3,4 \tag{13.20}$$

Projections of the four outward normal vectors on the body basis triad, $\mathbf{b}_1$, $\mathbf{b}_2$, $\mathbf{b}_3$ are

$$\begin{Bmatrix} \mathbf{n}_1 \\ \mathbf{n}_2 \\ \mathbf{n}_3 \\ \mathbf{n}_4 \end{Bmatrix} = \begin{bmatrix} \alpha_1 & \beta_1 & \gamma_1 \\ \alpha_2 & \beta_2 & \gamma_2 \\ \alpha_3 & \beta_3 & \gamma_3 \\ \alpha_4 & \beta_4 & \gamma_4 \end{bmatrix} \begin{Bmatrix} \mathbf{b}_1 \\ \mathbf{b}_2 \\ \mathbf{b}_3 \end{Bmatrix} \tag{13.21}$$

With the help of Eq. (13.21), Eq. (13.18) can be seen as the four velocity constraint equations:

$$\alpha_1 u_{10} + \beta_1 u_{11} + \gamma_1 u_{12} = 0$$

$$\alpha_2 u_{13} + \beta_2 u_{14} + \gamma_2 u_{15} = 0$$

$$\alpha_3 u_{16} + \beta_3 u_{17} + \gamma_3 u_{18} = 0 \tag{13.22}$$

$$\alpha_4 u_{19} + \beta_4 u_{20} + \gamma_4 u_{21} = 0$$

Differentiating the terms in Eq. (13.22) with respect to time, yields the acceleration constraint equations, as the coordinates of the crawling particles change with time:

$$\alpha_1 \dot{u}_{10} + \beta_1 \dot{u}_{11} + \gamma_1 \dot{u}_{12} + \dot{\alpha}_1 u_{10} + \dot{\beta}_1 u_{11} + \dot{\gamma}_1 u_{12} = 0$$

$$\alpha_2 \dot{u}_{13} + \beta_2 \dot{u}_{14} + \gamma_2 \dot{u}_{15} + \dot{\alpha}_2 u_{13} + \dot{\beta}_2 u_{14} + \dot{\gamma}_2 u_{15} = 0$$

$$\alpha_3 \dot{u}_{16} + \beta_3 \dot{u}_{17} + \gamma_3 \dot{u}_{18} + \dot{\alpha}_3 u_{16} + \dot{\beta}_3 u_{17} + \dot{\gamma}_3 u_{18} = 0 \tag{13.23}$$

$$\alpha_4 \dot{u}_{19} + \beta_4 \dot{u}_{20} + \gamma_4 \dot{u}_{21} + \dot{\alpha}_4 u_{19} + \dot{\beta}_4 u_{20} + \dot{\gamma}_4 u_{21} = 0$$

Note that Eq. (13.20) yields

$$\dot{\alpha}_k = \frac{\dot{x}_k}{a_k^2}; \quad \dot{\beta}_k = \frac{\dot{y}_k}{b_k^2}; \quad \dot{\gamma}_k = \frac{\dot{z}_k}{b_k^2} \quad k = 1,2,3,4 \tag{13.24}$$

This acceleration form of the four constraint equations in Eq. (13.23), corresponding to the four k's, that is, the four tanks, can be written in the sparse matrix form shown below, where the zeros in the four rows are not shown.

$$\begin{bmatrix} \alpha_1 & \beta_1 & \gamma_1 & & & & & & & \\ & & & \alpha_2 & \beta_2 & \gamma_2 & & & & \\ & & & & & & \alpha_3 & \beta_3 & \gamma_3 & \\ & & & & & & & & & \alpha_4 & \beta_4 & \gamma_4 \end{bmatrix} \begin{pmatrix} \dot{u}_{10} \\ \dot{u}_{11} \\ \dot{u}_{12} \\ \dot{u}_{13} \\ \dot{u}_{14} \\ \dot{u}_{15} \\ \dot{u}_{16} \\ \dot{u}_{17} \\ \dot{u}_{18} \\ \dot{u}_{19} \\ \dot{u}_{20} \\ \dot{u}_{21} \end{pmatrix}$$

$$+ \begin{pmatrix} \dot{\alpha}_1 u_{10} + \dot{\beta}_1 u_{11} + \dot{\gamma}_1 u_{12} \\ \dot{\alpha}_2 u_{13} + \dot{\beta}_2 u_{14} + \dot{\gamma}_2 u_{15} \\ \dot{\alpha}_3 u_{16} + \dot{\beta}_3 u_{17} + \dot{\gamma}_3 u_{18} \\ \dot{\alpha}_4 u_{19} + \dot{\beta}_4 u_{20} + \dot{\gamma}_4 u_{21} \end{pmatrix} = \begin{pmatrix} 0 \\ 0 \\ 0 \\ 0 \end{pmatrix}$$

(13.25)

or, in shorthand notation:

$$[A]\{\dot{U}\} + \{B\} = \{0\} \tag{13.26}$$

Now we form the generalized active forces due to the physical forces that are required to enforce the kinematical constraints involving the generalized speeds in Eq. (13.18). Force at the contact point of the k-th particle with the tank wall is given along the negative of the outward gradient vector at the point, with λ_k as its magnitude, and a friction force opposite to the velocity of the particle relative to the tank can be assumed friction (due to fuel viscosity) with the coefficient of friction, μ:

$$\mathbf{F}^{B/P_k} = -\lambda_k \mathbf{n}_k - \sum_{i=1}^{3} \mu\,(u_{6+3k+i}\mathbf{b}_i) \quad k = 1,\dots,4 \tag{13.27}$$

The partial velocity vectors of P_k in B are obtained by inspection of Eq. (13.11) as

$$^{B}\mathbf{v}_{6+3k+i}^{P_k} = \mathbf{b}_i \quad i = 1,2,3 \tag{13.28}$$

The generalized active force due to the physical constraint forces [9] λ_k on the particle, P_k, representing the fuel in the k-th tank along with the friction forces, is given for this action–reaction of forces between wall and particles:

$$f_{6+3k+i} = \mathbf{F}^{B/P_k} \cdot {}^{B}\mathbf{v}^{P_k}_{6+3k+1} \quad i = 12,3; \quad k = 1,\dots,4 \tag{13.29}$$

By referring to the direction cosines in Eq. (13.21) and using Eqs. (13.27) and (13.28) in Eq. (13.29) one gets the generalized active forces corresponding to the generalized speeds $u_{10}, \dots, u_{21}$ in the matrix form shown below, where the zeros in the four columns are not shown:

$$
\begin{pmatrix} f_{10} \\ f_{11} \\ f_{12} \\ f_{13} \\ f_{14} \\ f_{15} \\ f_{16} \\ f_{17} \\ f_{18} \\ f_{19} \\ f_{20} \\ f_{21} \end{pmatrix} = -
\begin{bmatrix}
\alpha_1 & & & \\
\beta_1 & & & \\
\gamma_1 & & & \\
& \alpha_2 & & \\
& \beta_2 & & \\
& \gamma_2 & & \\
& & \alpha_3 & \\
& & \beta_3 & \\
& & \gamma_3 & \\
& & & \alpha_4 \\
& & & \beta_4 \\
& & & \gamma_4
\end{bmatrix}
\begin{pmatrix} \lambda_1 \\ \lambda_2 \\ \lambda_3 \\ \lambda_4 \end{pmatrix} - \mu
\begin{pmatrix} u_{10} \\ u_{11} \\ u_{12} \\ u_{13} \\ u_{14} \\ u_{15} \\ u_{16} \\ u_{17} \\ u_{18} \\ u_{19} \\ u_{20} \\ u_{21} \end{pmatrix} \tag{13.30}
$$

Note that the coefficient matrix of λ_k ($k = 1,2,3,4$) in Eq. (13.30) is precisely the transpose of the coefficient matrix of $\dot{u}_{6+3k+i}$ in the acceleration constraints, Eq. (13.25).

Assembling generalized inertia forces and generalized active forces and adding the generalized constraint forces, the overall dynamical equations can be expressed in matrix form, and written as Kane's dynamical equations with undetermined multipliers [9]:

$$\mathbf{M}\dot{\mathbf{U}} + \mathbf{A}^{t}\Lambda = \mathbf{C} + \mathbf{F} \tag{13.31}$$

Here, elements of the column matrix Λ represent the force required to maintain the constraints; $\mathbf{C}$ is the generalized inertia force terms associated with the remainder acceleration; and $\mathbf{F}$ represents the generalized active force terms that include contributions from the reaction wheel torques, the elasticity of the solar panels, and even gravity, however small as in nearly zero-g condition. Equation (13.31), together with the acceleration constraints, Eq. (13.26), can be solved for the undetermined multipliers, which represent the constraint forces and derivatives of the generalized speeds as follows:

$$\Lambda = [\mathbf{A}\mathbf{M}^{-1}\mathbf{A}^{t}]^{-1}\{\mathbf{B} + \mathbf{A}\mathbf{M}^{-1}(\mathbf{C} + \mathbf{F})\} \tag{13.32}$$

$$\dot{\mathbf{U}} = \mathbf{M}^{-1}\{\mathbf{C} + \mathbf{F} - \mathbf{A}^{t}\Lambda\} \tag{13.33}$$

13.3 SIMULATION OF SPACECRAFT MOTION WITH FUEL SLOSH FOR VARIOUS FILL-FRACTIONS OF THE TANK

To validate the reliability of this apparently simplistic theory of the crawling particle model in predicting the total slosh force and slosh torque under complex excitations, comparisons of the results predicted by the foregoing theory with those from the internationally available commercial code, FLOW-3D, were undertaken. Results are given for a zero-g operating condition to analyze the effects of fuel slosh in tanks, for various fraction-fills, to demonstrate the rigid–liquid–flexibility coupling effects of the spacecraft during an in-orbit maneuver with a three-axial-stabilized proportional-derivative controller.

13.3.1 COMPUTATION OF THE TOTAL SLOSH FORCE AND TORQUE

The environment, a description of the physical properties of the propellant in the tanks, and the location coordinates of the tank mass centers are given as follows:
Zero-g environment, i.e., $g = 0$.

i. The radius of the circular propellant tanks is $R_k = 0.3796$ m; $k = 1,\ldots, 4$.

ii. For Tank-1 and Tank-2, the fuel contained is the propellant; Tank-3 and Tank-4 contain the oxidizer propellant. The physical properties of the two kinds of propellant are listed in Table 13.1.

iii. Coordinates of the geometric center of the tanks in the B-frame are in Table 13.2.

The rigid body motion of the spacecraft is prescribed as shown in Figures 13.3 and 13.4. In detail, Figure 13.3 shows the time history of the acceleration components of B* in N described in the B-frame, with $^{N}\mathbf{a}^{B*} = a_1\mathbf{b}_1 + a_2\mathbf{b}_2 + a_3\mathbf{b}_3$.

TABLE 13.1
Physical Properties of Propellant

Propellant (N/m)	Mass density (kg/m³)	Kinematic viscosity (m²/s)	Surface tension (N/m)
Fuel	0.8744e3	9.572e-6	0.03392
Oxidizer	1.446e3	2.839e-6	0.02561

TABLE 13.2
Coordinates of the Center of the Tanks

Tank number	X_k (m)	Y_k (m)	Z_k (m)
1	0.1805	0	0.4085
2	0.1805	0	−0.4085
3	−0.4545	−0.4085	0
4	−0.4545	0.4085	0

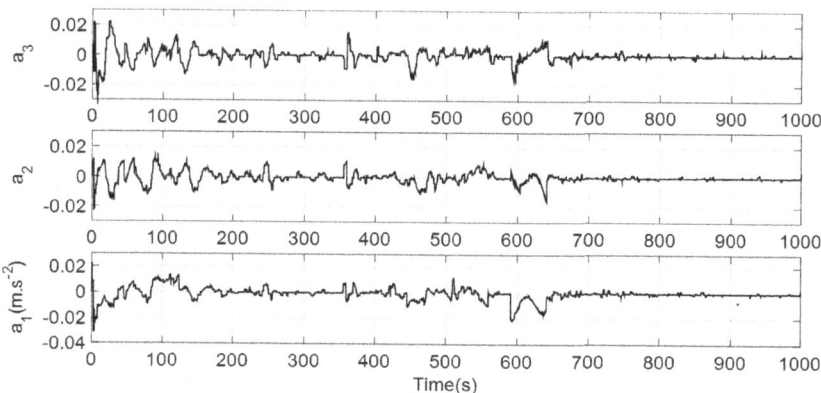

FIGURE 13.3   Time history of the acceleration components for the mass center of the spacecraft.

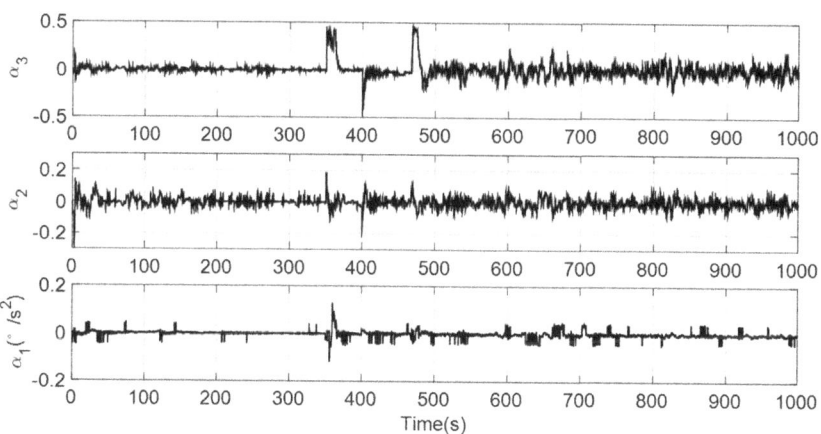

FIGURE 13.4   Time history of the angular acceleration components.

Figure 13.4 shows the time history of the angular acceleration of the rigid hub B in N, $^{N}\alpha^{B} = \alpha_1\mathbf{b}_1 + \alpha_2\mathbf{b}_2 + \alpha_3\mathbf{b}_3$. Finally, note that the percentage fill-fractions of the fuel correspond to the equivalent mass fractions of the particles used in the model. Cases A and B represented fuel fill-fractions indicated in the figures:

Case A: 40% fill-fraction (results in Figures 13.5 and 13.6).
Case B: 80% fill-fraction (results in Figures 13.7 and 13.8).

The total equivalent slosh force on B* and the torque on B are calculated by summing up the physical constraint forces, the four λ's for the forces normal to the tank surface and the wall friction force on the particles P_k, see Eq. (13.26); force and torque about B* is computed as shown below:

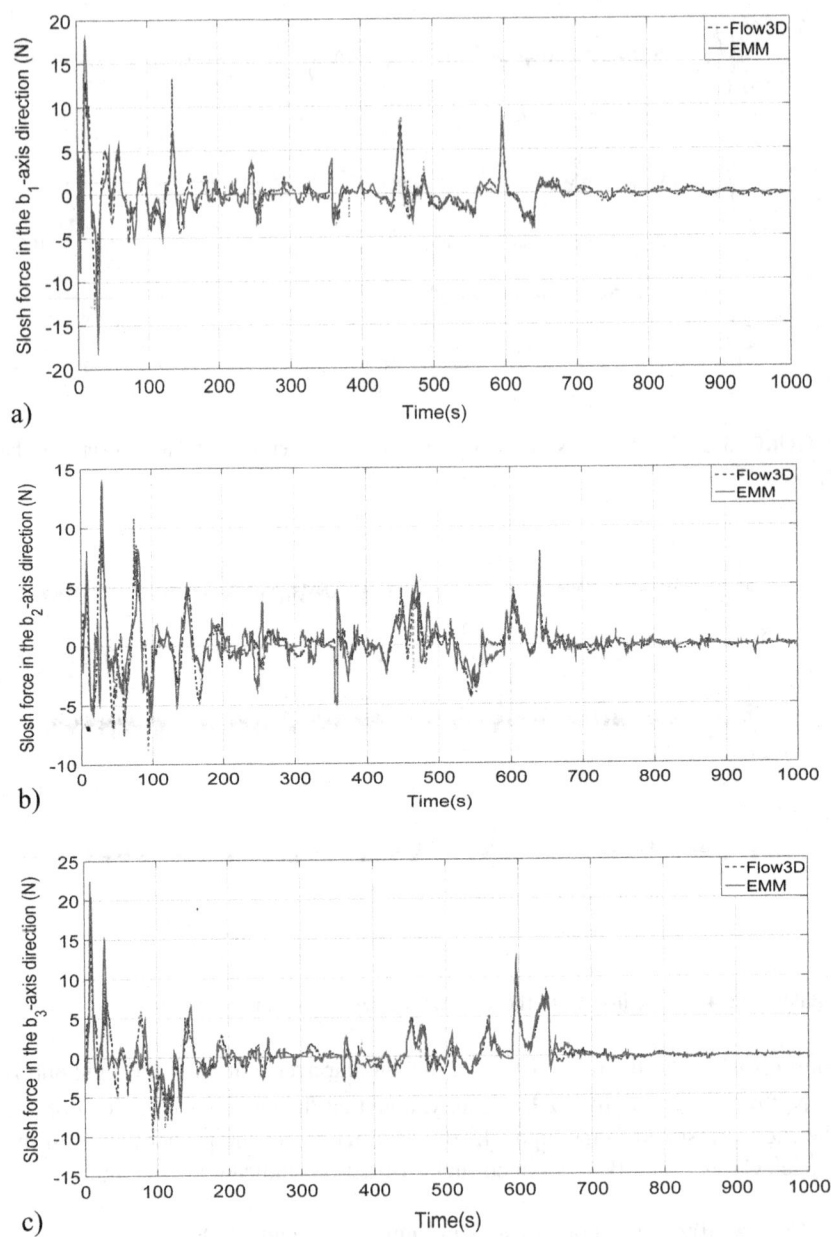

FIGURE 13.5   Total slosh force for tank 40% full: model results vs. Flow3D results.

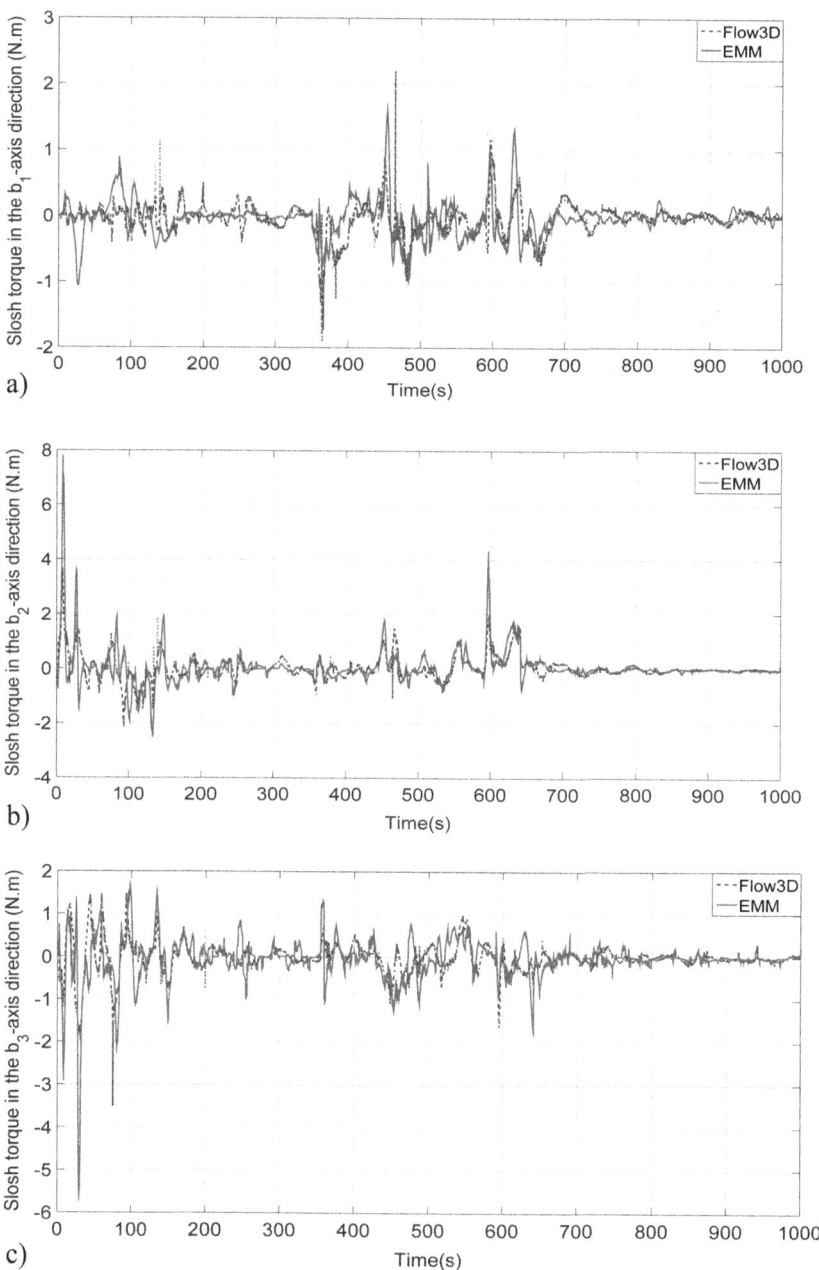

FIGURE 13.6 Total slosh torque with tank 40% full: model results vs. Flow3D results; comparisons in Case B (tank 80% full).

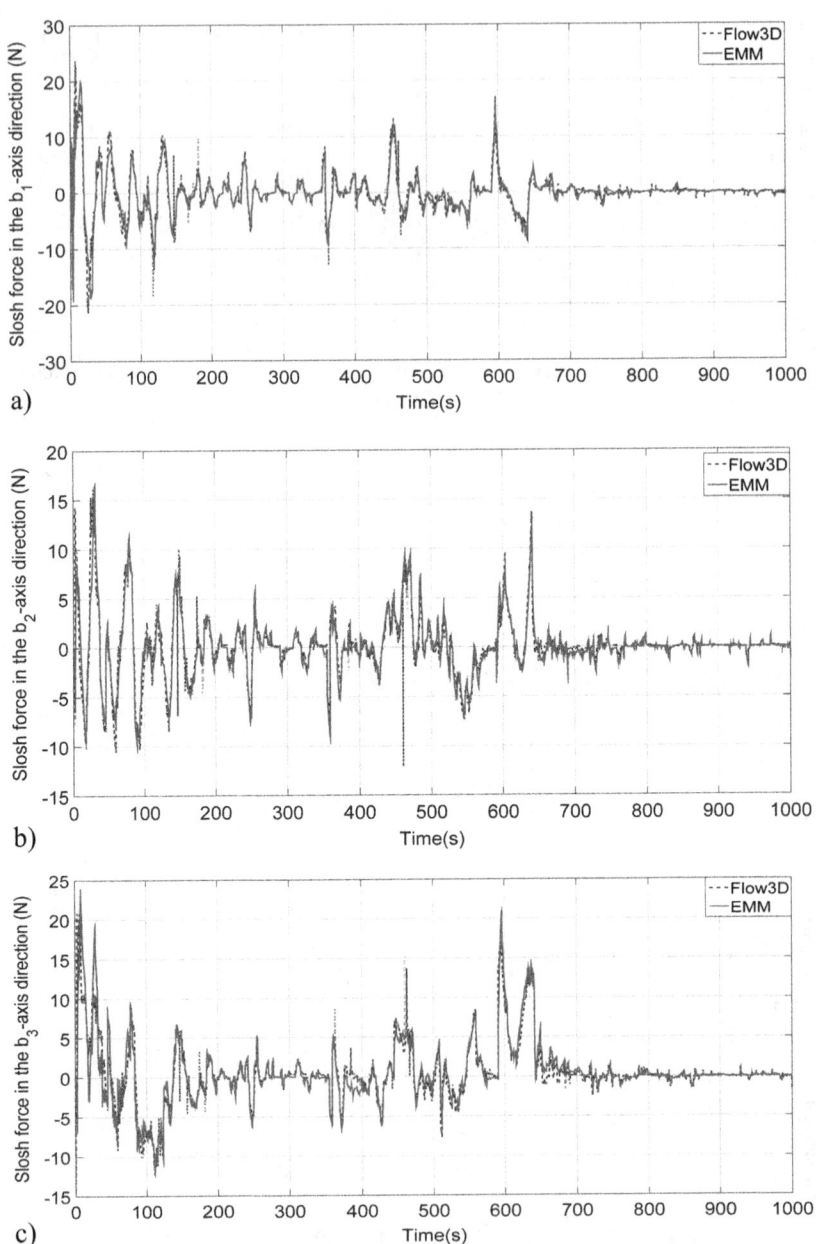

FIGURE 13.7   Total slosh force for tank 80% full: model results vs. Flow3D results.

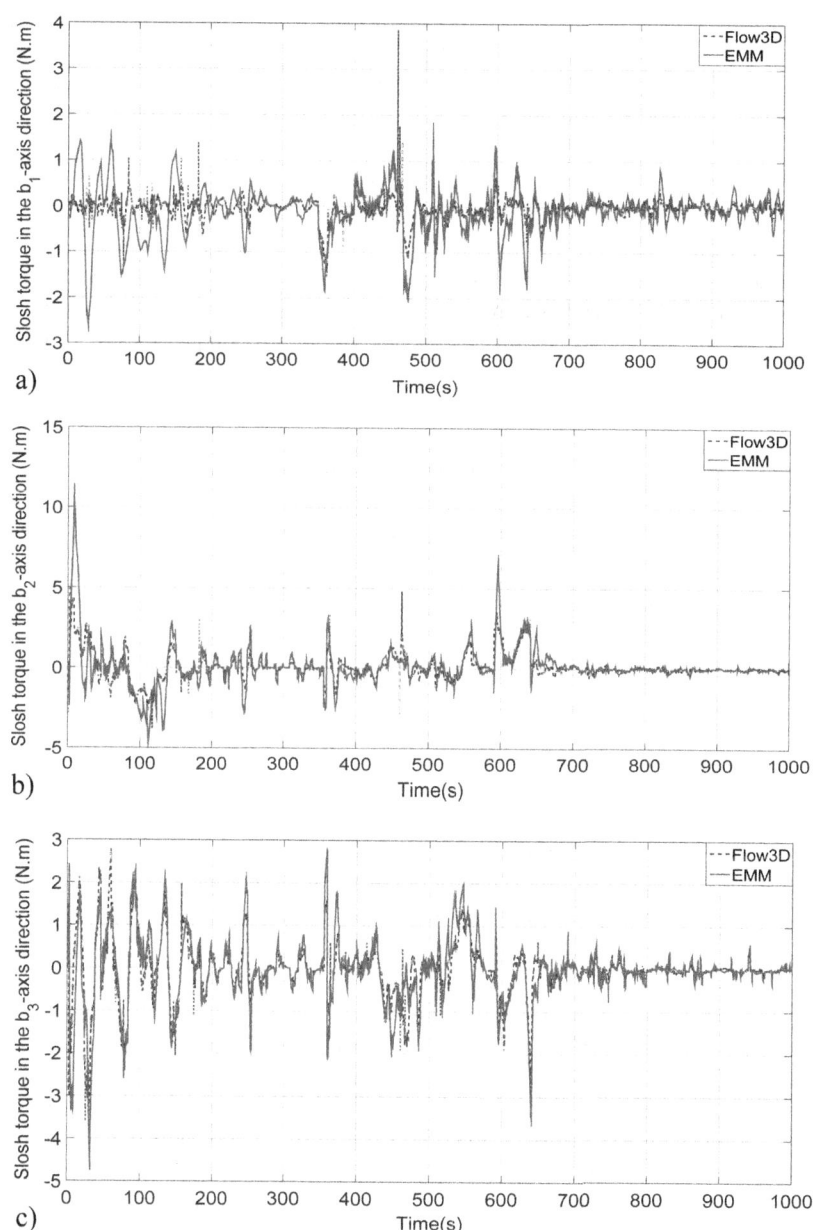

FIGURE 13.8 Total slosh torque for tank 80% full: model vs. Flow3D results.

$$\mathbf{F}_s^{B*} = \sum_{k=1}^{4} \mathbf{F}_c^{P_k} \tag{13.34}$$

$$\mathbf{T}_s^{B} = \sum_{k=1}^{4} \mathbf{p}^{B*P_k} \times \mathbf{F}_c^{P_k} \tag{13.35}$$

To validate the total equivalent slosh force and torque calculated by the equivalent mechanical model (particles crawling on the tank walls) used in this chapter, the following two cases are simulated and the results are compared with Flow3D computations:

Comparisons in Case A (tank 40% full):

a. Component of the total slosh force in $\mathbf{b}_1$ direction.
b. Component of the total slosh force in $\mathbf{b}_2$ direction.
c. Component of the total slosh force in $\mathbf{b}_3$ direction.

a. Component of the total slosh torque in $\mathbf{b}_1$ direction.
b. Component of the total slosh torque in $\mathbf{b}_2$ direction.
c. Component of the total slosh torque in $\mathbf{b}_3$ direction

a. Component of the total slosh force in $\mathbf{b}_1$ direction.
b. Component of the total slosh force in $\mathbf{b}_2$ direction.
c. Component of the total slosh force in $\mathbf{b}_3$ direction

a. Component of the total slosh torque in $\mathbf{b}_1$ direction.
b. Component of the total slosh torque in $\mathbf{b}_2$ direction.
c. Component of the total slosh torque in $\mathbf{b}_3$ direction

As mentioned before, Figures 13.5–13.8 show the time history of the components of the total equivalent slosh force and torque responses in two liquid-fill ratio cases, 40% and 80%, with a comparison of the results from the present theory and those of the code Flow3D. It can be seen that the simplified mechanical model, of a particle crawling on the surface of the tank wall, to represent fuel slosh, becomes more accurate if the liquid-fill ratio is higher. The main reason is that the slosh amplitude of the low liquid-fill ratio case is larger than that of the high liquid-fill ratio case under the same excitation level. Actually, the Flow3D computations display liquid disintegration under lower liquid-fill ratio. In addition, the slosh torque response level obtained by the crawling particle model is uniformly higher than that simulated by Flow3D software. This may be due to the fact that the distributed liquid mass is approximated as a lumped-mass particle in the model. However, the comparisons show good agreement, and the results indicate that the dynamics model given here is useful to predict the total slosh force and total slosh torque under complex excitations.

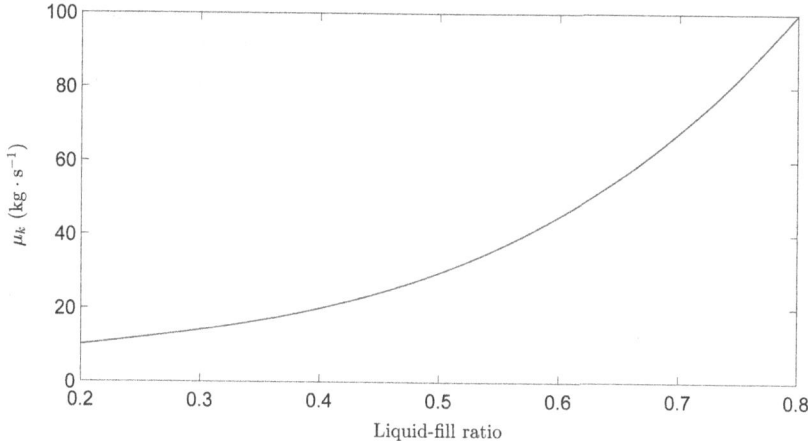

FIGURE 13.9 Empirical relationship between equivalent damping parameter μ_k and liquid-fill ratio.

To conclude, Figure 13.9 shows an empirical relationship between the equivalent damping parameter μ_k and the liquid-fill ratio. The result is obtained from additional simulations in different liquid-fill ratio cases, and the empirical curve can be approximated to a quadratic function curve. It should be noted that the equivalent damping parameter of the equivalent mechanical model depends on the shape and the size of the tank, the gravity environment, and the physical properties of the liquid.

APPENDIX 13 LINEARIZED EQUATIONS OF MOTION OF FUEL SLOSH IN A TANK, FOR SPACECRAFT CONTROL DESIGN

Here, we use the procedure for deriving linearized dynamical equations without deriving the non-linear dynamical equations for the system. The procedure is given in Ref. [8]. A classic reference work for fuel slosh is Ref. [7], where it was shown that a pendulum model is quite adequate for representing slosh. Here, a compound pendulum with an axial offset hinge, locating the mass center of the fuel, is used to describe partially filled fuel slosh, in three dimensions. When the offset is zero, it means the tank is full. A sketch of the system is given in Figure A.1. Let the various kinematical variables be defined as follows, where B is the tank with its mass center B*, and R is the rod of the pendulum:

$$^N\mathbf{v}^{B*} = \varepsilon_1\mathbf{b}_1 + \varepsilon_2\mathbf{b}_2 + \varepsilon_3\mathbf{b}_3 \tag{A13.1}$$

$$^N\boldsymbol{\omega}^B = u_1\mathbf{b}_1 + u_2\mathbf{b}_2 + u_3\mathbf{b}_3 \tag{A13.2}$$

$$^N\boldsymbol{\omega}^R = {}^N\boldsymbol{\omega}^B + {}^B\boldsymbol{\omega}^R \tag{A13.3}$$

Here, ε_1, ε_2, ε_3 are prescribed velocity components of the mass center B^* of the tank. Beyond the angular velocity generalized speeds, introduce two more generalized speeds for the rate of change of orientation of the pendulum frame R, see Figure A13.1.

$$u_4 = \dot{\theta}_1 \tag{A13.4}$$

$$u_5 = \dot{\theta}_2 \tag{A13.5}$$

For a body 3-2 rotation, [10], the angular velocity of the rod R with respect to B is:

$$^B\omega^R = -u_4 s_2 \mathbf{r}_1 + u_5 \mathbf{r}_2 + u_4 c_2 \mathbf{r}_3 \tag{A13.6}$$

Here, we have used the abbreviations $s_2 = \sin\theta_2$, $c_2 = \cos\theta_2$.
The direction cosine matrix between the B- and R-basis vectors are [10]

$$
\begin{bmatrix} \mathbf{b1} \\ \mathbf{b2} \\ \mathbf{b3} \end{bmatrix} =
\begin{bmatrix} c1 & -s1 & c1s2 \\ s1c2 & c1 & s1s2 \\ -s2 & 0 & c2 \end{bmatrix}
\begin{bmatrix} \mathbf{r1} \\ \mathbf{r2} \\ \mathbf{r3} \end{bmatrix} \tag{A13.7}
$$

From Eqs. (A13.6) and (A13.7), we get, after simplification:

$$^B\omega^R = \mathbf{b}_1(-u_5 s_1) + \mathbf{b}_2 u_5 c_1 + \mathbf{b}_3 u_4 \tag{A13.8}$$

From Eqs. (A13.3) and (A13.8), we get

$$^N\omega^R = \mathbf{b}_1(u_1 - u_5 s_1) + \mathbf{b}_2(u_2 + u_5 c_1) + \mathbf{b}_3(u_3 + u_4) \tag{A13.9}$$

The velocity of the mass center of R^* of the compound pendulum R is given by

$$^N\mathbf{v}^{R^*} = {}^N\mathbf{v}^H + {}^N\omega^R \times (L_s \mathbf{r}_1) = \varepsilon_1 \mathbf{b}_1 + \varepsilon_2 \mathbf{b}_2 + \varepsilon_3 \mathbf{b}_3 + (u_1 \mathbf{b}_1 + u_2 \mathbf{b}_2 + u_3 \mathbf{b}_3)$$

$$\times (-h\mathbf{b}_3) + [\mathbf{b}_1(u_1 - u_5 s_1) + \mathbf{b}_2(u_2 + u_5 c_1) + \mathbf{b}_3(u_3 + u_4)]$$

$$\times L_s(c_1 c_2 \mathbf{b}_1 + s_1 c_2 \mathbf{b}_2 - s_2 \mathbf{b}_3)$$

$$= \varepsilon_1 \mathbf{b}_1 + \varepsilon_2 \mathbf{b}_2 + \varepsilon_3 \mathbf{b}_3 + h(u_1 \mathbf{b}_2 - u_2 \mathbf{b}_1) + \tag{A13.10}$$

$$L_s\{\mathbf{b}_1[(u_2 + u_5 c_1)(-s_2) - (u_3 + u_4)s_1 c_2] +$$

$$\mathbf{b}_2[(u_3 + u_4)c_1 c_2 - (u_1 - u_5 s_1)(-s_2)] +$$

$$\mathbf{b}_3[(u_1 - u_5 s_1)s_1 c_2 - (u_2 + u_5 c_1)c_1 c_2]\}$$

This is rewritten as

$$
\begin{aligned}
{}^N\mathbf{v}^{R*} = &\ \mathbf{b}_1\{\varepsilon_1 - hu_2 - L_s[(u_2 + u_5c_1)(s_2) + (u_3 + u_4)s_1c_2]\} \\
&+ \mathbf{b}_2\{\varepsilon_2 + hu_1 + L_s[(u_3 + u_4)c_1c_2 + (u_1 - u_5s_1)s_2]\} \qquad \text{(A13.11)} \\
&+ \mathbf{b}_3\{\varepsilon_3 + L_s[u_1s_1c_2 - u_5c_2 - u_2c_1c_2]\}
\end{aligned}
$$

Using Kane's method of deriving linearized equations without deriving non-linear equations [8], we obtain first the non-linear partial velocity/partial angular velocity expressions and linearize them, and then linearize all non-linear terms in the generalized inertia force and generalized active force acceleration expressions, to form Kane's dynamical equations. For the problem of a container body B and a compound pendulum rod R, the non-linear partial velocity/partial angular velocity expressions are as follows:

$$
\begin{bmatrix}
i & {}^N\mathbf{v}_i^{R*} & {}^N\boldsymbol{\omega}_i^B & {}^N\boldsymbol{\omega}_i^R \\
1 & \mathbf{b}_2(h + L_s s_2) + \mathbf{b}_3 L_s s_1 c_2 & \mathbf{b}_1 & \mathbf{b}_1 \\
2 & -\mathbf{b}_1(h + L_s s_2) - \mathbf{b}_3 L_s c_1 c_2 & \mathbf{b}_2 & \mathbf{b}_2 \\
3 & -\mathbf{b}_1 L_s s_1 c_2 + \mathbf{b}_2 L_s c_1 c_2 & \mathbf{b}_3 & \mathbf{b}_3 \\
4 & -\mathbf{b}_1 L_s s_1 c_2 + \mathbf{b}_2 L_s c_1 c_2 & 0 & \mathbf{b}_3 \\
5 & -\mathbf{b}_1 L_s c_1 s_2 - \mathbf{b}_2 L_s s_1 s_2 - \mathbf{b}_3 L_s c_2 & 0 & -\mathbf{b}_1 s_1 + \mathbf{b}_2 c_1
\end{bmatrix} \quad \text{(A13.12)}
$$

To remove multiplications between two variables in the analysis, as will be required later in the analysis, we linearize the partial velocity/partial angular velocity matrix further as

$$
\begin{bmatrix}
i & {}^N\tilde{\mathbf{v}}_i^{R*} & {}^N\tilde{\boldsymbol{\omega}}_i^B & {}^N\tilde{\boldsymbol{\omega}}_i^R \\
1 & \mathbf{b}_2 h & \mathbf{b}_1 & \mathbf{b}_1 \\
2 & -\mathbf{b}_1 h - \mathbf{b}_3 L_s & \mathbf{b}_2 & \mathbf{b}_2 \\
3 & \mathbf{b}_2 L_s & \mathbf{b}_3 & \mathbf{b}_3 \\
4 & \mathbf{b}_2 L_s & 0 & \mathbf{b}_3 \\
5 & -\mathbf{b}_3 L_s & 0 & \mathbf{b}_2
\end{bmatrix} \quad \text{(A13.13)}
$$

Linearized angular acceleration of B and the compound pendulum R are:

$$
{}^N\tilde{\boldsymbol{\alpha}}^B = \dot{\tilde{u}}_1 \mathbf{b}_1 + \dot{\tilde{u}}_2 \mathbf{b}_2 + \dot{\tilde{u}}_3 \mathbf{b}_3 \qquad \text{(A13.14)}
$$

$$
{}^N\tilde{\boldsymbol{\alpha}}^R = \dot{\tilde{u}}_1 \mathbf{b}_1 + (\dot{\tilde{u}}_2 + \dot{\tilde{u}}_5)\mathbf{b}_2 + (\dot{\tilde{u}}_3 + \dot{\tilde{u}}_5)\mathbf{b}_3 \qquad \text{(A13.15)}
$$

$$
\begin{aligned}
{}^N\tilde{\mathbf{a}}^{R*} = &\ \dot{\varepsilon}_1 \mathbf{b}_1 + \dot{\varepsilon}_2 \mathbf{b}_2 + \dot{\varepsilon}_3 \mathbf{b}_3 + + \mathbf{b}_2 h\dot{u}_1 \\
&- (\mathbf{b}_1 h + \mathbf{b}_3 L_s)\dot{u}_2 + \mathbf{b}_2 L_s \dot{u}_3 + \mathbf{b}_2 L_s \dot{u}_4 - \mathbf{b}_3 L_s \dot{u}_5
\end{aligned}
\qquad \text{(A13.16)}
$$

This leads to the linearized generalized inertia force expressions [8]:

$$-\tilde{F}_i^* = {}^N\tilde{v}_i^{R*} \cdot [m_s \, {}^N\tilde{a}^{R*}]$$

$$+ {}^N\tilde{\omega}_i^R \cdot [\overline{I}^{R/R*} \cdot {}^N\tilde{\omega}^R] + {}^N\tilde{\alpha}_i^B \cdot [\overline{I}^{B/B*} \cdot {}^N\tilde{\omega}^B] \quad i=1,\ldots,5 \tag{A13.17}$$

These are explicitly expressed for the five generalized speeds, defined for this problem.

$$-\tilde{F}_1^* = m_s \left\langle \dot{\varepsilon}_2 h + h^2 \dot{\tilde{u}}_1 + hL_s \dot{\tilde{u}}_3 + hL_s \dot{\tilde{u}}_4 \right\rangle + I_{11}^B \dot{\tilde{u}}_1 + I_{12}^B \dot{\tilde{u}}_2 + I_{13}^B \dot{\tilde{u}}_3 \tag{A13.18}$$

$$-\tilde{F}_2^* = \left\langle -m_s(\dot{\varepsilon}_1 h + \dot{\varepsilon}_3 L_s) + m_s(h^2 + L_s^2)\dot{\tilde{u}}_2 + m_s L_s^2 \dot{\tilde{u}}_5 \right\rangle$$

$$+ I_{21}^B \dot{\tilde{u}}_1 + I_{22}^B \dot{\tilde{u}}_2 + I_{23}^B \dot{\tilde{u}}_3 + \frac{m_s L_s^2}{3}(\dot{\tilde{u}}_2 + \dot{\tilde{u}}_5) \tag{A13.19}$$

$$-\tilde{F}_3^* = L_s \varepsilon_2 + [m_s L_s h + I_{31}]\dot{\tilde{u}}_1 + I_{32}\dot{\tilde{u}}_2$$

$$+ (m_s L_s^2 + I_{33})\dot{\tilde{u}}_3 + m_s L_s^2 \dot{\tilde{u}}_4 + \frac{m_s L_s^2}{3}(\dot{\tilde{u}}_2 + \dot{\tilde{u}}_5) \tag{A13.20}$$

$$-\tilde{F}_4^* = m_s[\dot{\varepsilon}_2 L_s + hL_s \dot{\tilde{u}}_1 + L_s^2 \dot{\tilde{u}}_3 + L_s^2 \dot{\tilde{u}}_4] + \frac{m_s L_s^2}{3}(\dot{\tilde{u}}_2 + \dot{\tilde{u}}_5) \tag{A13.21}$$

$$-\tilde{F}_5^* = -m_s L_s \dot{\varepsilon}_3 + m_s L_s^2(\dot{\tilde{u}}_2 + \dot{\tilde{u}}_5) + \frac{m_s L_s^2}{3}(\dot{\tilde{u}}_2 + \dot{\tilde{u}}_5) \tag{A13.22}$$

These equations are written involving a "mass matrix" as follows:

$$\begin{Bmatrix} -\tilde{F}_1^* \\ -\tilde{F}_2^* \\ -\tilde{F}_3^* \\ -\tilde{F}_4^* \\ -\tilde{F}_5^* \end{Bmatrix} = \begin{bmatrix} (m_s h^2 + I_{11}^B) & I_{12}^B & (m_s hL_s + I_{13}^B) & m_s hL_s & 0 \\ I_{12}^B & [I_{22}^B + m_s(h^2 + 4L_s^2/3)] & I_{23}^B & 0 & 4m_s L_s^2/3 \\ (m_s hL_s + I_{13}^B) & I_{23}^B & (I_{33}^B + 4m_s L_s^2/3) & 4m_s L_s^2/3 & 0 \\ m_s hL_s & 0 & 4m_s L_s^2/3 & 4m_s L_s^2/3 & 0 \\ 0 & 4m_s L_s^2/3 & 0 & 0 & 4m_s L_s^2/3 \end{bmatrix} \begin{Bmatrix} \dot{\tilde{u}}_1 \\ \dot{\tilde{u}}_2 \\ \dot{\tilde{u}}_3 \\ \dot{\tilde{u}}_4 \\ \dot{\tilde{u}}_5 \end{Bmatrix}$$

$$\tag{A13.23}$$

Note the symmetry in the mass matrix.

13A.1 GENERALIZED ACTIVE FORCES

There are four types of forces on the system: interaction forces like springs and dashpots restraining the rod R representing the fuel as a compound pendulum,

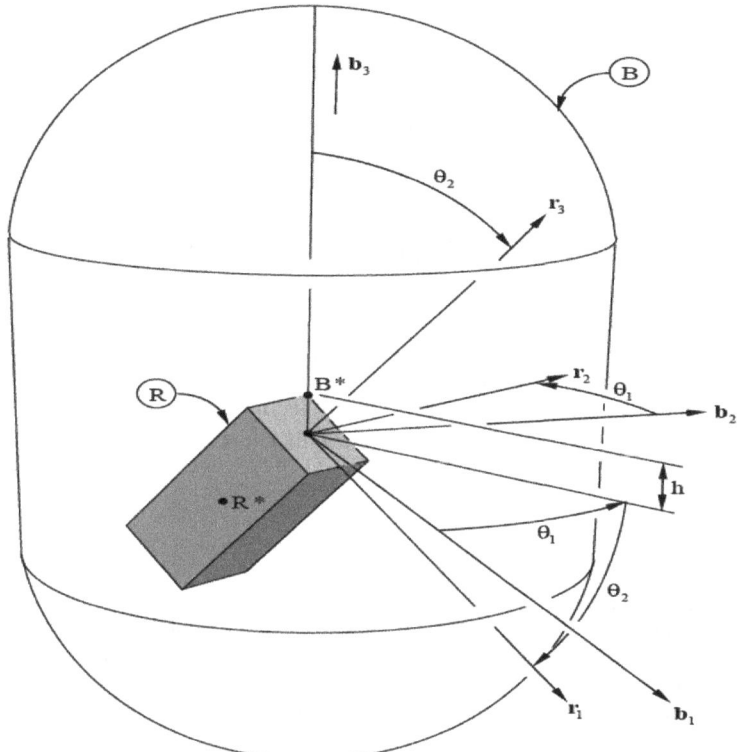

FIGURE A13.1 Sketch of a hinged compound pendulum with an axial offset, representing fuel slosh in a partially filled tank rotating in three dimensions. (Drawn by D. Levinson.)

force due to gravity on the pendulum, and control torques on the container (spacecraft). The linearized equations are

$$
\begin{Bmatrix} F_1^{sd} \\ F_2^{sd} \\ F_3^{sd} \\ F_4^{sd} \\ F_5^{sd} \end{Bmatrix} = -m_s \begin{Bmatrix} \dot{\varepsilon}_2 h \\ -\dot{\varepsilon}_1 h - \dot{\varepsilon}_3 L_s \\ 0 \\ \dot{\varepsilon}_2 L_s \\ -\dot{\varepsilon}_3 L_s \end{Bmatrix} - \begin{Bmatrix} 0 \\ 0 \\ 0 \\ k\theta_1 + cu_4 \\ k\theta_2 + cu_5 \end{Bmatrix} - m_s g \begin{Bmatrix} 0 \\ -L_s \\ 0 \\ 0 \\ -L_s \end{Bmatrix} + \begin{Bmatrix} T_1^c \\ T_2^c \\ T_3^c \\ 0 \\ 0 \end{Bmatrix} \qquad (A13.24)
$$

Note that setting $g = 0$ takes care of the zero-g condition in space. The spring constant k and damping are expected to be determined from a test. Finally, following Kane's method, linearized equations for fuel slosh obtained without deriving the non-linear equations are given by

$$-\begin{Bmatrix} \tilde{F}_1^* \\ \tilde{F}_2^* \\ \tilde{F}_3^* \\ \tilde{F}_4^* \\ \tilde{F}_5^* \end{Bmatrix} = \begin{Bmatrix} F_1^{sd} \\ F_2^{sd} \\ F_3^{sd} \\ F_4^{sd} \\ F_5^{sd} \end{Bmatrix} \qquad\qquad (A13.25)$$

These equations can be used to design the control laws, for T_1^c, T_2^c, T_3^c.

ACKNOWLEDGEMENT

The work reported in this chapter was done by the students of the Aerospace Engineering department, Beijing Institute of Technology, under the supervision of Prof. Baozeng Yue, who had invited the author to teach from the first edition of the book.

REFERENCES

1. Feng, L., Baozeng, Y., Banerjee, A.K., Yong, T., Wenjun, W., and Zhengyong, L., (December 2019), "Large Motion Dynamics of In-Orbit Rigid-Liquid-Flexible Coupling of Spacecraft Undergoing Large Overall Motion", *Journal of Guidance Control, and Dynamics*, **43**, pp. 438–450.
2. Schiehlen, W., (1997), "Multibody System Dynamics: Roots and Perspectives", *Multibody System Dynamics*, **1**, pp. 149–188.
3. Banerjee, A.K., (2003), "Contribution of Multi-body Dynamics to Space Flight: A Brief Review", *Journal of Guidance, Control, and Dynamics*, **26**(3), pp. 385–394.
4. Zhou, Z.C. and Huang, H., (2015), "Constraint Surface Model for Large Amplitude Sloshing of the Spacecraft with Multiple Tanks", *Acta Astronautica*, **111**, pp. 222–229.
5. Wei, C., Wang, L., and Shabana, A.A., (2015), "A Total Lagrangian ANCF Liquid Sloshing Approach for Multibody System Applications", *Journal of Computational and Nonlinear Dynamics*, **10**(5), 51014.
6. Berry, R.L. and Tegart, J.R., (1975), "Experimental Study of Transient Liquid Motion in Orbiting Spacecraft", *NASA Report* NAS8-30690.
7. Dodge, F., (2000), "The New Dynamic Behavior of Liquids in Moving Containers", *NASA Report* SP-106.
8. Kane, T.R. and Levinson, D.A., (1985), *Dynamics: Theory and Applications*, McGraw-Hill.
9. Wang, J.T. and Huston, E.L., (1987), "Kane's Equations with Undetermined Multipliers: Application to Constrained Multibody Systems", *Journal of Applied Mechanics*, **54**, pp. 424–429.
10. Kane, T.R., Likine, P.W., and Levinson, D.A., (1983), *Spacecraft Dynamics*, McGraw-Hill.
11. Singh, R.P., van der Voort, R.T., and Likins, P.W., (1985), "Dynamics of Flexible Bodies in Tree Topology: A Computer Oriented Approach", *Journal of Guidance, Control, and Dynamics*, **8**(5), pp. 584–590.
12. Banerjee, A.K. and Lemak, M.E., (1991), "Multi-Flexible Body Dynamics Capturing Motion-Induced Stiffness", *Journal of Applied Mechanics*, **58**, pp. 766–775.

Appendix A: Modal Integrals for an Arbitrary Flexible Body

The following modal integrals arise pervasively in flexible multibody dynamics equations of motion. The notations follow from the figure shown below, where the elastic deflection δ at a point from a point O to a point $\hat{G}$ defined by the position vector $\mathbf{r}^j$ in the j-th body, $j = 1, \ldots, n$ in the undeformed configuration. Local deformation δ is described in terms of a linear combination of component modes ϕ_k^j:

$$\delta = \sum_{k=1}^{v} \phi_k^j \eta_k^j \qquad (A.1)$$

The integrals are evaluated over the whole mass of the j-th body, B_j, (the component) of the flexible multibody system. The notations follow from the modal integrals [1] of the j-th flexible body of mass m^j, $j = 1, \ldots, n$ for an n-body with vibration modes, φ_k^j, $k = 1, \ldots, v^j$ are defined below. These vector-dyadic relations are used in Ref. [2], and in discrete form in Chapter 12.

1.
$$\mathbf{b}_k^j = \int_{B_j} \varphi_k^j \, dm \qquad (A.2)$$

2.
$$\mathbf{s}^j = \int_{B_j} \mathbf{r}^j dm + \sum_{k=1}^{\nu^j} \mathbf{b}_k^j \eta_k^j; \quad \mathbf{r}^j = \mathbf{p}^{OG_j} \qquad (A.3)$$

3.
$$\mathbf{c}_k^j = \int_{B_j} \mathbf{r}^j \times \varphi_k^j \, dm \qquad (A.4)$$

4.
$$\mathbf{d}_{ik}^j = \int_{B_j} \varphi_i^j \times \varphi_k^j \, dm \qquad (A.5)$$

5.
$$\mathbf{g}_k^j = \mathbf{c}_k^j + \sum_{i=1}^{\nu^j} \mathbf{d}_{ik}^j \eta_i^j \qquad (A.6)$$

6.
$$\mu_{ik}^j = \int_{B_j} \varphi_i^j \cdot \varphi_k^j \, dm \qquad (A.7)$$

In the following four cases, bold notation with an overbar indicates an inertia dyadic:

7.
$$\overline{\mathbf{I}}_0^{j/O} = \int_{B_j} \left[(\mathbf{r}^j \cdot \mathbf{r}^j) \overline{\mathbf{U}} - \mathbf{r}^j \mathbf{r}^j \right] dm \qquad (A.8)$$

8.
$$\overline{\mathbf{N}}_n^{j/O} = \int_{B_j} \left[(\mathbf{r}^j \cdot \varphi_n^j) \overline{\mathbf{U}} - \mathbf{r}^j \varphi_n^j \right] dm \qquad (A.9)$$

9.
$$\overline{\mathbf{I}}_{total}^{j/O} = \overline{\mathbf{I}}_0^{j/O} + \sum_{n=1}^{\nu^j} \left\langle \overline{\mathbf{N}}_n^{j/O} + [\overline{\mathbf{N}}_n^{j/O}]^t \right\rangle \eta_n^j \qquad (A.10)$$

10.
$$\overline{\mathbf{D}}_i^{j/O} = \overline{\mathbf{N}}_i^{j/O} + \sum_{k=1}^{\nu} \eta_k^j \int_{B_j} \left[(\varphi_k^j \cdot \varphi_i^j) \overline{\mathbf{U}} - \varphi_k^j \varphi_i^j \right] dm \qquad (A.11)$$

Note that in the case of integral 6, Eq. (A.5), the result is non-zero from the orthogonality condition of vibration modes; if modal truncation vectors are used, these are orthogonal between themselves and the regular vibration modes. Note that $\overline{\mathbf{U}}$ in Eqs. (A.8), (A.9), and (A.11) is a unity dyadic. The integrals in (A.7), (A.8), and (A.10) produce dyadics that are developed by using the following identity:

$$\mathbf{a} \times (\mathbf{b} \times \mathbf{c}) = [(\mathbf{c} \cdot \mathbf{a})\overline{\mathbf{U}} - \mathbf{c}\mathbf{a}] \cdot \mathbf{b}$$

In the modal integrals given below, certain dyadics, used in the text in connection with rotary inertia of flexible bodies, use nodal rigid bodies in the finite element models. These are marked with a tilde. Superscript * on a dyadic denotes its transpose. Also used is a skew-symmetric matrix $\tilde{\psi}_k^j$ formed out of the column vector matrix ψ_k^j for small elastic rotation vector ψ_k^j.

$$\tilde{\mathbf{W}}_{1j} = \int_{B_j} d\tilde{\mathbf{I}}_0 \quad \text{nominal inertia dyadic for rigid body rotation} \tag{A.12}$$

$$\tilde{\mathbf{W}}_{2k}^j = \int_{B_j} d\tilde{\mathbf{I}}_k^j, \text{ where the total dyadic for rotation is } d\tilde{\mathbf{I}}_k^j = d\tilde{\mathbf{I}}_0^j + \eta_i \int_{B_j} \sum_{i=1}^{v_j} d\tilde{\mathbf{I}}_0^j \psi_i^j \{\psi_i^j\}^t \tag{A.13}$$

$$\tilde{\mathbf{W}}_{3k}^j = \int_{B_j} \mathbf{\Psi}_k^j \times d\tilde{\mathbf{I}}_0^{j*} \tag{A.14}$$

$$\bar{\mathbf{W}}_{4k}^j = \int_{B_j} d\tilde{\mathbf{I}}_0^j \cdot \mathbf{\Psi}_k^j \tag{A.15}$$

$$\tilde{\mathbf{W}}_{5k}^j = \int_{B_j} d\mathbf{I}_0^j \times \mathbf{\Psi}_k^j \tag{A.16}$$

$$\bar{\mathbf{W}}_{6k}^j = \int_{B_j} \psi_k^j \cdot d\mathbf{I}_0^j \tag{A.17}$$

$$\bar{\mathbf{W}}_{7lk}^j = \int_{B_j} \psi_l^j \cdot d\mathbf{I}_k \tag{ }$$

$$\bar{\mathbf{W}}_{8lk}^j = \int_{B_j} \psi_l^j \times d\tilde{\mathbf{I}}_0^j \cdot \mathbf{\Psi}_k^j \tag{A.18}$$

$$\mathbf{W}_{9lk}^j = \int_{B_j} \psi_l^j \cdot d\mathbf{I}_0^j \cdot \mathbf{\Psi}_k^j \tag{A.19}$$

$$\tilde{\mathbf{W}}_{10lk}^j = \int_{B_j} \mathbf{\Psi}_l^j \times d\mathbf{I}_k^* \tag{A.20}$$

$$\bar{\mathbf{W}}_{11lk}^j = \int_{B_j} d\mathbf{I}_0^j \cdot \mathbf{\Psi}_l^j \times \mathbf{\Psi}_k^j \tag{A.21}$$

$$\bar{\mathbf{W}}_{12lk}^j = \int_{B_j}^j [d\mathbf{I}_0^j \times \psi_l^j] \cdot \psi_l^j \tag{A.22}$$

The dyadic algebra results shown in Eqs. (A.14)–(A.22) are based on the rules given in Ref. [2].

REFERENCES

1. Banerjee, A.K. and Lemak, M.E., (1991), "Multi-flexible dynamics capturing motion-induced stiffness", *Journal of Applied Mechanics*, 58(3), pp. 766–775.
2. Thomson, W.T., (1963), *Introduction to Spacecraft Dynamics*, Wiley.

Appendix B: Flexible Multibody Dynamics for Small Overall Motion

While all of the preceding chapters in this book deal with large overall motion of flexible multibody systems, there are practical applications to spacecraft [1], where the main body of the spacecraft and its appendages undergo only small rotational and translational motions at slow speeds. In such cases, the material in Chapter 6 applies with all non-linearities removed. This means that products of rigid body velocity and angular velocity terms, products of such terms with modal coordinates and their rates and, of course, terms representing geometric stiffness are removed from the equations of motion. Refer to Figure 6.2, reproduced below:

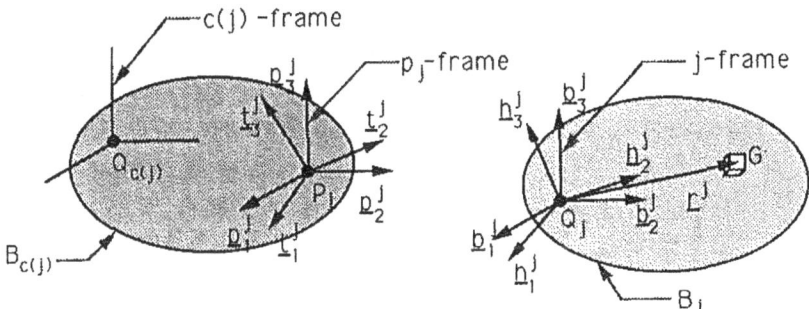

Define the generalized speeds:

$$u_i^j = {}^{Pj}\omega^j \cdot \mathbf{h}_i^j = \dot{\theta}_i^j \quad (j=1,\dots,n; \ i=1,\dots,R_j)$$

$$u_{R_j+i}^j = {}^{Pj}\mathbf{v}^j \cdot \mathbf{t}_i^j = \dot{\delta}_i^j \quad (j=1,\dots,n; \ i=1,\dots,T_j) \tag{B.1}$$

$$u_{R_j+T_j+i}^j = \dot{\eta}_i^j \quad (j=1,\dots,n; \ i=1,\dots,M_j)$$

The total number of degrees of freedom is then, $\mathrm{ndof} = \sum_{j=1}^{n}(R_j + T_j + M_j)$.

In Eq. (B.1) the modal coordinate comes from the assumption that the deformation at a point is described in terms of the j-th body vibration modes:

$$\mathbf{d}^j = \sum_{i=1}^{M_j} \varphi_i^j \eta_i^j \tag{B.2}$$

375

The linearized form of the velocity of a generic point G with a position vector $\mathbf{r}^j$ from Q_j of the j-th flexible body is then given by

$$^N\mathbf{v}^G = {}^N\mathbf{v}^{Q_j} + {}^N\boldsymbol{\omega}^j \times \mathbf{r}^j + \sum_{i=1}^{M_j} \varphi_i^j u_{R_j+T_j+i}^j \tag{B.3}$$

The partial velocities of point G are then

$$^N\mathbf{v}_i^G = {}^N\mathbf{v}_i^{Q_j} + {}^N\boldsymbol{\alpha}_i^j \times \mathbf{r}^j + \varphi_k^j \delta_{k,R_j+T_j+i} \tag{B.4}$$

The linearized acceleration of the generic particle G in the j-th body follows from Eq. (B.3) after removing centripetal and Coriolis terms:

$$^N\mathbf{a}^G = {}^N\mathbf{a}^{Q_j} + {}^N\boldsymbol{\alpha}^j \times \mathbf{r}^j + \sum_{k=1}^{M_j} \varphi_k^j \dot{u}_{R_j+T_j+k}^j \tag{B.5}$$

The i-th linearized (negative) generalized inertia force [3] is expressed by using the modal integrals:

$$-F_i^{j*} = \int_{B_j} {}^N\mathbf{v}_i^G \cdot {}^N\mathbf{a}^G dm$$

$$= \int_{B_j} [\,{}^N\mathbf{v}_i^{Q_j} + {}^N\boldsymbol{\omega}_i^j \times \mathbf{r}^j + \varphi_k^j \delta_{k,R_j+T_j+i}] \cdot [\,{}^N\mathbf{a}^{Q_j} + {}^N\boldsymbol{\alpha}^j \times \mathbf{r}^j + \sum_{n=1}^{M_j} \varphi_n^j \dot{u}_{R_j+T_j+n}^j] dm$$

$$= {}^N\mathbf{v}_i^{Q_j} \cdot \left[m^j\,{}^N\mathbf{a}^{Q_j} - \mathbf{s}^j \times {}^N\boldsymbol{\alpha}^j + \sum_{k=1}^{M_j} \mathbf{b}_k^j \dot{u}_{R_j+T_j+k}^j \right] \tag{B.6}$$

$$+ {}^N\boldsymbol{\omega}_i^j \cdot \left[\mathbf{s}^j \times {}^N\mathbf{a}^{Q_j} + \overline{\mathbf{I}}^{j/Q_j} \cdot {}^N\boldsymbol{\alpha}^j + \sum_{k=1}^{M_j} \mathbf{c}_k^j \dot{u}_{R_j+T_j+k}^j \right]$$

$$+ \delta_{k,R_j+T_j+i} \left[\mathbf{b}_k^j \cdot {}^N\mathbf{a}^{Q_j} + \mathbf{c}^j \cdot {}^N\boldsymbol{\alpha}^j + \sum_{k=1}^{M_j} \mu_{ik}^j \ddot{\eta}_k \right]$$

The i-th generalized active force [1] is written as usual:

$$F_i^j = \int_{B_j} {}^N\mathbf{v}_i^G \cdot \mathbf{df}^G \tag{B.7}$$

Finally, Kane's equations of motion are

$$-\sum_{j=1}^{n} F_i^{j*} = \sum_{n=1}^{n} F_i \quad i=1,\ldots,\text{ndof} \tag{B.8}$$

REFERENCE

1. Kane, T.R. Likins, P.W. and Levinson, D.A., (1983), *Spacecraft Dynamics*, McGraw-Hill, pp. 318–343.

Appendix C: A FORTRAN Code of the Order-n Algorithm: Application to an Example

```
C ORDER-N FORMULATION FOR EXTRUSION OF A BEAM WITH
  LARGE
C FRAME ROTATION, BENDING AND TORSION, CHAPTER 10
C CODE LINES SHOULD START AT 6TH COLUMNS. THIS IS
  ONLY A GUIDE
C IT WILL NEED ADJUSTMENT FOR USER'S COMPUTER
  EXTERNAL EQNS
  LOGICAL STPSZ
  DIMENSION X(144),Q(72),DFLN(72)
  COMMON/DFQLST/T,STEP,RELERR,ABSERR,NCUTS,NEQNS,STP
  SZ
  COMMON/PRESCR/OMG,OMGDT,DIST,DVEL,DACC
  COMMON/PARAM/NPMAX,NP,EM(72),EL(72),AI(3,3,72),AK(72)
  DATA T,TMAX,STEP/0.0,120.0,0.1/
  DATA    NCUTS,ABSERR,RELERR,STPSZ/20,1.0E-6,1.0E-6,.F
  ALSE./
  DATA  NPMAX,EM1,EL1,AK1,AI3/72,  3.124352E-3,  5.42,
  0.02, 4.0357/
C
  ATIME=SECOND( )
  NP=NPMAX
  DO 1 K=1,NPMAX
  EM(K)=EM1
  EL(K)=EL1
  AK(K)=AK1
  DO 2 I=1,3
  DO 2 J=1,3
2 AI(I,J,K)=0.0
1 AI(3,3,K)=AI3
  NEQNS=2*NPMAX
  DO 5 I=1,NEQNS
```

```
  5 X (I)=0.0
    X(31)=1.0
    NSTEPS=INT(TMAX/STEP+0.1)+1
    DO 50 JLOOP =1,NSTEPS
    IF(MOD(JLOOP,2).EQ.1)THEN
    WRITE(6,20)  T,X(NEQNS)
    END IF
    IF (JLOOP.EQ.NSTEPS)GO TO 50
    CALL DEQS(EQNS,X,*99)
 50 CONTINUE
    BTIME=SECOND( )
    CPU=BTIME-ATIME
    WRITE(6,101) CPU
101 FORMAT (1X,'CPU TIME FOR ORDER-N FORMULATION',E13.5)
 20 FORMAT(1X,6E12.4)
 99 WRITE(6,100)T
100 FORMAT(1X,'STEPSIZE HALVED 20 TIMES AT T=',E13.5)
101 STOP
    END
CCCCCCCCCCCCCCCCCCCCCCCCCCCCCCCCCCCCCCCCCCCCCCCCCCCCCCCCC
    SUBROUTINE EQNS(T,X,XDT)
    SAVE
    COMMON/PARAM/NPMAX,NP,EM(72),EL(72),AI(3,3,72),AK(72)
    COMMON/PRESCR/OMG,OMGDT,DIST,DVEL,DACC
    DIMENSION X(144),XDT(144)
    DIMENSION W0(3),ALF0(3),A0(3)
    DIMENSION  WKK(3,72),WKI(3,72),CKI(3,3,73),WK(3,72),ALFK
    T(3,72),
  2 BIGAKT(3,3,72),AKT(3,72),VKK(3,72),FSTKK(3,72),
  3 TSTKK(3,72),FXTK(3,72),TXTK(3,72),TAUK(72),FSTKT(3,72).
  4 TSTKT(3,72),BIGMK(6,6,72),VEC(6),A(6),B(6),
  5 BIGXK(6,72),BIGM(6,6),BIGX(6),YK(6,72),ZK(6,72),EMKK(
    72),
  6 FK(72),PK(6),WC(6,6,72),DUM(6,6)
    DATA I1ST/0/
    EL2=EL(1)/2.0
    DO 5 I=1,NPMAX
    NI=NPMAX+I
    XDT(NI)=X(I)
  5 XDT(I)=0.0
    DO 100 K=1,NP
    WKK(1,K)=0.0
    WKK(2,K)=0.0
    WKK(3,K)=0.0
    WKI(1,K)=0.0
```

```
   WKI(2,K)=0.0
   WKI(3,K)=X(K)
 C CKI
   NPK=NPMAX+K
   SK=SIN(X(NPK))
   CK=COS(X(NPK))
   CKI(1,1,K)=CK
   CKI(1,2,K)=SK
   CKI(1,3,K)=0.0
   CKI(2,1,K)=-SK
   CKI(2,2,K)=CK
   CKI(2,3,K)=0.0
   CKI(3,1,K)=0.0
   CKI(3,2,K)=0.0
   CKI(3,3,K)=1.0
   IF(K.EQ.1)THEN
   WK(1,K)=WKI(1,K)
   WK(2,K)=WKI(2,K)
   WK(3,K)=WKI(3,K)
 C
   ALFKT(1,K)=0.0
   ALFKT(2,K)=0.0
   ALFKT(3,K)=0.0
   AKT(1,K)= -WK(3,K)**2*EL2
   AKT(2,K)=0.0
   AKT(3,K)=0.0
   BIGAKT(1,1,K)= -(WK(3,K)**2+WK(2,K)**2)
   BIGAKT(2,2,K)= -(WK(1,K)**2+WK(3,K)**2)
   BIGAKT(3,3,K)= -(WK(1,K)**2+WK(2,K)**2)
   BIGAKT(1,2,K)= -ALFKT(3,K)+WK(1,K)*WK(2,K)
   BIGAKT(1,3,K)= ALFKT(2,K)+WK(1,K)*WK(3,K)
   BIGAKT(2,1,K)= ALFKT(3,K)+WK(2,K)*WK(1,K)
   BIGAKT(2,3,K)= -ALFKT(1,K)+WK(2,K)*WK(3,K)
   BIGAKT(3,1,K)= -ALFKT(2,K)+WK(3,K)*WK(1,K)
   BIGAKT(3,2,K)= ALFKT(1,K)+WK(3,K)*WK(2,K)
   END IF
 C
   IF(K.GT.1)THEN
   I=K-1
   WK(1,K)=WKI(1,K)+CK*WK(1,I)+SK*WK(2,I)
   WK(2,K)=WKI(2,K)-SK*WK(1,I)+CK*WK(2,I)
   WK(3,K)=WKI(3,K)+WK(3,I)
   ALFKT(1,K)=0.0
   ALFKT(2,K)=0.0
   ALFKT(3,K)=0.0
```

```
BIGAKT(1,1,K)= -(WK(3,K)**2+WK(2,K)**2)
BIGAKT(2,2,K)= -(WK(1,K)**2+WK(3,K)**2)
BIGAKT(3,3,K)= -(WK(1,K)**2+WK(2,K)**2)
BIGAKT(1,2,K)= -ALFKT(3,K)+WK(1,K)*WK(2,K)
BIGAKT(1,3,K)= ALFKT(2,K)+WK(1,K)*WK(3,K)
BIGAKT(2,1,K)= ALFKT(3,K)+WK(1,K)*WK(2,K)
BIGAKT(2,3,K)= -ALFKT(1,K)+WK(3,K)*WK(2,K)
BIGAKT(3,1,K)= -ALFKT(2,K)+WK(1,K)*WK(3,K)
BIGAKT(3,2,K)= ALFKT(1,K)+WK(3,K)*WK(2,K)
VEC(1)=AKT(1,I)+BIGAKT(1,1,I)*EL2
VEC(2)=AKT(2,I)+BIGAKT(2,1,I)*EL2
VEC(3)=AKT(3,I)+BIGAKT(3,1,I)*EL2
 A K T ( 1 , K ) = C K I ( 1 , 1 , K ) * V E C ( 1 ) +
CKI(1,2,K)*VEC(2)+CKI(1,3,K)*
2 VEC(3)+BIGAKT(1,1,K)*EL2
 A K T ( 2 , K ) = C K I ( 2 , 1 , K ) * V E C ( 1 ) +
CKI(2,2,K)*VEC(2)+CKI(2,3,K)*
2 VEC(3)+BIGAKT(2,1,K)*EL2
 A K T ( 3 , K ) = C K I ( 3 , 1 , K ) * V E C ( 1 ) +
CKI(3,2,K)*VEC(2)+CKI(3,3,K)*
2 VEC(3)+BIGAKT(3,1,K)*EL2
END IF
VKK(1,K)=0.0
VKK(2,K)=WKK(3,K)*EL2
VKK(3,K)=0.0
CALL EXTFOR (FXTK,TXTK,K)
FXTK(1,K)=EM(K)*AKT(1,K)-FXTK(1,K)
FXTK(2,K)=EM(K)*AKT(2,K)-FXTK(2,K)
FXTK(3,K)=EM(K)*AKT(3,K)-FXTK(3,K)
C
Z1=AI(1,1,K)*WK(1,K)+AI(1,2,K)*WK(2,K)+AI(1,3,K)*WK(3,K)
Z2=AI(2,1,K)*WK(1,K)+AI(2,2,K)*WK(2,K)+AI(2,3,K)*WK
 (3,K)
Z3=AI(3,1,K)*WK(1,K)+AI(3,2,K)*WK(2,K)+AI(3,3,K)*WK(3,K)
TSTKT(1,K)=AI(1,1,K)*ALFKT(1,K)+ AI(1,2,K)*ALFKT(2,K)+
AI(1,3,K)*
2 ALFKT(3,K)-WK(3,K)*Z2+ WK(2,K)*Z3-TXTK(1,K)
TSTKT(2,K)=AI(2,1,K)*ALFKT(1,K)+ AI(2,2,K)*ALFKT(2,K)+
AI(2,3,K)*
2 ALFKT(3,K)+WK(3,K)*Z1- WK(1,K)*Z3-TXTK(2,K)
TSTKT(3,K)=AI(3,1,K)*ALFKT(1,K)+ AI(3,2,K)*ALFKT(2,K)+
AI(3,3,K)*
2 ALFKT(3,K)-WK(2,K)*Z1+ WK(1,K)*Z2-TXTK(3,K)
DO 10 I=1,6
DO 10 J=1,6
```

```
 10 BIGMK(I,J,K)=0.0
    BIGMK(1,1,K)=EM(K)
    BIGMK(2,2,K)=EM(K)
    BIGMK(3,3,K)=EM(K)
    DO 15 I=4,6
    I3=I-3
    BIGXK(I3,K)=FSTKT(I3,K)
    BIGXK(I,K)=TSTKT(I3,K)
    DO 15 J=4,6
 15 BIGMK(I,J,K)=AI(I3,J-3,K)
    DO 20 I=1,3
    YK(I,K)=VKK(I,K)
 20 YK(I+3,K)=WKK(I,K)
  100 CONTINUE
  C BEGIN BACKWARD PASS
    K=NP
    DO 110 I=1,6
    BIGX(I)=BIGXK(I,K)
    DO 110 J=1,6
  110 BIGM(I,J)=BIGMK(I,J,K)
  200 CONTINUE
    DO 215 I=1,6
    ZK(I,K)=0.0
    DO 215 J=1,6
  215 ZK(I,K)=ZK(I,K)+BIGM(I,J)*YK(J,K)
    EMKK(K)=YK(1,K)*ZK(1,K)+              YK(2,K)*ZK(2,K)+
    YK(3,K)*ZK(3,K)
  2 +YK(4,K)*ZK(4,K)+ YK(5,K)*ZK(5,K)+ YK(6,K)*ZK(6,K)
    CALL HINGE(TAUK,X,K)
    FK(K)=YK(1,K)*BIGX(1)+YK(2,K)*BIGX(2)+YK(3,K)*BIGX(3)+Y
    K(4,K)
  2 *BIGX(4)+YK(5,K)*BIGX(5)+YK(6,K)*BIGX(6)-TAUK(K)
    IF(K.EQ.1)GO TO 300
    DO 220 I=1,6
  220 PK(I)=ZK(I,K)/EMKK(K)
    DO 230 I=1,6
    BIGX(I)=BIGX(I)-PK(I)*FK(K)
    DO 230 J=1,6
  230 BIGM(I,J)=BIGM(I,J)-ZK(I,K)*PK(J)
  C SHIFT TRANSFORM
    DO 240 I=1,3
    I3=I+3
    DO 240 J=1,3
    WC(I,J,K)=CKI(I,J,K)
    WC(I3,J,K)=0.0
```

```
      J3=J+3
      WC(I,J3,K)=0.0
  240 WC(I3,J3,K)=WC(I,J,K)
      VEC(1)=EL2+CKI(1,1,K)*EL2
      VEC(2)=CKI(2,1,K)*EL2
      VEC(3)=0.0
      WC(1,4,K)=0.0
      WC(2,4,K)=CKI(3,1,K)*VEC(1)
      WC(3,4,K)= -CKI(2,1,K)*VEC(1)+CKI(1,1,K)*VEC(2)
      WC(1,5,K)=0.0
      WC(2,5,K)=CKI(3,2,K)*VEC(1)
      WC(3,5,K)= -CKI(2,2,K)*VEC(1)+CKI(1,2,K)*VEC(2)
      WC(1,6,K)= -CKI(3,3,K)*VEC(2)
      WC(2,6,K)= CKI(3,3,K)*VEC(1)
      WC(3,6,K)= -CKI(2,3,K)*VEC(1)
      DO 250 I=1,6
      DO 250 J=1,6
      DUM(I,J)=0.0
      DO 250 L=1,6
  250 DUM(I,J)=DUM(I,J)+BIGM(I,L)*WC(L,J,K)
      KM1=K-1
      DO 260 I=1,6
      DO 260 J=1,6
      BIGM(I,J)=BIGMK(I,J,KM1)
      DO 260 L=1,6
  260 BIGM(I,J)=BIGM(I,J)+WC(L,I,K)*DUM(L,J)
      DO 270 I=1,6
      VEC(I)=0.0
      DO 270 J=1,6
  270 VEC(I)=VEC(I)+WC(J,I,K)*BIGX(J)
      DO 280 I=1,6
  280 BIGX(I)=BIGXK(I,KM1)+VEC(I)
      K=KM1
      GO TO 200
  300 CONTINUE
C BEGIN FORWARD PASS
      XDT(1)=-FK(1)/EMKK(1)
      IF(NP.EQ.1) RETURN
      DO 310 I=1,6
  310 A(I)=YK(I,1)*XDT(1)
      K=2
  320 CONTINUE
      DO 330=0 I=1,6
      B(I)=0.0
      DO 330 J=1,6
```

```
  330 B(I)=B(I)+WC(I,J,K)*A(J)
      XDT(K)= -(ZK(1,K)*B(1) + ZK(2,K)*B(2)+ ZK(3,K)*B(3)
  2 +ZK(4,K)*B(4) +ZK(5,K)*B(5) +ZK(6,K)*B(6)
  3 +FK(K))/EMKK(K)
      IF(K.EQ.NP) RETURN
      DO 340 I=1,6
  340 A(I)=B(I)+YK(I,K)*XDT(K)
      K=K+1
      GP TO 320
      END
CCCCCCCCCCCCCCCCCCCCCCCCCCCCCCCCCCCCCCCCCCCCC
      SUBROUTINE EXTFOR (FXTK,TXTK,K)
      SAVE
      DIMENSION FXTK(3.72),TXTK(3,72)
C SPECIFY EXTERNAL FORCES & TORQUES; BELOW IS A
      PLACE HOLDER
      DO 10 I=1,3
      FXTK(I,K)=0.0
  10 TXTK(I,K)=0.0
      RETURN
      END
CCCCCCCCCCCCCCCCCCCCCCCCCCCCCCCCCCCCCCCCCCCCC
      SUBROUTINE HINGE (TAUK,X,K)
      SAVE
      COMMON/PARAM/NPMAX,NP,EM(72),EL(72),AI(3,3,72),AK(72)
      DIMENSION TAUK(1),X(1)
      TAUK(K)= -AK(K)*X(NPMAX+K)
      RETURN
      END
CCCCCCCCCCCCCCCCCCCCCCCCCCCCCCCCCCCCCCCCCCCCCCCCCCC
      CCCC
      SUBROUTINE DEQS(F,Y,*)
C VARIABLE STEP, FOURTH ORDER RUNGE-KUTTA-MERSON
      INTEGRATOR
      IMPLICIT REAL (A-Z)
      INTEGER I,NCUTS,NEQ
      LOGICAL DBL,STPSZ
      SAVE
      EXTERNAL F
      COMMON/DFQLST/T,STEP,REL,ABSERR,NCUTS,NEQ,STPSZ
      DIMENSION F0(200),F1(200),F2(200),Y1(200),Y2(200),Y(NEQ)
      DATA HC /0.0/
C CHECK FOR INITIAL ENTRY AND ADJUST HC, IF NECESSARY
      IF (NEQ.NE.0) GO TO 10
      HC=STEP
```

```
      RETURN
   10 IF(STEP.EQ.0.0) RETURN 1
    C CHANGE DIRECTION, IF REQUIRED
      IF (HC*STEP) 20,30,40
   20 HC=-HC
      GO TO 40
   30 HC=STEP
    C SET LOCAL VARIABLES
   40 EPSL=REL
      FINAL=T+STEP
      H=HC
      TT=T+H
      T=FINAL
      H2=H/2.0
      H3=H/3.0
      H6=H/6.0
      H8=H/8.0
    C MAIN KUTTA-MERSON LOOP
   50 IF ((H.GT.0.0 .AND. TT.GT.FINAL) .OR.
      2 (H.LT.0.0 .AND. TT.LT.FINAL)) GO TO 190
   60 CALL F(TT-H,Y,F0)
      DO 70 I=1,NEQ
   70 Y1(I)=F0(I)*H3+Y(I)
      CALL F(TT-2.0*H3,Y1,F1)
      DO 80 I=1,NEQ
   80 Y1(I)=(F0(I)+F1(I))*H6+Y(I)
      CALL F(TT-2.0*H3,Y1,F1)
      DO 90 I=1,NEQ
   90 Y1(I)=(F1(I)*3.0+F0(I))*H8+Y(I)
      DO 100 I=1,NEQ
  100 Y1(I)=(F2(I)*4.0-F1(I)*3.0+F0(I))*H2+Y(I)
      CALL F(TT,Y1,F1)
      DO 110 I=1,NEQ
  110 Y2(I)=(F2(I)*4.0+F1(I)+F0(I))*H6+Y(I)
    C ** DOES THE STEPSIZE H NEED TO BE CHANGED
      IF(EPSL.LE.0.0) GO TO 170
      DBL=.TRUE.
      DO 160 I=1,NEQ
      ERR=ABS(Y1(I)-Y2(I))*0.2
      TEST=ABS(Y1(I))*EPSL
      IF (ERR.LT.TEST .OR. ERR.LT.ABSERR) GO TO 150
    C ** HALVE THE STEPSIZE
      H=H2
      TT=TT-H2
      IF (.NOT.STPSZ) GO TO 120
```

```
      TEMP=TT-H2
      WRITE(6,200)H,TEMP
C HAS THE STEPSIZE BEEN HALVED TOO MANY TIMES ?
120 NCUTS=NCUTS-1
      IF (NCUTS.GE.0) GO TO 130
      T=TT-H2
      WRITE(6,210)T
      RETURN 1
C ** IF STEPSIZE IS TOO SMALL RELATIVE TO TT TAKE
      RETURN 1
130 IF(TT+H .NE. TT) GO TO 140
      T=TT
      RETURN 1
140 H2=H/2.0
      H3=H/3.0
      H6=H/6.0
      H8=H/8.0
      GO TO 60
150 IF (DBL .AND. 64.0*ERR .GT. TEST
    2 .AND. 64.0*ERR .GT. ABSERR) DBL=.FALSE.
160 CONTINUE
C DOUBLE THE STEPSIZE, MAYBE.
      IF(.NOT. DBL .OR. ABS(2.0*H) .GT. ABS(STEP) .OR.
    2 ABS(TT+2.0*H) .GT. ABS(FINAL) .AND.
    3 ABS(TT-FINAL) .GT. ABS(FINAL)*1.0E-7) GO TO 170
      H2=H
      H=H+H
      IF (STPSZ) WRITE(6,200)H,TT
      H3=H/3.0
      H6=H/6.0
      H8=H/8.0
      NCUTS=NCUTS+1
170 DO 180 I=1,NEQ
180 Y(I)=Y2(I)
      TT=TT+H
      GO TO 50
190 IF(EPSL.LT.0.0) RETURN
C * NOW BE SURE TO HAVE T=FINAL
      HC=H
      H=FINAL-(TT-H)
      IF (ABS(H).LE.ABS(FINAL)*1.0E-7)RETURN
      TT=FINAL
      EPSL= -1.0
      H2=H/2.0
      H3=H/3.0
```

```
      H6=H/6.0
      H8=H/8.0
      GO TO 60
  200 FORMAT(1X,'THE     STEPSIZE     IS     NOW',1E12.4,'AT
     T=',1E12.4)
  210 FORMAT(1X,'THE STEPSIZE HAS BEEN HALVED TOO MANY
     TIMES;',
     2  'T=',1E12.4)
      END
```

Index